PC Based Instrumentation and Control

PC Based Instrumentation and Control

Mike Tooley

ELSEVIER
BUTTERWORTH
HEINEMANN

AMSTERDAM • BOSTON • HEIDELBERG • LONDON • OXFORD • NEW YORK
PARIS • SAN DIEGO • SAN FRANCISCO • SINGAPORE • SYDNEY • TOKYO

Elsevier Butterworth-Heinemann
Linacre House, Jordan Hill, Oxford OX2 8DP
30 Corporate Drive, Burlington, MA 01803

First published 2005

British Library Cataloguing in Publication Data
A catalogue record for this book is available from the British Library

Library of Congress Cataloguing in Publication Data
A catalogue record for this book is available from the Library of Congress

ISBN 0 7506 4716 7

For information on all Elsevier Butterworth-Heinemann publications visit our website at: www.books.elsevier.com

Typeset by Charon Tec Pvt. Ltd, Chennai, India
Printed and bound in Great Britain

Contents

Preface

Ask any production engineer, control or instrumentation specialist to define his objectives and his reply will probably include increasing efficiency without compromising on quality or reliability. Ask him what his most pressing problems are and lack of suitably trained personnel will almost certainly be high on the list. Happily, both of these perennial problems can be solved with the aid of a PC (or PC-compatible) acting as an intelligent controller. All that is required is sufficient peripheral hardware and the necessary software to provide an interface with the production/test environment.

As an example, consider the procedure used for testing and calibrating an item of electronic equipment. Traditional methods involve the use of a number of items of stand-alone test equipment (each with its own peculiarities and set-up requirements). A number of adjustments may then be required and each will require judgment and expertise on the part of the calibration technician or test engineer. The process is thus not only time consuming but also demands the attention of experienced personnel. Furthermore, in today's calibration laboratory and production test environment, the need is for a cluster of test equipment rather than for a number of stand-alone instruments. Such an arrangement is an ideal candidate for computer control.

The computer (an ordinary PC or PC-compatible) controls each item of external instrumentation and automates the test and calibration procedure, increasing throughput, consistency, and reliability, freeing the test engineer for higher level tasks. A PC-based arrangement thus provides a flexible and highly cost-effective alternative to traditional methods. Furthermore, systems can be easily configured to cope with the changing requirements of the user.

In general, PC-based instrumentation and control systems offer the following advantages:

- Flexible and adaptable: the system can be easily extended or reconfigured for a different application.
- The technology of the PC; is well known and understood, and most companies already have such equipment installed in a variety of locations.
- Low-cost PC-based systems can be put together at a faction of the cost associated with dedicated controllers.
- Rugged embedded PC controllers are available for use in more demanding applications. Such systems can be configured for a wide range of instrumentation and control applications with the added advantage that they use the same familiar operating system environment and programming software that runs on a conventional PC.
- Availability of an extensive range of PC-compatible expansion cards from an increasingly wide range of suppliers.
- Ability to interface with standard bus systems (including the immensely popular IEEE-488 General Purpose Instrument Bus).

- Support for a variety of popular network and asynchronous data communications standards (allowing PC-based systems to become fully integrated within larger manufacturing and process control systems).
- Internationally accepted standards, including ISA, PCI, PC/104, and USB bus systems.

Typical applications for PC-based instrumentation and control systems include:

- Data acquisition and data logging.
- Automatic component and QA acceptance testing.
- Signal monitoring.
- Production monitoring and control.
- Environmental control.
- Access control.
- Security and alarm systems.
- Control of test and calibration clusters.
- Process control systems.
- Factory automation systems.
- Automated monitoring and performance measurement.
- Simple machine-vision systems.
- Small-scale production management systems.
- A 'virtual' replacement for conventional laboratory test equipment.

Aims

The book aims to provide readers with sufficient information to be able to select the necessary hardware and software to implement a wide range of practical PC-based instrumentation and control systems. Wherever possible the book contains examples of practical configurations and working circuits (all of which have been rigorously tested). Representative software is also included in a variety of languages including x86 assembly language, BASIC, Visual BASIC, C, and C++. In addition, a number of popular software packages for control, instrumentation and data analysis have been described in some detail.

Information has been included so that circuits and software routines can be readily modified and extended by readers to meet their own particular needs. Overall, the aim has been that of providing the reader with sufficient information so that he or she can solve a wide variety of control and instrumentation problems in the shortest possible time and without recourse to any other texts.

Readership

This book is aimed primarily at the professional control and instrumentation specialist. It does not assume any previous knowledge of microprocessors or microcomputer systems and thus should appeal to a wide audience (including mechanical and production engineers looking for new solutions to control and instrumentation problems). The book is also ideal for students at undergraduate and post-graduate level who need a 'source book' of practical ideas and solutions.

Chapter 1 This chapter provides an introduction to microcomputer systems and the IBM PC compatible equipment. The Intel range of microprocessors is introduced as the 'legacy' chipsets and VLSI support devices found in the generic PC.

Chapter 2 This chapter describes various expansion systems which can be used to extend the I/O capability of the PC. These systems include the original Industry Standard Architecture 8- and 16-bit PC expansion bus, the Peripheral Component Interconnect (PCI), and the PC/104 architecture. Representative expansion cards and bus configurations are discussed in some detail. The chapter concludes with a detailed examination of the Universal Serial Bus (USB).

Chapter 3 This chapter is devoted to the facilities offered by the PC's operating system whether it be a basic DOS-based system or one operating under Windows 9x, NT, or XP. Each of the most popular MS-DOS commands is described and details are provided which should assist readers in creating batch files (which can be important in unattended systems which must be capable of initializing themselves and automatically executing an appropriate control program in the event of power failure) as well as executing and debugging programs using the MS-DOS debugger, DEBUG. The chapter also describes the facilities offered by the Windows operating system as a platform for the development and execution of control, instrumentation and data acquisition software.

Chapter 4 Programming techniques are introduced in this chapter. This chapter is intended for those who may be developing programs for their own specialized applications and for whom no 'off-the-shelf-' software is available. The virtues of- modular and structured programming are stressed and various control structures are discussed in some detail. Some useful pointers are included for those who need to select a language for control, instrumentation and data acquisition applications.

Chapter 5 This chapter deals with assembly language programming. The x86 instruction set is briefly explained and several representative assembly language routines written using the original Microsoft Macro Assembler (MASM) and its 32-bit reincarnation (MASM32) are included.

Chapter 6 The BASIC programming language is introduced in this chapter. Generic BASIC programming techniques and control structures are introduced, and sample routines are provided in QBASIC, PowerBASIC, and the ever-popular MS Visual BASIC.

Chapter 7 This chapter is devoted to C and C++ programming. As with the two preceding chapters, this chapter aims to provide readers with a brief introduction to programming techniques and numerous examples are included taken from applications within the general field of control, instrumentation, and data acquisition.

Chapter 8 The ever-popular IEEE-488 instrument bus is introduced in this chapter. A representative PC adapter card is described which allows a PC to be used as an IEEE-488 bus controller.

Chapter 9 This chapter deals with the general principles of interfacing analogue and digital signals to PC expansion bus modules, analogue-to-digital and digital-to-analogue conversion. A variety of sensors, transducers, and practical interface circuits have been included.

Chapter 10 Virtual instruments can provide a flexible low-cost alternative to the need to have a variety of dedicated test instruments available. This chapter provides an introduction to virtual test instruments and describes, in detail, the use of a high-performance digital storage oscilloscope.

Chapter 11 Commercial software packages are frequently used in industry to deal with specific data acquisition and instrumentation requirements. This chapter provides details of several of the most popular packages and has been designed to assist the newcomer in the selection of a package which will satisfy his or her needs.

Chapter 12 The general procedure for selection and specification of system hardware and software is described in this chapter. Eight practical applications of PC-based data acquisition, instrumentation, and control are described in detail complete with specifications, circuit diagrams, screen shots and code where appropriate.

Chapter 13 This chapter deals with reliability and fault tolerance. Basic quality procedures are described together with diagnostic and benchmarking software, and detailed fault-location charts.

A glossary is included in Appendix A while Appendices B and C deal with fundamental SI units, multiples, and sub-multiples. A binary, hexadecimal, and ASCII conversion table appears in Appendix D. A further nine appendices provide additional reference information including an extensive list of manufacturers, suppliers and distributors, useful web sites and a bibliography.

The third edition includes:

- Updated information on PC hardware and bus systems (including PCI, PC/104 and USB).
- A new chapter on PC instruments complete with examples of measurement and data logging applications.
- An introduction to software development in a modern 32-bit environment with the latest software tools that make it possible for applications running in a Windows NT or Windows XP environment to access system I/O.
- New sections on MASM32, C++, and Visual BASIC including examples of the use of visual programming languages and integrated development environments (IDE) for BASIC, assembly language and Visual Studio applications.
- New sections on LabVIEW, DASYLab, Matlab with an updated section on DADiSP.
- An expanded chapter with eight diverse PC applications described in detail.
- A revised and expanded chapter on reliability and fault-finding including detailed fault-location charts, diagnostic and benchmarking software.
- Considerably extended and updated reference information.
- A companion web site with downloadable executables, source code, links to manufacturers and suppliers, and additional reference material.

Companion website

The companion website, www.key2control.com, has a variety of additional resources including downloadable source code and executable programs. A visit to the site is highly recommended!

This book is the end result of several thousand hours of research and development and I should like to extend my thanks and gratitude to all those, too numerous to mention, who have helped and assisted in its production. May it now be of benefit to many!

Mike Tooley

1 The PC

Ever since IBM entered the personal computer scene, it was clear that its 'PC' (first announced in 1981) would gain an immense following. In a specification that now seems totally inadequate, the original PC had an 8088 processor, 64–256 kilobyte (KB) of system board RAM (expandable to 640 KB with 384 KB fitted in expansion slots). It supported two 360 KB floppy disk drives, an 80 columns × 25 lines display, and 16 colours with an IBM colour graphics adapter.

The original PC was quickly followed by the PC-XT. This machine, an improved PC, with a single $5\frac{1}{4}$ in. 360 KB floppy disk drive and a 10 megabyte (MB) hard disk, was introduced in 1983. In 1984, the PC-XT was followed by a yet further enhanced machine, the PC-AT (where XT and AT stood for eXtended and Advanced Technology, respectively). The PC-AT used an 80286 microprocessor and catered for a $5\frac{1}{4}$ in. 1.2 MB floppy drive together with a 20 MB hard disk.

While IBM were blazing a trail, many other manufacturers were close behind. The standards set by IBM attracted much interest from other manufacturers, notable among whom were Compaq and Olivetti. These companies were not merely content to produce machines with an identical specification but went on to make further significant improvements. Other manufacturers were happy to simply 'clone' the PC; indeed, one could be excused for thinking that the highest

Photo 1.1 *Setting up a PC requires access to both hardware and software*

Table 1.1 *Typical PC specifications from 1981 to the present day*

Standard	Approximate year of introduction	Processor	RAM	Cache	Floppy disk	Hard disk	Graphics	Parallel port(s)	Serial port(s)	Clock speed	Bus
PC	1981	8088	16–256 KB	Nil	1 or 2 5¼ in. 360 KB	None	Text or CGA	1 or 2	1 or 2	8 MHz	ISA
XT	1982	8088 or 80286	640 KB	Nil	1 or 2 5¼ in. 360 KB	10 MB	Text or CGA	1 or 2	1 or 2	8 or 10 MHz	ISA
AT	1984	80286	1 MB	Nil	1 5¼ in. 1.2 MB	20 MB	Text, CGA, or EGA	1 or 2	1 or 2	12 or 16 MHz	EISA
386SX based	1986	80386SX	1–8 MB	64 KB	1 3½ in. 1.44 MB	80 MB	Text, VGA, or SVGA	1 or 2	1 or 2	16 or 20 MHz	EISA
386DX based	1986	80386DX	1–16 MB	128 KB	1 3½ in. 1.44 MB	120 MB	Text, VGA, or SVGA	1 or 2	1 or 2	25 or 33 MHz	EISA
486SX based	1991	80486SX	4–16 MB	256 KB	1 3½ In. 1.44 MB	230 MB	Text, VGA, or SVGA	1 or 2	1 or 2	25 or 33 MHz	ISA and VL
486DX based	1991	80486DX	4–64 MB	256 KB	1 3½ in. 1.44 MB	340 MB	Text, VGA, or SVGA	1 or 2	1 or 2	33, 50, or 66 MHz	ISA and VL
PS/2		80286 or 80386	1–16 MB	Nil	1 3½ in. 1.44 MB	44, 70, or 117 MB	Text, EGA, or VGA	1 or 2	1 or 2	8, 10, 16, or 20 MHz	MCA
PS/1	1986	80286 or 80386	1–16 MB	Nil	1 3½ in. 1.44 MB	85 or 130 MB	Text, VGA, or SVGA	1 or 2	1 or 2	8, 10, 16, or 20 MHz	MCA
Early Pentium	1993	Pentium	8–64 MB	512 KB	1 3½ in. 1.44 MB	640 MB or 1.2 GB	Text, VGA, or SVGA	1 or 2	1 or 2	66 or 133 MHz	EISA and VL
Current	2004	Pentium 4, Celeron, Athlon, etc.	256 MB to 1 GB	512 KB	1 3½ in. 1.44 MB	60, 80, or 120 GB	Text, VGA, SVGA, or XGA	1 or 2	1 or 2	2.1, 2.8, or 3.2 GHz	PCI and USB

Photo 1.2 *A modern high-specification dual-BIOS PC motherboard*

accolade that could be offered by the computer press was that a machine was 'IBM compatible'.

This chapter sets out to introduce the PC and provide an insight into the architecture, construction, and operation of a 'generic PC'. It should, perhaps, be stated that the term 'PC' now applies to such a wide range of equipment that it is difficult to pin down the essential ingredients of such a machine. However, at the risk of oversimplifying matters, a 'PC' need only satisfy two essential criteria:

- Be based upon an Intel 16-, 32-, or 64-bit processor, such as a 'x86, Pentium, or a compatible device (such as a Celeron, Athlon, or Duron processor).
- Be able to support the Microsoft MS-DOS operating system, Microsoft Windows, or a compatible operating system.

Other factors, such as available memory size, disk capacity, and display technology remain secondary.

To illustrate the progress in technology over the last 20 or so years, Table 1.1 shows typical specifications for various types of PC. However, before considering PC architecture in more detail, we shall begin by briefly describing the basic elements of a microcomputer system.

Microcomputer systems

The principal elements within a microcomputer system consist of a central processing unit (CPU), read/write memory (RAM), read-only memory (ROM), together with one (or more) input/output (I/O) devices. These elements are

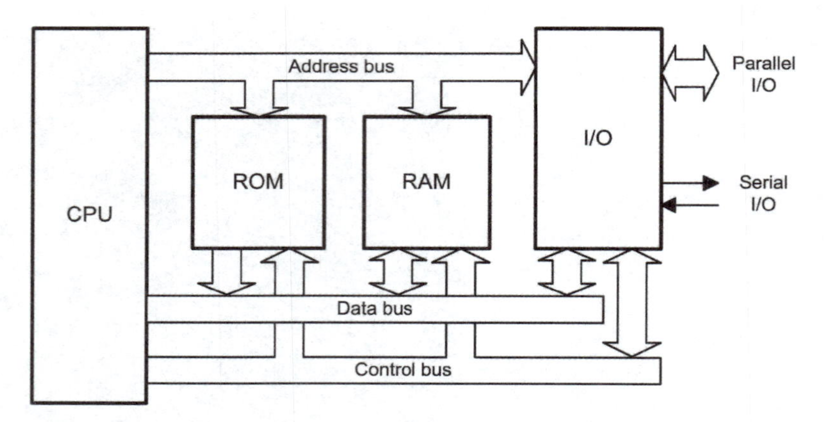

Figure 1.1 *Elements of a microcomputer system*

connected together by a bus system along which data, address, and control signals are passed, as shown in Figure 1.1.

The CPU is the microprocessor itself (e.g. a x86 or Pentium device), whilst the read/write and read-only memory are implemented using a number of semiconductor memory devices (RAM and ROM, respectively). The semiconductor ROM provides non-volatile storage for part of the operating system code (the code remains intact when the power supply is disconnected, whereas the semiconductor RAM provides storage for the remainder of the operating system code, applications programs, and transient data. It is important to note that this memory is volatile, and any program or data stored within it will be lost when the power supply is disconnected.

The operating system is a collection of programs and software utilities that provide an environment in which applications software can easily interact with system hardware. The operating system also provides the user with a means of carrying out general housekeeping tasks, such as disk formatting, disk copying, etc. In order to provide a means of interaction with the user (via keyboard entered commands and onscreen prompts and messages), the operating system incorporates a shell program (e.g. the COMMAND.COM program provided within MSDOS).

Part of the semiconductor RAM is reserved for operating system use and for storage of a graphic/text display (as appropriate). In order to optimize the use of the available memory, most modern operating systems employ memory management techniques which allocate memory to transient programs and then release the memory when the program is terminated. A special type of program (known as a 'terminate and stay resident' program) can, however, remain resident in memory for immediate execution at some later stage (e.g. when another application program is running).

I/O devices provide a means of connecting external hardware, such as keyboards, displays, and disk controllers. I/O is usually handled by a number of specialized VLSI devices, each dedicated to a particular I/O function (such as disk control, graphics control, etc.). Such I/O devices are, in themselves, very complex and are generally programmable (requiring software configuration during system initialization).

Photo 1.3 *Arcom's Pegasus embedded PC controller (photo courtesy of Arcom)*

The elements within the microcomputer system shown in Figure 1.1 (CPU, ROM, RAM, and I/O) are connected together by three distinct bus systems:

1 The *address bus* along which address information is passed.
2 The *data bus* along which data is passed.
3 The *control bus* along which control signals are passed.

Data representation

The information present on the bus lines is digital and is represented by the two binary logic states: logic 1 (or *high*) and logic 0 (or *low*). All addresses and data values must therefore be coded in binary format with the *most significant bit* (MSB) present on the uppermost address or data line and the *least significant bit* (LSB) on the lowermost address or data line (labelled A0 and D0, respectively).

The bus lines (whether they be address, data, or control) are common to all four elements of the system. Data is passed via the data bus line in parallel groups of either 8, 16, 32, or 64 bits. An 8-bit group of data is commonly known as a *byte* whereas a 16-bit group is usually referred to as a *word*.

As an example, assume that the state of the eight data bus lines in a system at a particular instant of time is as shown in Figure 1.2.

The binary value (MSB first, LSB last) is 10100111 and its decimal value (found by adding together the decimal equivalents wherever a '1' is present in the corresponding bit position) is 167.

It is often more convenient to express values in hexadecimal (base 16) format (see Appendix D). The value of the byte (found by grouping the binary digits into two 4-bit *nibbles* and then converting each to its corresponding hexadecimal character, is A7 (variously shown as A7h, A7H, HA7, or $A7_{16}$ in order to indicate that the base is 16).

The data bus invariably comprises 8, 16, or 32 separate lines labelled D0 to D7 (or D0 to D16, etc.), whilst the address bus may have as few as 20 lines in

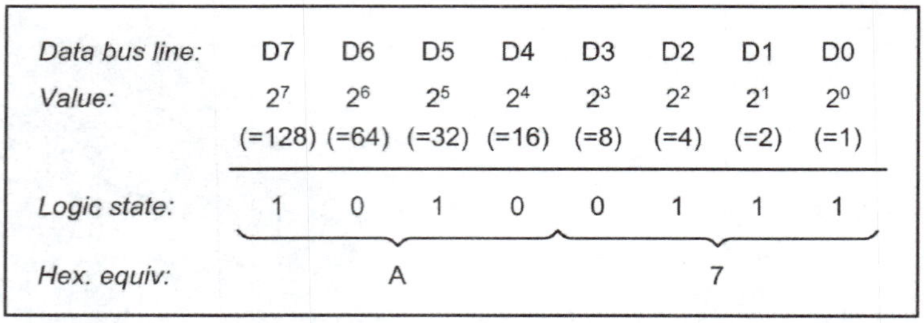

Data bus line:	D7	D6	D5	D4	D3	D2	D1	D0
Value:	2^7 (=128)	2^6 (=64)	2^5 (=32)	2^4 (=16)	2^3 (=8)	2^2 (=4)	2^1 (=2)	2^0 (=1)
Logic state:	1	0	1	0	0	1	1	1
Hex. equiv:	A				7			

Figure 1.2 *Data representation in a microcomputer system*

Table 1.2 *Relationship between data bus size and largest data value*

Number of data lines	Number of bytes	Largest data value
8	1	255
16	2	65 535
32	4	4 294 967 295
64	8	Approximately 1.84×10^{19}

Table 1.3 *Relationship between address bus lines and linear addressable memory*

Number of address lines	Linear addressable memory
16	64 KB
20	1 MB
22	4 MB
24	16 MB
32	4 GB

early PC, XT, and AT models (labelled A0 to A19) and as many as 32 bits in modern equipment (where the address lines are labelled A0 to A31).

The relationship between data bus lines and the largest data value possible that can be conveyed *at any particular instant* is shown in Table 1.2.

Similarly, with more address lines it is possible to address a larger memory. The relationship between address bus lines and *linear addressable* memory is shown in Table 1.3.

Bus expansion

The system shown in Figure 1.1 can be expanded by making the three bus systems accessible to a number of expansion modules, as shown in Figure 1.3. These modules (which invariably take the form of plug-in printed circuit cards) provide additional functionality associated with input/output (I/O), graphics, or disk control. Expansion cards are often referred to as 'option cards' or 'adapter cards', and they provide a means of extending a basic microcomputer system for a particular application.

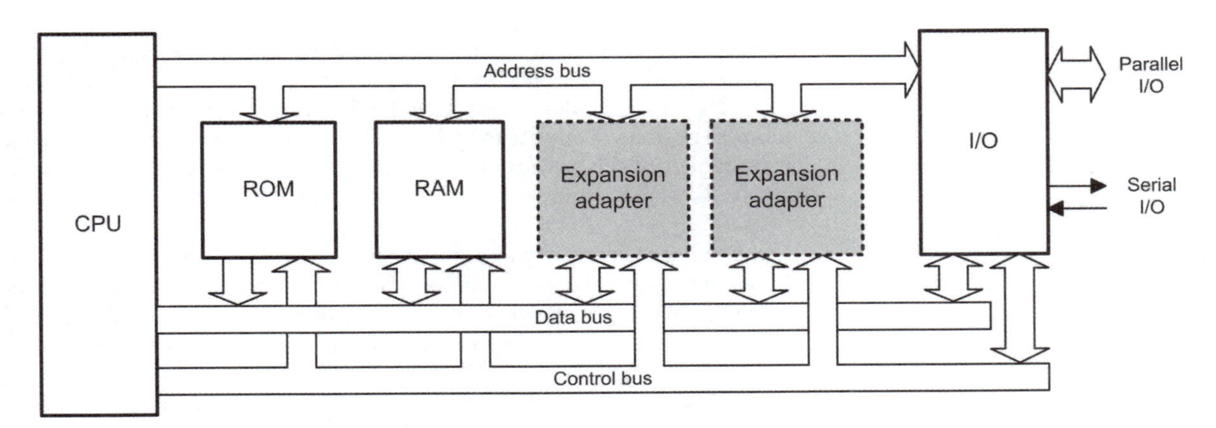

Figure 1.3 *Microcomputer system with bus expansion capability*

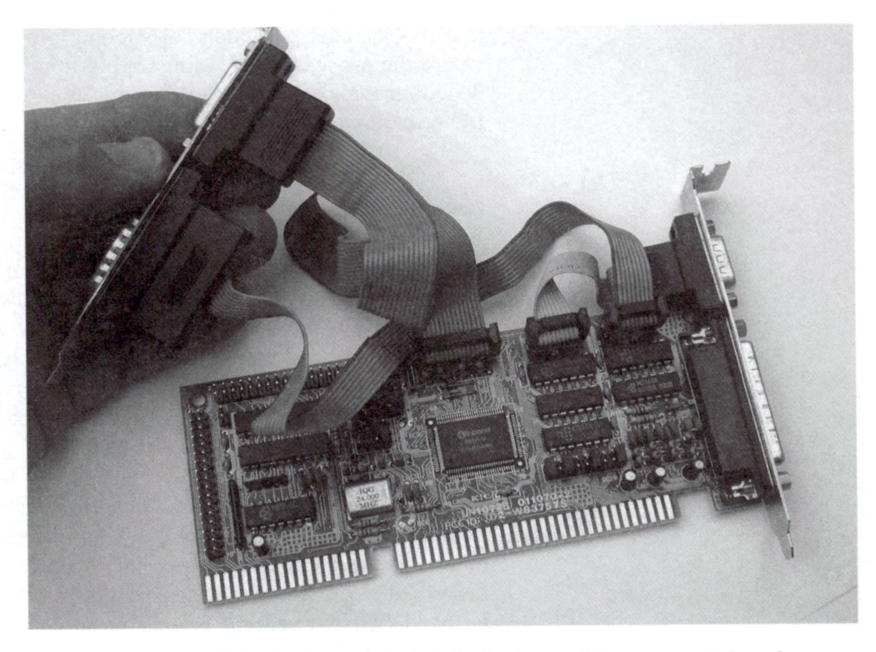

Photo 1.4 *A typical ISA expansion card which provides two serial and two parallel ports*

Microprocessor operation

The majority of operations performed by a microprocessor involve the movement of data. Indeed, the program code (a set of instructions stored in ROM or RAM) must itself be fetched from memory prior to execution. The microprocessor thus performs a continuous sequence of instruction fetch and execute cycles. The act of fetching an instruction code or operand or data value from memory involves a read operation whilst the act of moving data from the microprocessor to a memory location involves a write operation.

Microprocessors determine the source of data when it is being read (and the destination of data when it is being written) by placing a unique address on the address bus. The address at which the data is to be placed (during a *write operation*) or from which it is to be fetched (during a *read operation*) can either constitute part of the memory of the system (in which case it may be within ROM or RAM) or it can be considered to be associated with an input/output (I/O) port.

Since the data bus is connected to a number of VLSI devices, an essential requirement of such chips (e.g. ROM or RAM) is that their data outputs should be capable of being isolated from the bus whenever necessary. These VLSI devices are fitted with select or enable inputs which are driven by address decoding logic (not shown in Figures 1.1 and 1.3). This logic ensures that several ROM, RAM, and I/O devices never simultaneously attempt to place data on the bus!

The inputs of the address decoding logic are derived from one, or more, of the address bus lines. The address decoder effectively divides the available memory into blocks, each of which correspond to one (or more VLSI device). Hence, where the processor is reading and writing to RAM, for example, the address decoding logic will ensure that only the RAM is selected whilst the ROM and I/O remain isolated from the data bus.

Data transfer and control

The transfer of data to and from I/O devices (such as hard drives) can be arranged in several ways. The simplest method (known as *programmed I/O*, involves moving all data through the CPU. Effectively, each item of data is first read into a CPU register and then written from the CPU register to its destination. This form of data transfer is straightforward but relatively slow, particularly where a large volume of data has to be transferred. The method is also somewhat inflexible as the transfer of data has to be incorporated specifically within the main program flow.

An alternative method allows data to be transferred 'on demand' in response to an *interrupt request*. Essentially, an interrupt request (IRQ) is a signal that is sent to the CPU when a peripheral device requires attention (this topic is described in greater detail later in this chapter). The advantage of this method is that CPU intervention is only required when data is actually ready to be transferred or is ready to be accepted (the CPU can thus be left to perform more useful tasks until data transfer is necessary).

The final method, *direct memory access* (DMA), provides a means of transferring data between I/O and memory devices without the need for direct CPU intervention. Direct memory access provides a means of achieving the highest possible data transfer rates, and it is instrumental in minimizing the time taken to transfer data to and from the hard disk or another mass storage device. Additional *DMA request* (DRQ) and *DMA acknowledge* (DACK) signals are necessary so that the CPU is made aware that other devices require access to the bus. Furthermore, as with IRQ signals, several different DMA channels must be provided in order to cater for the needs of several devices that may be present within a system. This topic is dealt with in greater detail later in this chapter.

Photo 1.5 *Serial (RS-232 and USB) and parallel port I/O on a modern motherboard. The DIP switch is used for setting various I/O options*

Parallel versus serial I/O

Most microcomputer systems (including the PC) have provision for both parallel (e.g. a parallel printer) and serial (e.g. an RS-232 port) I/O. Parallel I/O involves transferring data one (or more) bytes at a time between the microcomputer and peripheral along multiple wires; usually eight plus a common *ground* connection). Serial I/O, on the other hand, involves transferring 1-bit after another along a pair of lines (one of which is usually a ground connection).

In order to transmit a byte (or group of bytes) the serial method of I/O must comprise a sequence or stream of bits. The stream of bits will continue until all of the bytes concerned have been transmitted and additional bits may be added to the stream in order to facilitate decoding and provide a means of error detection.

Since data present on a microprocessor data bus exists in parallel form, it should be apparent that a means of parallel-to-serial and serial-to-parallel conversion will be required in order to implement a serial data link between microcomputers and peripherals (see Figure 1.4).

Serial data may be transferred in either synchronous or asynchronous mode. In the former case, all transfers are carried out in accordance with a common clock signal (the clock must be available at both ends of the transmission path). Asynchronous operation involves transmission of data in packets: each packet containing the necessary information required to decode the data which it contains. Clearly this technique is more complex, but it has the considerable advantage that a separate clock signal is not required.

As with parallel I/O, signals from serial I/O devices are invariably TTL compatible. It should be noted that, in general, such signals are unsuitable for anything other than the shortest of transmission paths (e.g. between a keyboard

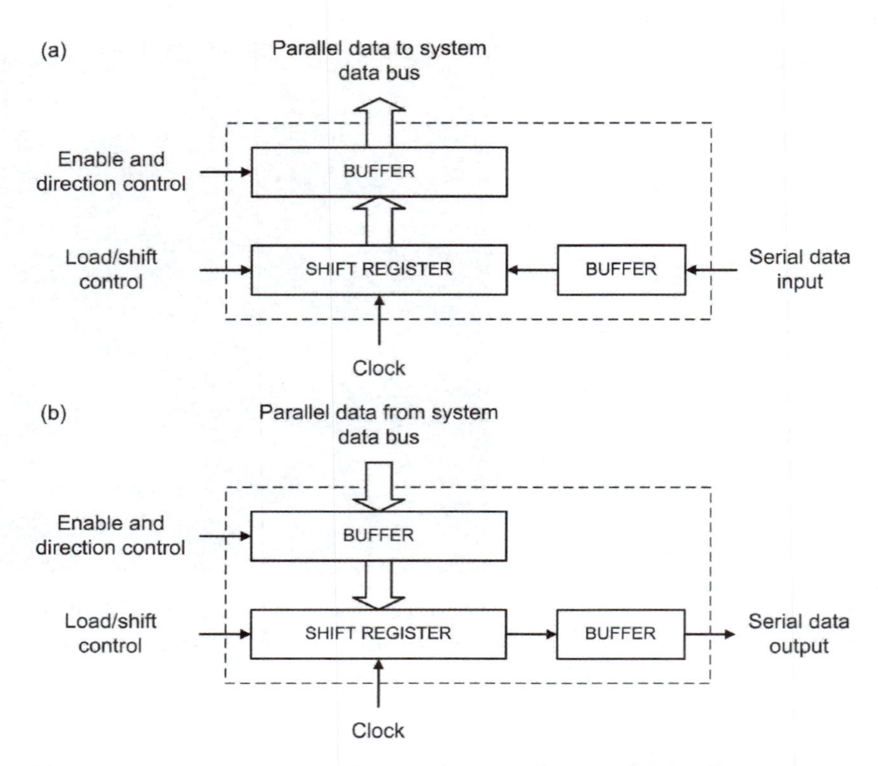

Figure 1.4 *Data conversion: (a) serial-to-parallel and (b) parallel-to-serial*

and a computer system enclosure). Serial data transmission over any appreciable distance requires additional line drivers to provide buffering and level shifting between the serial I/O device and the physical medium. In addition, line receivers are required to condition and modify the incoming signal to TTL levels.

The processor

The processor, or *central processing unit* (CPU), is crucial in determining the performance of a PC and processors (see Table 1.4) have been consistently upgraded since the first PC arrived on the scene in 1981. Not surprisingly given the advances in semiconductor technology, the latest processors offer vastly improved performance when compared with their predecessors. Despite this, it is important to remember that a core of common features has been retained in order to preserve compatibility. Hence all current CPU devices are based on a superset of the basic 8088/8086 registers. For this reason it is worth spending a little time looking at the development of processor technology over the last two decades.

The x86 processor family

The original member of the x86 family was Intel's first true 16-bit processor which had 20 address lines that could directly address up to 1 MB of RAM. The chip was available in 5, 6, 8, and 10 MHz versions. The 8086 was designed

Table 1.4 *Common processors used in modern PC equipment (see Appendix G for more detailed information)*

CPU type	Manufacturer	Socket (see Appendix F)	Speeds (MHz)	Effective front side bus (FSB) – MHz	L2 Cache	Internal bus (bit)	Year of introduction
Pentium	Intel	Socket 4	60–66	60–66		64	March 1993
Pentium	Intel	Socket 5	75–120	60–66		64	March 1994
Pentium	Intel	Socket 7	120–200	60–66		64	March 1995
6x86	Cyrix/IBM	Socket 7	PR90–PR200	40–75		64	October 1995
K5	AMD	Socket 7	PR75–PR166	60–66		64	June 1996
6x86L	Cyrix/IBM	Socket 7	PR120–PR200	50–75		64	January 1997
Pentium MMX	Intel	Socket 7	133–233	60–66		64	January 1997
K6	AMD	Socket 7	166–233	66		64	April 1997
6x86MX/MII	Cyrix/IBM	Socket 7	PR166–PR366	66–83		64	May 1997
K6-III	AMD	Socket 7	400–450	100	256 KB 4-way	64	February 1999
K6-2+	AMD	Socket 7	450–550	100	128 KB	64	April 2000
K6-III+	AMD	Socket 7	450–500	95–100	256 KB 4-way	64	April 2000
Pentium Pro	Intel	Socket 8	150–200	60/66	256, 512, and 1024 KB	64	November 1995
Pentium II	Intel	Slot 1	233–300	66	512 KB	64	May 1997
Celeron	Intel	Slot 1	266–300	66		64	April 1998
Pentium II Xeon	Intel	Slot 2	400–450	100	512, 1024, and 2048 KB	64	June 1998
Celeron	Intel	Slot 1/ Socket 370	300–533	66	128 KB	64	August 1998
Pentium III	Intel	Slot 1	450–600	100/133	512 KB	64	February 1999
Pentium III Xeon	Intel	Slot 2	500–550	100	512, 1024, and 2048 KB	64	March 1999
Pentium III Xeon	Intel	Slot 2	600–1000	100/133	256, 1024, and 2048 KB	256	October 1999
Celeron II	Intel	Socket 370	533–1100	66/100	128 KB	256	March 2000
Athlon	AMD	Slot A	500–700	200	512 KB	64	August 1999
Duron	AMD	Socket A	600–950	200	64 KB	64	June 2000
Athlon	AMD	Slot A/ Socket A	650–1400	200/266	256 KB	64	June 2000
Athlon XP	AMD	Socket A	1333–1733 (XP1500+ to XP2100+)	266	256 KB	64	October 2001
Pentium 4	Intel	Socket 423/ Socket 478	1300–2000	400	256 KB	256	November 2000
Pentium 4	Intel	Socket 478	1600–2533	400–533	512 KB	256	January 2002

Table 1.5 *8088/8086 signal lines*

Signal	Function	Notes
AD0–AD7 (8088)	Address/data bus	Multiplexed 8-bit address/data lines
A8–A19 (8088)	Address bus	Non-multiplexed address lines
AD0–AD15 (8086)	Address/data bus	Multiplexed 16-bit address/data bus
A16–A19 (8086)	Address bus	Non-multiplexed address lines
S0–S7	Status lines	S0–S2 are only available in Maximum Mode and are connected to the 8288 Bus Controller. Note that status lines S3–S7 all share pins with other signals.
INTR	Interrupt line	Level-triggered, active high interrupt request input
NMI	Non-maskable interrupt line	Positive edge-triggered non-maskable interrupt input
RESET	Reset line	Active high reset input
READY	Ready line	Active high ready input
TEST	Test	Input used to provide synchronization with external processors. When a WAIT instruction is encountered in the instruction stream, the CPU examines the state of the TEST line. If this line is found the to be high, processor waits in an 'idle' state until the signal goes low.
QS0, QS1	Queue status lines	Outputs from the processor which may be used to keep track of the internal instruction queue.
LOCK	Bus lock	Output from the processor which is taken low to indicate that the bus is not currently available to other potential bus masters.
RQ/GT0–RQ/GT1	Request/grant	Used for signalling bus requests and grants placed in the CL register.

with modular internal architecture. This approach to microprocessor design has allowed Intel to produce a similar microprocessor with identical internal architecture but employing an 8-bit external bus. This device, the 8088, shares the same 16-bit internal architecture as its 16-bit bus counterpart. Both devices were packaged in 40-pin DIL encapsulations. The CPU signal lines are described in Table 1.5 while the pin connections for the legacy processor family will be found later in this chapter in Figure 1.12.

The 8086/8088 can be divided internally into two functional blocks comprising an Execution Unit (EU) and a Bus Interface Unit (BIU), as shown in Figure 1.5. The EU is responsible for decoding and executing instructions, whilst the BIU pre-fetches instructions from memory and places them in an instruction queue where they await decoding and execution by the EU.

The EU comprises a general and special purpose register block, temporary registers, arithmetic logic unit (ALU), a Flag (Status) Register, and control logic. It is important to note that the principal elements of the 8086 EU remain common to each of the subsequent members of the x86 family, but with additional registers with the more modern processors.

The BIU architecture varies according to the size of the external bus. The BIU comprises four Segment Registers and an Instruction Pointer, temporary storage for instructions held in the instruction queue, and bus control logic.

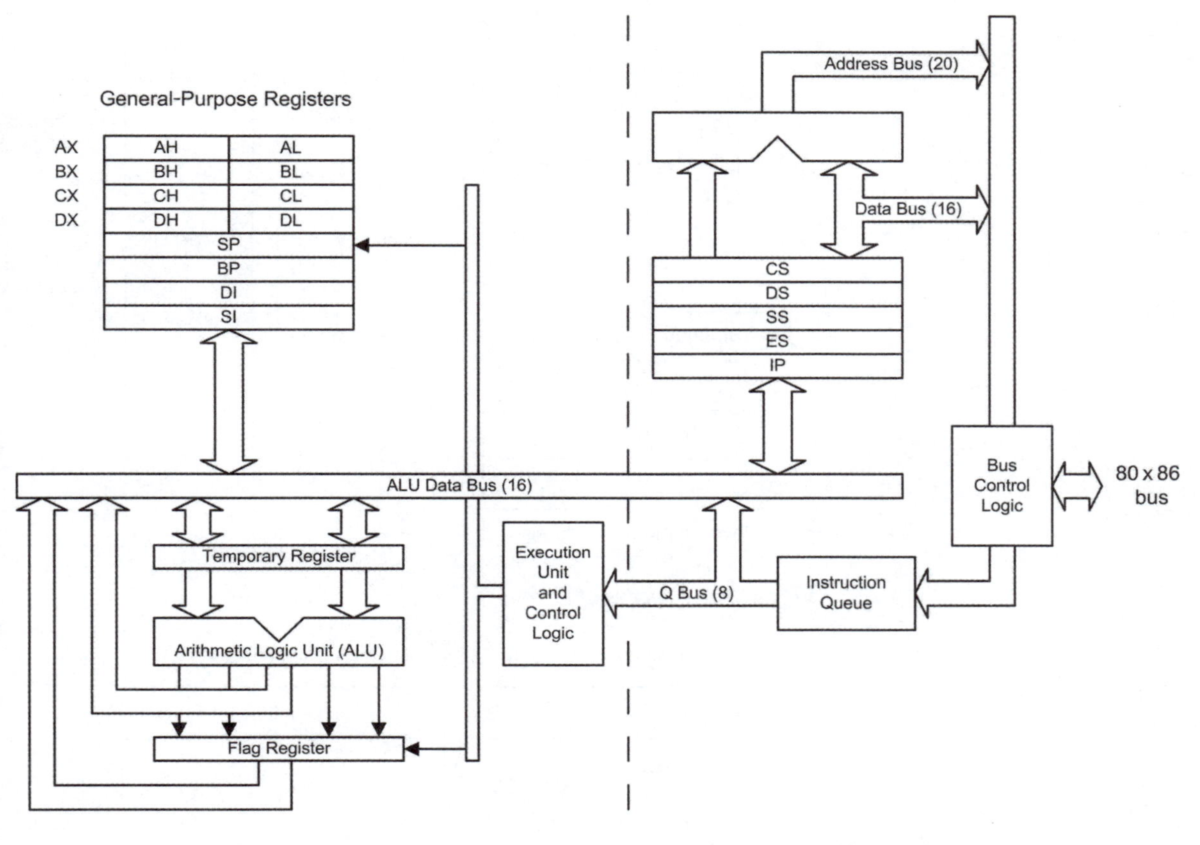

Figure 1.5 *Internal architecture of the 8086*

Addressing

The 8086 has 20 address lines and thus provides for a physical 1 MB memory address range (memory address locations 00000 to FFFFF hex.). The I/O address range is 64 KB (I/O address locations 0000 to FFFF hex.).

The actual 20-bit physical memory address is formed by shifting the segment address four 0-bits to the left (adding four least significant bits), which effectively multiplies the Segment Register contents by 16. The contents of the Instruction Pointer (IP), Stack Pointer (SP), or other 16-bit memory reference are then added to the result. This process is illustrated in Figure 1.6.

As an example of the process of forming a physical address reference, Table 1.6 shows the state of the 8086 registers after the RESET signal is applied. The instruction referenced (i.e. the first instruction to be executed after the RESET signal is applied) will be found by combining the Instruction Pointer (offset address) with the Code Segment Register (paragraph address). The location of the instruction referenced is FFFF0 (i.e. F0000 + FFF0). Note that the PC's ROM physically occupies addresses F0000 to FFFFF and that, following

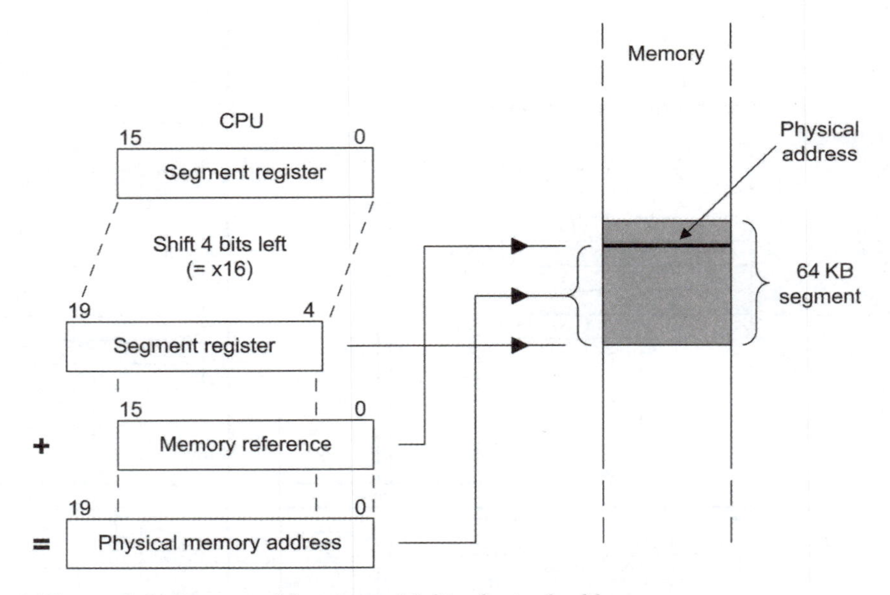

Figure 1.6 *Process of forming a 20-bit physical address*

Table 1.6 *Contents of the 8086 registers after a reset*

Register	Contents (hex.)
Flag	0002
Instruction Pointer	FFF0
Code Segment	F000
Data Segment	0000
Extra Segment	0000
Stack Segment	0000

a power-on or hardware reset, execution commences from address FFFF0 with a jump to the initial program loader.

80286, 80386, and 80486 processors

Intel's 80286 CPU was first employed in the PC-AT and PS/2 Models 50 and 60. The 80286 offers a 16 MB physical addressing range but incorporates memory management capabilities that can map up to a gigabyte of virtual memory. Depending upon the application, the 80286 is up to six times faster than the standard 5 MHz 8086 while providing upward software compatibility with the 8086 and 8088 processors.

The 80286 had 15 16-bit registers, of which 14 are identical to those of the 8086. The additional machine status word (MSW) register controls the operating mode of the processor and also records when a task switch takes place.

The bit functions within the MSW are summarized in Table 1.7. The MSW is initialized with a value of FFF0H upon reset, the remainder of the 80286

Table 1.7 *Bit functions in the 80286 machine status word*

Bit	Name	Function
0	Protected mode (PE)	Enables protected mode and can only be cleared by asserting the RESET signal.
1	Monitor processor (MP)	Allows WAIT instructions to cause a 'processor extension not present' exception (Exception 7).
2	Emulate processor (EP)	Causes a 'processor extension not present' exception (Exception 7) on ESC instructions to allow emulation of a processor extension.
3	Task switched (TS)	Indicates that the next instruction using a processor extension will cause Exception 7 (allowing software to test whether the current processor extension context belongs to the current task).

registers being initialized as shown in Table 1.6. The 80286 is packaged in a 68-pin JEDEC type-A plastic leadless chip carrier (PLCC), see Figure 1.12.

The 80386 (or '386) was designed as a *full* 32-bit device capable of manipulating data 32 bits at a time and communicating with the outside world through a 32-bit address bus. The 80386 offers a 'virtual 8086' mode of operation in which memory can be divided into 1 MB chunks with a different program allocated to each partition.

The 80386 was available in two basic versions. The 80386SX operates internally as a 32-bit device but presents itself to the outside world through only 16 data lines. This has made the CPU extremely popular for use in low-cost systems which could still boast the processing power of a 386 (despite the obvious limitation imposed by the reduced number of data lines, the 'SX' version of the 80386 runs at approximately 80% of the speed of its fully fledged counterpart).

The 80386 comprises a Bus Interface Unit (BIU), a Code Pre-fetch Unit, an Instruction Decode Unit, an Execution Unit (EU), a Segmentation Unit, and a Paging Unit. The Code Pre-fetch Unit performs the program 'look-ahead' function. When the BIU is not performing bus cycles in the execution of an instruction, the Code Pre-fetch Unit uses the BIU to fetch sequentially the instruction stream. The pre-fetched instructions are stored in a 16-byte 'code queue' where they await processing by the Instruction Decode Unit. The pre-fetch queue is fed to the Instruction Decode Unit which translates the instructions into micro-code. These micro-coded instructions are then stored in a three-deep instruction queue on a first-in first-out (FIFO) basis. This queue of instructions awaits acceptance by the EU. Immediate data and op-code offsets are also taken from the pre-fetch queue.

The 80486 processor was not merely an upgraded 80386 processor; its redesigned architecture offers significantly faster processing speeds when running at the *same* clock speed as its predecessor. Enhancements include a built-in maths coprocessor, internal cache memory, and cache memory control. The internal cache is responsible for a significant increase in processing speed. As a result, a '486 operating at 25 MHz can achieve a faster processing speed than a '386 operating at 33 MHz.

The '486 uses a large number of additional signals associated with parity checking (PCHK) and cache operation (AHOLD, FLUSH, etc.). The cache comprises a set of four 2-KB blocks (128 × 16 bytes) of high-speed internal memory. Each 16-byte line of memory has a matching 21-bit 'tag'. This tag comprises a 17-bit linear address together with four protection bits. The cache control block contains 128 sets of 7 bits. Three of the bits are used to implement the least recently used (LRU) system for replacement and the remaining 4 bits are used to indicate valid data.

Interrupt handling

Interrupt service routines are subprograms stored away from the main body of code that are available for execution whenever the relevant interrupt occurs. However, since interrupts may occur at virtually any point in the execution of a main program, the response must be automatic; the processor must suspend its current task and save the return address so that the program can be resumed at the point at which it was left. Note that the programmer must assume responsibility for preserving the state of any registers which may have their contents altered during execution of the interrupt service routine.

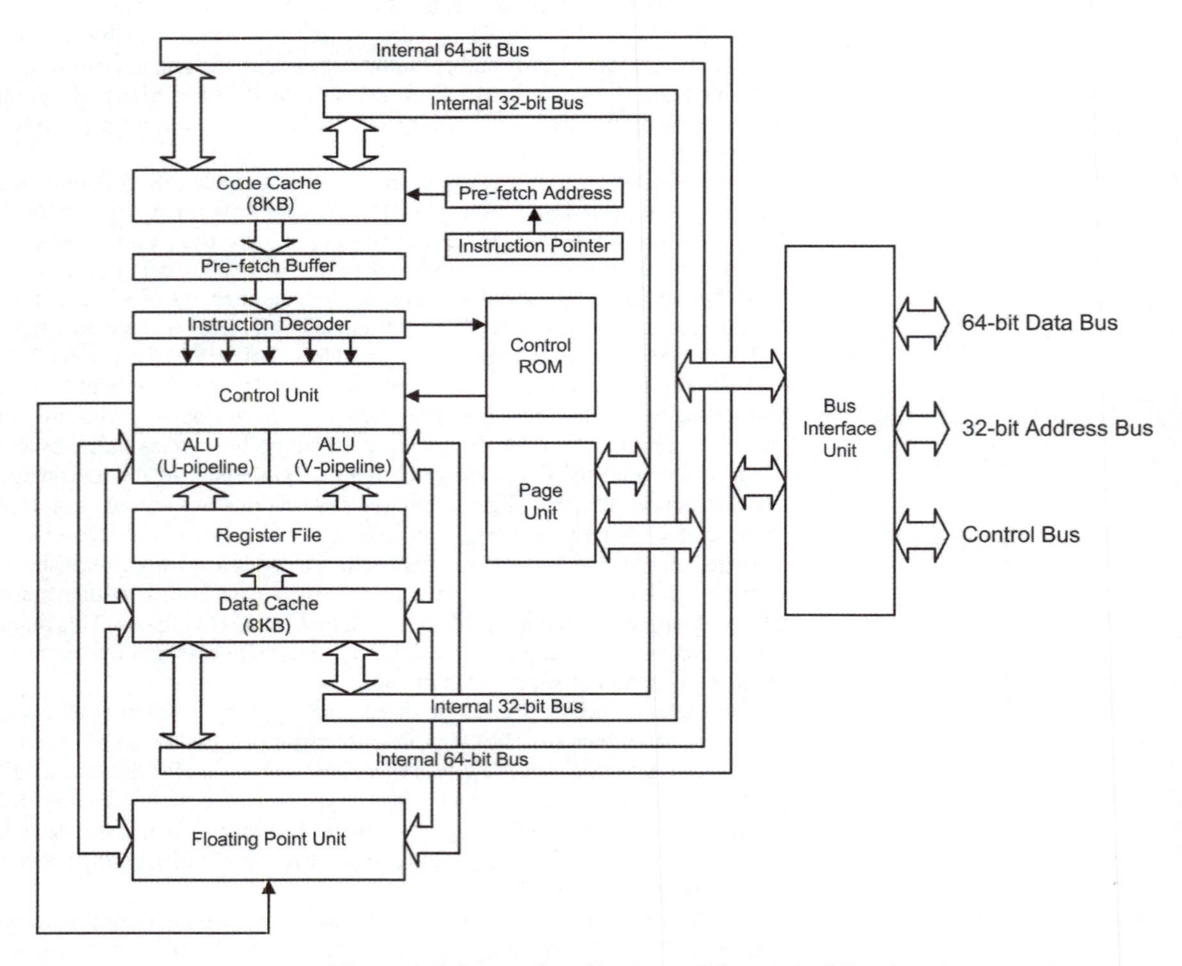

Figure 1.7 *Internal architecture of the original Pentium processor*

The Intel processor family uses a table of 256 4-byte pointers stored in the bottom 1 KB of memory (addresses 0000H to 03FFH). Each of the locations in the Interrupt Pointer Table can be loaded with a pointer to a different interrupt service routine. Each pointer contains 2 bytes for loading into the Instruction Pointer (IP). This allows the programmer to place his/her interrupt service routines in any appropriate place within the 1 MB physical address space.

The Pentium family of processors

Initially running at 60 MHz, the Pentium could achieve 100 MIPS. The original Pentium had an architecture based on 3.2 million transistors and a 32-bit address bus like the 486 but a 64-bit external data bus. The chip was capable of operation at twice the speed of its predecessor, the '486 (Figure 1.7).

The first generation Pentium was eventually to become available in 60, 66, 75, 90, 100, 120, 133, 150, 166, and 200 MHz versions. The first ones fitted Socket 4 boards whilst the rest fitted Socket 7 boards (see Photo 1.6). The Pentium was *super-scalar* and could execute two instructions per clock cycle. With two separate 8 KB caches it was much faster than a '486 with the same clock speed.

The Pentium Pro incorporated a number of changes over the Pentium which made the chip run faster for the same clock speeds. Three instead of two instructions can be decoded in each clock cycle and instruction decoding and execution are decoupled, meaning that instructions can still be executed if one pipeline stops. Instructions could also be executed out of strict order. The Pentium Pro had an 8 KB level 1 cache for data and a separate cache for instructions. The chip was available with up to 1 MB of onboard level 2 cache which further

Photo 1.6 *Socket 7 (with lever raised ready to accept a processor)*

Photo 1.7 *A modern Slot 1 Pentium processor*

increased data throughput. The architecture of the Pentium Pro was optimized for 32-bit code, but the chip would only run 16-bit code at the same speed as its predecessor.

Originally released in 1997, the Pentium MMX was intended to improve multimedia performance although software had to be specially written for it to have an effect. This software had to make use of the new MMX instruction set that was an extension off the normal 8086 instruction set. Other improvements produced a chip that could run faster than previous Pentiums.

Optimized for 32-bit applications, the Pentium 2 had 32 KB of level 1 cache (16 KB each for data and instructions) and had a 512 KB of level 2 cache on package. To discourage competitors from making direct replacement chips, this was the first Intel chip to make use of its patented 'Slot 1'. The Intel Celeron was a cut down version of Pentium II aimed primarily at the laptop market. The chip was slower as the level 2 cache had been removed. Later versions were supplied with 128 KB of level 2 cache.

The Pentium III was released in February 1999 and first made available in a 450 MHz version supporting 100 MHz bus. As a means of further improving the multimedia performance of the processor (particularly for 3D graphics), the Pentium III supports extensions to the MMX instruction set.

The latest Pentium 4 architecture is based on new 'NetBurst' architecture that combines four technologies: Hyper Pipelined Technology, Rapid Execution Engine, Execution Trace Cache, and a 400 MHz system bus. The Pentium 4 processor (see Photo 1.7) is available at speeds ranging from 1.70 to 2.80 GHz with system bus speeds of 400 and 533 MHz (the latter delivering a staggering 4.2 GB of data-per-second into and out of the processor). This performance is accomplished through a physical signalling scheme of quad pumping the data transfers over a 133-MHz clocked system bus and a buffering scheme allowing for sustained 533 MHz data transfers.

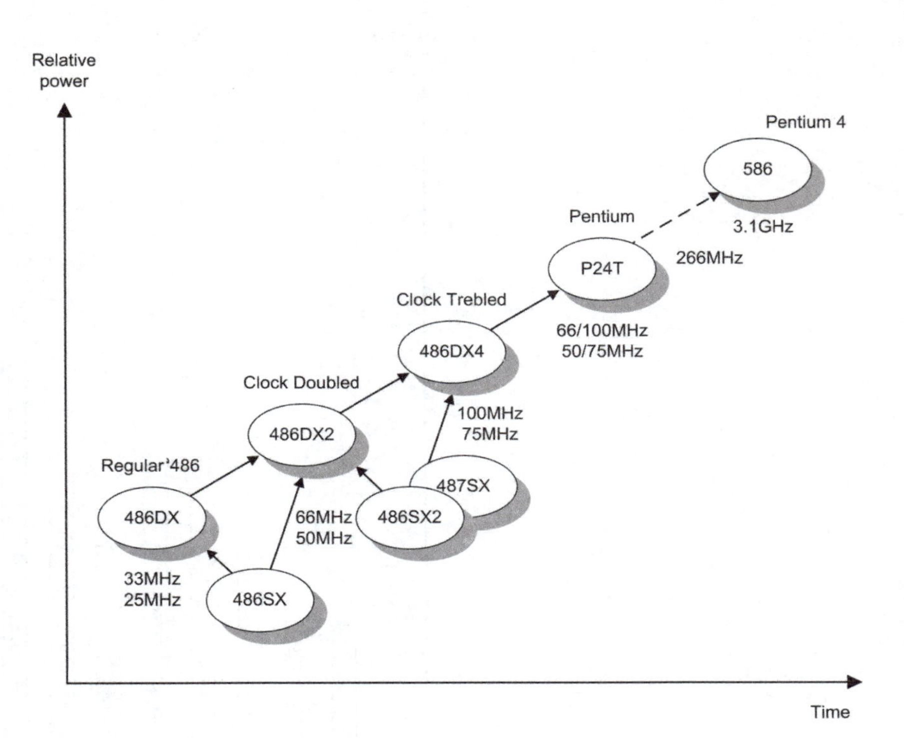

Figure 1.8 *Development of the 'x86 Intel processor family*

Figure 1.8 shows the development of x86 processor technology into the modern Pentium family of processors whilst Figure 1.9 shows how the relative power of PC processors has increased over the last two decades.

PC architecture

The generic PC, whether a 'desktop' or 'tower' system, comprises three units: system unit, keyboard, and display. The system unit itself comprises three items: system board, power supply, and floppy/hard disk drives.

The original IBM PC System Board employed approximately 100 IC devices including an 8088 CPU, an 8259A Interrupt Controller, an optional 8087 Maths Coprocessor, an 8288 Bus Controller, an 8284A Clock Generator, an 8253 Timer/Counter, an 8237A DMA Controller, and an 8255A Parallel Interface together with a host of discrete logic (including bus buffers, latches, and transceivers). Figure 1.10 shows the simplified bus architecture of the system.

Much of this architecture was carried forward to the PC-XT and the PC-AT. This latter machine employed an 80286 CPU, 80287 Maths Coprocessor, two 8237A DMA Controllers, 8254-2 Programmable Timer, 8284A Clock Generator, two 8259A Interrupt Controllers, and a 74LS612N Memory Mapper.

In order to significantly reduce manufacturing costs as well as to save on space and increase reliability, more recent AT-compatible microcomputers are based on a significantly smaller number of devices (many of which may be surface mounted types). This trend has been continued with today's powerful 386- and 486-based systems. However, the functions provided by the highly

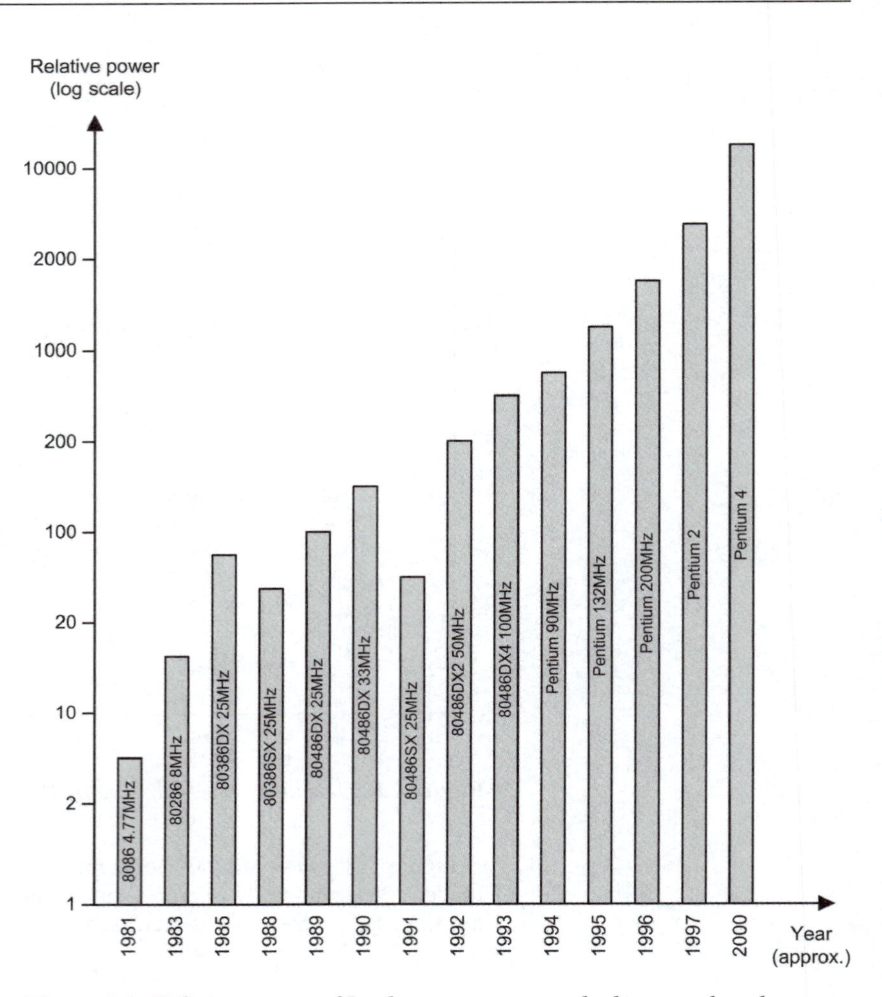

Figure 1.9 *Relative power of Intel processors over the last two decades*

integrated chipsets are usually a superset of those provided by the much larger number of devices found in their predecessors.

There is more to Figure 1.10 than mere historical interest might indicate as modern PCs can still trace their origins to this particular arrangement. It is, therefore, worth spending a few moments developing an understanding of the configuration before moving on to modern systems that employ a much faster multiple bus structure.

The 'CPU bus' (comprising lines A8 to A19 and AD0 to AD7 on the left side of Figure 1.10) is separated from the 'system bus' which links the support devices and expansion cards.

The eight least significant address and all eight of the data bus lines share a common set of eight CPU pins. These lines are labelled AD0 to AD7. The term used to describe this form of bus (where data and address information take turns to be present on a shared set of bus lines) is known as 'multiplexing'. This saves pins on the CPU package and it allowed Intel to make use of standard 40-pin packages for the 8088 and 8086 processors.

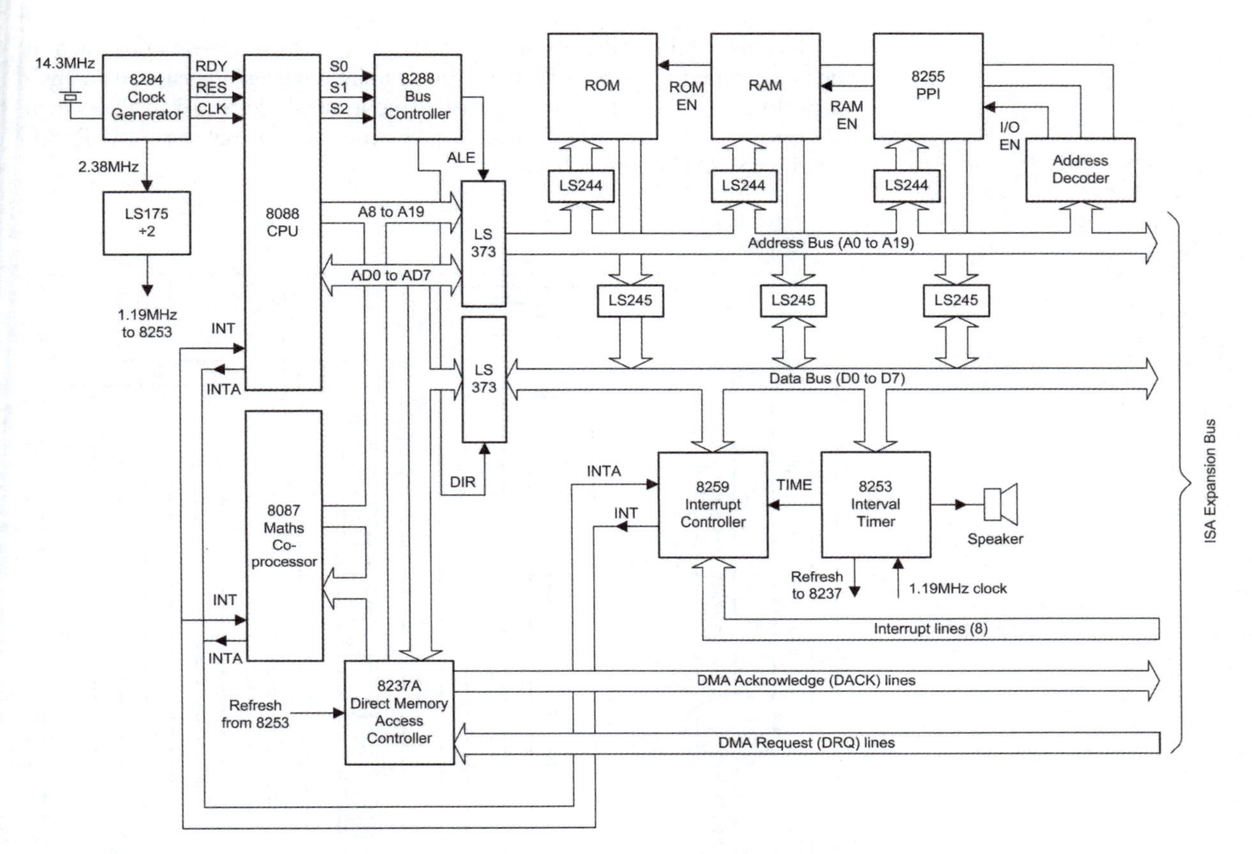

Figure 1.10 *Internal architecture of the original IBM PC*

The system address bus (available on each of the expansion connectors) comprises 20 address lines, A0 to A19. The system data bus comprises eight lines, D0 to D7. Address and data information are alternately latched onto the appropriate set of bus lines by means of the four 74LS373 8-bit data latches. The control signals, ALE (address latch enable), and DIR (direction) derived from the 8288 bus controller are used to activate the two pairs of data latches.

The CPU bus is extended to the 8087 numeric data processor (maths coprocessor). This device is physically located in close proximity to the CPU in order to simplify the PCB layout.

The original PC required a CPU clock signal of 4.773 MHz from a dedicated Intel clock generator chip. The basic timing element for this device is a quartz crystal which oscillates at a fundamental frequency of 14.318 MHz. This frequency is internally divided by three in order to produce the CPU clock.

The CPU clock frequency is also further divided by two internally and again by two externally in order to produce a clock signal for the 8253 Programmable Interrupt Timer. This device provides three important timing signals used by the system. One (known appropriately as TIME) controls the 8259 Programmable Interrupt Controller, another (known as REFRESH) provides a timing input for the 8237 DMA Controller, whilst the third is used (in conjunction with some extra logic) to produce an audible signal at the loudspeaker.

74LS244 8-bit bus drivers and 74LS245 8-bit bus transceivers link each of the major support devices with the 'system address bus' and 'system data bus', respectively. Address decoding logic (with input signals derived from the system address bus) generates the chip enable lines which activate the respective ROM, RAM, and I/O chip select lines.

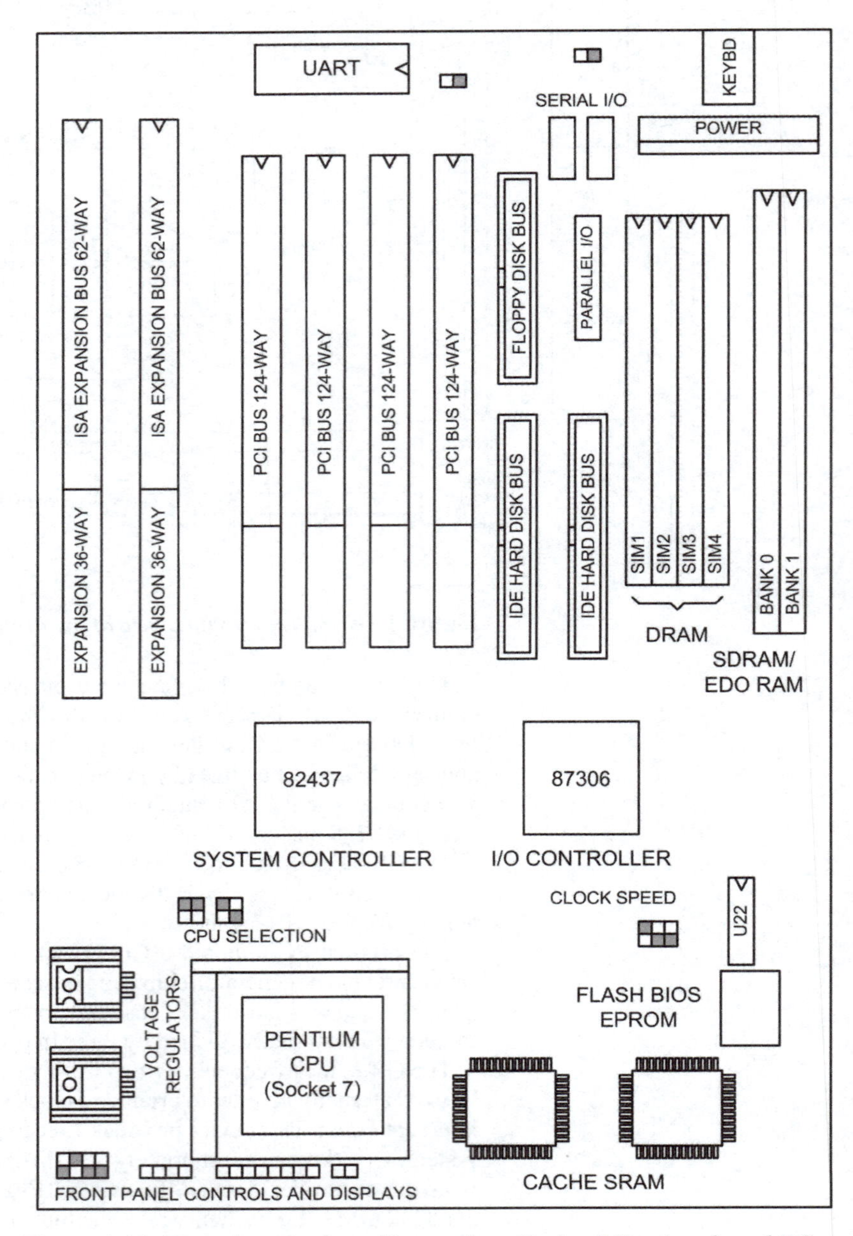

Figure 1.11 *Typical motherboard layout for a Socket 7 Pentium-based PC*

The basic system board incorporates a CPU, provides a connector for the addition of a maths coprocessor, incorporates bus and DMA control, and provides the system clock and timing signals. The system board also houses the BIOS ROM, main system RAM, and offers some limited parallel I/O. It does not, however, provide a number of other essential facilities including a video interface, disk, and serial I/O. These important functions must normally be provided by means of adapter cards (note that some systems which offer only limited expansion may have some or all of these facilities integrated into the system board).

Adapter cards are connected to the expansion bus by means of a number of expansion slots (see Chapter 2). The expansion cards are physically placed so that any external connections required are available at the rear (or side) of the unit. Connections to internal subsystems (such as hard and floppy disk drives) are usually made using lengths of ribbon cables and PCB connectors (see later).

A typical Pentium system motherboard layout is shown in Figure 1.11. This system board provides five and a single AGP card slot. Two three-terminal integrated circuit voltage regulators provide the low-voltage 3.3 V supply required by the faster Pentium processors. The 296-pin ZIF socket ('Socket 7') is suitable for a wide variety of devices, including all 6x86 and Pentium chips (including MMX) as well as the AMD K5 and K6. 512 KB of surface mounted cache memory is fitted. Two 168-pin sockets accept up to two dual-inline memory modules (DIMM) carrying fast (6–7 ns) synchronous DRAM or EDO DIMMs. Once again, standard 'primary' and 'secondary' IDE hard disk drive and/or CD-ROM ports are provided by means of two 40-way connectors.

Various combinations of DRAM can be fitted, with 128, 256 and 512 MB being the most popular. Standard IDE hard disk drive and/or CD-ROM ports are provided by means of two 40-way connectors (these are the 'primary' and 'secondary' IDE ports). Note that the floppy disk interface is provided as part

Photo 1.8 *AMD Athlon processor fitted with a CPU fan*

of the 'multi-function I/O' adapter card. The 'bare' system has provision for the following I/O facilities for:

- one or two floppy disk drives (via a 34-way ribbon cable header);
- six USB ports (two on the front panel and four on the rear panel);
- a first serial port (with its 9-way D-connector fitted to the rear bracket);
- an optional second serial port connector (via an 8-way header);
- a parallel port (with its 25-way D-connector fitted to the rear bracket);
- a game/joystick port (via a 16-way ribbon cable header);
- an optional IDE device (via a 40-way ribbon cable header which is not normally used if IDE facilities are available on the motherboard);
- firewire (a high-speed serial bus).

Cooling

All PC systems produce heat and some systems produce more heat than others. Adequate ventilation is thus an essential consideration and fans are included within the system unit to ensure that there is adequate air flow. Furthermore, internal air flow must be arranged so that it is unrestricted as modern processors and support chips run at high temperatures. These devices are much more prone to failure when they run excessively hot than when they run cool or merely warm.

Legacy support devices

Each of the major support devices present within a PC has a key role to play in off-loading a number of routine tasks that would otherwise have to be performed by the CPU. This section provides a brief introduction to each generic device together with internal architecture and, where appropriate, pin connecting details (Table 1.8).

Maths coprocessors

Maths coprocessors, 'numeric data processors' (NDP) or 'floating point units' (FPU) as they are variously called, provide a means of carrying out mathematical

Table 1.8 *Intel legacy support chips originally used with original x86 processors*

Processor type	8086	8088	80186	80286	80386
Clock generator	8284A	8284A	On-chip	82284	82384
Bus controller	8288	8288	On-chip	82288	82288
Integrated support chips				82230/82231, 82335	82230/82231, 82335
Interrupt controller	8259A	8259A	On-chip	8259A	8259A
DMA controller	8089/82258	8089/8237/82258	On-chip/82258	8089/82258	8237/82258
Timer/counter	8253/8254	8253/8254	On-chip	8253/8254	8253/8254
Maths coprocessor	8087	8087	8087	80287	80287/80387
Chip select/ wait state logic	TTL	TTL	On-chip	TTL	TTL

operations on large, 'floating point' numbers. A floating point number comprises three parts: the *sign* which may be positive or negative, the significant digits (or *mantissa*), and an *exponent* (which effectively fixes the position of the decimal point within the number). Hence, floating point numbers are essentially numbers in which the decimal point 'floats' rather than occupies a fixed position. The manipulation of floating point numbers is exclusively the province of the maths coprocessor – the ALU of a normal CPU is not equipped to operate with such numbers.

The 8087 was the original maths coprocessor which was designed to be active when mathematics related instructions were encountered in the instruction stream of an 8086 or 8088 CPU. The 8087, which is effectively wired in parallel with the 8086 or 8088 CPU, adds eight 80-bit floating point registers to the CPU register set. The 8087 maintains its own instruction queue and executes only those instructions which are specifically intended for it. The 8087 is supplied in a 40-pin DIL package, the pin connections for which are shown in Figure 1.12.

The active low TEST input of the 8086/8088 CPU is driven from the BUSY output of the 8087 NDP. This allows the CPU to respond to the WAIT instruction (inserted by the assembler/compiler) which occurs before each coprocessor instruction. An FWAIT instruction follows each coprocessor instruction which deposits data in memory for immediate use by the CPU. The instruction is then translated to the requisite 8087 operation (with the preceding WAIT) and the FWAIT instruction is translated as a CPU WAIT instruction.

During coprocessor execution, the BUSY line is taken high and the CPU (responding to the WAIT instruction) halts its activity until the line goes low. The two Queue Status (QS0 and QS1) signals are used to synchronize the instruction queues of the two processing devices. 80287 and 80387 chips provide maths co-processing facilities within AT and '386-based PC's, respectively. In '486DX (and later systems) there is no need for a maths coprocessor as these facilities have been incorporated within the CPU itself.

The 80287 and 80387 Maths Coprocessors operate in conjunction with 80286 and 80386 CPU, respectively. The '287 coprocessor was introduced in 1985 whilst the '387 made its debut in 1987. Each device represented a significant upgrade on its predecessor – the most notable factor being an increase in speed from 5 MHz (the original 8087) to 33 MHz (the fastest version of the 80387).

With the advent of the 80486, Intel placed the floating point unit *inside* the CPU (the floating point units was actually based on the 33 MHz version of the 80387). Since not all applications demand the power of a maths coprocessor, Intel developed a 'cut down' version of the '486 CPU *without* the internal floating point unit. This processor was designated the '486SX (to upgrade a system based on such a device so that it can take advantage of maths coprocessor instructions it is merely necessary to add a '487 coprocessor). The logic behind Intel's approach was apparent that users could later upgrade their systems if they found that the addition of a maths coprocessor was necessary for the software that they intended to run.

This approach could hardly be described as cost effective since the falling cost of CPUs meant that a full '486DX soon cost less than the two chips it could replace (i.e. a '486SX plus a '487SX). Happily, all modern processors incorporate internal floating point units and there is thus no further need for separate coprocessors.

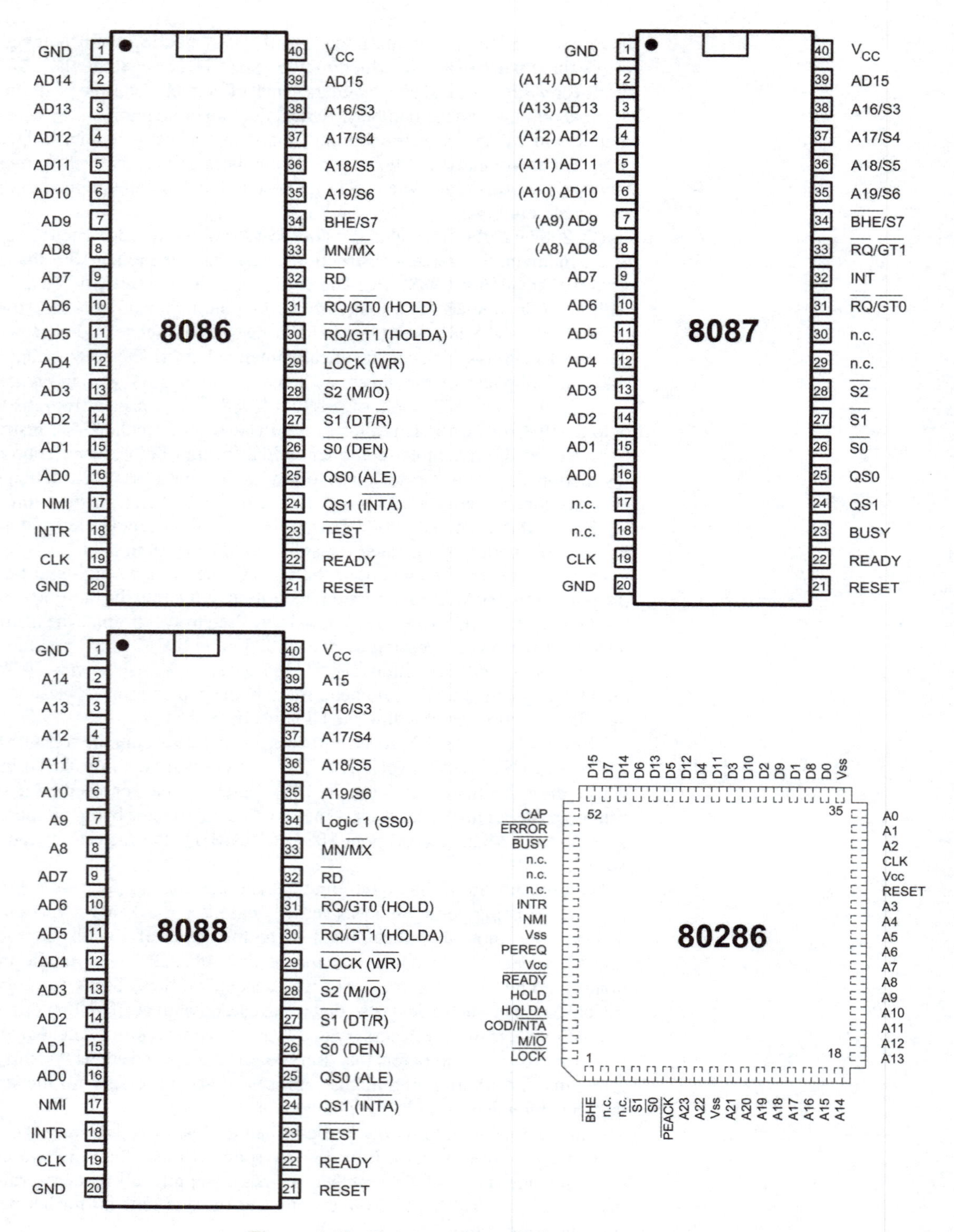

Figure 1.12 *Pin connections for legacy processors*

8237A Direct Memory Access Controller

The 8237A DMA Controller (DMAC) can provide service for up to four independent DMA channels, each with separate registers for Mode Control, Current Address, Base Address, Current Word Count, and Base Word Count. The DMAC is designed to improve system performance by allowing external devices to directly transfer information to and from the system memory. The 8237A offers a variety of programmable control features to enhance data throughput and allow dynamic reconfiguration under software control.

The 8237A provides four basic modes of transfer: Block, Demand, Single Word, and Cascade. These modes may be programmed as required, however, channels may be auto-initialize to their original condition following an End Of Process (EOP) signal.

The 8237A is designed for use with an external octal address latch such as the 74LS373. A system's DMA capability may be extended by cascading further 8237A DMAC chips and this feature is exploited in the PC-AT which has two such devices.

The least significant four address lines of the 8237A are bi-directional: when functioning as inputs, they are used to select one of the DMA controllers' 16 internal registers. When functioning as outputs, on the other hand, a 16-bit address is formed by taking the eight address lines (A0 to A7) to form the least significant address byte whilst the most significant address byte (A8 to A15) is multiplexed onto the data bus lines (D0 to D7). The requisite address latch enable signal (ADSTB) is available from pin-8. The upper four address bits (A16 to A19) are typically supplied by a 74LS670 4×4 register file. The requisite bits are placed in this device (effectively a static RAM) by the processor before the DMA transfer is completed.

DMA channel 0 (highest priority) is used in conjunction with the 8253 Programmable Interval Timer (PIT) in order to provide a memory refresh facility for the PC's dynamic RAM. DMA channels 1–3 are connected to the expansion slots for use by option cards.

The refresh process involves channel 1 of the PIT producing a negative going pulse with a period of approximately 15 μs. This pulse sets a bistable which, in turn, generates a DMA request at the channel-0 input of the DMAC (pin-19). The processor is then forced into a wait state, and the address and data bus buffers assume a tri-state (high impedance) condition. The DMAC then outputs a row refresh address and the row address strobe (RAS) is asserted. The 8237 increments its refresh count register and control is then returned to the processor. The process then continues such that all 256 rows are refreshed within a time interval of 4 ms. The pin connections for the 8237A are shown in Figure 1.13.

8253 Programmable Interval Timer

The 8253 is a Programmable Interval Timer (PIT) which has three independent presettable 16-bit counters each offering a count rate of up to 2.6 MHz. The pin connections for the 8253 are shown in Figure 1.13. Each counter consists of a single 16-bit presettable down counter. The counter can function in binary or BCD and its input, gate, and output are configured by the data held in the Control

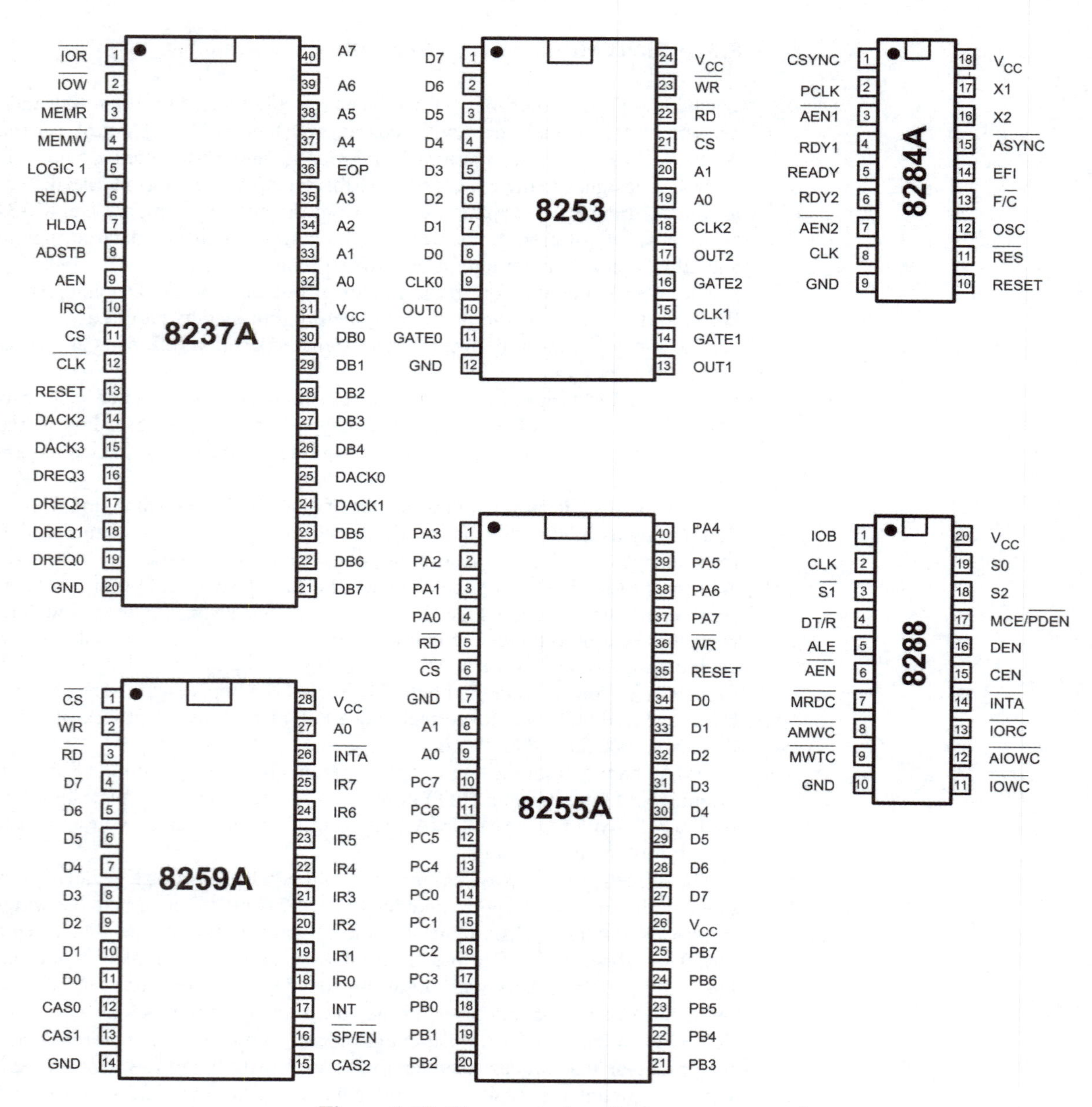

Figure 1.13 *Pin connections for legacy support chips*

Word Register. The down counters are negative edge triggered such that, on a falling clock edge, the contents of the respective counter is decremented.

The three counters are fully independent and each can have separate mode configuration and counting operation, binary or BCD. The contents of each 16-bit count register can be loaded or read using simple software referencing the relevant port addresses shown in Table 1.10. The truth table for the chip's active low chip select (CS), read (RD), write (WR) and address lines (A1 and A0) is shown in Table 1.9.

Table 1.9 *Truth table for the 8253*

CS	RD	WR	A1	A0	Function
0	1	0	0	0	Load counter 0
0	1	0	0	1	Load counter 1
0	1	0	1	0	Load counter 2
0	1	0	1	1	Write mode word
0	0	1	0	0	Read counter 0
0	0	1	0	1	Read counter 1
0	0	1	1	0	Read counter 2
0	0	1	1	1	No-operation (tri-state)
1	x	x	x	x	Disable tri-state
0	1	1	x	x	No-operation (tri-state)

8255A Programmable Peripheral Interface

The 8255A Programmable Peripheral Interface (PPI) is a general purpose I/O device which provides no less than 24 I/O lines arranged as three 8-bit I/O ports. The pin connections and internal architecture of the 8255A are shown in Figures 1.13 and 1.14, respectively. The Read/Write and Control Logic block manages all internal and external data transfers. The port addresses used by the 8255A are given in Table 1.10.

The functional configuration of each of the 8255's three I/O ports is fully programmable. Each of the control groups accepts commands from the Read/Write Control Logic, receives Control Words via the internal data bus, and issues the requisite commands to each of the ports. At this point, it is important to note that the 24 I/O lines are, for control purposes, divided into two logical groups (A and B). Group A comprises the entire eight lines of Port A together with the four upper (most significant) lines of Port B. Group B, on the other hand, takes in all eight lines from Port B together with the four lower (least significant) lines of Port C. The upshot of all this is simply that Port C can be split into two in order to allow its lines to be used for status and control (handshaking) when data is transferred to or from Ports A or B.

8259A Programmable Interrupt Controller

The 8259A Programmable Interrupt Controller (PIC) was designed specifically for use in real-time interrupt driven microcomputer systems. The device manages eight levels of request and can be expanded using further 8259A devices.

The sequence of events which occurs when an 8259A device is used in conjunction with an 8086 or 8088 processor is as follows:

(a) One or more of the interrupt request lines (IR0–IR7) are asserted (note that these lines are active high) by the interrupting device(s).
(b) The corresponding bits in the IRR register become set.
(c) The 8259A evaluates the requests on the following basis:
   (i) If more than one request is currently present, determine which of the requests has the highest priority.

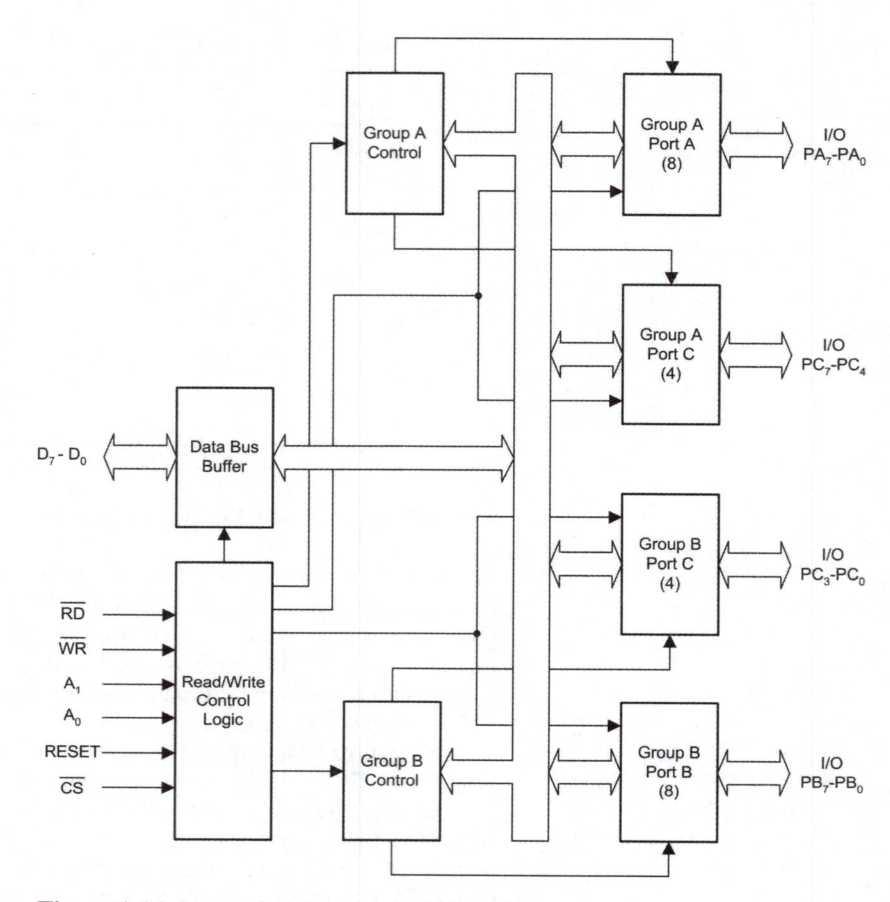

Figure 1.14 *Internal architecture of the 8255A*

(ii) Ascertain whether the successful request has a higher priority than the level currently being serviced.

(iii) If the condition in (ii) is satisfied, issue an interrupt to the processor by asserting the active high INT line.

(d) The processor acknowledges the interrupt signal and responds by pulsing the interrupt acknowledge (INTA) line.

(e) Upon receiving the INTA pulse from the processor, the highest priority ISR bit is set and the corresponding IRR bit is reset.

(f) The processor then initiates a second interrupt acknowledge (INTA) pulse. During this second period for which the INTA line is taken low, the 8259 outputs a pointer on the data bus which is then read by the processor.

The pin connections for the 8259A are shown in Figure 1.13.

8284A Clock generator

The 8284A is a single chip clock generator/driver designed specifically for use by the 8086 family of devices. The chip contains a crystal oscillator, divide-by-3

Table 1.10 *Port addresses (hexadecimal) used in the PC family*

Device	PC-XT	PC-AT
8237A DMA Controller	000-00F	000-01F
8259A Interrupt Controller	020-021	020-03F
8253/8254 timer	040-043	040-05F
8255 Parallel Interface	060-063	n.a.
8042 keyboard controller	n.a.	060-06F
DMA page register	080-083	080-09F
NMI mask register	0A0-0A7	070-07F
Second 8259A Interrupt Controller	n.a.	0A0-0BF
Second 8237A DMA Controller	n.a.	0C0-0DF
Maths Coprocessor (8087, 80287)	n.a.	0F0-0FF
Games controller	200-20F	200-207
Expansion unit	210-217	n.a.
Second parallel port	n.a.	278-27F
Second serial port	2F8-2FF	2F8-2FF
Prototype card	300-31F	300-31F
Fixed (hard) disk	320-32F	1F0-1F8
First parallel printer	378-37F	378-37F
SDLC adapter	380-38F	380-38F
BSC adapter	n.a.	3A0-3AF
Monochrome adapter	3B0-3BF	3B0-3BF
Enhanced graphics adapter	n.a.	3C0-3CF
Colour graphics adapter	3D0-3DF	3D0-3DF
Floppy disk controller	3F0-3F7	3F0-3F7
First serial port	3F8-3FF	3F8-3FF

counter, ready, and reset logic. On the original PC, the quartz crystal is a series mode fundamental device which operates at a frequency of 14.312818 MHz. The output of the divide-by-3 counter takes the form of a 33% duty cycle square wave at precisely one-third of the fundamental frequency (i.e. 4.77 MHz). This signal is then applied to the processor's clock (CLK) input. The clock generator also produces a signal at 2.38 MHz which is externally divided to provide a 5.193 MHz 50% duty cycle clock signal for the 8253 Programmable Interval Timer (PIT).

8288 Bus Controller

The 8288 Bus Controller decodes the status outputs from the CPU (S0 and S1) in order to generate the requisite bus command and control signals. These signals are used as shown in Table 1.11. The 8288 issues signals to the system to strobe addresses into the address latches, to enable data onto the buses, and to determine the direction of data flow through the data buffers. The internal architecture and pin connections for the 8288 are shown in Figure 1.13.

Table 1.11 *8288 Bus Controller status inputs*

Processor status line			
S2	S1	S0	Condition
0	0	0	Interrupt acknowledge
0	0	1	I/O read
0	1	0	I/O write
0	1	1	Halt
1	0	0	Memory read
1	0	1	Memory read
1	1	0	Memory write
1	1	1	Inactive

Chipsets

In modern PCs, the overall device count has been significantly reduced by integrating several of the functions associated with the original PC chipset within one or two VLSI devices or within the CPU itself.

Early examples of integrated chipsets include the Chips and Technology 82C100 XT Controller found in older 'XT-compatible systems', provides the functionality associated with no less than six of the original XT chipset and effectively replaces the following devices: one 8237 DMA Controller, one 8253 Counter/ Timer, one 8255 Parallel Interface, one 8259 Interrupt Controller, one 8284 Clock Generator, and one 8288 Bus Controller. In order to ensure software compatibility with the original PC, the 82C100 contains a superset of the registers associated with each of the devices which it is designed to replace. The use of the chip is thus completely transparent as far as applications software is concerned.

Another example is OPTi's 82C206 and 82C495XLC 'AT controller' chipset found in many early '486 and Pentium-based systems. The 82C206 provides the functions of two 82437 DMA Controllers, two 8259 Interrupt Controllers, one 8254 Counter/Timer, one 146818-compatible Real-Time Clock, and one 74LS612 Memory Mapper. In addition, the chip provides 114 bytes of CMOS RAM (used for storing the BIOS configuration settings). The matching 82C495XLC device provides cache memory control and shadow RAM support for system, video, and adapter card BIOS. The chip also contains on-chip hardware that provides direct support for up to two VL-bus master devices.

Modern PCs use chipsets supplied by a number of different manufacturers. The chipsets provide an interface between the processor, memory and graphics controllers (which must all operate at this highest possible speed), and the various expansion buses (PCI, ISA, etc.). One of the functions of the chipset is to act as a *bridge* between the various bus systems, managing the data flow and ensuring the efficient transfer of data Table 1.12. Figure 1.15 shows the typical architecture of a system that supports both PCI and ISA expansion bus systems. The *front side bus* (FSB) allows data to be transferred at high speed between the processor, memory controller, and graphics controller whilst the *back side bus* (BSB) allows the processor to be fed with an instruction stream from the level 2 *cache memory* (see page 39).

Table 1.12 *Representative chipset data*

Chipsets	Typical CPUs	Supported DRAM types	Supported DRAM density (Mbit)	Maximum memory size supported	ECC/ parity	AGP	Bus speeds (MHz)	PCI clock dividers
Intel 850 (Tehama) [82850] (MCH) [82801BA] (ICH2) [82802] (FWH)	Pentium 4	RDR DC PC800	128 256	2 GB	ECC	1x 2x 4x 1.5v	100 (×4)	1/3 1/4 PCI 2.2
SiS 645 [645] [961]	Pentium 4	SDRAM PC133 DDR PC2700 Mem = 4/3 Bus Mem = 5/3 Bus	16 64 128 256 512	3 GB	No	1x 2x 4x	100 (×4)	1/3 PCI 2.2
ALi MAGiK 1 [M1647] [M1535D+]	Athlon Duron	SDRAM PC133 DDR PC2100 Asynch Mem	16 64 128 256 512	3 GB	?	1x 2x 4x	100 (×2) 133 (×2)	1/3 1/4 Asynchronous PCI 2.2
AMD 750 [751] (Irongate) [756] (Viper)	Athlon Duron	SDRAM PC100	16 64 128	768 MB	ECC	1x 2x	100 (×2)	1/3 PCI 2.2
AMD 760 [761] (Irongate-4) [766]	Athlon Duron	DDR PC2100 Reg DDR	64 128 256 512	2 GB 4 GB Reg.	ECC	1x 2x 4x	100 (×2) 133 (×2)	1/3 1/4 PCI 2.2
VIA KT-266 [VT8366] [VT8233]	Athlon Duron	SDRAM PC133 Reg. SDRAM VC SDRAM DDR PC2100 Reg. DDR Mem = 3/4 Bus Mem = 4/3 Bus Mem = AGP	64 128 256 512	3 GB S 4 GB Reg.	Both	1x 2x 4x	100 (×2) 133 (×2)	1/3 1/4 Pseudosynchronous PCI 2.2

Another arrangement is shown in Figure 1.16. This architecture uses a *North Bridge* and *South Bridge* (both separate chips within the chipset). The North Bridge provides the processor with an interface to the memory bus, advanced graphics port bus (AGP) – see Chapter 2 – and the PCI expansion bus. The South

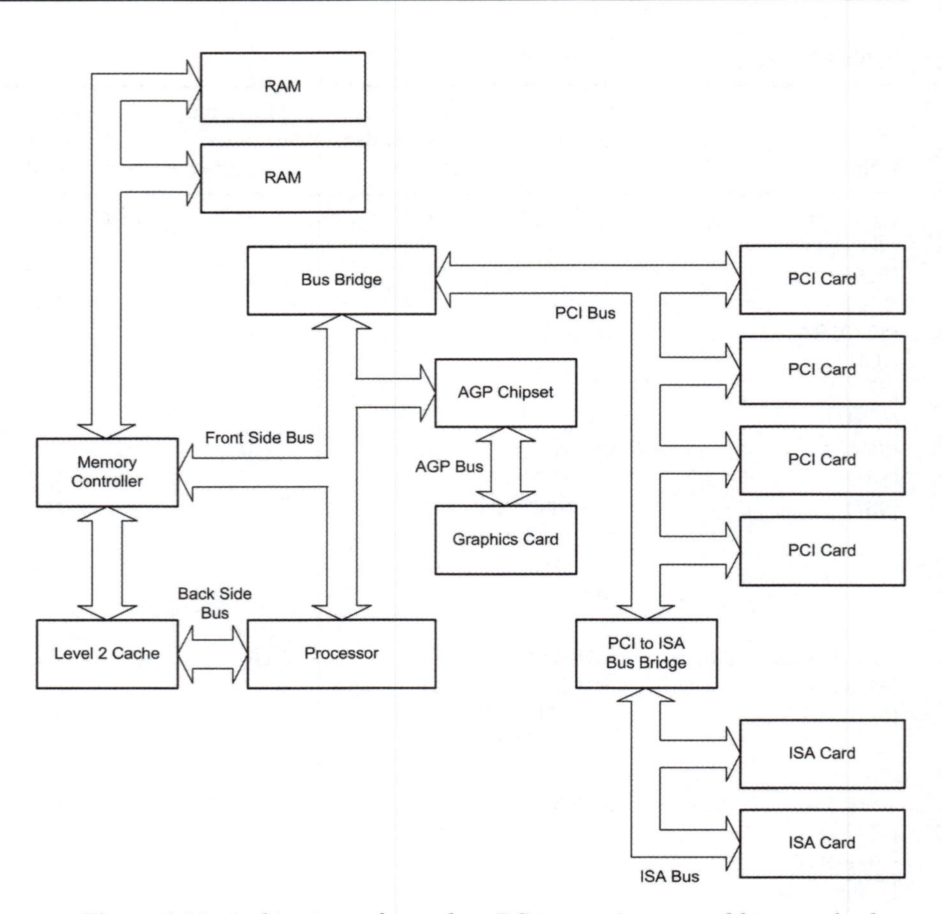

Figure 1.15 *Architecture of a modern PC supporting several bus standards (AGP, PCI, and ISA).*

Bridge handles all of the system I/O, including an interface to the IDE/ATA bus (see page 53).

Figure 1.17 shows the typical layout of a modern PC motherboard. This system employs an architecture which is based on a bus controller (North Bridge) and an I/O controller (South Bridge), and an AMD Socket 7 processor. Four ISA and three PCI expansion slots are provided. By contrast, an example of an embedded PC controller is shown in Figure 1.18. This system is based on an AMD processor (designed specifically for embedded controller applications) and uses the PC/104 expansion architecture (see Chapter 2).

PC memory

The PC system board's read/write memory provides storage for the memory resident parts of the operating system (e.g. Windows, Linux, or DOS) as well as user applications programs and transient data. Read/write memory is also used to store data that is displayed on the screen. On some systems this memory is separate from the system board's read/write memory (and usually fitted to a specialized graphics card) whilst on others it is 'mapped' into the main

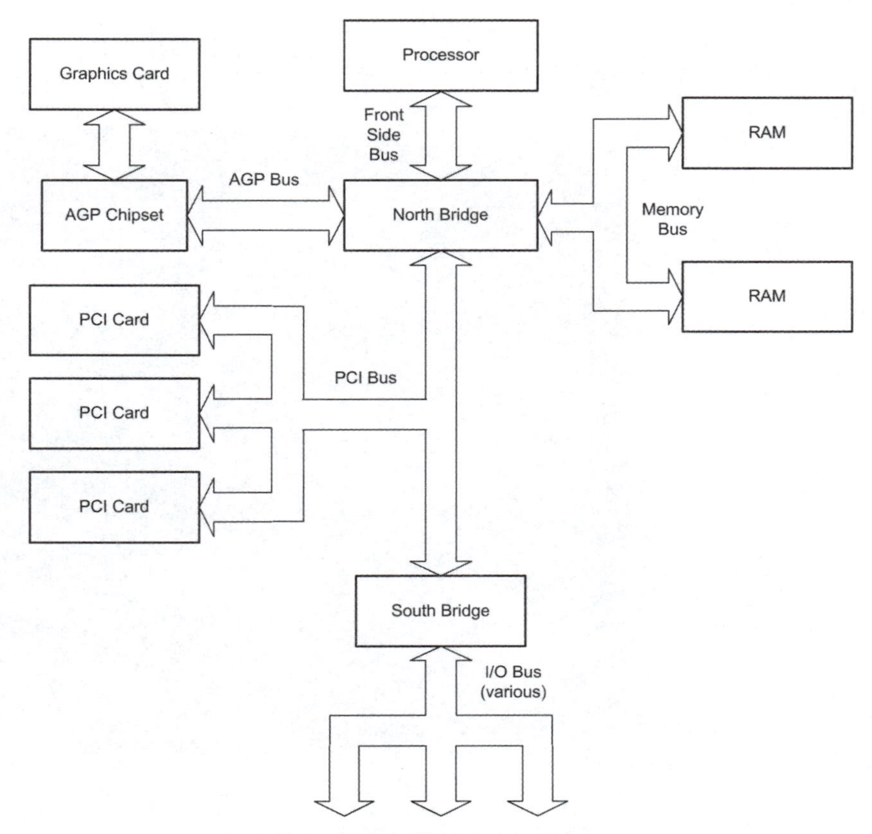

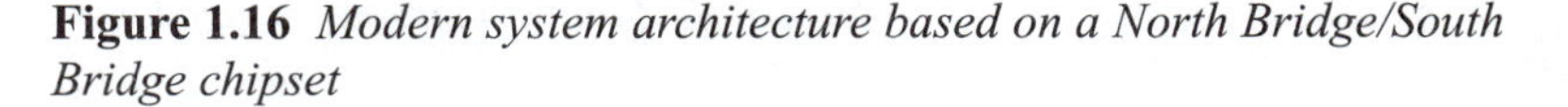

Figure 1.16 *Modern system architecture based on a North Bridge/South Bridge chipset*

read/write memory of the system. What makes all this possible is the availability of fast semiconductor random access memory (RAM) devices. This section explains what these devices are and how they are incorporated into a PC system.

Modern PCs require large amounts of RAM in order to run increasingly powerful software applications. Today, memory capacities of 64 MB or 128 MB are commonplace. Early PCs, on the other hand, were designed to operate with a mere 640 KB or 1 MB of memory.

Memory operation

Unfortunately, it takes a finite time in order to access data stored in a memory device. Since program execution involves constantly reading and writing data from/to memory the amount of time taken to transfer data has an important bearing on the time that it takes to execute a program. *Access time* is the average time (usually specified in nanoseconds) for a RAM device to complete one data access. Access time itself is comprises the initial address setup time and the time it takes to initiate a request for data and prepare access (this is known as *latency*).

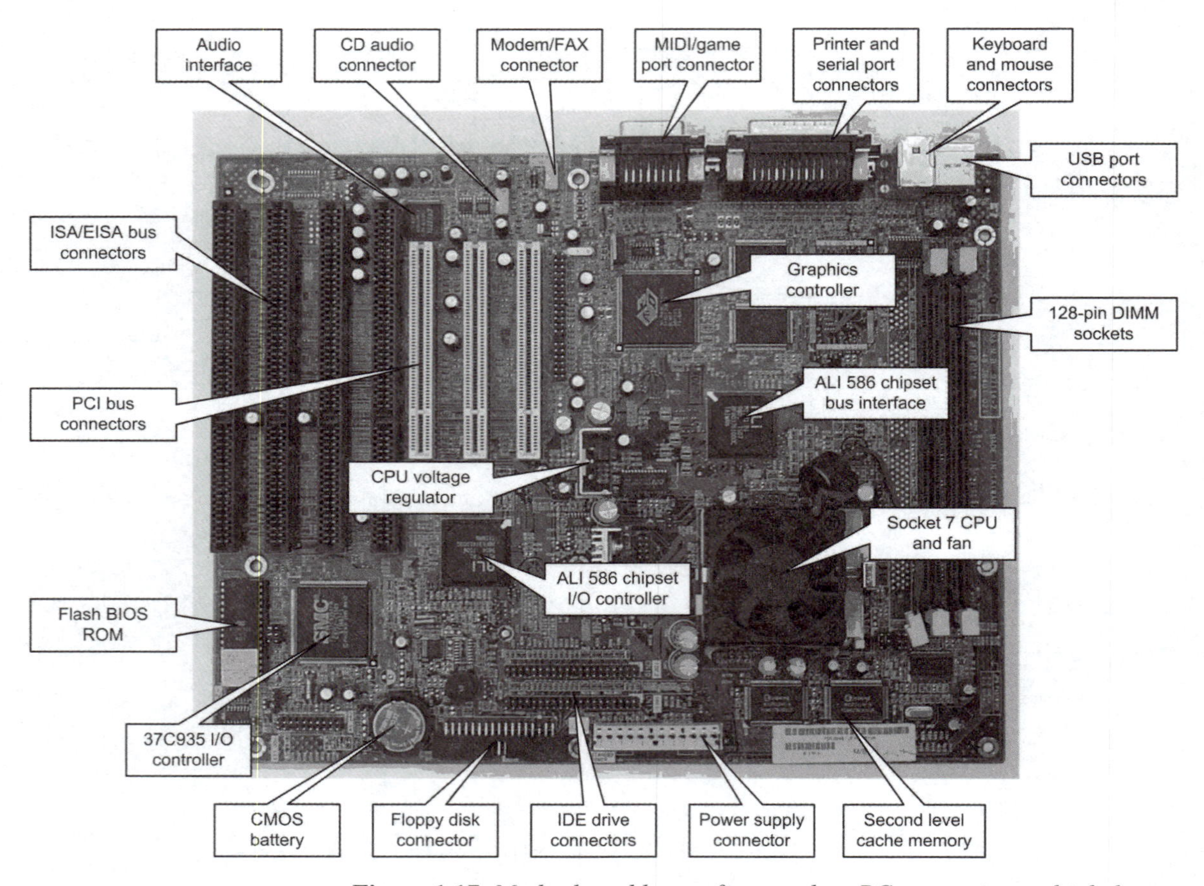

Figure 1.17 *Motherboard layout for a modern PC supporting multiple bus standards*

Most memory device consist of a matrix of cells arranged on the basis of rows and columns. A *row address strobe* (RAS) signal is used to latch the row address of a particular memory location whilst a *column address strobe* (CAS) signal is used to latch the column address of a particular memory location into the row–column matrix of a RAM device. *CAS latency* is the ratio of column access time to clock cycle time.

In addition, modern large-scale memories are based on dynamic RAM (DRAM) technology in which the data is stored as a tiny electric charge which will leak away if it is not periodically *refreshed*. The process of reading and then writing back the data stored in a DRAM device is known as *refreshing*, and this process must operate continuously otherwise data will be lost.

Memory organization

The memory in a PC is usually arranged in *banks*. Many modern PCs have two or more memory banks (i.e. Bank A, Bank B, and so on) and each bank comprises a group of adjacent sockets or modules. The banks are usually easily

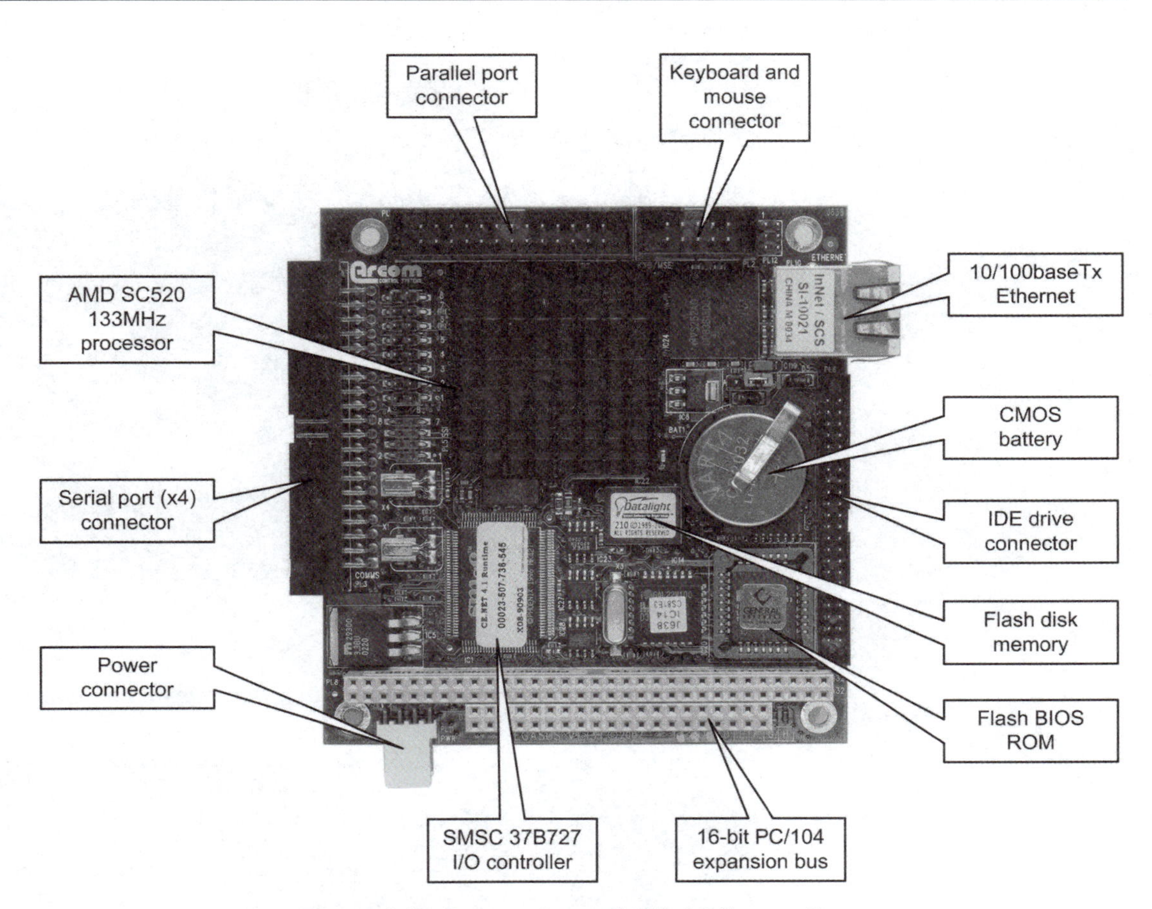

Figure 1.18 *Layout of an embedded PC controller*

identified on the system board but are also described in the system board manual. Furthermore, because memory bank configurations can vary from system to system, it is important to refer to manufacturers' data before attempting to fit memory modules. Some PCs require *all* the sockets in one bank to be filled with the same capacity module, some computers require the first bank to house the highest capacity modules, and others require the banks to be filled in a particular order!

Most of today's PCs use 168-pin DIMMs, which support 64-bit data paths. Earlier 72-pin SIMMs supported 32-bit data paths, and were originally used with 32-bit CPUs. It is important to note that, when 32-bit SIMMs were used with 64-bit processors, they had to be installed in pairs, with each pair of modules making up one memory bank.

Data integrity

With early PCs, data integrity checking was based on the use of a simple parity check of each byte of data. The *parity bit* (stored separately) is used to detect errors in the other 8 bits. Parity checking may be either *odd* or *even*. In the

Photo 1.9 *Various PC memory devices with capacities varying from 1 to 256 MB*

Photo 1.10 *DIMM memory module with heatsinks fitted to each memory device*

former case, the parity bit is set when there is an odd number of 1's in the byte of data. In the latter case, the parity bit is set when there is an even number of 1's in the byte of data. Other, more powerful, data integrity checking methods are now available, such as error correction code (ECC) methods. ECC provides more elaborate error detection than simple parity checking. Note that ECC can detect multiple-bit errors and can locate *and* correct single-bit errors.

Memory terminology

The following terminology is commonly used to describe the various types of memory present within a PC or PC-compatible system.

Buffered memory

A buffered memory module contains buffers that are used to interface the module to the external memory bus. So that more memory devices can be included in the module itself, the built-in buffers provide additional drive capability and also regularize the logic levels employed. It is important to note that buffered and un-buffered memory cannot be mixed. See also Registered memory.

BEDO RAM

BEDO (burst extended data output) RAM can process four memory addresses in one burst. BEDO bus speeds range from 50 to 66 MHz compared with up to 33 MHz for EDO RAM and 25 MHz for FPM RAM.

Cache memory

Cache memory comprises a limited amount (often 256 or 512 KB) of high-speed read/write memory in close physical and electrical proximity to the CPU. Instead of having to fetch instructions and data from the relatively slow main system board RAM, the cache memory provides the CPU with rapid access to the most recent and frequently requested instructions. The primary cache or *level 1 cache* is the cache memory closest to the processor core. Secondary cache (*level 2 cache*) may also be provided. This cache is normally fitted to the system board.

CMOS memory

See page 44.

Double data rate (DDR) memory

The latest generation of synchronous dynamic random access memory (SDRAM) operates at double the data rate (DDR). With DDR SDRAM, data is read on *both* the rising and the falling edge of the PC clock, thereby delivering twice the bandwidth of standard SDRAM. With DDR SDRAM, memory speed doubles *without* increasing the clock frequency.

Dual-inline memory module (DIMM)

Dual-inline memory modules (DIMM) are similar to single inline memory modules but the contacts on each side of the DIMM are differently connected (unlike the SIMM in which the contacts on each side of the module are electrically connected). See also SIMM.

Direct Rambus

Direct Rambus is the name of a third generation memory technology that offers a completely new DRAM architecture for high-performance PCs. With Direct Rambus data transfers are made at speeds of up to 800 MHz over a relatively narrow 16-bit data bus compared with current SDRAM technology that operates at 100 MHz on a relatively wide 64-bit data bus.

DIP memory

Early PCs were fitted with DRAM devices supplied in conventional dual-inline packages (DIP). These chips were either fitted in sockets (16-or 18-pin DIL) or permanently soldered into the system board. This type of memory is now obsolete.

Dynamic random access memory (DRAM)

Dynamic random access memory is the most commonly used form of PC RAM. Because of its cell architecture (in which charge is stored in a semiconductor junction capacitance) data can only be stored for a very short time. In order to retain the data, DRAM devices must be refreshed (i.e. read and then written back) on a regular basis.

Dual-ported memory

See VRAM.

Extended data-output (EDO) memory

Extended data-output is a DRAM technology that shortens the read cycle between the memory and the CPU. EDO memory allows a CPU to access memory up to 20% faster than comparable fast-page mode (FPM) memory. Note that EDO RAM can only be fitted to a system board that supports its use.

Enhanced synchronous dynamic random access memory (ESDRAM)

Enhanced synchronous DRAM is a type of memory that replaces expensive SRAM in embedded systems and offers comparable speed with less power consumption and lower cost.

Fast-page mode (FPM) RAM

Fast-page mode RAM is a technology that was used to improve the performance of early DRAM devices. Compared with conventional page mode technology,

FPM provides faster access to data that is stored in the same row of a memory matrix.

Non-volatile random access memory (NVRAM)

See CMOS memory on page 44.

Registered memory

Registered memory is SDRAM memory that contains registers directly on the module. The registers re-drive the signals through the memory chips and allow the module to be built with more memory chips. Registered (buffered) and un-buffered memory cannot be mixed. The design of the computer memory controller dictates which type of memory the computer requires.

Synchronous dynamic random access memory (SDRAM)

Synchronous DRAM (SDRAM) is a DRAM technology that uses a memory clock to synchronize signal input and output on a memory chip. The memory clock is synchronized with the CPU clock so the timing of the memory chips and the timing of the CPU are locked together. Synchronous DRAM saves time in executing commands and transmitting data, thereby increasing the overall performance of the computer. SDRAM allows the CPU to access memory approximately 25% faster than EDO memory.

Self-refreshing RAM

Self-refreshing is a memory technology that enables DRAM to refresh itself independently of the CPU or external refresh circuitry. Self-refresh technology is built into the DRAM chip itself and reduces power consumption dramatically. Notebook and laptop computers use this technology.

Synchronous graphics random access memory (SGRAM)

Synchronous graphics RAM is video memory that includes graphics-specific read/write features. SGRAM allows data to be retrieved and modified in blocks instead of individually. Blocking reduces the number of reads and writes the memory must perform and increases the performance of the graphics controller.

Single inline memory module (SIMM)

Single inline memory modules are small printed circuit boards populated by semiconductor memory devices and fitted with gold or lead/tin contacts. A SIMM plugs into a memory expansion connector on the system board. SIMMs offer various advantages over DIP packaged RAM including ease of installation and minimal footprint. See also DIMM.

Small-outline dual-inline memory module (SODIMM)

Small-outline dual-inline memory modules are enhanced versions of standard DIMM devices. A 72-pin small-outline DIMM is about half the length of a 72-pin SIMM.

Small-outline J-lead (SOJ) packaged memory

Small-outline J-lead packages are commonly used for surface-mounted DRAM devices. The package is rectangular with J-shaped connecting pins on the two long sides.

Static random access memory (SRAM)

Static RAM (SRAM) is a type of RAM that requires no refreshing and retains its data as long as power is applied. Provided that the data is not changing (i.e. remains *static*), SRAM devices require very little power. SRAM is frequently used to provide cache memory.

Thin small-outline packaged (SOP) memory

Thin small-outline packages are an alternative to SOJ packages for surface-mounting DRAM devices. TSOP packages are approximately one-third of the thickness of an SOJ. TSOP components are often found in small-outline DIMMs and credit card memory.

Un-buffered memory

An un-buffered memory device does not have internal buffers or registers. See Buffered memory and Registered memory.

Video random access memory (VRAM)

Video RAM is special dual-ported memory (two separate data ports are provided) fitted to a video or graphics card.

Zero wait state memory

Zero wait state memory offers fast access times because older and slower memory devices may require between one and five wait states (i.e. do-nothing cycles) to slow a CPU down to match the access time of the RAM.

Memory size

The amount of memory required by a PC depends not only on the software applications that are installed but also on the operating system that is used.

The following are minimum memory sizes recommended for use with the most common PC operating systems over the last 20 years:

Operating system	Minimum RAM	Recommended RAM (MB)
MS-DOS 3.3	640 KB	4
MS-DOS 5	1 MB	16
Windows 3.1	3 MB	32
Windows 95	8 MB	64
Windows 98	24 MB	128
Windows ME	32 MB	128
Windows 2000 Professional	64 MB	128
Windows 2000 Server	128 MB	256
Windows 2000 Advanced Server	256 MB	512
Windows XP	256 MB	512

It is worth mentioning that adding more memory to a PC can have a *very* significant effect on its performance, particularly if the memory is at, or near, the minimum recommended for the type of operating system. The reason for this is that, when insufficient RAM is available the PC's operating system will create virtual memory on the hard disk which will replace the physical memory which would otherwise be needed. Unfortunately, writing to and reading from the hard disk takes significantly longer than performing the same operation to a semiconductor memory. Frequent accesses to the hard disk impose will cause a program to run much slower than if the hard disk was not in regular use. To put this into context, it takes typically less than 200 ns to access physical RAM and around 10 ms to access a reasonably hard disk drive!

Memory speed

The speed of memory is one of the most important factors in defining the performance of a system. Furthermore, memory speed forms (or the speed of memory components) forms an essential part of specification of every PC. Memory fitted to the PC must comply with this specification and failure to observe this prerequisite may cause a wide variety of problems including lock-ups, re-booting, and failure to boot. Some of the most significant milestones in the development of memory devices are listed below:

Year first introduced	Memory technology	Access time/speed
1981	DIL RAM	100 ns
1987	FPM RAM	70 ns
1995	EDO RAM	50 ns
1997	SDRAM (PC66)	66 MHz
1998	SDRAM (PC100)	100 MHz
1999	RDRAM	800 MHz
1999/2000	SRAM (PC133)	133 MHz (VCM)
2000	DDR SDRAM	266 MHz

In many cases you can fit a memory module rated at the same speed or faster than that at which a PC's memory system is rated. This means that you should be able to replace a 70 ns module with one rated at either 70 or 60 ns *but not* one rated at 80 ns. It is, however, worth noting that some older systems check the module speed at boot-up and will only accept a module that has the same speed rating as that of the system to which it is fitted. This explains why some systems will refuse to accept faster memory modules than those being replaced!

CMOS memory

The PC-AT and later machine's CMOS memory is 64 byte of battery-backed memory contained within the real-time clock chip (a Motorola MC146818).

Table 1.13 *CMOS memory organization*

Offset (hex.)	Contents
00	Seconds
01	Seconds alarm
02	Minutes
03	Minutes alarm
04	Hours
05	Hours alarm
06	Day of week
07	Day of the month
08	Month
09	Year
0A	Status Register A
0B	Status Register B
0C	Status Register C
0D	Status Register D
0E	Diagnostic status byte
0F	Shutdown status byte
10	Floppy disk type (drives A and B)
11	Reserved
12	Fixed disk type (drives 0 and 1)
13	Reserved
14	Equipment byte
15	Base memory (low byte)
16	Base memory (high byte)
17	Extended memory (low byte)
18	Extended memory (high byte)
19	Hard disk 0 extended type
1A	Hard disk 1 extended type
1B-2D	Reserved
2E-2F	Check-sum for bytes 10 to 1F
30	Actual extended memory (low byte)
31	Actual extended memory (high byte)
32	Date century byte (in BCD format)
33-3F	Reserved

Sixteen byte of this memory are used to retain the real-time clock settings (date and time information), whilst the remainder contains important information on the configuration of the system. When the CMOS battery fails or when power is inadvertently removed from the real-time clock chip, all data becomes invalid and the set-up program has to be used to restore the settings of the system. This can be a real problem *unless* you know what the settings should be! The organization of the CMOS memory is shown in Table 1.13 (note that locations marked 'reserved' may have different functions in different systems). CMOS memory is also sometimes referred to as non-volatile random access memory (NVRAM).

BIOS ROM

The BIOS ROM is programmed during manufacture. The programming data is supplied to the semiconductor manufacturer by the BIOS originator. This process is cost-effective for large-scale production. However, programming of the ROM is irreversible; once programmed, devices cannot be erased in preparation for fresh programming. Hence, the only way of upgrading the BIOS is to remove and discard the existing chips, and replace them with new ones. This procedure is fraught with problems, not least of which is compatibility of the BIOS upgrade with existing DOS software (see page 398 for further information relating to BIOS upgrading).

The BIOS ROM invariably occupies the last 64 or 128 KB of memory (from F0000 to FFFFF or E0000 to FFFFF, respectively) within the first megabyte of physical memory. Additional BIOS extensions are provided for other I/O functions (see Figure 1.19).

Photo 1.11 *CMOS battery. The link adjacent to the battery can be used to clear the memory*

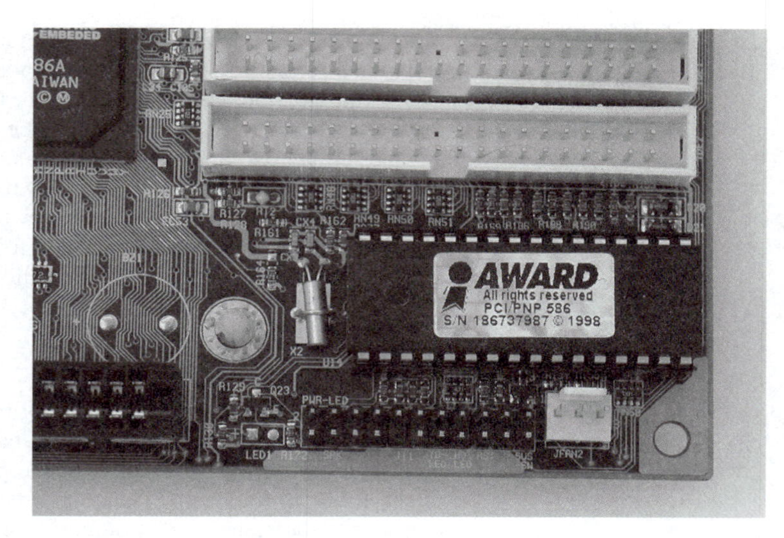

Photo 1.12 *Award BIOS ROM (note the real-time clock crystal adjacent to the chip)*

PC memory allocation

The allocation of memory space within a PC can he usefully illustrated by means of a memory map. An 8086 microprocessor can address any one of 048 576 different memory locations with its 20 address lines. It thus has a memory which ranges from 00000 (the lowest address) to FFFFF (the highest address). We can illustrate the use of memory using a diagram known as a 'memory map'. Figure 1.19 shows a memory map for the first megabyte of PC memory.

BIOS data area

The memory region starting at address 0400H (see Figure 1.19) contains data that is maintained by the BIOS. A number of memory locations within this space can provide useful information about the current state of a PC. You can easily display the contents of these memory locations (summarized in Table 1.13) using the MS-DOS DEBUG utility (see page 136) or using a short routine written in QuickBASIC.

As an example of the first method, the following DEBUG command can be used to display the contents of 10 bytes of RAM starting at memory location 0410:

```
D0:0410 L 0A
```

A rather more user-friendly method of displaying the contents of RAM is shown in the following QuickBASIC code fragment:

```
DEF SEG = 0
CLS
INPUT "Start address (in hex) "; address$
```

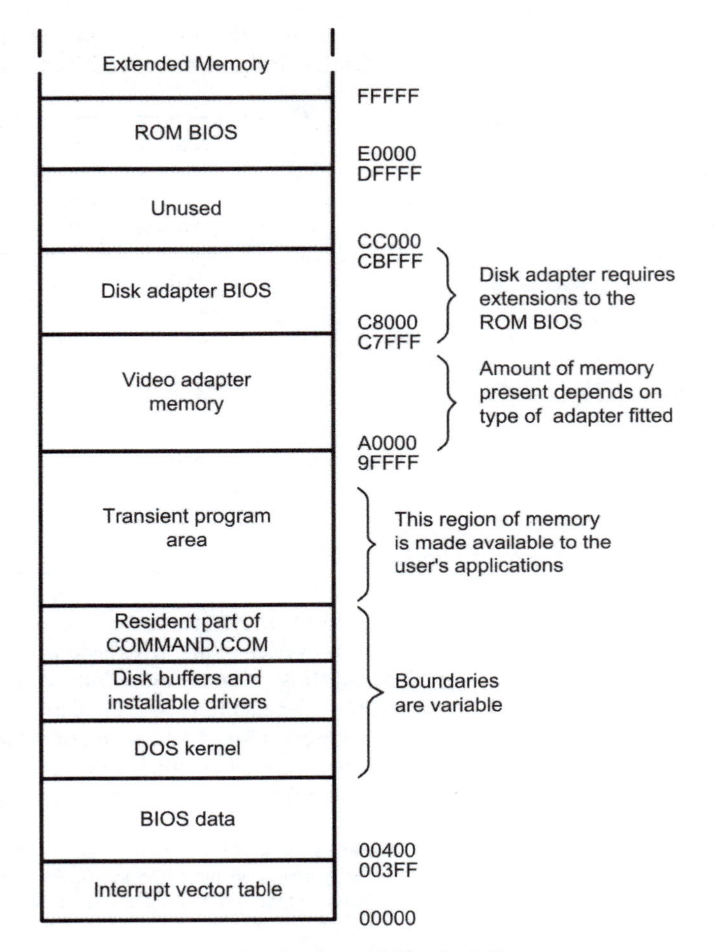

Figure 1.19 *Memory map for the first 1 MB of a PC memory*

```
address$ = "&H" +address$
address = VAL(address$)
INPUT "Number of bytes to display "; number
PRINT
PRINT "Address", "Byte"
PRINT " (hex) " , " (hex) "
PRINT
FOR 1% = 0 TO number - 1
  v = PEEK(address + i%)
  PRINTHEX$(address+ i%), HEX$(v)
NEXT i%
PRINT
END
```

This simple QuickBASIC program prompts the user for a start address (expressed in hexadecimal) and the number of bytes to display. A typical example of running the program is shown below. The program has been used to display the contents of 10 bytes of RAM from address 0410 onwards:

Start address (in hex.)?	410
Number of bytes to display?	10
Address (hex.)	**Byte (hex.)**
410	63
411	44
412	BF
413	80
414	2
415	0
416	18
417	20
418	0
419	0

Unfortunately, the above information is not particularly useful unless you know how each of the bytes is constructed and what the data actually represents! Despite this, it is possible to interrogate the BIOS data area with simple software in order, for example, to display the port addresses used on a system Table 1.14. The PowerBASIC 3.5 program (available for downloading from the companion web site) shows how the BIOS data area can be accessed and useful information extracted from it. The result of running the program is shown in Figure 1.20.

```
' Name: biosdata.bas      Version: 0.5    Modified: 25/08/04
' Language: PowerBASIC 3.5
' Function: Display BIOS data
'
' Initialise
'
dim high as integer
dim low as integer
division$ = String$(40, Chr$(205))
Color 15, 1
Cls
'
' Get equipment list word at offset &H10
'
Def Seg = &H40
high% = peekl(&H10) \ 256       ' low byte
low% = peekl(&H10) MOD 256      ' high byte
'
' Print title and version number
'
print division$
print "BIOS Data 0.5"
'
' Get BIOS date
'
```

Table 1.14 *BIOS data area*

Offset (hex.)	Offset (dec.)	BIOS service	Field size (bytes)	Function
00h	0	Int 14h	2	Base I/O address for serial port 1 (COM 1)
02h	2	Int 14h	2	Base I/O address for serial port 2 (COM 2)
04h	4	Int 14h	2	Base I/O address for serial port 3 (COM 3)
06h	6	Int 14h	2	Base I/O address for serial port 4 (COM 4)
08h	8	Int 17h	2	Base I/O address for parallel port 1 (LPT 1)
0Ah	10	Int 17h	2	Base I/O address for parallel port 2 (LPT 2)
0Ch	12	Int 17h	2	Base I/O address for parallel port 3 (LPT 3)
0Eh	14	POST	2	Base I/O address for parallel port 4 (LPT 4)
10h	16	Int 11h	2	Equipment Word
12h	18	POST	1	Interrupt flag – Manufacturing test
13h	19	Int 12h	2	Memory size in KB
15h	21		2	Error codes for AT; adapter memory size for PC and XT
17h	22	Int 16h	1	Keyboard shift flag 1
18h	23	Int 16h	1	Keyboard shift flag 2
19h	24	Int 09h	1	Alt Num pad work area
1Ah	26	Int 16h	2	Pointer to the address of the next character in the keyboard buffer
1Ch	28	Int 16h	2	Pointer to the address of the last character in the keyboard buffer
1Eh	60	Int 16h	32	Keyboard buffer
3Eh	61	Int 13h	1	Floppy disk drive calibration status
3Fh	62	Int 13h	1	Floppy disk drive motor status
40h	63	Int 13h	1	Floppy disk drive motor time-out
41h	64	Int 13h	1	Floppy disk drive status
42h	65	Int 13h	1	Hard disk and floppy controller Status Register 0
43h	66	Int 13h	1	Floppy drive controller Status Register 1
44h	67	Int 13h	1	Floppy drive controller Status Register 2
45h	68	Int 13h	1	Floppy disk controller: cylinder number
46h	69	Int 13h	1	Floppy disk controller: head number
47h	70	Int 13h	1	Floppy disk controller: sector number
48h	71		1	Floppy disk controller: number of written
49h	72	Int 10h	1	Active video mode setting
4Ah	74	Int 10h	2	Number of text columns per row for the active video mode
4Ch	76	Int 10h	2	Size of active video in page s
4Eh	78	Int 10h	2	Offset address of the active video page relative to the start of video RAM
50h	80	Int 10h	2	Cursor position for video page 0
52h	82	Int 10h	2	Cursor position for video page 1
54h	84	Int 10h	2	Cursor position for video page 2
56h	86	Int 10h	2	Cursor position for video page 3
58h	88	Int 10h	2	Cursor position for video page 4
5Ah	90	Int 10h	2	Cursor position for video page 5
5Ch	92	Int 10h	2	Cursor position for video page 6
5Eh	94	Int 10h	2	Cursor position for video page 7
60h	96	Int 10h	2	Cursor shape
62h	97	Int 10h	1	Active video page
63h	99	Int 10h	2	I/O port address for the video display adapter
65h	100	Int 10h	1	Video display adapter internal mode register
66h	101	Int 10h	1	Colour palette
67h	103		2	Adapter ROM offset address

(*continued*)

Table 1.14 *(Continued)*

Offset (hex.)	Offset (dec.)	BIOS service	Field size (bytes)	Function
69h	106		2	Adapter ROM segment address
6Bh	107		1	Last interrupt (not PC)
6Ch	111	Int 1Ah	4	Counter for Interrupt 1Ah
70c	112	Int 1Ah	1	Timer 24-h flag
71h	113	Int 16h	1	Keyboard Ctrl-Break flag
72h	115	POST	2	Soft reset flag
74h	116	Int 13h	1	Status of last hard disk operation
75h	117	Int 13h	1	Number of hard disk drives
76h	118	Int 13h	1	Hard disk control
77h	119	Int 13h	1	Offset address of hard disk I/O port (XT)
78h	120	Int 17h	1	Parallel port 1 timeout
79h	121	Int 17h	1	Parallel port 2 timeout
7Ah	122	Int 17h	1	Parallel port 3 timeout
7Bh	123		1	Parallel port 4 timeout support for virtual DMA services
7Ch	124	Int 14h	1	Serial port 1 timeout
7Dh	125	Int 14h	1	Serial port 2 timeout
7Eh	126	Int 14h	1	Serial port 3 timeout
7Fh	127	Int 14h	1	Serial port 4 timeout
80h	129	Int 16h	2	Starting address of keyboard buffer
82h	131	Int 16h	2	Ending address of keyboard buffer
84h	132	Int 10h	1	Number of video rows (minus 1)
85h	134	Int 10h	2	Number of scan lines per character
87h	135	Int 10h	1	Video display adapter options
88h	136	Int 10h	1	Video display adapter switches
89h	137	Int 10h	1	VGA video flag 1
8Ah	138	Int 10h	1	VGA video flag 2
8Bh	139	Int 13h	1	Floppy disk configuration data
8Ch	140	Int 13h	1	Hard disk drive controller status
8Dh	141	Int 13h	1	Hard disk drive error
8Eh	142	Int 13h	1	Hard disk drive task complete flag
8Fh	143	Int 13h	1	Floppy disk drive information
90h	144	Int 13h	1	Disk 0 media state
91h	145	Int 13h	1	Disk 1 media state
92h	146	Int 13h	1	Disk 0 operational starting state
93h	147	Int 13h	1	Disk 1 operational starting state
94h	148	Int 13h	1	Disk 0 current cylinder
95h	149	Int 13h	1	Disk 1 current cylinder
96h	150	Int 16h	1	Keyboard status flag 3
97h	151	Int 16h	1	Keyboard status flag 4
98h	155		4	Address of user wait flag pointer
9Ch	159		4	User wait count
A0h	160		1	User wait flag
A1h	167		7	Local area network (LAN)
A8h	171		4	Address of video parameter control block
ACh	239		68	Reserved
F0h	255		16	Intra-applications communications area

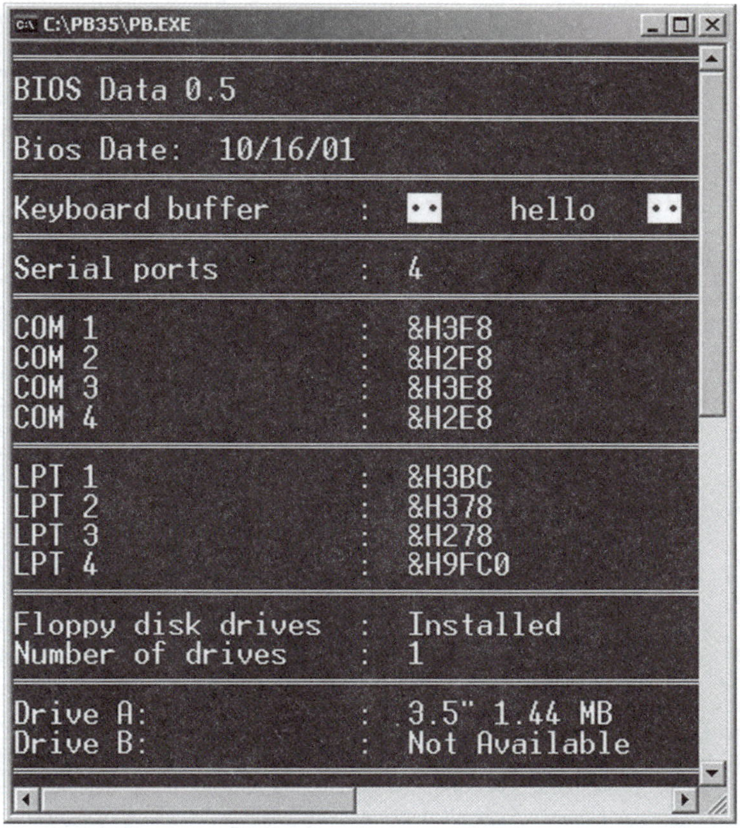

```
C:\PB35\PB.EXE

BIOS Data 0.5

Bios Date:  10/16/01

Keyboard buffer       :  ••    hello   ••

Serial ports          :  4

COM 1                 :  &H3F8
COM 2                 :  &H2F8
COM 3                 :  &H3E8
COM 4                 :  &H2E8

LPT 1                 :  &H3BC
LPT 2                 :  &H378
LPT 3                 :  &H278
LPT 4                 :  &H9FC0

Floppy disk drives    :  Installed
Number of drives      :  1

Drive A:              :  3.5" 1.44 MB
Drive B:              :  Not Available
```

Figure 1.20 *Output of the BIOS data program*

```
print division$
Def Seg=&HF000
Print "Bios Date:    " Peek$(&H0FFF5,8)
'
' Retrieve and print the keyboard buffer
'
Def Seg = &H40
c$ = ""
for i% = 0 to 31 step 2
cb? = peek(&H1E + i%)
c$ = c$ + chr$(cb?)
next i%
print division$
print "Keyboard buffer         :   "; c$
print division$
'
' Get and display COM port addresses
'
num% = val("&B"+left$(right$(bin$(high%),4),3))
print "Serial ports            :   "num%
print division$
for temp=1 to 4
```

```
    ' Address of COM1 is at offset &H00
    ' Address of COM2 is at offset &H02
    ' Address of COM3 is at offset &H04
    ' Address of COM4 is at offset &H06
    if peeki(&H0+(temp-1)*2)<>0 then
       Print "COM"temp"          :    &H";
       print hex$(peeki(&H0+(temp-1)*2))
    end if
next temp
'
' Get and display LPT port addresses
'
print division$
for temp=1 to 4
    ' Address of LPT1 is at offset &H08
    ' Address of LPT2 is at offset &H0A
    ' Address of LPT3 is at offset &H0C
    ' Address of LPT4 is at offset &H0E
    if peeki(&H08+(temp-1)*2)<>0 then
       print "LPT"temp"          :    &H";
       print hex$(peeki(&H08+(temp-1)*2))
    end if
next temp
'
' Determine floppy disk drives installed
'
print division$
high% = peekl(&H10) \ 256
low%  = peekl(&H10) MOD 256
if bit(low%,0)=1 then
   print "Floppy disk drives   :  Installed    "
else
   print "Floppy disk drives   :  Not Available"
end if
if bit(low%,6)=0 and bit(low%,7)=0 then print"Number of
    drives    :  1"
if bit(low%,6)=0 and bit(low%,7)=1 then print"Number of
    drives    :  2"
if bit(low%,6)=1 and bit(low%,7)=0 then print"Number of
    drives    :  3"
if bit(low%,6)=1 and bit(low%,7)=1 then print"Number of
    drives    :  4"
print division$
Out &H70,&H10
x=Inp(&H71)
Print "Drive A:              :    ";
if x\16=0 Then Print "Not Available"
if x\16=1 Then Print "5.25"chr$(34)" 360 KB"
if x\16=2 Then Print "5.25"chr$(34)" 1.2 MB"
if x\16=3 Then Print "3.5"chr$(34)"  720 KB"
if x\16=4 Then Print "3.5"chr$(34)" 1.44 MB"
if x\16=5 Then Print "3.5"chr$(34)" 2.88 KB"
Print "Drive B:              :    ";
if (x And &H0F)=0 Then Print "Not Available"
if (x And &H0F)=1 Then Print "5.25"chr$(34)" 360 KB"
if (x And &H0F)=2 Then Print "5.25"chr$(34)" 1.2 MB"
if (x And &H0F)=3 Then Print "3.5"chr$(34)"  720 KB"
```

```
if (x And &H0F)=4 Then Print "3.5"chr$(34)" 1.44 MB"
if (x And &H0F)=5 Then Print "3.5"chr$(34)" 2.88 KB"
print division$
end
```

Disk drives

Disk drives provide low-cost high-capacity storage for data and programs. Standard floppy disk drives operate at 300 rpm and use an 80-track format with 135 tracks per inch. The standard data transfer rate is around 250 KB/s while the formatted storage capacity is 1.44 MB.

Like floppy disks, the data stored on a hard disk takes the form of a magnetic pattern stored in the oxide-coated surface of a disk. Unlike floppy disks, hard disk drives are sealed in order to prevent the ingress of dust, smoke and dirt particles. This is important since hard disks work to much finer tolerances (track spacing, etc.) than do floppy drives. Furthermore, the read/write heads of a hard disk 'fly' above the surface of the disk when the platters arc turning. The speed of data transfer greatly exceeds that of a floppy disk drive because the hard disk rotates at speeds of typically between 4200 and 7200 rpm (around 20 times faster than a floppy drive).

Modern Integrated Drive Electronics (IDE) hard drives are designed to interface very easily with the PC bus by means of one, or more, 40-way IDC connectors on the motherboard. The 40-way bus extension is sometimes known as an *AT attachment* (ATA). This system interface is simply a subset of the original ISA bus signals and it can support up to two IDE drives in a *daisy chain* fashion (i.e. similar to that used originally with floppy disk drives).

IDE drives are *low-level formatted* with a pattern of tracks and sectors already in place. This allows drives to be more efficiently formatted than would otherwise be possible. The actual *physical* layout of the data on the disk is hidden

Photo 1.13 *Interior of a hard disk drive*

from the BIOS which only sees the *logical* format of tracks and sectors presented to it by the integrated electronics. This means that the disk can have a much larger number of sectors on the outer tracks than on the inner tracks. Consequently, a much greater proportion of the disk space is available for data storage.

Photo 1.14 *DIMM, ATA/IDE, and power connectors on modern motherboard*

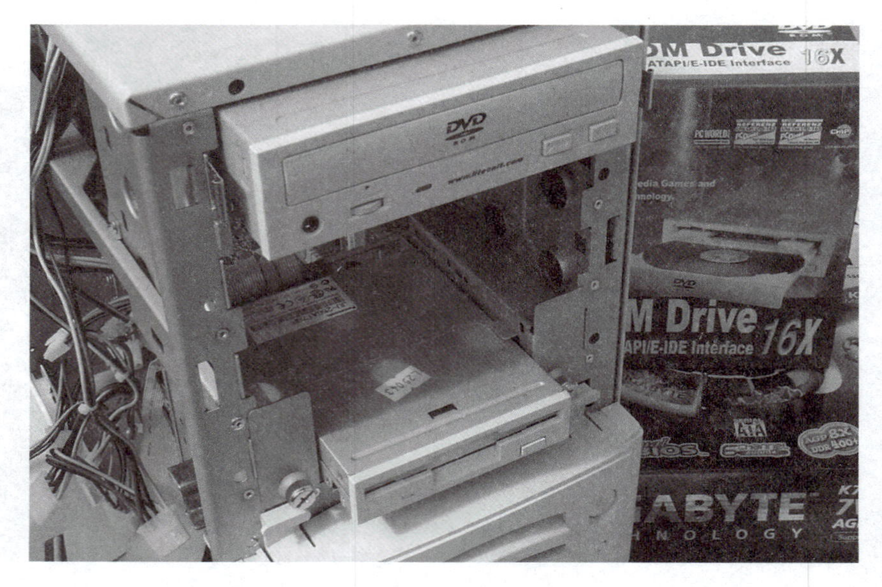

Photo 1.15 *Drive bays in a tower PC (the hard drive has been removed)*

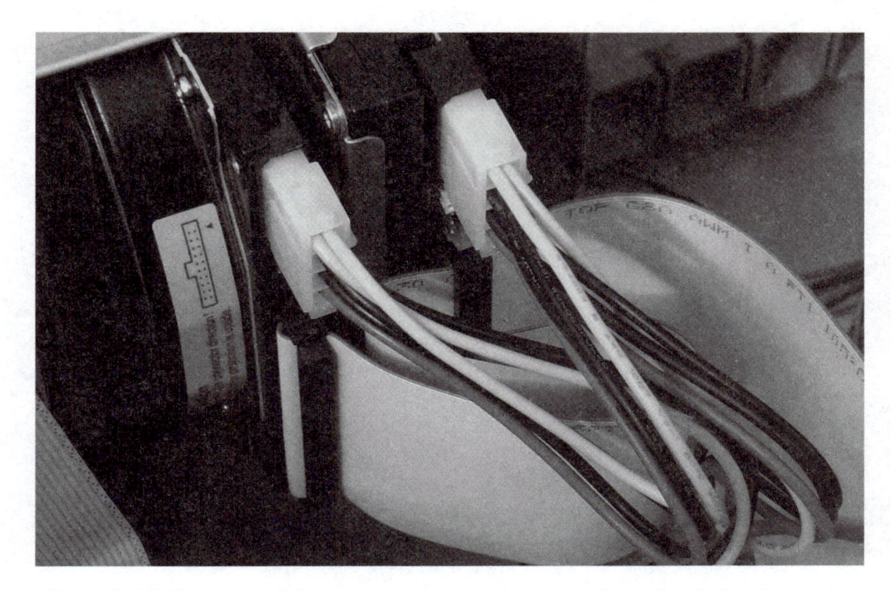

Photo 1.16 *Power and data connectors at the rear of two hard drives*

The next generation of hard drives are set to use the newly introduced Serial ATA (SATA) interface which is now becoming widely available. Existing parallel ATA drives transfer data concurrently on multiple parallel wires within an 80-wire cable. In contrast, SATA drives transfer data at high speeds over a thin 7-wire cable.

Serial ATA drives offer several advantages over IDE drives, not the least of which is speed. The maximum data transfer rate (or burst rate) for most parallel drives is between 100 and 133 MB/s whilst drives using the first generation of the SATA interface can often reach 150 MBps. SATA drive speeds are expected to increase significantly over the next few years.

2 PC expansion bus systems

The availability of a variety of standard expansion bus systems within the PC environment must surely be the single most crucial factor in harnessing the power of the machine. Having decided upon the platform for your application, whether it be a conventional PC, an industrial PC, or some form of embedded PC controller, there is a need to find an effective means of connecting your hardware via an appropriate interface.

For many applications the internally available expansion (ISA/EISA, PCI, or PCI-X) provides a means of connecting a wide range of external hardware devices. Happily, a large number of manufacturers have recognized this fact and have developed expansion cards specifically for control, data acquisition, and instrumentation applications. For other applications it may be necessary to make use of an interface to an external bus via the USB, serial, or parallel ports. Alternatively, specialized PC controller/bus standards (such as PC/104) may be appropriate. This chapter discusses a variety of different solutions to the problem of connecting a PC to external hardware.

Expansion methods

PC expansion can be readily achieved by means of cards connected to the PC bus by any one or more of the following general methods:

- connectors available on the system motherboard (e.g. ISA/EISA, PCI, or PCI-X);
- an external *backplane bus* or a *stacking bus* system (e.g. PC/104 and PC/104-Plus);
- a high-speed serial interface to the external hardware (e.g. USB);
- serial and/or parallel ports available on the motherboard.

The first two of these methods provide a more direct route to the system bus which is based on connection to the motherboard bus signals. The second two methods are less direct and may require substantial buffering as well as serial-to-parallel conversion before external data can reach the system bus.

Development of PC expansion bus architectures

The signals present on an expansion bus can be divided into the following general categories:

- Address bus lines
- Data bus lines
- Read and write control signals
- Interrupt request signals
- DMA request and DMA acknowledge signals
- Miscellaneous control signals
- Clock signals
- Power rails.

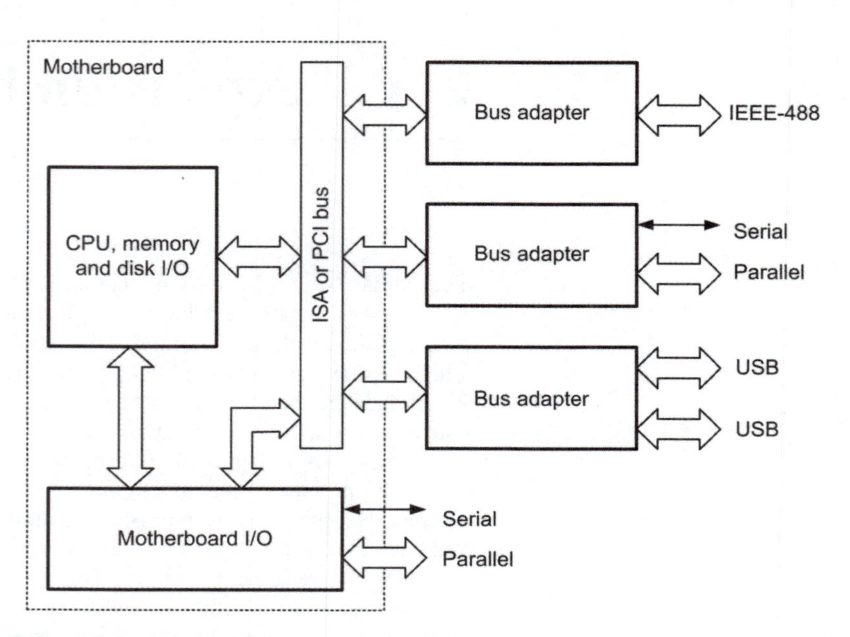

Figure 2.1 *A typical PC expansion scheme*

The most obvious method of expanding the PC bus is simply to provide a number of access points to the bus on the system motherboard. This approach was followed by IBM (and countless manufacturers of clones and compatibles) as a means of connecting essential items of peripheral hardware (such as displays and disk drives) via controllers fitted to *adapter* (or *option*) *cards.* This same method of connection can also be employed for more specialized applications such as analogue data acquisition, IEEE-488 bus control, etc.

Several PC expansion bus schemes have evolved over the past two decades. The original and most widely used standard is based on *Industry Standard Architecture* (ISA). This standard is also referred to as the *8-bit expansion bus* or simply as the *PC expansion bus.* This original PC expansion scheme was based on a single 62-way direct edge connector which provided access to the 8-bit data bus and the majority of control bus signals and power rails (Figure 2.1).

In order to provide access to a full 16-bit data path available from the PC/AT standard a further 36-way direct edge connector was later added. This provided access to the remaining data bus lines together with some additional control bus signals. Cards that required only an 8-bit data path and a subset of the PC's standard control signals were able to make use of *only* the first 64-way connector. Cards that needed the full 16-bit data path (not available on the early PC and XT machines) required *both* connectors. This enhanced standard is often referred to as *Extended Industry Standard Architecture* (EISA), the *16-bit expansion bus*, or simply the *PC/AT expansion bus.*

With the advent of PS/2, a more advanced expansion scheme has become available. This expansion standard was known as *Micro Channel Architecture* (MCA) and it provided access to the 16-bit data bus in the IBM PS/2 Models 50 and 60 whereas access to a full 32-bit data bus was made available in the Model 80 (which was fitted with an 80386 CPU).

An important advantage of MCA was that it permitted data transfer at significantly faster rates than was possible with ISA. In fairness, the increase in data transfer rate may be unimportant in many applications and also tends to vary somewhat from machine to machine. As a rough guide, when a standard AT machine is compared with a PS/2 Model 50, data transfer rates could be expected to increase by around 25% for conventional memory transfers and by 100% (or more) for DMA transfers.

Since MCA interrupt signals were shared between expansion cards, MCA interrupt structure tends to differ from that employed within ISA where interrupt signals tend to remain exclusive to a particular expansion card. More importantly, MCA provided a scheme of bus arbitration in order to decide which of the 'feature cards' had rights to exercise control of the MCA bus at any particular time. The arbitration mechanism provided for up to 15 bus masters, each one able to exercise control of the bus. As a further bonus, MCA provided an auxiliary video connector and programmable option configuration to relieve the tedium of setting DIP switches on system boards and expansion cards.

Despite its advantages over ISA/EISA, MCA was a relatively short-lived standard and it was never widely adopted by the industry. Instead, a new (and much enhanced) standard was introduced. This bus expansion standard is referred to as *Peripheral Component Interconnect* (PCI) and it quickly became the dominant standard leading to the rapid obsolescence of the PC and PC/AT bus standards. That said, many ISA/EISA cards are still in use today and so we shall begin by describing these standards in some detail before moving on to more modern bus standards.

PC ISA/EISA expansion bus

The PC ISA/EISA expansion bus is based upon a number of *expansion slots*, each of which is fitted with a 62-way direct edge connector together with an optional subsidiary 36-way direct edge connector. Expansion or option cards may be designed to connect only to the 62-way connector or may, alternatively, mate with both the 62- and 36-way connectors. Since only the 62-way connector was fitted on early machines (which had an 8-bit data bus), cards designed for use with this connector are sometimes known as *8-bit expansion cards* or *PC expansion cards*. The AT standard, however, provides access to a full 16-bit data bus together with additional control signals and hence requires the additional 36-way connector. Cards that are designed to make use of *both* connectors are generally known as *16-bit expansion cards* or *AT expansion cards*.

The original PC was fitted with only five expansion slots (spaced approximately 25 mm apart). The standard XT provided a further three slots to make a total of eight (spaced approximately 19 mm apart). Some cards, particularly those providing hard disk storage, required the width occupied by two expansion slot positions on the PC-XT. This was unfortunate, particularly where the number of free slots was often at a premium!

All of the XT expansion slots provided identical signals with one notable exception; the slot nearest to the power supply was employed in a particular IBM configuration (the IBM 3270 PC) to accept a keyboard/timer adapter. This particular configuration employed a dedicated *card select* signal (B8 on the connector) which was required by the system motherboard. Other cards which *would* operate in this position included the IBM 3270 Asynchronous Communications adapter.

Photo 2.1 *ISA/EISA and PCI expansions slots*

Like the PC-XT, the standard PC/AT also provided eight expansion slots. Six of these slots were fitted with two connectors (62- and 36-way) while two positions (slots 1 and 7) only had 62-way connectors. Slot positions 1 and 7 were designed to accept earlier 8-bit expansion cards which made use of the maximum allowable height throughout their length. If a 36-way connector *had* been fitted to the system motherboard, this would have fouled the lower edge of the card, preventing effective insertion of the card!

Finally, it should be noted that boards designed for AT systems (i.e. those specifically designed to take advantage of the availability of the full 16-bit data bus) will usually offer a considerable speed advantage over those which were based upon the 8-bit PC expansion bus. In some applications, this speed advantage was critical.

PC expansion cards

Expansion cards for PC systems tend to vary slightly in their outline and dimensions (see Figure 2.2). However, the *maximum* allowable dimensions for the adapter and expansion cards fitted to PC (and PS/2) equipment is usually quoted as follows:

		Height		Length		Width	
Standard	System type	in.	mm	in.	mm	in.	mm
ISA	8-bit PC and PC-XT	4.2	107	13.3	335	0.5	12.7
EISA	16-bit PC/AT	4.8	122	13.2	335	0.5	12.7
MCA	PS/2	3.8	96	13.2	335	0.5	12.7
PCI	PC/AT	4.8	122	7.3	185	0.5	12.7

With the exception of slot 8 in the PC-XT, the position in which an adapter or expansion card is placed *should* be unimportant. In most cases, this does hold

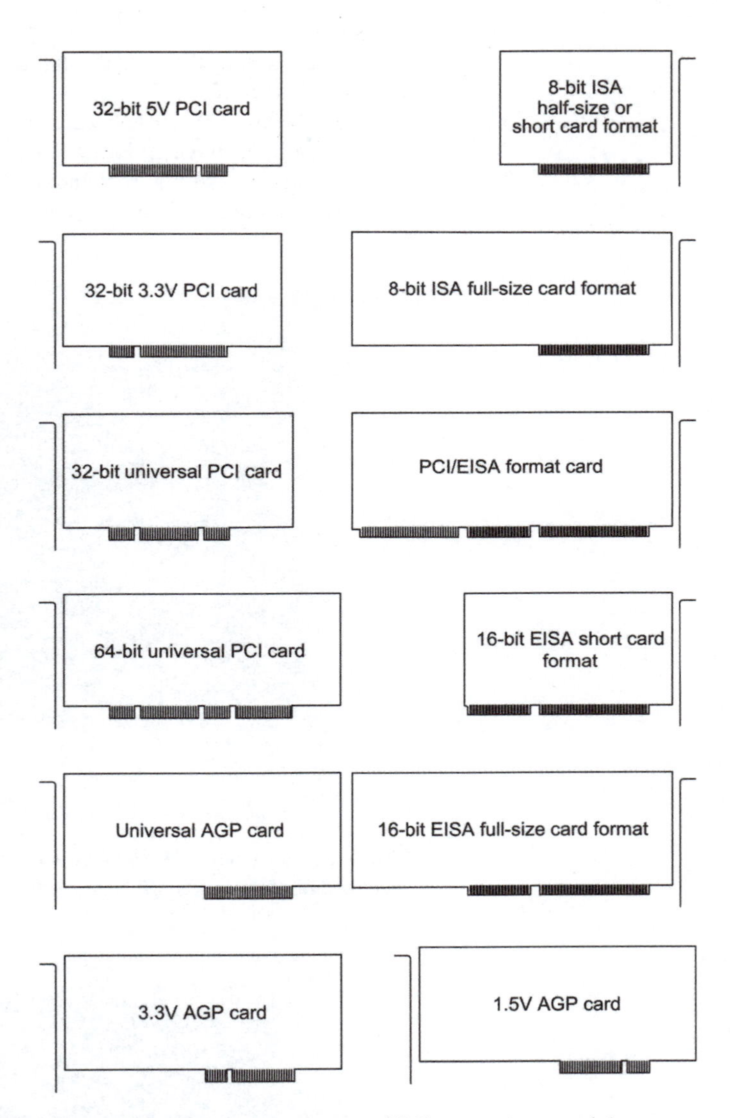

Figure 2.2 *Outlines for various types of PC expansion card (component side view)*

true however; in certain circumstances it is worth considering in which slot one should place a card.

The most important factor that should be taken into account is ventilation. Where cards are tightly packed together (particularly where ribbon cables may reduce airflow in the space between expansion cards) it is wise to optimize arrangements for cooling. Boards that are tightly packed with heat-producing components should be located in the positions around which airflow can be expected to be the greatest. This generally applies to the higher numbered slots in a system. Furthermore, when introducing a new card to a system, it may be worth re-arranging those cards that are already fitted in order to promote the unimpeded flow of air.

Accessibility of ISA/EISA cards (as well as later PCI cards) is also a point which is well worth considering. This is particularly important when the card in question is a prototype card that may require adjustment or alignment when the system is running. The card placed in an end slot is usually very much more

Photo 2.2 *An early ISA SCSI interface card (note the use of links along the upper edge of the card for base address selection)*

Photo 2.3 *A low-cost ISA parallel interface card*

accessible than any of the others. Furthermore, measurements are often more easily taken from a board fitted in this slot position without having to resort to the use of a *bus extender*. This point is also worth bearing in mind when fault finding becomes necessary.

To avoid the possibility of induced noise and glitches on the supply rails, it is usually beneficial to place boards that make large current demands or switch rapidly, in close proximity to the power supply (e.g. in slots 6, 7, and 8 of an ISA/EISA PC system). This precaution can be instrumental in reducing supply-borne disturbances (glitches) and can also help to improve overall system integrity and reliability. If, however, effective decoupling precautions have been observed, this precaution will be of minor importance.

Photo 2.4 *EISA dual serial/parallel port interface card (note the use of two mounting brackets for the external port connections)*

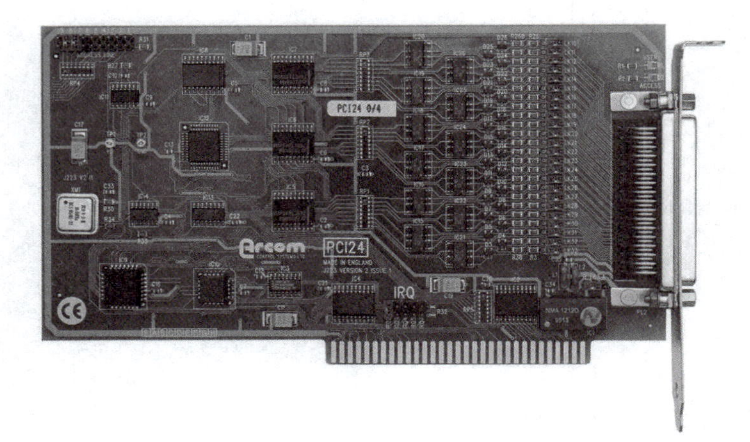

Photo 2.5 *A modern ISA card which provides 24 optically-isolated digital inputs (photo courtesy of Arcom)*

Finally, whilst timing is rarely a critical issue, some advantages can accrue from placing cards in older ISA/EISA-based systems closer to the processor. A particular case in point is the memory expansion cards that may be fitted to older ISA/EISA systems. These should ideally be fitted in slot positions 6, 7, and 8 in preference to positions 1, 2, and 3. In some cases this precaution could be instrumental in improving overall memory access times and avoiding parity errors. We continue this chapter by examining the ISA/EISA, PCI, and AGP bus standards in greater detail.

Industry Standard Architecture (ISA) bus

The original PC expansion bus supported an 8-bit data path (ISA) but the bus was soon extended to support the full 16-bit bus (EISA). Despite the emergence of PCI as an enhanced bus standard, many ISA and EISA cards are still in current use in control and instrumentation systems, and are still available from a number of suppliers.

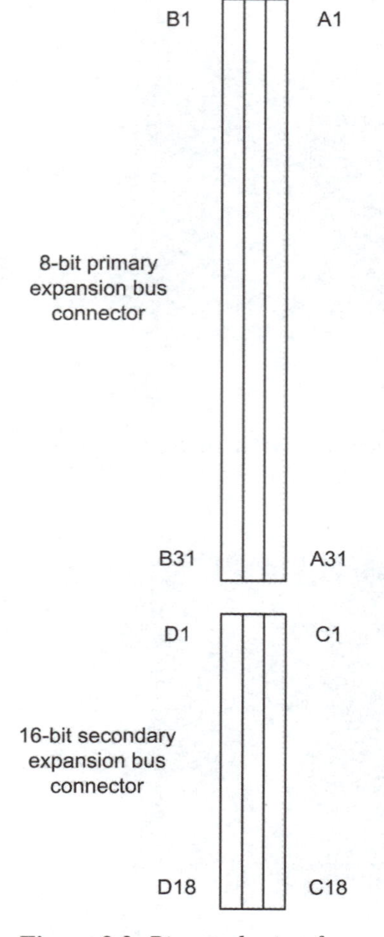

Figure 2.3 *Pin numbering for PC/PC-AT ISA and EISA expansion cards (viewed from above)*

The 62-way ISA (PC expansion bus) connector

The 62-way ISA expansion bus connector was based on a number of direct edge connectors fitted to the system motherboard. One side of the connector is referred to as A (lines as numbered Al to A31) while the other is referred to as B (lines are numbered B1 to B31). The address and data bus lines are grouped together on the A-side of the connector while the control bus and power rails occupy the B-side (see Figure 2.3).

It is, however, important to be aware that some early PC expansion bus pin-numbering systems did not use letters A and B to distinguish the two sides of the expansion bus connector. In such cases, odd-numbered lines (1 to 61) formed one side of the connector whilst even-numbered lines (2 to 62) formed the other. Here we shall, however, adopt the more commonly used pin-numbering convention described earlier.

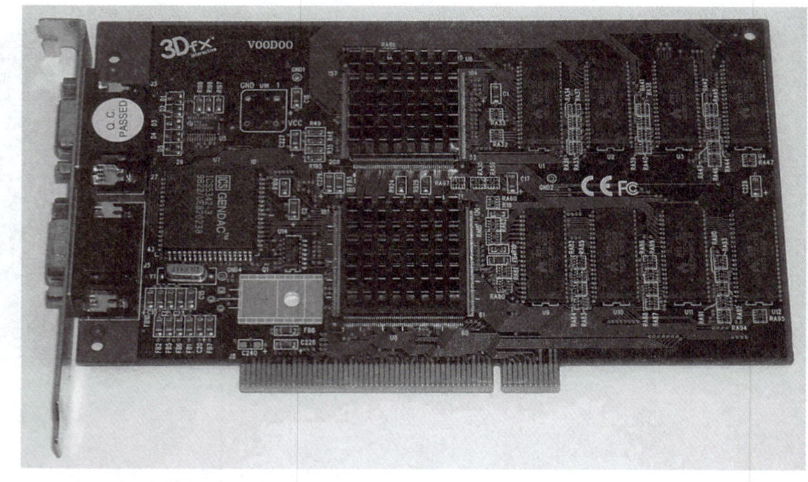

Photo 2.6 *A PCI 3D graphics adapter card*

The following table describes each of the signals present on the 62-way ISA expansion bus connector:

Pin number	Abbreviation	Direction	Signal	Function
A1	/IOCHK	I	I/O channel check	Taken low to indicate a parity error in a memory or I/O device
A2	D7	I/O	Data 7	Data bus line
A3	D6	I/O	Data 6	Data bus line
A4	D5	I/O	Data 5	Data bus line
A5	D4	I/O	Data 4	Data bus line
A6	D3	I/O	Data 3	Data bus line
A7	D2	I/O	Data 2	Data bus line
A8	D1	I/O	Data 1	Data bus line
A9	D0	I/O	Data 0	Data bus line
A10	/IOCHRDY	I	I/O channel ready	Pulsed low by a slow memory or I/O device to signal that it is not ready for data transfer
A11	AEN	O	Address enable	Issued by the DMA controller to indicate that a DMA cycle is in progress. Disables port I/O during a DMA operation in which /IOR and /IOW may be asserted
A12	A19	I/O	Address 12	Address bus line
A13	A18	I/O	Address 13	Address bus line
A14	A17	I/O	Address 14	Address bus line
A15	A16	I/O	Address 15	Address bus line
A16	A15	I/O	Address 16	Address bus line
A17	A14	I/O	Address 17	Address bus line
A18	A13	I/O	Address 18	Address bus line
A19	A12	I/O	Address 19	Address bus line
A20	A11	I/O	Address 20	Address bus line
A21	A10	I/O	Address 21	Address bus line
A22	A9	I/O	Address 22	Address bus line
A23	A8	I/O	Address 23	Address bus line
A24	A7	I/O	Address 24	Address bus line
A25	A6	I/O	Address 25	Address bus line
A26	A5	I/O	Address 26	Address bus line
A27	A4	I/O	Address 27	Address bus line
A28	A3	I/O	Address 28	Address bus line
A29	A2	I/O	Address 29	Address bus line
A30	A1	I/O	Address 30	Address bus line
A31	A0	I/O	Address 31	Address bus line
B1	GND	n.a.	Ground	Ground/common 0 V
B2	RESET	O	Reset	When taken high this signal resets all expansion cards
B3	+5 V	n.a.	+5 V DC	+5 V supply voltage
B4	IRQ2	I	Interrupt request level 2	Interrupt request (highest priority)
B5	−5 V	n.a.	−5 V DC supply	−5 V supply voltage
B6	DRQ2	I	Direct memory access request level 2	Taken high when a DMA transfer is required. The signal remains high until the corresponding /DACK line goes low

(*continued*)

Pin number	Abbreviation	Direction	Signal	Function
B7	−12 V	n.a.	−12 V DC	−12 V supply voltage
B8	0WS	I	Zero wait state	Indicates to the processor that the present bus cycle can be completed without any additional wait cycles
B9	+12 V	n.a.	+12 V DC	+12 V supply voltage
B10	GND	n.a.	Ground	Ground/common 0 V
B11	/MEMW	O	Memory write	Taken low to signal a memory write operation
B12	/MEMR	O	Memory read	Taken low to signal a memory read operation
B13	/IOW	O	I/O write	Taken low to signal an I/O write operation
B14	/IOR	O	I/O read	Taken low to signal an I/O read operation
B15	/DACK3	O	Direct memory access acknowledge level 3	Taken low to acknowledge a DMA request on the corresponding level (see notes)
B16	DRQ3	I	Direct memory access request level 3	Taken high when a DMA transfer is required. The signal remains high until the corresponding /DACK line goes low
B17	/DACK1	O	Direct memory access acknowledge level 1	Taken low to acknowledge a DMA request on the corresponding level (see notes)
B18	DRQ1	I	Direct memory access request level 1	Taken high when a DMA transfer is required. The signal remains high until the corresponding /DACK line goes low
B19	/DACK0	O	Direct memory access acknowledge level 0	Taken low to acknowledge a DMA request on the corresponding level (see notes)
B20	CLK4	O	4.77 MHz clock	Processor clock divided by 3 with 210 ns period and 33% duty cycle
B21	IRQ7	I	Interrupt request level 7	Asserted by an I/O device when it requires service (see notes)
B22	IRQ6	I	Interrupt request level 6	Asserted by an I/O device when it requires service (see notes)
B23	IRQ5	I	Interrupt request level 5	Asserted by an I/O device when it requires service (see notes)
B24	IRQ4	I	Interrupt request level 4	Asserted by an I/O device when it requires service (see notes)
B25	IRQ3	I	Interrupt request level 3	Asserted by an I/O device when it requires service (see notes)
B26	/DACK2	O	Direct memory access acknowledge level 2	Taken low to acknowledge a DMA request on the corresponding level (see notes)
B27	TC	O	Terminal count	Pulse high to indicate that a DMA transfer terminal count has been reached
B28	ALE	O	Address latch enable	A falling edge indicates that the address latch is to be enables. The signal is taken high during DMA transfers

(*continued*)

Pin number	Abbreviation	Direction	Signal	Function
B20	+5 V	n.a.	+5 V DC	+5 V supply voltage
B30	OSC	O	14.31818 MHz clock	Fast clock with 70 ns period and 50% duty cycle
B31	GND	n.a.	Ground	Ground/common 0 V

Notes:

1 Signal directions are quoted relative to the system motherboard; I represents input, O represents output, and I/O represents a bidirectional signal used both for input and also for output (n.a. indicates 'not applicable').
2 IRQ4, IRQ6, and IRQ7 are generated by the motherboard serial, disk, and parallel interfaces, respectively.
3 DACK0 (sometimes labelled REFRESH) is used to refresh dynamic memory while DACK1 to DACK3 are used to acknowledge other DMA requests.
4 A / indicates a signal line that is active low (or asserted low).

The 36-way EISA (PC-AT expansion bus) connector

The PC-AT is fitted with an additional expansion bus connector which provides access to the upper eight data lines, D8 to Dl5, as well as further control bus lines. The AT-bus employs an additional 36-way direct edge-type connector. One side of the connector is referred to as C (lines are numbered C1 to C18) whilst the other is referred to as D (lines are numbered Dl to D18), as shown in Figure 2.3. The upper eight data bus lines and latched upper address lines are grouped together on the C-side of the connector (together with memory read and write lines) while additional interrupt request, DMA request, and DMA acknowledge lines occupy the D-side.

The following table describes each of the signals present on the 32-way EISA expansion bus connector:

Pin number	Abbreviation	Direction	Signal	Function
C1	SBHE	I/O	System bus high enable	When asserted this signal indicates that the high byte (D8 to D15) is present on the data bus
C2	LA23	I/O	Latched address 23	Address bus line
C3	LA22	I/O	Latched address 22	Address bus line
C4	LA21	I/O	Latched address 21	Address bus line
C5	LA20	I/O	Latched address 20	Address bus line
C6	LA19	I/O	Latched address 19	Address bus line
C7	LA18	I/O	Latched address 18	Address bus line
C8	LA17	I/O	Latched address 17	Address bus line
C9	/MEMW	I/O	Memory write	Taken low to signal a memory write operation
C10	/MEMR	I/O	Memory read	Taken low to signal a memory read operation
C11	D8	I/O	Data 8	Data bus line
C12	D9	I/O	Data 9	Data bus line
C13	D10	I/O	Data 10	Data bus line

(*continued*)

Pin number	Abbreviation	Direction	Signal	Function
C14	D11	I/O	Data 11	Data bus line
C15	D12	I/O	Data 12	Data bus line
C16	D13	I/O	Data 13	Data bus line
C17	D14	I/O	Data 14	Data bus line
C18	D15	I/O	Data 15	Data bus line
D1	/MEMCS16	I	Memory chip select 16	Taken low to indicate that the current data transfer is a 16-bit (single wait state) memory operation
D2	/IOCS16	I	I/O chip select 16	Taken low to indicate that the current data transfer is a 16-bit (single wait state) I/O operation
D3	IRQ10	I	Interrupt request level 10	Asserted by an I/O device when it requires service
D4	IRQ11	I	Interrupt request level 11	Asserted by an I/O device when it requires service
D5	IRQ12	I	Interrupt request level 12	Asserted by an I/O device when it requires service
D6	IRQ13	I	Interrupt request level 10	Asserted by an I/O device when it requires service
D7	IRQ14	I	Interrupt request level 10	Asserted by an I/O device when it requires service
D8	/DACK0	O	Direct memory access acknowledge level 0	Taken low to acknowledge a DMA request on the corresponding level
D9	DRQO	I	Direct memory access request level 0	Taken high when a DMA transfer is required. The signal remains high until the corresponding DACK line goes low
D10	/DACK5	O	Direct memory access acknowledge level 5	Taken low to acknowledge a DMA request on the corresponding level
D11	DRQ5	I	Direct memory access request level 5	Taken high when a DMA transfer is required. The signal remains high until the corresponding DACK line goes low
D12	/DACK6	O	Direct memory access acknowledge level 6	Taken low to acknowledge a DMA request on the corresponding level
D13	DRQ6	I	Direct memory access request level 6	Taken high when a DMA transfer is required. The signal remains high until the corresponding DACK line goes low
D14	/DACK7	O	Direct memory access acknowledge level 7	Taken low to acknowledge a DMA request on the corresponding level
D15	DRQ7	I	Direct memory access request level 7	Taken high when a DMA transfer is required. The signal remains high until the corresponding DACK line goes low
D16	+5 V	n.a.	+5 V DC	+5 V supply voltage
D17	/MASTER	I	Master	Taken low by the I/O processor when controlling the system address, data and control bus lines
D18	GND	n.a.	Ground	Ground/common 0 V

n.a. indicates 'not applicable.'

Photo 2.7 *Motherboard with ISA/EISA bus connectors (note that two out of the eight slots only provide access to the 8-bit bus)*

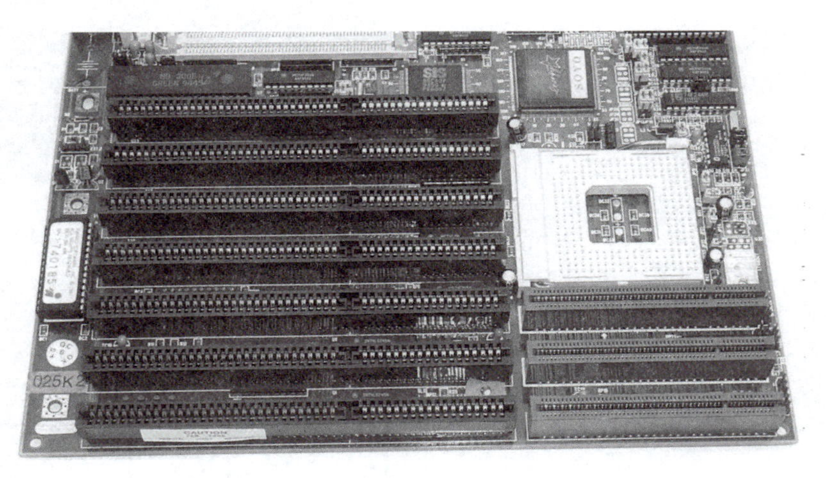

Photo 2.8 *Motherboard with ISA/EISA slots and PCI (CCA) combined slots*

Electrical characteristics

All of the signal lines present on the expansion connector(s) are TTL compatible. In the case of output signals from the system motherboard, the maximum loading imposed by an expansion card adapter should be limited to no more than two low-power (LS) TTL devices. The following expansion bus lines are open-collector: /MEMCSI6, /IOCS16, and 0WS. Note that the '/' indicates that the signal in question is *active low* (or *asserted low*).

The /IOCHRDY line is available for interfacing slow memory or I/O devices. Normal processor generated read and write cycles use four clock (CLK) cycles per byte transferred. The standard PC clock frequency of 4.77 MHz results in a single clock cycle of 210 ns. Thus each processor read or write cycle requires 840 ns at the standard clock rate. DMA transfers, I/O read and write cycles, on the other hand, require five clock cycles (1050 μs). When the /IOCHRDY line is asserted, the processor machine cycle is extended for an integral number of clock cycles.

Finally, when an I/O processor wishes to take control of the bus, it must assert the /MASTER line. This signal should not be asserted for more than 15 μs as it may otherwise impair the refreshing of system memory.

Photo 2.9 *Motherboard with four ISA/EISA slots and three PCI slots*

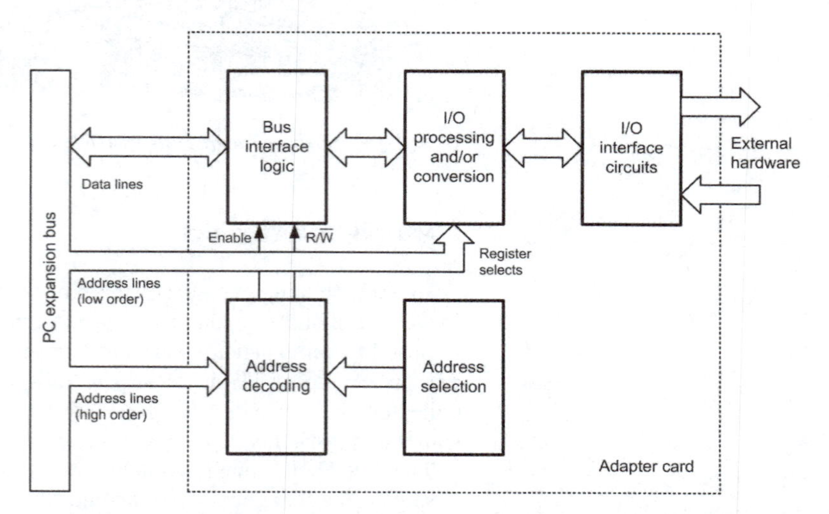

Figure 2.4 *Generalized block schematic for an ISA/EISA expansion card*

Design of PC expansion cards

Several factors need to be taken into account when designing PC expansion cards. These include power-supply requirements, power-supply rail distribution and decoupling, and address decoding (see Figure 2.4). In addition, access to the more specialized bus control signals (such as /IOCHK, /IOCHRDY, DRQ, and IRQ) may be required in the case of cards which are fitted with slow I/O devices, require DMA transfer or need to be interrupt driven. The following pointers are given for the benefit of those involved with the design and development of PC expansion cards.

Power rails

The available power for additional expansion cards depends upon the rating of the system power-supply, the requirements of the motherboard, and the demands of other adapter cards which may be fitted. When designing ISA/EISA expansion cards, the recommended limit (per card) for each of the four power rails is as follows:

Voltage rail (V)	Connection	Maximum current
+5	B3 and B29	1.5 A
−5	B5	100 mA
+12	B9	500 mA
−12	B7	100 mA

Where several adapter cards are fitted, the current demand for each supply rail should be estimated and the total power requirements calculated. It should go without saying that the total demand should not exceed the spare capacity rating of the system power supply. In some cases this may be less than 25 W!

As a guide, the following data refers to the power supplies fitted as 'standard' on most PCs with the nominal power ratings shown:

	Nominal power rating (W)			
	250	300	350	400
Maximum current rating (A)				
+3.3 V rail	20	28	28	28
+5 V rail	25	30	30	30
−5 V rail	0.5	0.5	0.5	0.5
+12 V rail	13	15	17	18
−12 V rail	0.8	0.8	0.8	0.8
+5 V standby	2	2	2	2

Whenever a system is built from scratch (or when expansion cards are added to a system) it is worth carrying out a *power audit* to ensure that the power supply is adequately rated. For example, assume that a motherboard with unexpanded

I/O is operated from a 250 W power supply and that the load on the various supply rails is as follows:

Supply rail (V)	Load (max) (A)	Power (W)
+3.3	15	50
+5	3	15
−5	0.5	2.5
+12	7	84
−12	0.5	6
+5 standby	1.5	7.5
	Total power:	165 W
	Remaining power available:	250 − 165 = 85 W

Supply rail distribution

In order to minimize supply-borne noise and glitches, the following recommendations should be observed when considering the design and layout of prototype expansion cards:

1 Ensure that the ground/common 0 V foil is adequate and that the three ground connections (B1, B10, and B31) are linked together via a substantial area of copper foil.
2 Include decoupling capacitors on each of the supply rails as follows:
   (a) 100 μF axial lead electrolytic to decouple the +5 V rail (locate close to pins B1 and B3 or B29 and B31).
   (b) 47 μF axial lead electrolytic to decouple the +12 V rail (locate close to pins B9 and B10).
   (c) 47 μF axial lead electrolytic to decouple the −12 V rail (locate close to pins B7 and B10).
   (d) 10 μF axial lead electrolytic to decouple the −5 V rail (locate close to pins B5 and B10).
   (*Note*: Capacitors can be omitted when the relevant voltage rail is not used within the expansion card.)
3 Fit 10 μF 16 V radial lead decoupling capacitors to the +5 V rail at the rate of one capacitor for every eight to ten TTL or CMOS logic devices. Capacitors should be distributed at regular points along the supply rail.
4 Fit 100 nF 16 V disk ceramic capacitors to the +5 V rail at the rate of one for every two to four TTL or CMOS logic devices. Capacitors should be placed at strategic points close to the supply pin connections of the integrated circuits.
5 Fit one 10 μF 16 V and one 100 nF 16 V capacitor to the +5 V rail for each VLSI device. Capacitors should be placed as close as possible to the supply pin connections of the devices in question.
6 Repeat (3), (4), and (5) for each of the other supply rails (where used).

Finally, it should go without saying that one should never attempt to insert or remove an expansion or adapter card when the power is connected and the system is running. Failure to observe this precaution may result in serious damage not only to the card in question but also to other cards that may be installed as well as to components on the system motherboard. If this all sounds rather obvious, no apologies are made for repeating it. In the heat of the moment it is all too easy to forget that a system is 'live'. You are only likely to make this mistake once – but the cost and frustration are likely to have a long-lasting effect!

Address decoding

The I/O provided by an expansion card will be mapped into either address or I/O space (the latter being conventionally used for digital and analogue I/O cards). The expansion card must, therefore, contain some address decoding logic which must be configured to avoid conflicts with other system hardware. Figure 2.5 shows some representative address decoding logic which provides access to eight *base addresses* within I/O space. Address lines A0 and A1 may then be used as optional *register select* lines for connection to VLSI devices (e.g. an 8255 Programmable Parallel Interface).

The address decoder shown in Figure 2.5 employs a three-to-eight line decoder (74LS138) in which the enable lines (G2A, G2B, and G1) are employed (note that G2A and G2B are active low, whilst G1 is an active-high input).

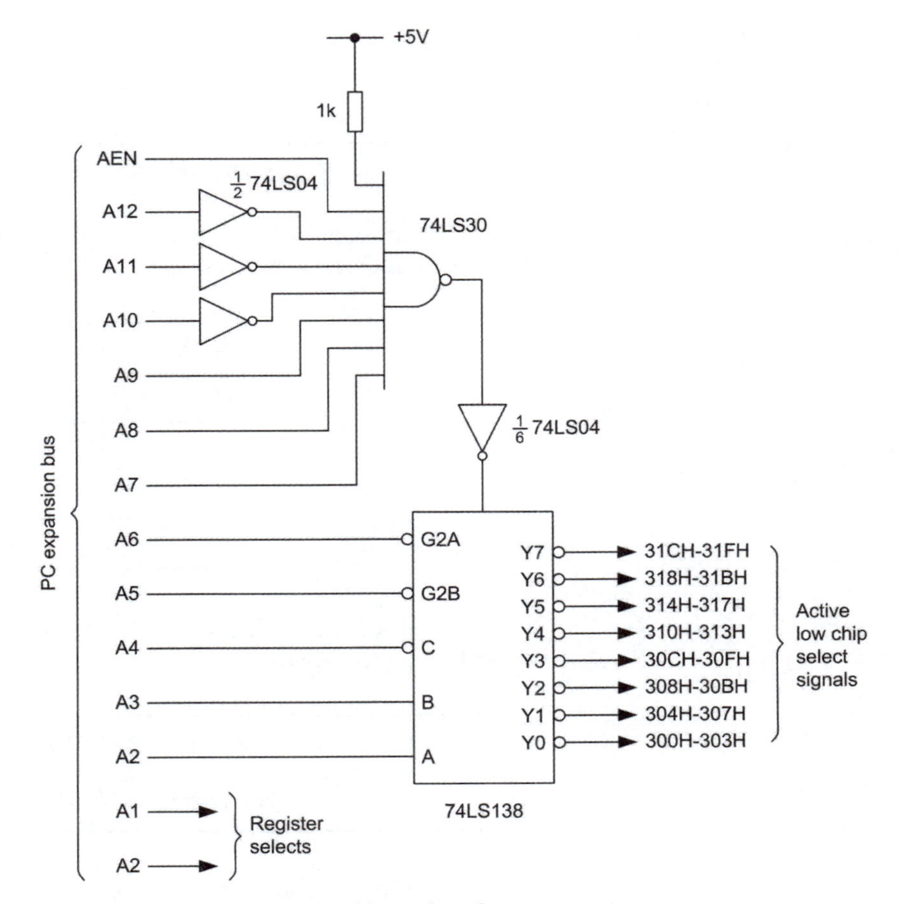

Figure 2.5 *Representative address decoder arrangement*

Outputs (Y0 to Y7) are active low and thus are ideal for use as chip select or enable signals. The truth table for the address decoder is as follows:

Address line											Base address	Output selected
A12	A11	A10	A9	A8	A7	A6	A5	A4	A3	A2	(hex.)	(taken low)
0	0	0	1	1	1	0	0	0	0	0	300	Y0
0	0	0	1	1	1	0	0	0	0	1	304	Y1
0	0	0	1	1	1	0	0	0	1	0	308	Y2
0	0	0	1	1	1	0	0	0	1	1	30C	Y3
0	0	0	1	1	1	0	0	1	0	0	310	Y4
0	0	0	1	1	1	0	0	1	0	1	314	Y5
0	0	0	1	1	1	0	0	1	1	0	318	Y6
0	0	0	1	1	1	0	0	1	1	1	31C	Y7
0	0	1	x	x	x	x	x	x	x	x	n.a.	None
0	1	0	x	x	x	x	x	x	x	x	n.a.	None
0	1	1	x	x	x	x	x	x	x	x	n.a.	None
1	0	0	x	x	x	x	x	x	x	x	n.a.	None
1	0	1	x	x	x	x	x	x	x	x	n.a.	None
1	1	0	x	x	x	x	x	x	x	x	n.a.	None
1	1	1	x	x	x	x	x	x	x	x	n.a.	None

x = don't care; n.a. = not applicable.

The remaining address lines (Al and A0) provide four *address offsets* from the base address, as follows:

Address lines		
A1	A0	Offset value
0	0	0
0	1	1
1	0	2
1	1	3

As an example, address 302 (hex.) will be selected when the following address pattern appears:

A12	A11	A10	A9	A8	A7	A6	A5	A4	A3	A2	Al	A0
0	0	0	1	1	1	0	0	0	0	0	1	0
Base address = 300 H											Offset = 2H	

It is, of course, quite permissible to use the chip select lines without making use of the register select lines, Al and A0. In such cases, it is important to remember that the I/O address will not be unique.

Photo 2.10 *Typical prototype card bus interface logic*

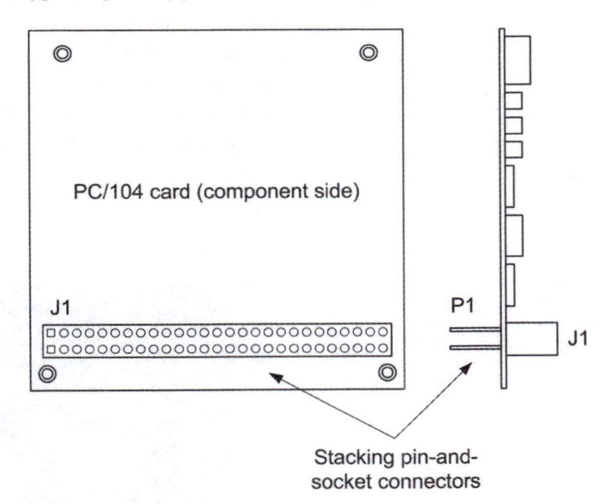

Figure 2.6 *Basic arrangement of the PC/104 bus*

The PC/104 bus

Despite the popularity of the PC and PC/AT bus architectures as a means of providing bus expansion for general purpose (desktop) and dedicated (non-desktop) applications their use on embedded systems has been limited by the relatively large size of expansion cards as well as the somewhat cumbersome method of interconnection. The PC/104 bus was developed in order to overcome these limitations.

The PC/104 bus offers the following advantages over the ISA/EISA standards:

- reduced card size;
- use of self-stacking system which has a small footprint and also eliminates the need for a backplane;
- reduced bus drive (and hence reduced power consumption overall).

The PC/104 bus is available in two versions, 8-bit and 16-bit, which correspond to the PC and PC/AT bus standards, respectively (Figure 2.6). To help

Photo 2.11 *Viper ultra-low power PC/104 format single board computer based on Intel's 400 MHz PXA255 XScale RISC processor. The board features a flat panel graphics controller, audio controller, 10/100baseT Ethernet, five serial ports, dual USB, digital I/O, onboard Flash memory and Compact Flash expansion (photo courtesy of Arcom)*

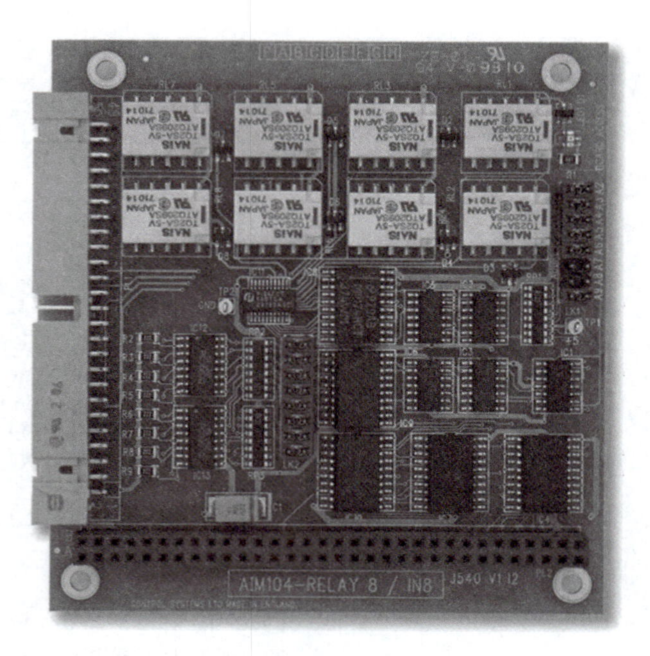

Photo 2.12 *AIM104 I/O card for PC/104 expansion. This card has eight changeover relays (each rated at 30 V DC, 1A) and eight optically isolated inputs (photo courtesy of Arcom)*

meet the tight space requirements of embedded systems, each of the two bus types (8-bit and 16-bit) offers two bus options, according to whether or not the P1 and P2 bus connectors extend through the module as *stack through* connectors.

Peripheral Component Interconnect/Interface (PCI) bus

The Peripheral Component Interconnect/Interface 'PCI' bus was originally developed as a local bus expansion for the ISA/EISA (PC/AT) bus. The first version of the PCI bus ran at 33 MHz with a 32-bit bus (133 MBps) but the current version runs at 66 MHz with a 64-bit bus. The PCI bus operates either synchronously or asynchronously with the motherboard bus rate. While operating asynchronously the bus will operate at any frequency up to the maximum (66 MHz). Flow control is used to allow the bus to operate with slower devices. The bus is unterminated and the bus clock operates at 133 MHz.

PCI supports full device *bus mastering*, and provides *bus arbitration* facilities through the system chipset. PCI architecture allows bus mastering of multiple devices on the bus simultaneously, with the arbitration circuitry working to ensure that no device on the bus (including the processor) locks out any other device. However, in the event that no other device requires access to the bus, PCI will allow a bus master to transfer data at the maximum permissible rate. Note that, with some early motherboards it might be possible that not all of the available PCI bus slots will be capable of bus mastering. When in doubt it is wise to check with the motherboard manual.

The PCI standard forms part of the *Plug and Play* standard developed by Intel, Microsoft, and many other companies in which the *PCI chipset* handles the identification of cards, working in conjunction with the BIOS and operating to automatically allocate resources for compatible peripheral cards.

The PCI bus uses its own internal interrupt system for dealing with requests from the cards on the bus. These interrupts are often called '#A', '#B', '#C', and '#D' to avoid confusion with the normal numbered system IRQs, though they are sometimes referred to by number. PCI interrupt levels are not generally seen by the user except in the PCI BIOS setup screen.

PCI interrupts are mapped to the normal system interrupts (usually IRQ9 to IRQ12). This imposes a limit of four interrupts available for PCI devices. Where more slots are provided (or where a USB controller is present) several PCI devices may be configured to share an IRQ.

Other variants and extensions of the basic PCI specification include:

PCI-X	The latest version 64 bits at 133 MHz
cPCI	Compact PCI is PCI in a VME form factor, using either 3U/6U modules and using 2 mm connectors
PC/104-Plus	PCI add-on to the PC/104 specification
PISA	PCI add-on for the PC/AT bus standard
P2CI	PCI on the VME64 P2 connector
PMC	PCI on a Mezzanine Card, 'PMC'
PXI	cPCI for Instrumentation
IPCI	Industrial PCI (another version of cPCI)
Serial PCI	PCI based on a serial link
Card Bus	32-bit PCI on the PC Card (PCMCIA) format

The pin connections and signals present on the PCI bus connector are summarized in the following table:

The PCI expansion bus connector

Pin	Name	Description	Pin	Name	Description
A1	TRST	Test logic reset	B1	−12 V	−12 V DC
A2	+12 V	+12 V DC	B2	TCK	Test Clock
A3	TMS	Test mode select	B3	GND	Ground
A4	TDI	Test data input	B4	TDO	Test data output
A5	+5 V	+5 V DC	B5	+5 V	+5 V DC
A6	INTA	Interrupt A	B6	+5 V	+5 V DC
A7	INTC	Interrupt C	B7	INTB	Interrupt B
A8	+5 V	+5 V DC	B8	INTD	Interrupt D
A9	—	Reserved	B9	PRSNT1	Present
A10	+5 V	Power (+5 V or +3.3 V)	B10	—	Reserved
A11	—	Reserved	B11	PRSNT2	Present
A12	GND03	Ground or keyway for 3.3 V/ universal cards	B12	GND	Ground or keyway for 3.3 V/ universal cards
A13	GND05	Ground or keyway for 3.3 V/ universal cards	B13	GND	Ground or open (key) for 3.3 V/ universal cards
A14	3.3 V aux		B14	RES	Reserved
A15	RESET	Reset	B15	GND	Ground
A16	+5 V	Power (+5 V or +3.3 V)	B16	CLK	Clock
A17	GNT	Grant PCI use	B17	GND	Ground
A18	GND08	Ground	B18	REQ	Request
A19	PME#	Power management event	B19	+5 V	Power (+5 V or +3.3 V)
A20	AD30	Address/Data 30	B20	AD31	Address/Data 31
A21	+3.3 V01	+3.3 V DC	B21	AD29	Address/Data 29
A22	AD28	Address/Data 28	B22	GND	Ground
A23	AD26	Address/Data 26	B23	AD27	Address/Data 27
A24	GND10	Ground	B24	AD25	Address/Data 25
A25	AD24	Address/Data 24	B25	+3.3 V	+3.3 VDC
A26	IDSEL	Initialization Device Select	B26	C/BE3	Command, Byte enable 3
A27	+3.3 V03	+3.3 V DC	B27	AD23	Address/Data 23
A28	AD22	Address/Data 22	B28	GND	Ground
A29	AD20	Address/Data 20	B29	AD21	Address/Data 21
A30	GND12	Ground	B30	AD19	Address/Data 19
A31	AD18	Address/Data 18	B31	+3.3 V	+3.3 V DC
A32	AD16	Address/Data 16	B32	AD17	Address/Data 17
A33	+3.3 V05	+3.3 V DC	B33	C/BE2	Command, Byte enable 2
A34	FRAME	Address or Data phase	B34	GND13	Ground
A35	GND14	Ground	B35	IRDY#	Initiator ready
A36	TRDY#	Target ready	B36	+3.3 V06	+3.3 V DC
A37	GND15	Ground	B37	DEVSEL	Device select
A38	STOP	Stop transfer cycle	B38	GND16	Ground
A39	+3.3 V07	+3.3 V DC	B39	LOCK#	Lock bus
A40	—	Reserved	B40	PERR#	Parity error

(*continued*)

Pin	Name	Description	Pin	Name	Description
A41	—	Reserved	B41	+3.3 V08	+3.3 V DC
A42	GND17	Ground	B42	SERR#	System Error
A43	PAR	Parity	B43	+3.3 V09	+3.3 V DC
A44	AD15	Address/Data 15	B44	C/BE1	Command, Byte enable 1
A45	+3.3 V10	+3.3 V DC	B45	AD14	Address/Data 14
A46	AD13	Address/Data 13	B46	GND18	Ground
A47	AD11	Address/Data 11	B47	AD12	Address/Data 12
A48	GND19	Ground	B48	AD10	Address/Data 10
A49	AD9	Address/Data 9	B49	GND20	Ground
A50	Keyway	Open or ground for 3.3 V cards	B50	Keyway	Open or ground for 3.3 V cards
A51	Keyway	Open or ground for 3.3 V cards	B51	Keyway	Open or Ground for 3.3 V cards
A52	C/BE0	Command, Byte Enable 0	B52	AD8	Address/Data 8
A53	+3.3 V11	+3.3 V DC	B53	AD7	Address/Data 7
A54	AD6	Address/Data 6	B54	+3.3 V12	+3.3 V DC
A55	AD4	Address/Data 4	B55	AD5	Address/Data 5
A56	GND21	Ground	B56	AD3	Address/Data 3
A57	AD2	Address/Data 2	B57	GND22	Ground
A58	AD0	Address/Data 0	B58	AD1	Address/Data 1
A59	+5 V	Power (+5 V or +3.3 V)	B59	VCC08	Power (+5 V or +3.3 V)
A60	REQ64	Request 64 bit	B60	ACK64	Acknowledge 64 bit
A61	VCC11	+5 V DC	B61	VCC10	+5 V DC
A62	VCC13	+5 V DC	B62	VCC12	+5 V DC
		64-bit spacer keyway			
A63	GND	Ground	B63	RES	Reserved
A64	C/BE[7]#	Command, Byte enable 7	B64	GND	Ground
A65	C/BE[5]#	Command, Byte enable 5	B65	C/BE[6]#	Command, Byte enable 6
A66	+5 V	Power (+5 V or +3.3 V)	B66	C/BE[4]#	Command, Byte enable 4
A67	PAR64	Parity 64	B67	GND	Ground
A68	AD62	Address/Data 62	B68	AD63	Address/Data 63
A69	GND	Ground	B69	AD61	Address/Data 61
A70	AD60	Address/Data 60	B70	+5 V	Power (+5 V or +3.3 V)
A71	AD58	Address/Data 58	B71	AD59	Address/Data 59
A72	GND	Ground	B72	AD57	Address/Data 57
A73	AD56	Address/Data 56	B73	GND	Ground
A74	AD54	Address/Data 54	B74	AD55	Address/Data 55
A75	+5 V	Power (+5 V or +3.3 V)	B75	AD53	Address/Data 53
A76	AD52	Address/Data 52	B76	GND	Ground
A77	AD50	Address/Data 50	B77	AD51	Address/Data 51
A78	GND	Ground	B78	AD49	Address/Data 49
A79	AD48	Address/Data 48	B79	+5 V	Power (+5 V or +3.3 V)
A80	AD46	Address/Data 46	B80	AD47	Address/Data 47
A81	GND	Ground	B81	AD45	Address/Data 45
A82	AD44	Address/Data 44	B82	GND	Ground
A83	AD42	Address/Data 42	B83	AD43	Address/Data 43
A84	+5 V	Power (+5 V or +3.3 V)	B84	AD41	Address/Data 41
A85	AD40	Address/Data 40	B85	GND	Ground

(*continued*)

Pin	Name	Description	Pin	Name	Description
A86	AD38	Address/Data 38	B86	AD39	Address/Data 39
A87	GND	Ground	B87	AD37	Address/Data 37
A88	AD36	Address/Data 36	B88	+5 V	Power (+5 V or +3.3 V)
A89	AD34	Address/Data 34	B89	AD35	Address/Data 35
A90	GND	Ground	B90	AD33	Address/Data 33
A91	AD32	Address/Data 32	B91	GND	Ground
A92	RES	Reserved	B92	RES	Reserved
A93	GND	Ground	B93	RES	Reserved
A94	RES	Reserved	B94	GND	Ground

Notes:

1 Signals on pins 63 to 94 are only used on 64-bit PCI bus cards.

2 The copper foil side of the card is side A whilst the component side is side B.

3 A # used after a signal name indicates that the signal in question is active low (or asserted low).

4 The time-multiplexed address and data bus may exist as either 0 to 31 bits (32 bits) or 0 to 63 bits (64 bits) using the 64-bit expansion bus. Both address and data signals use the same bus; addresses followed by data. A 32-bit PCI may also use 64-bit addressing by using two address cycles, referred to as Dual Address Cycles (DAC), in which the low order address is sent first. Additional control bits are used when the bus is used in 64-bit mode.

5 The bus connectors are labelled '+5 V *or* +3.3 V' in the case of +5 V systems and '+3.3 V' for 3.3 V systems. Note that the original PCI standard required that plug-in boards use +5 V supplies provided by the PCs motherboard. As the PCI standard evolved, the option was added for a + 3.3 V power source. Furthermore, the newer PCI 2.3 standard has now made the +5 V supply obsolete. This means that many of the most recent PCs can only accept 3.3 V or 'universal' PCI cards. Contacts on the PCI connector (keyways A12, B12, etc.) are used to determine the correct power rail voltages (see Figure 2.2).

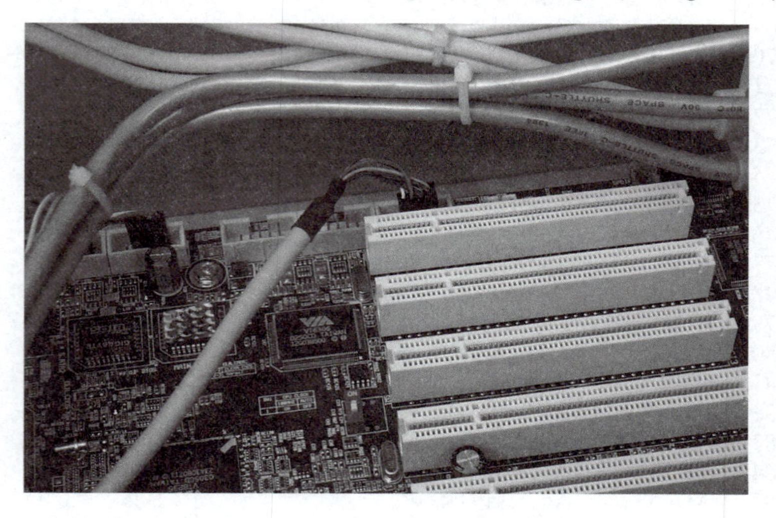

Photo 2.13 *Five PCI expansion connectors in a modern PC*

Accelerated Graphics Port (AGP)

Introduced in 1997, AGP was designed to provide a cost-effective means of improving the video performance of a PC and of reducing the burden that would otherwise be imposed on the PCI bus by having to cope with the fast throughput of video data. AGP enhances the interface between the video chipset and the processor, and also makes it possible for the video processor to have access to the main system memory. The AGP slot is physically similar to the PCI slots that may be fitted to a motherboard. However, the AGP slot is usually offset further from the edge of the motherboard. The AGP specification is based on the PCI 2.1 specification which includes a 66 MHz bus speed.

Like the PCI bus, the AGP bus is 32-bit wide however, instead of running at half of the system (memory) bus speed, AGP runs at the full speed of the bus (66 MHz). AGP also benefits from the fact that, as only one slot is present, there is no need to share the available bandwidth with any other devices!

In addition to doubling the speed of the bus, AGP has defined a double speed (2×) mode that allows twice as much data to be sent over the port at the same clock speed. In this mode, the hardware places data on the bus on both the rising and falling edges of the clock signal. In contrast, the PCI bus places data on only one of these transitions. The theoretical bandwidth is thus increased to a little over 500 MB/s.

In the context of data acquisition, control and instrumentation, and as a potential means for interfacing to external hardware the AGP has obvious limitations, not least of which is that there is only one slot available and this may already be occupied by a graphics card. Happily, for most applications the PCI bus is capable of providing sufficiently fast throughput with the added bonus that it is well supported by a huge range of I/O cards.

The Universal Serial bus

Offering true plug-and-play capability, the Universal Serial Bus (USB) has become the *de-facto* future standard for the interconnection of a host computer to a wide range of simultaneously accessible peripheral devices that share the available USB bandwidth through a host-scheduled, token-based protocol. Furthermore, unlike most other forms of computer bus, USB allows peripherals to be attached, configured, used, and detached while the host and other peripherals are in operation.

The Universal Serial Bus was originally specified as an industry-standard extension to the PC architecture with a focus on Computer Telephony Integration (CTI), consumer, and productivity applications. In framing the original specification, the following criteria were applied in defining the original USB specification:

- Ease-of-use for PC peripheral expansion
- Low-cost solution that supports transfer rates up to 12 MB/s
- Full support for real-time data for voice, audio, and compressed video
- Support for various types of data transfer
- Ability to cope with diverse system configurations, form factors, and host computers.

One of the principal advantages of USB is the speed at which it operates. USB supports two data transfer rates; 12 MB/s (described as *high-speed*) and an alternative (but still quite respectable) 1.5 MB/s (described as *low-speed*). Figure 2.7 shows how USB's two data transfer rates compare with those offered by other interface types and standards.

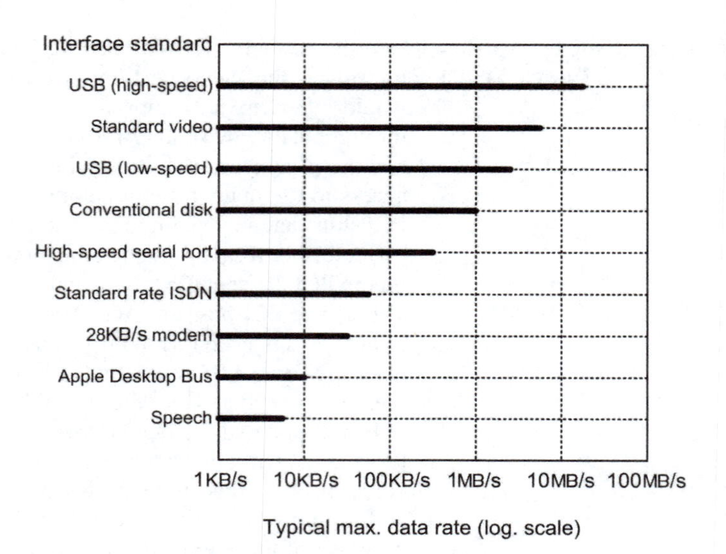

Figure 2.7 *Comparison of data rates for several interface standards*

USB applications and principal features

The Universal Serial Bus can be used in a very wide variety of applications. The ability to support data rates of up to 12 MB/s makes USB suitable for applications that are demanding in terms of data throughput whilst the simplicity of the USB interface makes it ideal for slower, low-cost peripheral devices such as keyboards, joysticks, and mice. Finally, the ability of USB to support multiple hubs and host controllers allows it to support more complex systems of computers and peripheral devices.

The following table, organized by speed of data transfer, summarizes the main applications envisaged for USB:

Speed of data transfer	Application	Examples
Low-speed (up to 128 KB/s)	Input devices	Mice (and other pointing devices) Keyboards Joysticks (and other game peripherals)
	Control systems	Plant and process control
	Communication	Low-speed modems
Medium-speed (128 KB/s to 2 MB/s)	Communication	High-speed modems and ISDN adapters
	Input devices	Scanners, video screen grabbers
	Storage devices	Removable disk drives (e.g. ZIP drives)
	Multimedia	Sound cards
	Instrumentation	Time, frequency, voltage measurement, etc.
	Data acquisition	Measurement and recording of temperature, humidity, stress, etc.
High-speed (greater than 2 MB/s)	Communication	Network adapters
	Storage devices	Optical drives
	Multimedia	Reduced bandwidth video

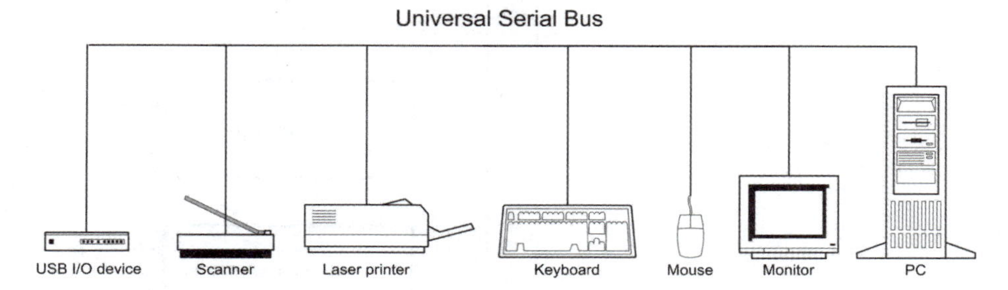

Figure 2.8 *USB arrangement for connecting a wide variety of peripheral devices*

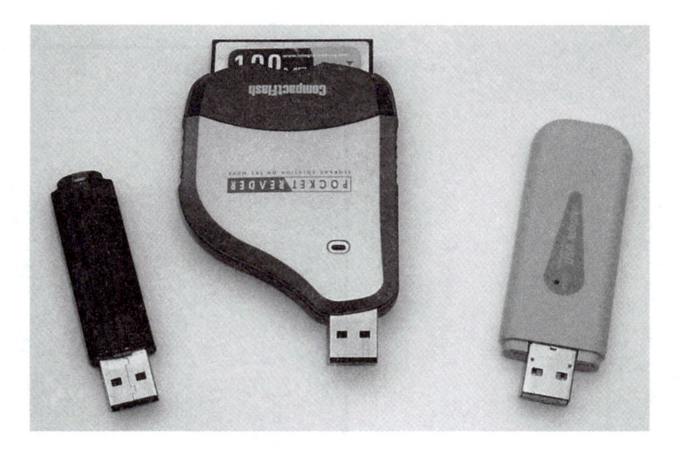

Photo 2.14 *Some typical USB devices; a 256 MB memory stick, a Flash memory card reader, and a wireless network adapter*

Figure 2.8 shows a typical range of peripherals that can be connected to a PC by means of the Universal Serial Bus. USB connectivity on a device can be easily recognized by the presence of the USB icon (see Figure 2.9).

The main features (and notable advantages compared with serial-port data transfer) of USB are as follows:

Figure 2.9 *The USB icon*

- Easy to set up and configure
- Simple cabling and connecting system
- Devices can be identified and configured automatically
- Peripheral devices can be 'hot-plugged' and 'hot un-plugged'
- Suitable for a wide range of device bandwidths
- Supports various types of data transfer (including isochronous)
- Supports concurrent operation of a large number of up to 127 devices
- Supports transfer of multiple data and message streams between the host and devices
- Efficient and transparent bus protocol
- Conforms with standard plug-and-play architecture
- Wide bandwidth
- Ability to use entire bus bandwidth in isochronous mode
- Flexible (easy to extend and modify)

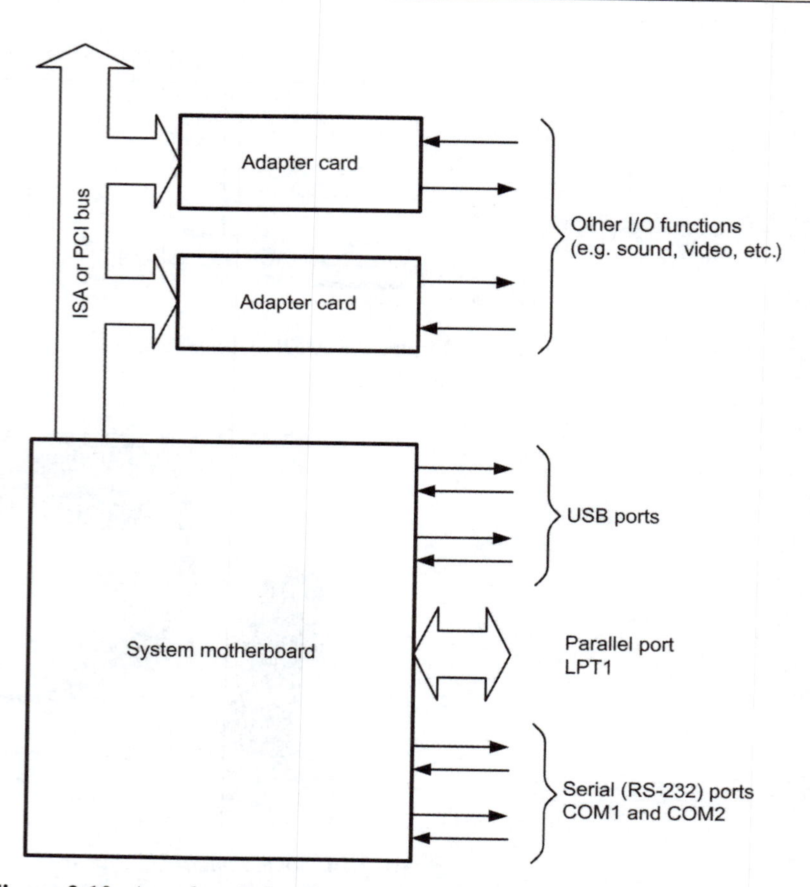

Figure 2.10 *A modern PC with motherboard USB ports and host controller circuitry*

- Allows a wide range of device data rates
- Flow control for buffer handling is built into the protocol
- Robust (incorporates error detection and fault recovery mechanisms)
- Relatively low-cost.

USB implementation

Most current desktop and tower PCs as well as Apple iMac computers are supplied with one or more USB ports. On PCs, these ports are additional to those that are normally associated with the original PC standard, such as the two serial ports (COM1: and COM2:) and the parallel port (LPT1:).

On most current PCs, the USB ports are functions provided by the system motherboard (see Figure 2.10). Older PCs can easily be fitted with USB ports by simply adding a low-cost adapter card (see Figure 2.11). However, in either case, the operating system must support the USB standard. This means that PC owners will have to upgrade to Windows 98 or later in order to have a system that *fully* implements the USB standard!

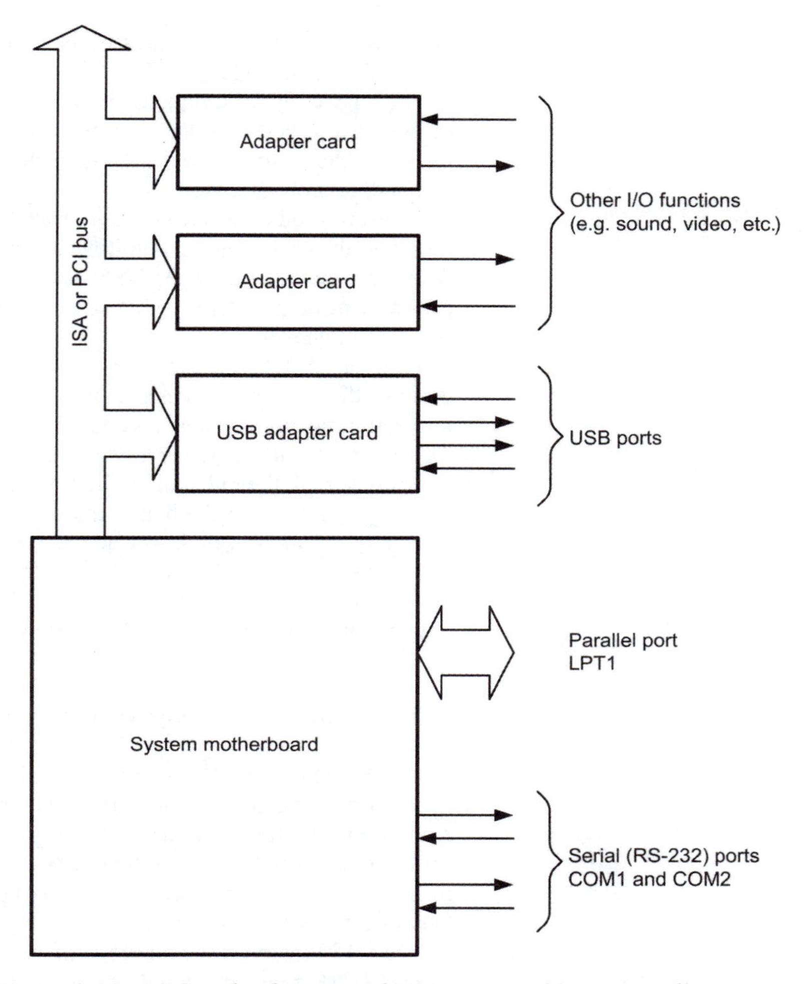

Figure 2.11 *A PC with adapter card USB ports and host controller circuitry*

The USB circuitry on the motherboard (or the adapter card) provides the functions of a USB host controller. This circuitry, in conjunction with buffers and drivers, provide the basic host interface to the USB.

Connection and disconnection of USB devices

One of the advantages of USB over other bus systems is its ability to support hot-connection and hot-disconnection from the bus. This important feature requires that the host's system software is not only able to recognize the connection and disconnection of devices but is able to reconfigure the system dynamically. Modern operating systems, such as Microsoft Windows 98, 2000 and XP have this facility.

USB devices attach to the USB through ports on hubs that incorporate status indicators to indicate the attachment or removal of a USB device. The host queries the hub to retrieve these indicators. In the case of an attachment, the host enables the port and addresses the USB device through the device's control pipe at the default address.

The host assigns a unique USB address to the device and then determines if the newly attached USB device is a *hub* or a *function*. The host then establishes its end of the control pipe for the USB device using the assigned USB address and endpoint number zero.

If the attached USB device is a *hub* and USB devices are attached to its ports, then the above procedure is followed for each of the attached USB devices. Alternatively, if the attached USB device is a *function*, then attachment notifications will be handled by host software that is appropriate for the particular function in question.

When a USB device has been removed from one of a hub's ports, the hub disables the port and provides an indication of device removal to the host. The removal indication is then handled by the appropriate USB system software. If the removed USB device is a hub, the USB system software must handle the removal of both the hub and all the USB devices that were previously attached to the system through the hub in question.

Finally, *enumeration* is the name given to process of allocating unique addresses to devices attached to a USB bus. Because the USB allows USB devices to attach to or detach from the USB at any time, bus enumeration has to be an on-going activity for the USB system software.

USB bus topology and physical connections

The USB connects USB devices with the USB host. The USB physical interconnect is a star topology that operates at a number of levels, extending downwards from the host. Hubs or nodes (i.e. peripheral USB devices) may be present at different levels but note that the nodes connected to any particular hub appear in the *next level down*. Each physical connection is a point-to-point connection between the relevant hub and node or between the relevant hub and another hub. Figure 2.12 illustrates the topology of the USB.

The USB transfers signal and power over a four-wire cable, shown in Figure 2.13. The signalling occurs over two wires on each point-to-point segment. In order to deliver power to devices, the cable also carries V_{BUS} (nominally +5 V) and GND wires on each segment. Cable segments may be of variable lengths (up to several metres) and the terminations allow rapid connection or disconnection at each port with differentiation between full-speed and low-speed devices.

It is important to note that each USB segment can provide only a *limited* amount of power. Furthermore, whilst the host can supply power for use by USB devices that are directly connected, any USB device may have its own power supply. USB devices that rely totally on power from the cable are called *bus-powered devices*. In contrast, those that have an alternate source of power are called *self-powered devices*. USB hubs supply power for any connected USB devices and this power may be derived from the host controller or may be externally derived. It is also worth noting that the mechanical specification for USB cables and connectors ensures that upstream and downstream connectors are not mechanically interchangeable, thus eliminating the possibility of *loopback connections* at hubs.

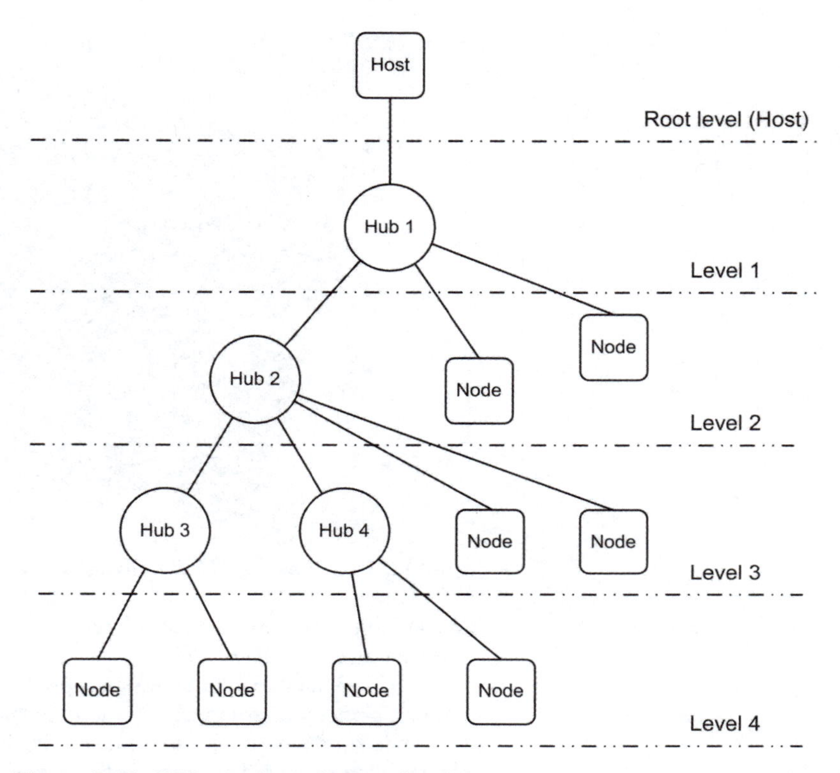

Figure 2.12 *USB topology showing levels or tiers*

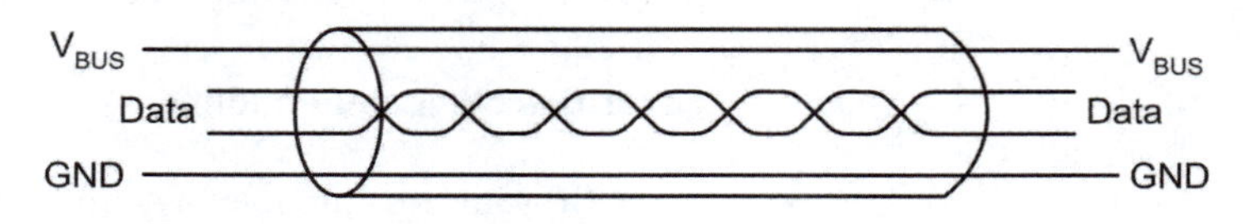

Figure 2.13 *USB cable (high-speed)*

As mentioned earlier, USB provides two basic data transfer rates; the full-speed bit rate of 12 MB/s and the reduced, low-speed, bit rate of 1.5 MB/s. Dynamic mode switching between transfers allows both modes to be supported in the same bus. The low-speed mode is defined to support a number of low-bandwidth devices, such as keyboards and pointing devices. The low-speed mode is also somewhat less demanding in terms of screening and EMI protection.

The USB clock signal is effectively encoded along with the differential data. The clock encoding scheme is NRZI with *bit stuffing* to ensure adequate transitions. In order to allow a receiver to synchronize its bit recovery clock, a SYNC field precedes each packet.

Note also that USB is a *polled bus* and that the *host controller* (i.e. the PC) initiates all of the data transfers. All bus transactions involve the transmission of up to three packets. Each transaction begins when the host controller, on a scheduled basis, sends a USB packet describing the type and direction of

Photo 2.15 *A low-cost USB four-port hub*

transaction, the USB device address, and endpoint number. This packet is referred to as the *token packet*.

Each USB device decodes the appropriate address fields to determine whether it is being selected. In a given transaction, data is transferred either from the host to a device or from a device to the host. The direction of data transfer is specified in the token packet. The source of the transaction then sends a data packet or indicates it has no data to transfer. The destination, in general, responds with a *handshake* packet indicating whether the transfer was successful.

Error detection and handling

The USB standard embodies a number of methods that contribute to the reduction of noise and data errors, and the overall enhancement of reliability. These include use of differential drivers, receivers, and shielding to improve signal integrity, *cyclic redundancy character* (CRC) checking of control and data fields, automatic detection of attachment and detachment of devices, and system-level configuration of resources. To provide protection against glitches and transients, each packet includes error protection fields. When a high level of data integrity is required, an error recovery procedure may be invoked in hardware or software. Hardware error handling includes reporting and retry of failed transfers, and a host controller will attempt retransmission three times before informing the client software of the failure. The client software can then recover in the most appropriate manner (according to the specific application and the particular device function).

USB data transfers

The USB architecture allows for four basic types of data transfers: *control transfers*, *bulk data transfers*, *interrupt data transfers*, and *isochronous data transfers*. We shall briefly describe each type.

Control transfers

Control data is used by the USB system software to configure devices when they are first attached. Other driver software can choose to use control transfers in implementation-specific ways. Data delivery is *lossless.*

Bulk data transfers

Bulk data typically consists of larger amounts of data, such as that used for printers or scanners. Bulk data is sequential. Reliable exchange of data is ensured at the hardware level by using hardware error and invoking a limited number of retries. Note also that the bandwidth taken up by bulk data can vary, depending on other bus activities.

Interrupt data transfers

A small, limited-latency transfer to or from a device is referred to as *interrupt data*. Such data may be presented for transfer by a device at *any* time. Interrupt data typically consists of event notification, characters, or coordinates that are organized as groups of one or more bytes. An example of interrupt data is the coordinates from a pointing device.

Isochronous data transfers

Isochronous data is continuous and delivered in real-time. Timing-related information is implied by the steady rate at which isochronous data is received and transferred. In order to maintain timing, isochronous data must be delivered at the rate that it is received. In addition to delivery rate, isochronous data may also be sensitive to delivery delays. For isochronous pipes, the bandwidth required is typically based upon the sampling characteristics of the associated function. The latency, on the other hand, is related to the buffering available at each endpoint.

A typical example of isochronous data is voice. If the delivery rate of this type of data stream is not maintained, *drop-outs* can occur due to buffer or frame *underruns* or *overruns*. Even if data is delivered at the appropriate rate by USB hardware, delivery delays introduced by software may degrade applications requiring real-time turn-around. To safeguard the delivery of data at the desired rate, USB isochronous data streams are allocated a dedicated portion of the USB bandwidth.

USB devices

As mentioned earlier, USB uses two major types of device: *hubs* and *functions*. The former class of device provides additional USB attachment points whilst the latter provides the host with additional capabilities. It is important to be clear about this distinction!

Hubs

Hubs are a key element in the plug-and-play architecture of the USB. Each hub converts a single *upstream port* into multiple *downstream ports*, each of which permits connection to another device or hub. Hubs can detect attachment and detachment at each downstream port and provide power to any downstream device that require it.

A hub consists of two elements: the *hub controller* and the *hub repeater*. The hub repeater is a protocol-controlled switch between the upstream port and downstream ports. It also has hardware support for reset and suspend/resume signalling. The host controller provides the interface registers to allow communication to/from the host. Hub-specific status and control commands permit the host to configure a hub and to monitor and control its ports.

Functions

A function is a USB device that is able to transmit or receive data or control information over the bus. A function is typically implemented as a separate peripheral device with a cable that plugs into a port on a hub. However, a physical package may implement multiple functions and an embedded hub with a single USB cable. This is known as a *compound device*. Such a device appears to the host as a hub with one or more non-removable USB devices.

Each function must incorporate configuration information that describes its capabilities and resource requirements. Before a function can be used, it must be configured by the host. This configuration includes allocating USB bandwidth and selecting function-specific configuration options. Examples of functions include:

- keyboards and keypads
- printers
- cameras
- graphics tablets
- mice
- trackballs.

USB data flow model

The USB host (normally part of the motherboard) interacts with USB devices and provides facilities for:

- Detecting the attachment and removal of devices
- Managing control flow between the host and devices
- Managing data flow between the host and devices
- Providing power to attached devices
- Collecting status information.

The simple view of communication between a host and a single attached USB device is shown in Figure 2.14. To account for the different layers, and the transactions that take place between them, Figure 2.15 provides a view of the underlying architecture of the interface.

For those who may be unfamiliar with the standard ISO model for Open System Interconnection, it is worth examining each of the major layers present within Figure 2.15. The *Bus interface* layer provides physical/signalling/packet connectivity between the host and a device whereas the *Device layer* is the view the USB system software has for performing generic USB operations with a

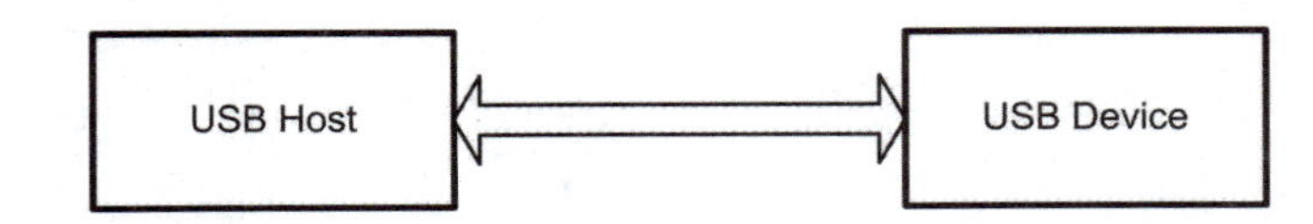

Figure 2.14 *Simple view of the USB interface*

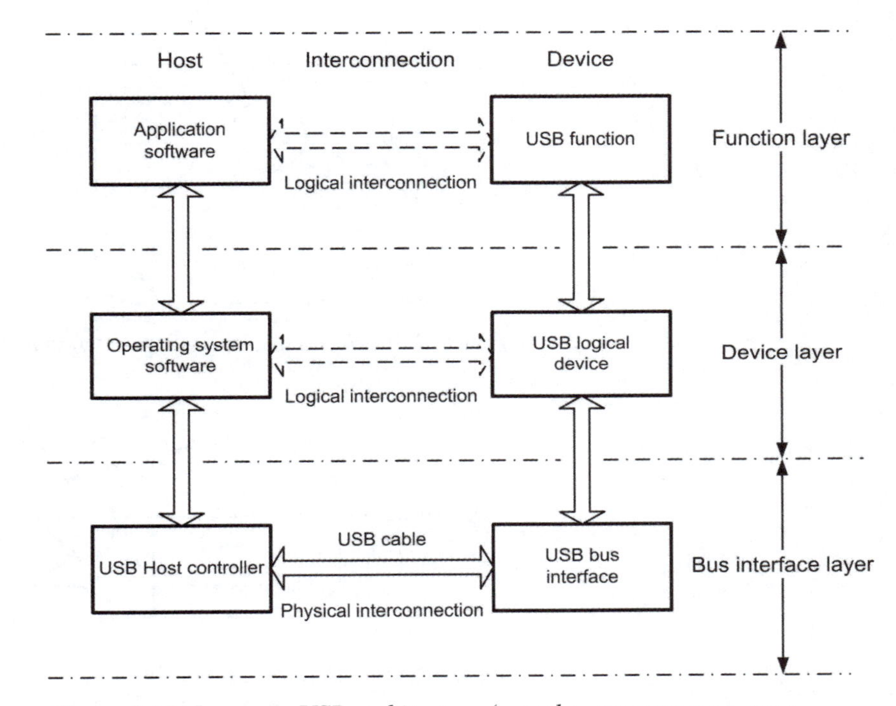

Figure 2.15 *Layers in USB architecture (note the apparent peer–peer logical interconnection)*

device. The *Function layer* provides additional capabilities to the host via an appropriate matched client software layer. The USB Device and Function layers each have a view of *logical* communication within their layer that actually uses the USB Bus interface layer to accomplish data transfer. What is important about this model is the apparent *peer-to-peer connectivity* that it provides!

Devices on the Universal Serial Bus are physically connected to the host via a tiered star topology. Figure 2.16 shows the topology of a typical USB arrangement. Note that host effectively incorporates its own embedded hub, called the *Root Hub*. The Root Hub, in turn, provides one or more attachment points.

Multiple functions may be packaged together in what appears to be a single physical device. For example, a keyboard and a trackball might be combined in a single package. Inside the package, the individual functions are permanently attached to a hub and it is the internal hub that is connected to the USB. When multiple functions are combined with a hub in a single package, they are referred to as a compound device. From the host's perspective, a compound device is the same as a separate hub with multiple functions attached.

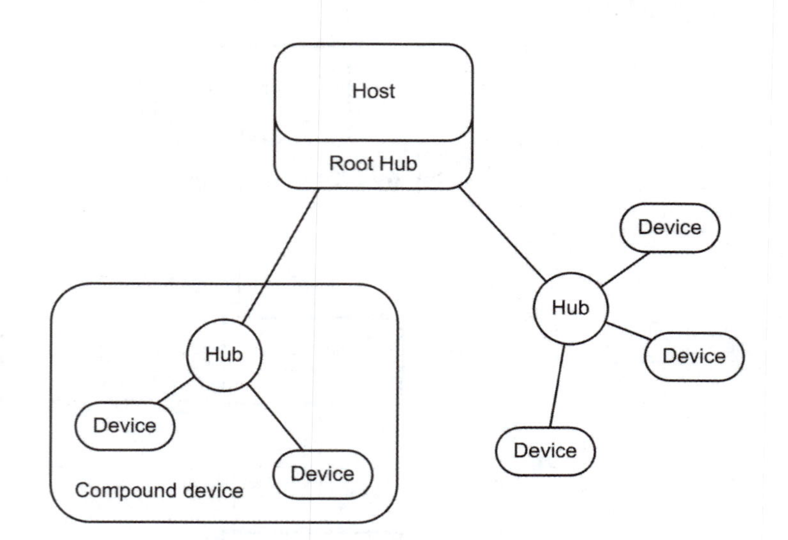

Figure 2.16 *Host, Hub, and device interconnections*

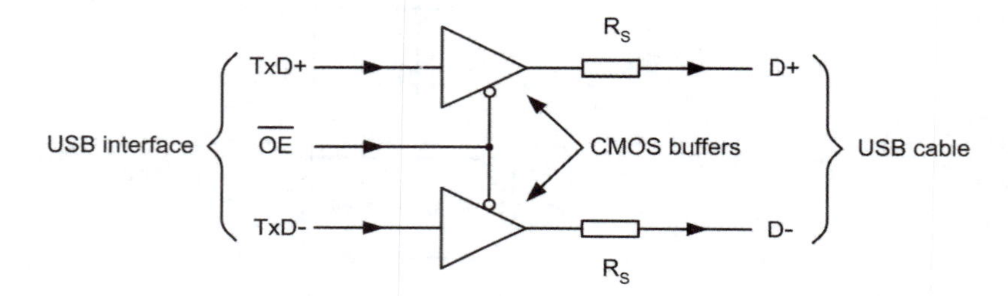

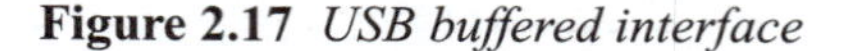

Figure 2.17 *USB buffered interface*

Client software for USB functions must use USB software programming interfaces to manipulate their functions instead of directly manipulating their functions via memory or I/O accesses as with other buses (e.g. PCI, EISA, PCMCIA, etc.). During operation, client software should be independent of other devices that may be connected to the USB. This allows the programmer and software to focus primarily on the interaction between hardware and software.

USB physical interface

The physical interface used in the Universal Serial Bus is quite straightforward. The interface specification involves electrical characteristics (voltage levels), cables, and connectors. We shall briefly describe each of these features:

Electrical interface

As mentioned earlier, USB uses just two differential data connections (D+ and D−) and two power connections. CMOS buffers are used to drive the relatively low impedance of the cable, as shown in Figure 2.17. The signal voltage present

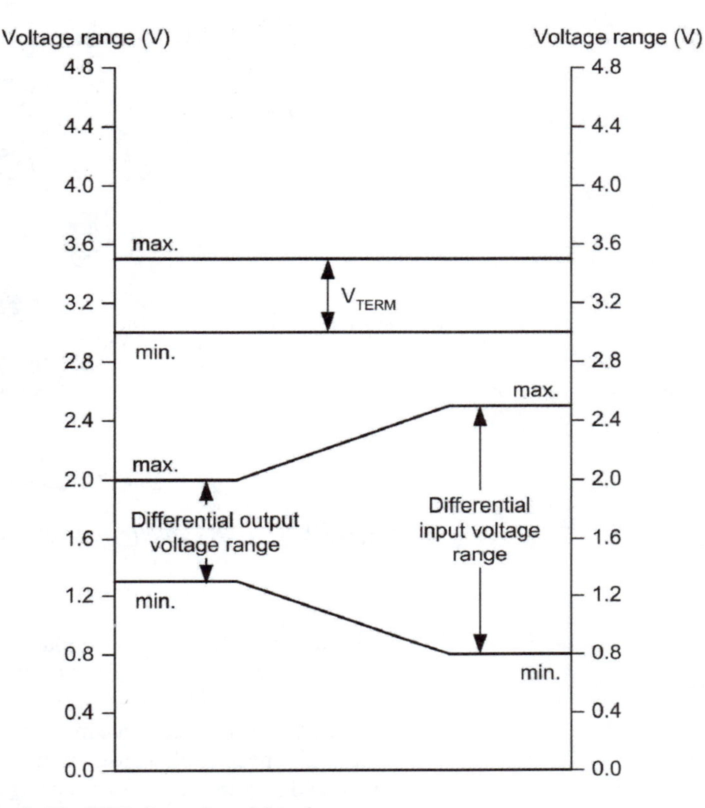

Figure 2.18 *USB data signal levels*

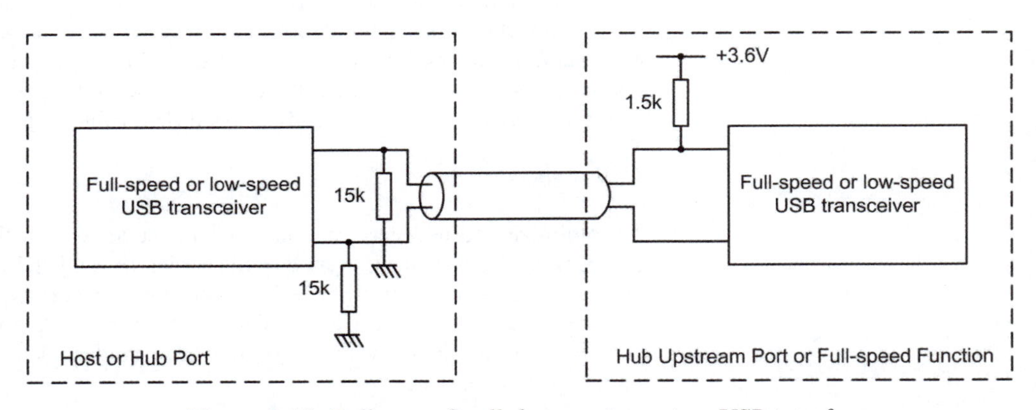

Figure 2.19 *Pull-up and pull-down resistors in a USB interface*

on the D+ and D− must be within the ranges shown in Figure 2.18. Note also that the terminating voltage (logic high) should be within the range 3.0–3.5 V.

Detection of device connection is accomplished by means of pull-up and pull-down resistors placed, respectively at the input or output of a port. USB pull-down resistors normally have a value of 15 kΩ whilst pull-up resistors have a value of 1.5 kΩ, as shown in Figure 2.19.

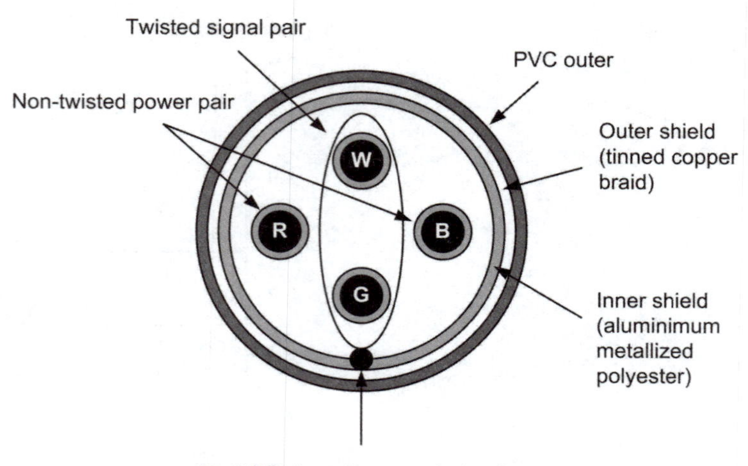

Figure 2.20 *USB cable (cross-sectional view)*

Cables

USB cables comprise four conductors, two power conductors and two signal conductors. Full-speed cable consists of a signalling twisted pair, V_{BUS}, GND, and an overall shield. Full-speed cable must be marked to indicate suitabillity for USB usage. Full-speed cable may be used with either low- or full-speed devices. When full-speed cable is used with low-speed devices, the cable must meet all low-speed requirements. Low-speed cable does not require twisted signalling conductors or the overall shield (since radiation of EMI is significantly reduced with low-speed data transmission).

The current USB specification describes three USB cable assemblies; *detachable cable*, *full-speed captive cable*, and *low-speed captive cable*. The recommended colours for the cable assembly are white, grey, or black.

A cross-sectional diagram of a full-speed USB cable is shown in Figure 2.20.

USB connectors

To minimize end user termination problems, USB uses a 'keyed connector' protocol. The physical difference in the Series 'A' and 'B' connectors insure proper end user connectivity. The 'A' connector is the principle means of connecting USB devices. All USB devices must have an 'A' connector. The 'B' connector allows device vendors to provide a standard detachable cable. It is important to note that:

- Series 'A' plugs are always oriented upstream towards the host system
- Series 'B' plugs are always oriented downstream towards the USB device.

The following list explains how the plugs and receptacles can be mated:

- The Series 'A' receptacle mates with a Series 'A' plug. Electrically, Series 'A' receptacles function as outputs from host systems and/or hubs.
- The Series 'A' plug mates with a Series 'A' receptacle. The Series 'A' plug always is oriented towards the host system.

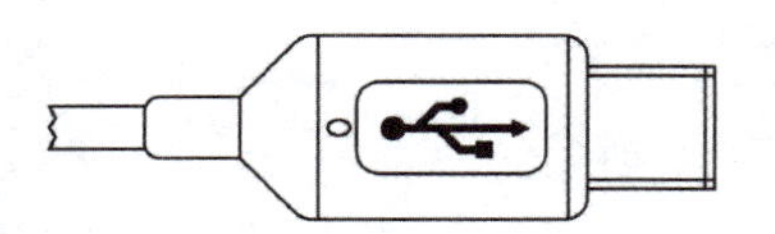

Figure 2.21 *USB cable connector*

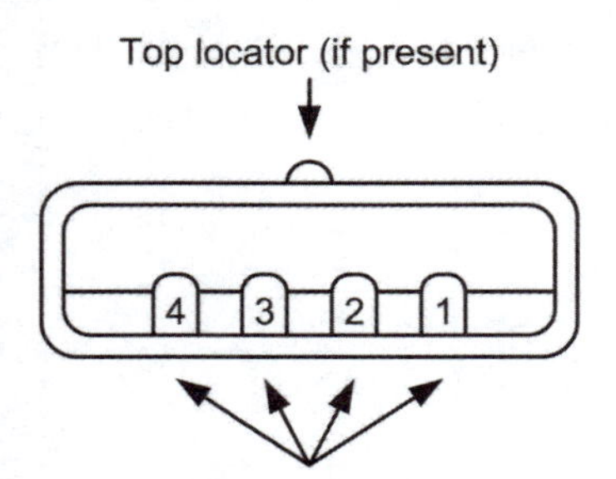

Figure 2.22 *USB connector (type A)*

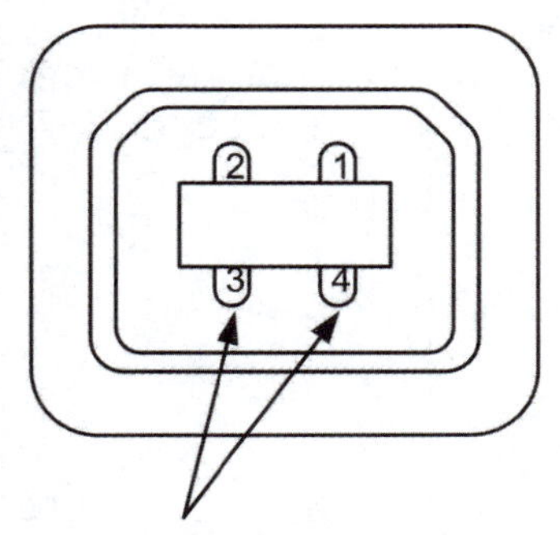

Figure 2.23 *USB connector (type B)*

Pin number	*Signal*	*Recommended colour*
1	V_{BUS}	Red
2	D-	White
3	D+	Green
4	GND	Black
Shell	Shield	Drain wire

Figure 2.24 *USB pin connections*

- The Series 'B' receptacle mates with a Series 'B' plug (male). Electrically, Series 'B' receptacles function as inputs to hubs or devices.
- The Series 'B' plug mates with a Series 'B' receptacle. The Series 'B' plug is always oriented towards the USB hub or device.

Full-speed devices can utilize the 'B' connector. This allows the device to have a detachable USB cable. This eliminates the need to build the device with a hardwired cable and minimizes end user problems if cable replacement is necessary. Figure 2.21 shows a typical Series 'A' connector showing the USB icon and the top locator (a small 'pip' located towards the cable end of the connector).

Devices utilizing the 'B' connector must be designed to work with worst case maximum length detachable cable. Detachable cable assemblies may be used only on full-speed devices. Note also that using a full-speed detachable cable on a low-speed device may exceed the maximum low-speed cable length.

Finally, Figures 2.22 and 2.23, respectively show the pin connections for connectors 'A' and 'B' whilst Figure 2.24 shows the pin assignment and recommended colour coding.

Representative I/O cards

The final part of this chapter describes some representative I/O interface cards and bus connected devices. These details have been included in order to provide readers with an insight into products that are currently available 'off-the-shelf' and that can be used for a wide variety of control, data acquisition and instrumentation applications.

Measurement Computing Corporation PDISO-8

The Measurement Computing Corporation PDISO-8 is an inexpensive eight-channel 500 V isolated input and 8-channel relay output interface card designed for control and sensing applications. The interface card is intended for use with a standard PCI bus and is Plug and Play compatible. Where necessary the input range may be extended by adding an additional fixed resistor in series with the existing 1.6 kΩ input resistor.

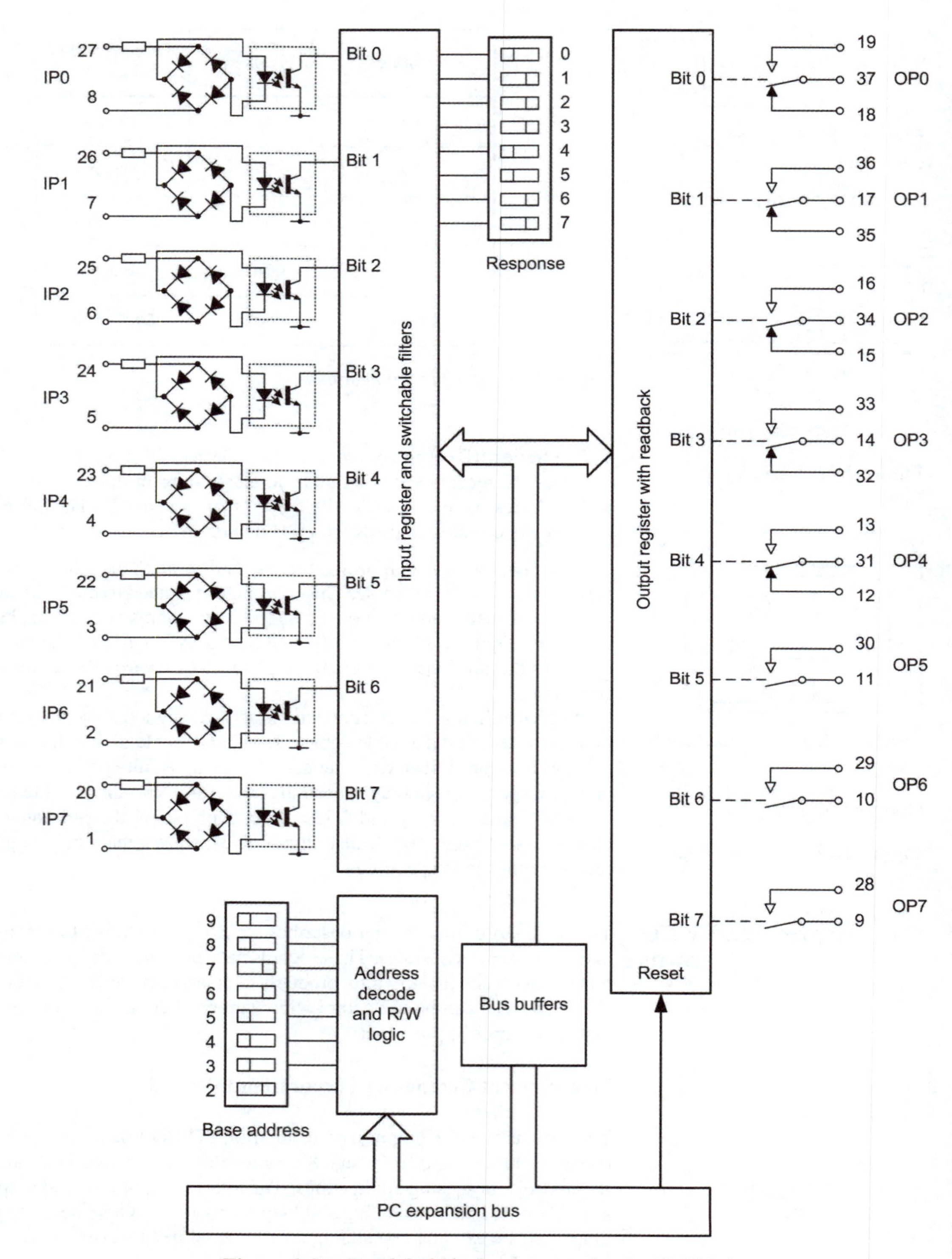

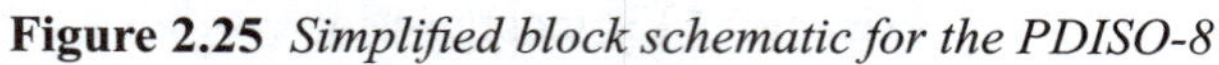

Figure 2.25 *Simplified block schematic for the PDISO-8*

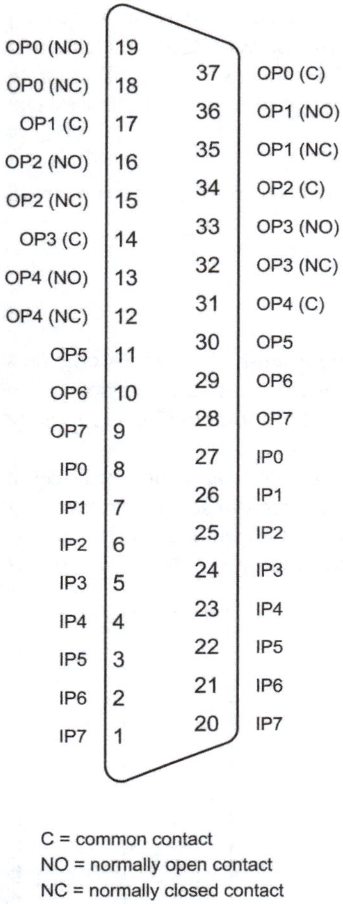

Signal	Pin	Pin	Signal
OP0 (NO)	19		
OP0 (NC)	18	37	OP0 (C)
OP1 (C)	17	36	OP1 (NO)
OP2 (NO)	16	35	OP1 (NC)
OP2 (NC)	15	34	OP2 (C)
OP3 (C)	14	33	OP3 (NO)
OP4 (NO)	13	32	OP3 (NC)
OP4 (NC)	12	31	OP4 (C)
OP5	11	30	OP5
OP6	10	29	OP6
OP7	9	28	OP7
IP0	8	27	IP0
IP1	7	26	IP1
IP2	6	25	IP2
IP3	5	24	IP3
IP4	4	23	IP4
IP5	3	22	IP5
IP6	2	21	IP6
IP7	1	20	IP7

C = common contact
NO = normally open contact
NC = normally closed contact

Figure 2.26 *Connector pin assignment for the PDISO-8*

Each of the eight inputs is optically isolated and fed via a bridge rectifier arrangement which allows for either AC or DC inputs of between 5 and 28 V. A fixed, current limiting, resistor of 1.6 kΩ is fitted to each input. The optoisolators provide electrical isolation of up to 500 V (channel–channel and channel–ground). A simplified block schematic for the PDISO-8 is shown in Figure 2.25.

The response time of each input may be individually selected using software control (the earlier ISA version of this card used dual-in-line switches to select the input filters). Input response time is typically 20 μs without the filter and 5 ms when the filter is switched in (note that filters are normally required with AC inputs in order to avoid the digital input pulsing on and off at twice the AC input frequency!).

The eight relay outputs each have contacts rated at 3 A at 120 V AC or 28 V DC (resistive loads). The maximum contact resistance is 100 mΩ and both SPDT (Channels 0 to 4) and SPST (Channels 5 to 7) contacts are available. Relay operating time is 20 ms (max.) and release time 10 ms (max.).

The PDISO-8 uses only the +5 V power rail from the PC and requires a typical supply current of 1 A (all relays energized). The I/O lines from the board connect via a standard 37-pin D-type male connector fitted to the rear metal bracket. The I/O connector pin assignment is shown in Figure 2.26.

The board address is selected by means of a dual-in-line switch. The PDISO-8 board occupies four consecutive addresses in the PC I/O address space of which only two addresses are actually used. The base address is selected by means of the dual-in-line switch and the two registers are located at (base address) and (base address +1). The I/O map for the board is as follows:

I/O address	Function	Mode
Base address	Relay outputs	Read/write
Base address +1	Isolated inputs	Read only

Each bit in the appropriate register corresponds to the equivalent I/O channel number. Bits are therefore allocated as follows:

	Data bit							
Address	D7	D6	D5	D4	D3	D2	D1	D0
Base	OP7	OP6	OP5	OP4	OP3	OP2	OP1	OP0
Base +1	IP7	IP6	IP5	IP4	IP3	IP2	IP1	IP0

As an example, assuming that the base address has been set to 300 hexadecimal, the relays can be operated by writing data to 0300H while the inputs can be sensed by reading data from 0301H. In the former case, a set bit (logic 1) will energize the relay connected to the channel in question while in the latter case, a set bit (logic 1) will indicate that an input has been asserted.

The state of the output register can be read by appropriate software in order to ascertain the current state of the relays. In some applications this can be useful since it avoids the need to preserve the state of the relay port within a variable. In order to operate a particular relay without disturbing any of the others, it is simply necessary to first read the data from (base address), bit-wise *or* the data with the bit to be set, and then write it back to (base address). Using the example addresses quoted earlier, the following single line of DOS BASIC will operate the relay connected to OP1 without altering the state of any of the other relays:

```
OUT &H300, INP(&H300) OR 2
```

Further information concerning programming this type of interface appears in Chapter 12.

Blue chip technology AIP-24

The Blue chip technology AIP-24 analogue input card provides 24 channels of single-ended or 12 channels of differential analogue input. The board is a 120 mm short format PC/AT compatible card and its simplified block schematic is shown in Figure 2.27.

The AIP-24 uses a 12-bit analogue-to-digital converter which provides a resolution of 0.025%. A sample and hold amplifier is used to capture fast moving analogue signals and freeze them in order to improve overall accuracy. The successive approximation ADC can operate in unipolar or bipolar modes

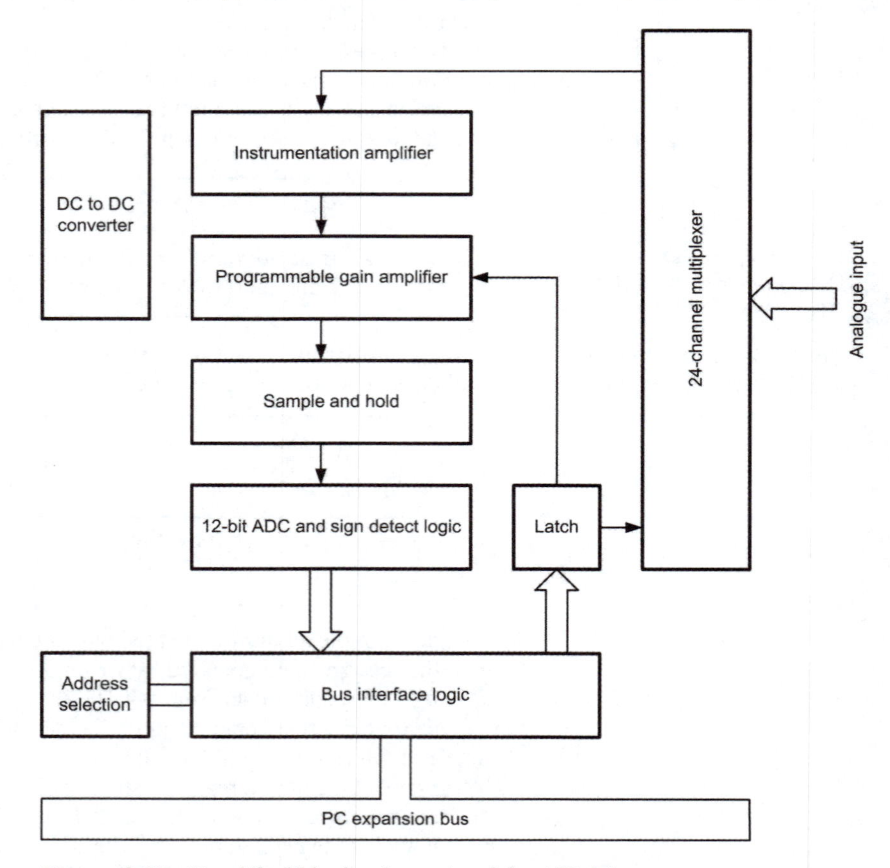

Figure 2.27 *Simplified block schematic of the AIP-24*

and handles signals in the range 0–±10 V. In order to cope with low amplitude input signals, an on-board programmable gain amplifier can be used to provide input gains of 1, 10, or 100. Conversion time is 25 μs but faster ADC chips may be fitted where conversion speed is critical.

Input connection is made via a 50-way IDC connector attached to the metal rear bracket. A ribbon cable or screw terminal may be fitted directly to the 50-way connector. An on-board DC–DC converter provides power for the analogue circuitry of the ADC.

Four addresses are used to set up and drive the card, and a set of links set the base address of the port within I/O address space. These provide gain selection (write), initiate conversion (write), and converted data (read). The base address of the card is selected by means of PCB links.

Programming the card is reasonably straightforward. The gain of the analogue input will normally be set by writing appropriate bytes during initialization. Thereafter, successive analogue-to-digital conversions are initiated by simply writing to the relevant port and then reading the value of the returned data.

The following BASIC program displays the inputs of the AIP-24 ports on the screen in decimal format. Note that the base address used for the program is 300H (768 decimal) which is the default factory setting:

```
REM Initialise
CLS: KEY OFF: LOCATE 1,1
p = 0
REM Main loop to print data
begin:
FOR y = 1 TO 3
  FOR x = 1 TO 80 STEP 10
    LOCATE y, x
    GOSUB getdata
    p = p + 1
    IF p > 23 THEN p = 0
    PRINT n; " ";
  NEXT
NEXT
GOTO begin
REM Get data from ports
getdata:
OUT &H300,p
OUT &H301,0
a = INP(&H302)
b = INP(&H303)
c = b AND &HF
n = (256 * c) + a
RETURN
```

Further examples of programming an analogue-to-digital converter appears in Chapter 12.

Measurement Computing Corporation Dual-422

The Measurement Computing Corporation Dual-422 is a two-channel RS-422 interface card (Figure 2.28). The half-size ISA/EISA card is compatible with

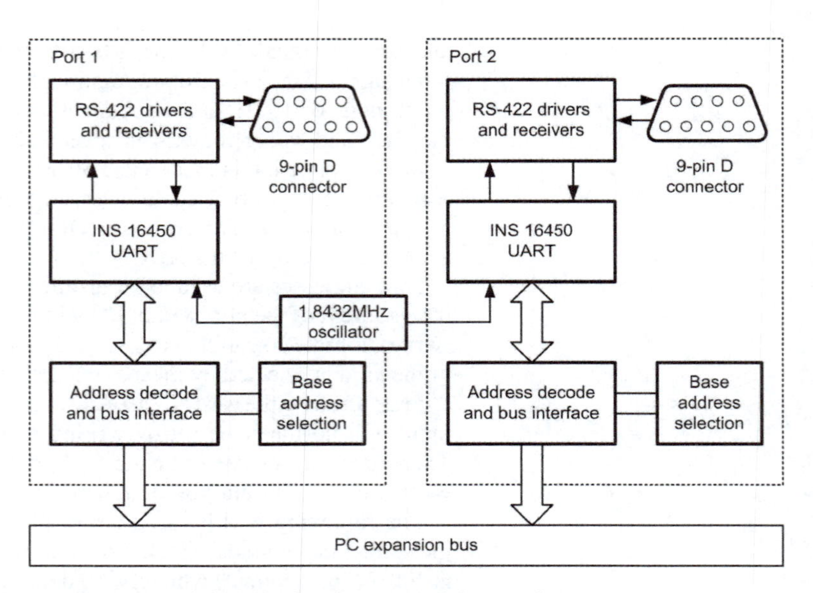

Figure 2.28 *Simplified block schematic of the Dual-422*

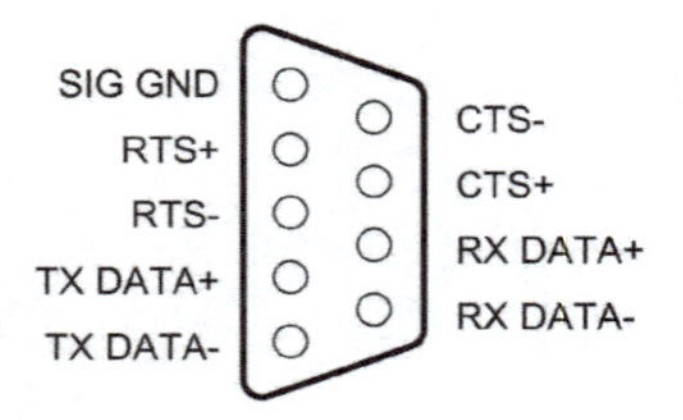

Figure 2.29 *I/O connector pin assignment for the Dual-422*

the PC/AT bus and permits serial communications at speeds up to 57.6 kilobaud at distances of up to 1.2 km (the 9.6 kilobaud limitation imposed by most PC communications routines can be overridden in most cases).

The simplified block schematic of the Dual-422 is shown in Figure 2.29. Both ports operate independently and each has its own case address and interrupt selection controls. A VLSI Universal Asynchronous Receiver/Transmitter (UART) device is used to form the basis of each channel and this device is augmented by external line drivers and receivers.

The UART employed is the National Semiconductor INS 16450 (an improved device which is compatible with the original 6250 device employed in the legacy PC). The INS 16450 is fully programmable and offers a choice of serial data word length (5, 6, 7, or 8 data bits) with selectable even, odd, or no parity checking. Baud rates are also selectable in the range 120 baud to 57.6 kilobaud.

Base address selection (for each port) is obtained via a dual-in-line switch (see Figure 2.30). Links are used to select the desired interrupt level (either channel can be configured as MS-DOS serial port COM1: or COM2: or any other interrupt level may be selected) whilst a further link is provided in order to enable or disable CTS/RTS data transfer control.

Programming the Dual-422 interface is extremely straightforward. Assuming that the ports have been configured as COM1: and COM2: (and that no other communication device has been configured to the same interrupt level), the following BASIC code transmits a test string (T$) output via COM1: for input via COM2: to the received string, R$:

```
REM Test string
T$="The quick brown fox jumps over the lazy dog"
REM Open the serial ports using 4800 baud
REM COM1 will be associated with channel 1
```

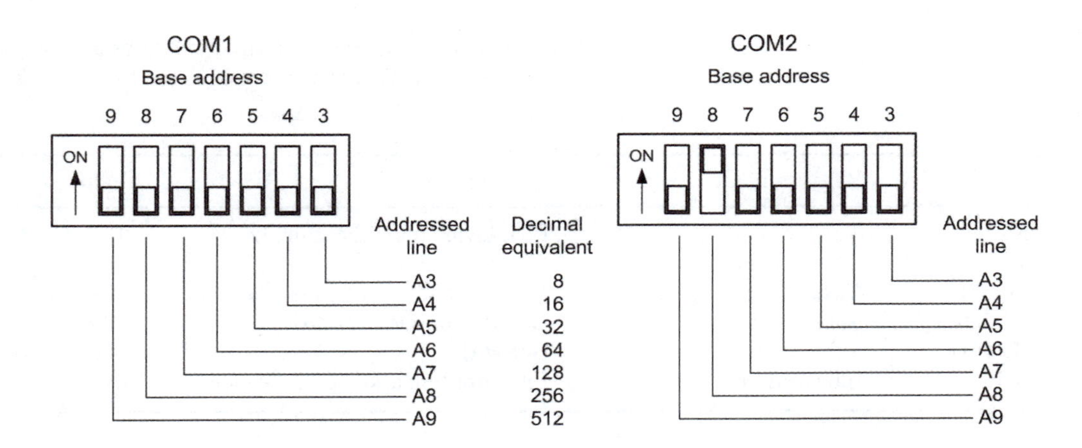

Note that a switch in the ON position corresponds to a zero and in the OFF position corresponds to the binary weight of the corresponding address bit (512, 256, 128, 64, 32, 14, 8). The base I/O address is the sum of all of the OFF switch address bits.

Figure 2.30 *Base address selection for the Dual-422*

```
REM COM2 will be associated with channel 2
OPEN "COM1:4800" AS #1
OPEN "COM2:4800" AS #2
REM Enable COM1: RS-485 driver
OUT &H3FF, 2
REM Enable COM2: RS-485 receiver
OUT &H2FF,1
REM Transmit data via COM1:
PRINT #1, T$
REM Receive data via COM2:
INPUT #2, R$
REM Close communication channels
CLOSE
```

Arcom APCI-ADADIO multifunction I/O card

The Arcom APCI-ADADIO is a 32-bit PCI card which provides eight differential (APCI-ADADIOCD) or 16 single-ended (APCI-ADADIOCS) multiplexed analogue inputs, two analogue outputs, 16 digital I/O lines, and three counter timer channels. All I/O signals are routed to a 50-way D-type connector which conforms to Arcom's standard signal conditioning system (SCS).

The range of features available from a single PCI card makes the APCI-ADADIO an excellent and highly versatile choice for use in modern control and data acquisition applications. The card is Plug and Play compatible and it uses a single chip PCI bus slave controller which is designed and manufactured by PLX Technology.

During power-up system initialization the PCI BIOS will detect the card and assign a unique I/O address and interrupt line. This ensures that there are no resource conflicts on the PCI bus. Multiple cards are supported by this method without the need for address decoding links.

The PLX Technology bus slave controller contains a standard configuration space header (type 00H). This header contains the following data:

Offset	Register name	Description	Value
00-01H	Vendor identification	PCI device manufacturer ID	10B5H (PLX Technology)
02-03H	Device identification	PCI device ID	9050H
18-1BH	Base address register	I/O base address of card	0000xxxx
2C-2DH	Subsystem Vendor ID	Board manufacturer ID	12ABH (Arcom)
2E-2FH	Subsystem ID	Board ID	0605H (APCI-ADADIO)
3CH	Interrupt line	Interrupt line assigned to device	0x

The above registers are accessed using PCI BIOS functions.

The APCI-ADADIO uses an indexed addressing scheme to access the on-board devices and special function registers. The addressing scheme is described in the following table:

I/O address	Function	Direction
Base	Index register	Write
Base+1	Control/Status	Read/Write
Base+2	ADC/DAC LSB data	Read/Write
Base+3	ADC/DAC MSB data	Read/Write

The APCI-ADADIO contains a single 12-bit successive approximation analogue-to-digital converter. The input to this device is connected to an 8-way multiplexer (APCI-ADADIOCD) or 16-way multiplexer (APCI-ADADIOCS). Prior to an analogue-to-digital conversion the appropriate channel can be selected by writing to the multiplexer channel select register. The ADC may be triggered by three different sources which are selected by links. These sources can be:

1 Software trigger, initiated by an I/O write sequence.
2 Hardware trigger from an external TTL input (approximately 1–2 μs low pulse).
3 Periodic timer programmed from the on-board counter/timer Channel 0.

The following sequence can be used to perform an analogue-to-digital conversion when using the software trigger mode:

1 Write 01H to the Base address
2 Write the appropriate multiplexer channel value to Base+1
3 Wait for approximately 50 μs for the input to settle
4 Write 00H to the Base address

Photo 2.16 *Arcom's APCI-ADADIO multifunction I/O card (photo courtesy of Arcom)*

5. Write any value to Base+1 in order to initiate conversion
6. Wait for approximately 20 μs for the conversion to complete
7. Read Base+1 and check that bit 0 is at logic 0 (i.e. conversion completed)
8. Read Base+2 ADC data low nibble (bits 0 to 3)
9. Read Base+3 ADC data high byte (bits 4 to 11).

The APCI-ADADIO contains two 12-bit digital-to-analogue converters. On-board links can be used to select between three possible output voltage ranges, ±5 V, 0–5 V, and 0–10 V. the DAC values are updated by writing to the data register at Base+2 (low nibble bits 0 to 3) and Base+3 (high byte bits 4 to 11). Prior to this the DAC channel must be selected by writing a value of 02H to the index register for DAC A and 03H for DAC B.

The APC-ADADIO provides 16 digital I/O lines grouped in four nibbles. Each nibble has a power-up/reset state link and can be programmed as either input or output via the digital I/O configuration register. Access to individual I/O lines are made possible via index registers 0AH and 0BH. Reading these provide the status of all I/O lines regardless of whether they are configured as input or output. With some careful programming it is also possible to use these lines in bi-directional mode. Note that, if a nibble is to be used as an input the corresponding reset state link must be set to the high position otherwise the lines will be driven low as outputs which may cause damage.

The APCI-ADADIO uses an 8254 compatible counter/timer. This provides three individual 16-bit counter/timers. Channel 0 can be used to trigger an analogue-to-digital conversion (as mentioned earlier) whilst Channel 1 may be used to generate an interrupt request sequence.

Photo 2.17 *Measurement Computing Corporation's PMD-1208LS Personal Measurement Device*

The PMD-1208LS USB device

The PMD-1208LS is an example of a modern and highly versatile USB device suitable for a wide variety of data acquisition and control applications. Whilst it was designed for slower USB 1.1 ports, the device is compatible with USB 2.0 ports and is supported under Microsoft Windows 98SE/ME/2000 and XP.

The PMD-1208LS features eight analogue inputs, two 10-bit analogue outputs, 16 digital I/O connections and one 32-bit event counter. The device is powered by the +5 V USB supply and does not require any external power source. The PMD-1208LS's analogue inputs are software configurable for either eight 11-bit single-ended inputs, or four 12-bit differential inputs. An on-board industry standard 82C55 programmable peripheral interface (see page 29) provides the 16 discrete digital I/O lines. Each digital channel can be configured for either input or output.

The block schematic diagram of the PMD-1208LS is shown in Figure 2.31. All I/O connections are made to the screw terminals located along each side of the device.

The PMD-1208LS is supplied with configuration software as well as Universal and OEM Software Libraries. Using these libraries it is a relatively simple matter to program applications using 32-bit Windows development software such as Microsoft Visual C++ and Microsoft Visual Basic. As an example, the following Visual basic code is all that is required to produce a simple digital frequency meter (see Figure 2.32).

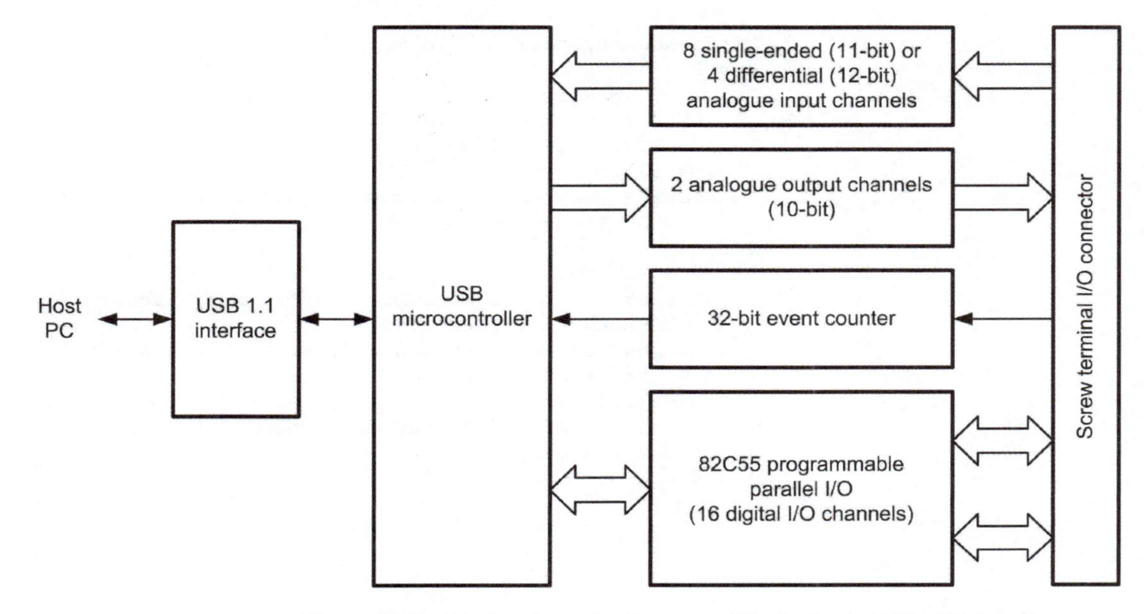

Figure 2.31 *Block schematic diagram of the PMD-1208LS USB device*

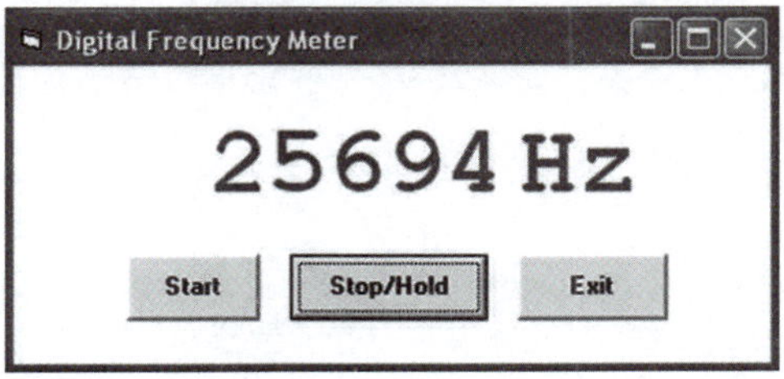

Digital Frequency Meter

25694 Hz

Start Stop/Hold Exit

Figure 2.32 *Output produced by the USB digital frequency meter*

```
'=============================================================
' File:                          pmdfreqy
' Library Calls:                 cbCLoad32%()
'                                cbCIn32%()
'                                cbErrHandling%()
' Purpose:                       Simple digital frequency meter
' Interface:                     PMD-1208LS USB HID
'=============================================================
Const BoardNum = 1               ' Board number
Const CounterNum% = 1            ' number of counter used
Const RegName% = LOADREG1        ' register name of counter 1
Private Sub cmdExit_Click()
   End
End Sub
Private Sub cmdStart_Click()
tmrReadCount.Enabled = True
End Sub
Private Sub cmdStopHold_Click()
```

```
tmrReadCount.Enabled = False
End Sub
Private Sub Form_Load()
   ULStat% = cbErrHandling(PRINTALL, DONTSTOP)
   If ULStat% <> 0 Then Stop
   LoadValue% = 0
   ULStat% = cbCLoad(BoardNum, RegName%, LoadValue%)
   If ULStat% <> 0 Then Stop
End Sub
Private Sub tmrReadCount_Timer()
   ULStat% = cbCIn32(BoardNum, CounterNum%, CBCount&)
   If ULStat% <> 0 Then Stop
   lblShowCountRead.Caption = Format$(CBCount&, "0")
   ' Reset count to zero
   LoadValue% = 0
   ULStat% = cbCLoad(BoardNum, RegName%, LoadValue%)
   If ULStat% <> 0 Then Stop
End Sub
```

3 Using the command line interface

This chapter outlines the facilities provided by the DOS operating system and the command line interpreter (CLI) in particular. Emphasis has been placed on those features which are of particular relevance to the engineer and software developer as well as those who may be unfamiliar with what lies below the Windows interface. If you are planning to develop applications that will run on minimal systems (without the overhead imposed and restriction imposed by Windows) or if you are developing *console applications* to run inside the Windows environment time spent in getting to know the CLI (including its peculiarities and foibles) can be instrumental in avoiding a variety of pitfalls.

The need for an operating system

Anyone who has made passing use of a microcomputer system will be aware of at least some of the facilities offered by its operating system. Such an awareness is developed by means of the interface between the operating system and the user; the system generates prompts and messages, and the user makes an appropriate response.

Within the familiar 'drag and drop' and 'point and click' interface provided by Windows where there is no need to use a command language. This, of course, is as it should be. As far as most end-users of computer systems are concerned, the operating system provides an environment from which it is possible to launch and run applications, and to carry out elementary maintenance of disk files. In such cases, the operating system is perhaps better described as a *microcomputer resource manager*. As such, the operating system provides an essential bridge between the user's application programs and the system hardware.

In order to provide a standardized environment (which will cater for a variety of different hardware configurations) and ensure a high degree of software portability, part of the operating system is hardware independent (DOS) whilst the hardware dependent (BIOS) provides the individual low-level routines required by the machine in question. Figure 3.1 illustrates this important point.

In the context of developing software for control and instrumentation applications, the software engineer needs to have a much deeper understanding of the role of the operating system as a means of accessing, configuring, and optimizing system resources. In addition, the software developer will need a variety of tools and utility programs (including items such as editors, assemblers, linkers, and debuggers). These *development tools* work together with the operating system to provide an environment which facilitates effective software development.

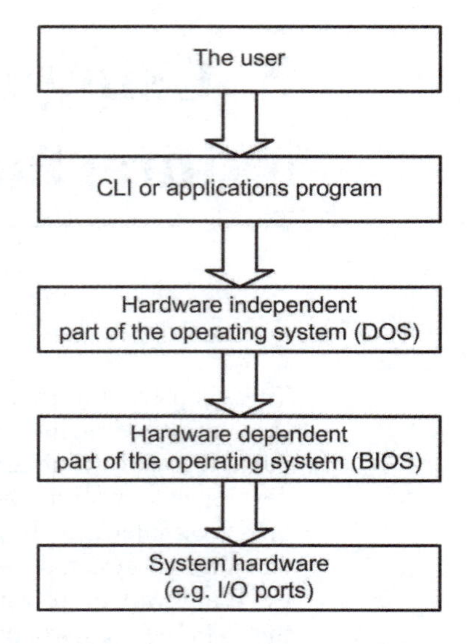

Figure 3.1 *Relationship between the different layers of the operating system (DOS and BIOS)*

Origins of DOS

Most microcomputer operating systems can trace their origins to the original control program for microprocessors (CP/M) developed by Garv Kildall as a software development environment for the Intel 8080 microprocessor. In its original form CP/M was supplied on an 8" IBM format floppy disk. CP/M was subsequently extensively developed and marketed by Digital Research in a variety of forms including those for use with Intel and Motorola microprocessor families.

In the last 1970s, CP/M rapidly became the *de facto* operating system for most 8080, 8085, and Z80-based 8-bit microcomputer systems. However, with the advent of 16-bit machines and the appearance of the IBM PC in 1981, a new and more powerful operating system was required. Digital Research produced an 8086-based version of CP/M (known as CP/M86); however, Microsoft produced a rival product (PC-DOS) which was adopted by IBM for use with the PC. Microsoft quickly also developed an operating system (MS-DOS) for use with PC-compatible machines. This operating system rapidly became the world's most popular microcomputer operating system. Windows was later added to the MS-DOS system in order to provide a more user-friendly graphical user interface (GUI).

Note that if the system is not configured to boot into Windows on start-up the user will normally be presented with the command line interface. This text-based interface permits the entry of DOS commands and the execution of programs by simply typing the executable filename at the DOS prompt. A comparison of the way in which similar information is presented DOS and Windows can be made from Figures 3.2 and 3.3.

The MS-DOS operating system can be configured for operation with a wide variety of peripheral devices including various types of monitor, serial and parallel printers, and modems. Each individual hardware configuration requires its own particular I/O provision and this is achieved by means of a piece of

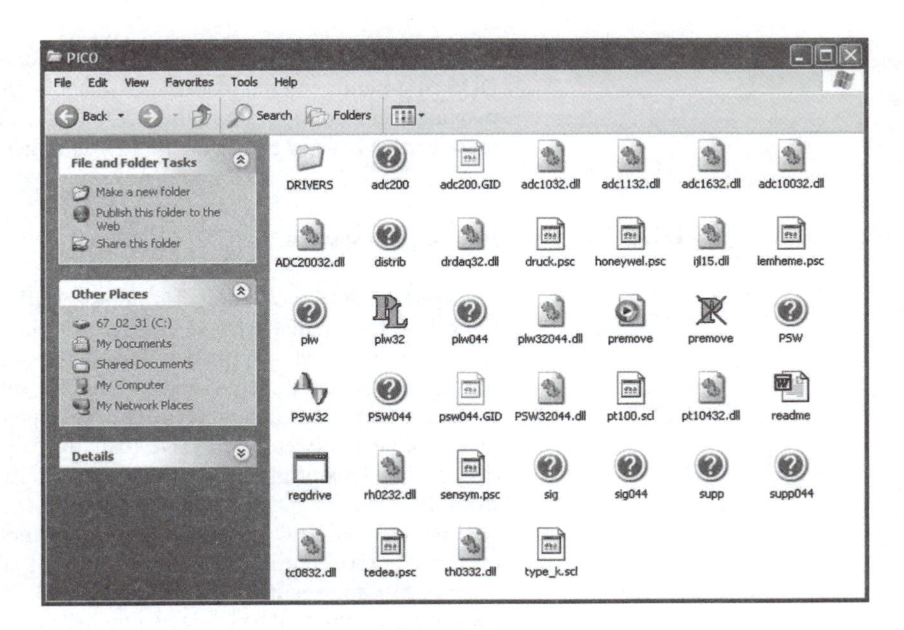

Figure 3.2 *The familiar Windows interface used to provide a graphical display of the files produced in the PICO directory of the C: drive*

```
C:\WINDOWS\System32\command.com
 Volume in drive C is 67_02_31
 Volume Serial Number is D402-D652

 Directory of C:\PICO

21/01/2004  23:42    <DIR>          .
21/01/2004  23:42    <DIR>          ..
31/07/2002  09:15            76,288 adc10032.dll
31/07/2002  08:50            69,120 adc1032.dll
31/07/2002  08:17            71,168 adc1132.dll
31/07/2002  15:21            58,880 adc1632.dll
31/05/2002  13:37           105,668 adc200.hlp
31/07/2002  09:59           160,256 ADC20032.dll
09/08/2002  15:47            53,617 distrib.hlp
31/07/2002  10:26            82,432 drdaq32.dll
21/01/2004  22:51    <DIR>          DRIVERS
13/11/2000  14:32             1,459 druck.psc
24/03/1999  09:43             1,335 honeywel.psc
30/05/2001  01:00           352,256 ijl15.dll
10/03/2000  17:05             2,151 lemheme.psc
15/07/2002  10:40           258,699 plw.hlp
15/07/2002  10:40           258,699 plw044.hlp
31/07/2002  13:41           906,240 plw32.exe
31/07/2002  13:41           114,688 plw32044.dll
21/01/2004  23:46            10,830 premove.dat
08/08/2001  15:25            40,960 premove.exe
15/07/2002  11:29           262,695 PSW.hlp
15/07/2002  11:29           262,695 PSW044.hlp
31/07/2002  13:01         1,264,640 PSW32.exe
Press any key to continue . . .
```

Figure 3.3 *The contents of the same directory as shown in Figure 3.2 but displayed using the DOS CLI*

software known as a *device driver*. A number of device drivers (e.g. those which deal with the standard serial and parallel ports) are resident within the BIOS ROM. Others which may be required must be loaded into RAM during system initialization.

DOS provides you with various methods for configuring a system: individual DOS commands entered from the CLI, batch files, hardware device drivers, and two important configuration files, CONFIG.SYS and AUTOEXEC.BAT. All of these can be instrumental in helping you to get the best out of a system.

DOS basics

Booting the system

The system (DOS or Windows) is automatically loaded from the hard disk (drive C:) or the floppy disk placed in drive A: whenever the system is *booted* (i.e. whenever the power is applied and the processor executes the BIOS code stored in the ROM). After successful loading, the title and version of the operating system is displayed on the screen. The message is then followed by a prompt that gives the currently selected drive (usually `C:\>` in a system fitted with a hard disk drive). This prompt shows that the system is ready to receive a command from the user.

If an AUTOEXEC batch file is present, the commands that it contains are executed before control is passed to the user. Furthermore, if such a file contains the name of an executable program (i.e. a file with a COM or EXE extension), then this program will be loaded from disk and executed. The program may take one of several forms including a program that simply performs its function and is then cleared from memory, a *terminate and stay-resident* (TSR) program, or a fully-blown control or data acquisition application.

It is important to remember that the currently selected drive remains the default drive unless explicitly changed by the user. As an example, consider a system that is booted with a system disk (floppy) placed in drive A:. The default drive will then be A: (unless an AUTOEXEC file is present that contains commands to change the current drive). The system prompt will indicate that A is the current drive. Thereafter, it is implicit that all commands which do not specify a drive refer, by default, to that drive. The `SET PATH` command (see page 118) can, however, be used to specify a directory path which will be searched if a command or filename does not appear in the current directory.

To return to the root directory from within any level of sub-directory you need only type `CD\` (followed, of course, by the `<ENTER>` key). To return to just one level of sub-directory towards the root you can simply type `CD..`. To help you navigate the system use the `PROMPT $p$g` command (see page 118).

I/O channels

In order to simplify the way in which DOS handles input and output, the system recognizes the names of its various I/O devices (see Table 3.1). This may, at first, appear to be unnecessarily cumbersome but it is instrumental in allowing DOS to redirect data. This feature can be extremely useful when, for example, output normally destined for the printer is to be redirected to an auxiliary serial port.

The `COPY` command (see page 115) can be used to transfer data from one device to another. As an example, the command `COPY CON: PRN:` copies data from the keyboard (console input device) to the printer, `COPY CON: COM1` copies data from the keyboard to the serial port. In either case, the end-of-file character, `<CTRL-Z>` or `<F6>`, must be entered to terminate input.

Table 3.1 *DOS I/O channels*

Channel	Meaning	Function	Notes
COM1:, COM2:, COM3:, and COM4:	Communications	Serial I/O	Via RS-232 ports
CON:	Console	Keyboard (input) and screen (output)	This channel combines the functions associated with the keyboard and the display (i.e. a 'terminal')
LPT1:, LPT2:, and LPT3:	Line printer	Parallel printer	This interface (output) conforms to the Centronics standard
PRN:	Printer	Serial or parallel printer (output)	
NUL:	Null device	Simulated I/O	Provides a means of simulating a physical I/O channel without data transfer taking place

DOS commands

DOS responds to command lines typed at the console and terminated with a `<RETURN>` or `<ENTER>` keystroke. A command line is thus composed of a command keyword, an optional command tail, and `<ENTER>`. The command keyword identifies the command (or program) to be executed. The command tail can contain extra information relevant to the command, such as a filename or other parameters. Each command line must be terminated using `<ENTER>` or `<ENTER>` (not shown in the examples which follow).

As an example, the following command can be used to display a directory of all BASIC source code (i.e. those with a BAS extension) within a directory named TEST in drive C:, indicating the size of each:

```
DIR C: \TEST\*.BAS
```

Note that, in this example and the examples that follow, we have omitted the prompt generated by the system (indicating the current drive).

It should be noted that the command line can be entered in any combination of upper- or lower-case characters. DOS converts all letters in the command line to upper-case before interpreting them. Furthermore, whilst a command line generally immediately follows the system prompt, DOS permits spaces between the prompt (e.g. `C:\>`) and the command word.

As characters are typed at the keyboard, the cursor moves to the right in order to indicate the position of the next character to be typed. Depending upon the keyboard used, a `<BACKSPACE>`, or `<DELETE>` key, can be used to delete the last entered character and move the cursor backwards one character position. Alternatively, a combination of the `CONTROL` and `H` keys (i.e. `<CTRL-H>`) may be used instead. Various other control characters are significant in DOS and these are shown in Table 3.2.

Table 3.2 *DOS control characters*

Control character	Hex. code	Function
<CTRL-C>	03	Terminates the current program (if possible) and returns control to the user.
<CTRL-G>	07	Sounds the audible warning device (bell) but can only be used as part of a program of batch file.
<CTRL-H>	08	Moves the cursor back by one space (i.e. the same as the <BACKSPACE> key) and deletes the character present at that position.
<CTRL-I>	09	Tabs the cursor right by a fixed number of columns (usually eight). Performs the same function as the <TAB> key.
<CTRL-J>	10	Issues a line feed and carriage return, effectively moving the cursor to the start of the next line.
<CTRL-L>	12	Issues a form feed instruction to the printer.
<CTRL-M>	13	Produces a carriage return (i.e. has the same effect as <RETURN>).
<CTRL-P>	16	Toggles screen output to the printer (i.e. after the first <CTRL-P> is issued, all screen output will be simultaneously echoed to the printer. A subsequent <CTRL-P> will disable the simultaneous printing of the screen output). Note that <CTRL-PRT.SC.> has the same effect as <CTRL-P>.
<CTRL-S>	19	Pauses screen output during execution of the TYPE command (<CTRL-NUM.LOCK> has the same effect).
<CTRL-Z>	26	Indicates the end of a file (can also be entered using <F6>).

Finally, the combination of three keys, <CTRL-ALT-DEL>, can be used to perform a 'warm' system reset. This particular combination should only be used in the last resort as it will clear system memory. Any unsaved data present in RAM will then be lost forever!

If it is necessary to repeat or edit the previous command, the <F1> (or right-arrow) key may be used to reproduce the command line, character by character, on the screen. The left-arrow key permits backwards movement through the command line for editing purposes. The <F3> key simply repeats the last command in its entirety.

File specifications

Many of the DOS commands make explicit reference to files. A file is simply a collection of related information stored on a disk. Program files comprise a series of instructions to be executed by the processor whereas data files simply contain a collection of records. A complete file specification has four distinct parts: a drive and directory specifier (known as a *pathname*), a *filename*, and a *filetype*.

The drive specifier is a single letter followed by a colon (e.g. C:). This is then followed by the directory and sub-directory names (if applicable) and the filename and filetype.

The filename comprises 1 to 8 characters whilst the filetype takes the form of a 1 to 3 character extension separated from the filename by means of a full-stop ('.'). A complete file specification (or *filespec*) thus takes the form:

```
[pathname]:[filename].[filetype]
```

As an example, the following file specification refers to a file named MOUSE and having a COM filetype found in the root directory of the disk in drive A:

```
A\:MOUSE.COM
```

DOS allows files to be grouped together within directories and sub-directories. Directory and sub-directory names are separated by means of the backslash (\) character. Directories and sub-directories are organized in an hierarchical (tree) structure and thus complete file specifications must include directory information.

The *root* or base directory (i.e. that which exists at the lowest level in the hierarchical structure) is accessed by default when we simply specify a drive name without further reference to a directory. Thus:

```
C:\MOUSE.COM
```

refers to a file in the root directory whilst:

```
C:\DOS\MOUSE.COM
```

refers to a identically named file resident in a sub-directory called 'DOS'.

Sub-directories can be extended to any practicable level. As an example:

```
C:\DOS\UTILS\MOUSE\MOUSE.COM
```

refers to a file named MOUSE.COM present in the MOUSE sub-directory which itself is contained within the UTILS sub-directory found within a directory named DOS.

When it is necessary to make explicit reference to the root directory, we can simply use a single backslash character as follows:

```
C:\
```

File extensions

The filetype extension provides a convenient mechanism for distinguishing different types of file and DOS provides various methods for manipulating groups of files having the same filetype extension. We could, for example, delete all of the back-up (BAK) present in the root directory of the hard disk (drive C:) using a single command of the form:

```
ERA C:\*.BAK
```

Alternatively, we could copy all of the executable (EXE) files from the root directory of the disk in drive A: to the root directory on drive C: using the command:

```
COPY A:\*.EXE C:\
```

Commonly used filetype extensions are shown in Appendix H on page 470.

Wildcard characters

DOS allows the user to employ wildcard characters when specifying files. The characters, '*' and '?', can be used to replace complete fields and individual characters, respectively, within a file specification. DOS will search then carry out the required operation on all files for which a match is obtained.

The following examples illustrate the use of wildcard characters:

```
A:\*.COM
```

refers to all files having a COM extension present in the root directory of drive A:.

```
C:\TOOLS\*.*
```

refers to all files (regardless of name or extension) present in the directory named TOOLS on drive C:.

```
B:\TURBO\PROG?.C
```

refers to all files having a C extension present in the TURBO directory on the disk in drive B which have PROG as their first three letters and any alphanumeric character in the fourth character place. A match will occur for each of the following files:

```
PROG1.C  PROG2.C PROG3.C PROGA.C PROGB.C, etc.
```

Internal and external commands

It is worth making a distinction between DOS commands which form part the resident portion of the operating system (internal commands) and those which involve other utility programs (external commands). *Intrinsic commands* are executed immediately whereas *extrinsic commands* require the loading of transient utility programs from disk, and hence there is a short delay before the command is acted upon.

In the case of external commands, DOS checks only the command keyword. Any parameters which follow are passed to the utility program without checking.

At this point we should perhaps mention that DOS only recognizes command keywords which are correctly spelled! Even an obvious typing error will result in the non-acceptance of the command and the system will respond with an appropriate error message.

As an example, suppose you attempt to format a disk but type `FORMATT` instead of `FORMAT`. Your system will respond with this message:

```
Bad command or file name
```

indicating that the command is unknown and that no file of that name (with a COM, BAT, or EXE extension) is present in the current directory.

To get online help from within DOS you can simple type the command name followed by `/?`. Hence `DIR /?` will bring you help before using the directory command. With later versions of DOS you can also type `HELP` followed by the command name (e.g. `HELP DIR`).

Internal DOS commands

We shall now briefly examine the function of each of the most commonly used internal DOS commands. Examples have been included wherever they can help to clarify the action of a particular command. The examples relate to the most common versions of DOS.

Command	Function

BREAK	The BREAK command disables the means by which it is possible to abort a running program. This facility is provided by means of the <CTRL-C> or <CTRL-BREAK> key combinations, and it normally only occurs when output is being directed to the screen or the printer. BREAK accepts two parameters: ON and OFF.
	Examples:
	`BREAK ON`
	enables full <CTRL-C> or <CTRL-BREAK> key checking (it is important to note that this will normally produce a dramatic reduction in the speed of execution of a program).
	`BREAK OFF`
	restores normal <CTRL-C> or <CTRL-BREAK> operation (i.e. the default condition).
	Note that BREAK ON will often result in a significant reduction in the speed of execution of a program. You should only use this command when strictly necessary!

CD	See CHDIR.

CHDIR	The CHDIR command allows users to display or change the current directory. CHDIR may be appreviated to CD.
	Examples:
	`CHDIR A:`
	displays the current directory path for the disk in drive A:.
	`CHDIR C:\APPS`
	changes the directory path to APPS on drive C:.
	`CD D:\DEV\PROCESS`
	changes the directory path to the sub-directory PROCESS within the directory named DEV on drive D:.
	`CD\`
	changes the directory path to the root directory of the *current* drive.
	`CD..`
	changes the directory path one level back towards the root directory of the *current* drive.

CLS	CLS clears the screen and restores the cursor position to the top left-hand corner of the screen.
COPY	The COPY command can be used to transfer a file from one disk to another using the same or a different filename. The COPY command is effective when the user has only a single drive. The COPY command must be followed by one or two file specifications. When only a single file specification is given, the command makes a single-drive copy of a file. The copied file

(*continued*)

Command	Function
	takes the same filename as the original and the user is prompted to insert the source and destination disks at the appropriate point. Where both source and destination file specifications are included, the file is copied to the specified drive and the copy takes the specified name. Where only a destination drive is specified (i.e. the destination filename is omitted) the `COPY` command copies the file to the specified drive without altering the filename. `COPY` may be used with the `*` and `?` wildcard characters in order to copy all files for which a match is found (see page 113). *Examples:* `COPY A\:ED.COM B:` copies the file `ED.COM` present in the root directory of the disk in drive A: to the disk present in drive B:. The copy will be given the name `ED.COM`. On a single-drive system the only available floppy drive can be used as both the source and destination when the `COPY` command is used. The single physical drive will operate as both drive A: and drive B:, and you will be prompted to insert the source and destination disks when required. `COPY` is unable to make copies files located within sub-directories. If you need this facility use `XCOPY` with the `/s` switch (see page 127).

`DATE`	The `DATE` command allows the date to be set or displayed.
	Examples:
	`DATE`
	displays the date on the screen and also prompts the user to make any desired changes. The user may press `<RETURN>` to leave the settings unchanged.
	`DATE 12-08-99`
	sets the date to 27th August 1999.

`DEL`	See `ERASE`.

`DIR`	The `DIR` command displays the names of files present within a directory. Variations of the command allow the user to specify the drive to be searched and the types of files to be displayed. Further options govern the format of the directory display.
	Examples:
	`DIR`
	displays all files in the current default directory.
	`A:\DIR`
	changes the default drive to A: (root directory) and then displays the contents of the root directory of the disk in drive A:.

(*continued*)

Command	Function

	`DIR *.BAS`
	displays all files with a `BAS` extension present in the current default directory drive.
	`DIR C:\DEV.*`
	displays all files named `DEV` (regardless of their type or extension) present in the root directory of drive C: (the hard disk).
	`DIR C:\MC\*.BIN`
	displays all files having a `BIN` extension present in the sub-directory named `MC` on drive C: (the hard disk).
	`DIR/W`
	displays a directory listing in 'wide' format (excluding size and creation date/time information) of the current default directory.
	To prevent directory listings scrolling off the screen use `DIR /P` or `DIR | MORE`. These commands will pause the listing at the end of each screen and wait for you to press a key before continuing.
	Later versions of DOS include many options for use with the `DIR` command including sorting the directory listing and displaying hidden system files.

`ERASE`	The `ERASE` command is used to erase a filename from the directory and release the storage space occupied by a file. The `ERASE` command is identical to the `DEL` command and the two may be used interchangeably. `ERASE` may be used with the `*` and `?` wildcard characters in order to erase all files for which a match occurs.
	Examples:
	`ERASE PROG1.ASM`
	erases the file named `PROG1.ASM` from the disk placed in the current (default) directory.
	`ERASE B:\TEMP.DAT`
	erases the file named `TEMP.DAT` from the root directory of the disk in drive B:.
	`ERASE C:\*.COM`
	erases all files having a `COM` extension present in the root directory of the hard disk (drive C:).
	`ERASE A:\PROG1.*`
	erases all files named `PROG1` (regardless of their type extension) present in the root directory of the disk currently in drive A:.

`MD`	See `MKDIR`.
`MKDIR`	The `MKDIR` command is used to make a new directory or sub-directory. The command may be abbreviated to `MD`.

(*continued*)

Command	Function

	Examples:
	`MKDIR APPS`
	creates a sub-directory named `APPS` within the *current* directory (note that the `CHDIR` command is often used after `MKDIR` – having created a new directory you will probably want to move to make it the current directory before doing something with it!).
	`MD C:\DOS\BACKUP`
	creates a sub-directory named `BACKUP` within the DOS directory of drive C:.

`PATH`	The `PATH` command may be used to display the current directory path. Alternatively, a new directory path may be established using the `SET PATH` command.
	Examples:
	`PATH`
	displays the current directory path (a typical response would be `PATH=C:\WINDOWS`).
	`SET PATH=C:\DOS`
	makes the directory path C:\DOS.

`PROMPT`	The `PROMPT` command allows the user to change the system prompt. The `PROMPT` command is followed by a text string which replaces the system prompt. Special characters may be inserted within the string, as follows:
	`$d` current date
	`$e` escape character
	`$g` >
	`$h` backspace and erase
	`$l` <
	`$n` current drive
	`$p` current directory path
	`$q` =
	`$t` current time
	`$v` DOS version number
	`$$` $
	`$` newline
	Examples:
	`PROMPT $t$g`
	changes the prompt to the current time followed by a >.
	`PROMPT Howard Associates PLC $?`
	changes the prompt to Howard Associates PLC followed by a carriage return and newline on which a ? is displayed.
	`PROMPT`
	restores the default system prompt (e.g. `C:\>`).

(continued)

Command	Function
	The most usual version of the PROMPT command is PROMPT pg which displays the current directory/sub-directory and helps to avoid confusion when navigating within DOS directories.
RD	See RMDIR.

RENAME	The RENAME command allows the user to rename a disk file. RENAME may be used with the * and ? wildcard characters in order to rename all files for which a match occurs. RENAME may be abbreviated to REN.
	Examples:
	`RENAME PROG2.ASM PROG1.ASM`
	renames PROG1.ASM to PROG2.ASM on the disk placed in the current (default) directory.
	`REN A:\HELP.DOC HELP.TXT`
	renames the file HELP.DOC to HELP.TXT in the root directory of the disk in drive A:.
	`REN B:\CONTROL.* PROG1.*`
	renames all files with name PROG1 (regardless of type extension) to CONTROL (with identical extensions) found in the root directory of the disk in drive B:.

RMDIR	The RMDIR command is used to remove a directory. RMDIR may be abbreviated to RD. The command cannot be used to remove the current directory and any directory to be removed must be empty and must not contain further sub-directories.
	Examples:
	`RMDIR ASSEM`
	removes the directory ASSEM from the current directory (note that DOS will warn you if the named directory is *not* empty!)
	`RD C:\DOS\BACKUP`
	removes the directory ASSEM from the current directory (once again, DOS will warn you if the named directory is *not* empty!)

SET	The SET command is use to set the environment variables (see PATH).

TIME	The TIME command allows the time to be set or displayed.
	Examples:
	`TIME`
	displays the time on the screen and also prompts the user to make any desired changes. The user may press <RETURN> to leave the settings unchanged.
	`TIME 14:30`
	sets the time to 2.30 p.m.

(*continued*)

Command	Function
TYPE	This useful command allows you to display the contents of an ASCII (text) file on the console screen. The TYPE command can be used with options which enable or disable paged mode displays. The <PAUSE> key or <CTRL-S> combination may be used to halt the display. You can press any key or use the <CTRL-Q> combination respectively to restart. <CTRL-C> may be used to abort the execution of the TYPE command and exit to the system. *Examples:* TYPE C\:AUTOEXEC.BAT will display the contents of the AUTOEXEC.BAT file stored in the root directory of drive C:. The file will be sent to the screen. TYPE B\:PROG1.ASM will display the contents of a file called PROG1.ASM stored in the root directory of the disk in drive B. The file will be sent to the screen. TYPE C:\WORK*.DOC will display the contents of *all* the files with a DOC extension present in the WORK directory of the hard disk (drive C:). You can use the TYPE command to send the contents of a file to the printer at the same time as viewing it on the screen. If you need to do this, press <CTRL-P> before you issue the TYPE command (but do make sure that the printer is 'online' and ready to go!). To disable the printer output you can use the <CTRL-P> combination a second time. The ability to redirect data is an extremely useful facility. DOS uses the < and > characters in conjunction with certain commands to redirect files. As an example: TYPE A:\README.DOC >PRN will redirect normal screen output produced by the TYPE command to the printer. This is usually more satisfactory than using the <PRT.SCREEN> key.
VER	The VER command displays the current DOS version.
VERIFY	The VERIFY command can be used to enable or disable disk file verification. VERIFY ON enables verification whilst VERIFY OFF disables verification. If VERIFY is used without ON or OFF, the system will display the state of verification (either 'on' or 'off').
VOL	The VOL command may be used to display the volume label of a disk.

External DOS commands

Unlike internal commands, these commands will not function unless the appropriate DOS utility program is resident in the current (default) directory. External commands are simply the names of utility programs (normally resident in the DOS sub-directory). If you need to gain access to these utilities from any directory or sub-directory, then the following lines should be included in your AUTOEXEC.BAT file (see page 135):

```
SET PATH=C:\DOS
```

The foregoing assumes that you have created a sub-directory called DOS on the hard disk and that this sub-directory contains the DOS utility programs. As with the internal DOS commands, the examples given apply to the majority of DOS versions.

Command	Function
`APPEND`	The `APPEND` command allows the user to specify drives, directories, and sub-directories which will be searched through when a reference is made to a particular data file. The `APPEND` command follows the same syntax as the `PATH` command (see page 118).

`ASSIGN`	The `ASSIGN` command allows users to redirect files between drives. `ASSIGN` is particularly useful when a RAM disk is used to replace a conventional disk drive.
	Examples:
	`ASSIGN A=E`
	results in drive E: being searched for a file whenever a reference is made to drive A:. The command may be subsequently countermanded by issuing a command of the form:
	`ASSIGN A=A`
	Alternatively, all current drive assignments may be overridden by simply using:
	`ASSIGN`
	`ASSIGN A=B` followed by `ASSIGN B=A` can be used to swap the drives over in a system which has two floppy drives. The original drive assignment can be restored using `ASSIGN`.

`ATTRIB`	The `ATTRIB` command allows the user to examine and/or set the attributes of a single file or a group of files. The `ATTRIB` command alters the file attribute byte (which appears within a disk directory) and which determines the status of the file (e.g. read-only).
	Examples:
	`ATTRIB A:\PROCESS.DOC`
	displays the attribute status of copies the file PROCESS.DOC contained in the root directory of the disk in drive A:.
	`ATTRIB +R A:\PROCESS.DOC`
	changes the status of the file PROCESS.DOC contained in the root directory of the disk in drive A: so that is a read-only file.

(*continued*)

Command	Function
	This command may be countermanded by issuing a command of the form: `ATTRIB -R A:\PROCESS.DOC` A crude but effective alternative to password protection is that of using `ATTRIB` to make all the files within a sub-directory hidden. As an example, `ATTRIB +H C:\PERSONAL` will hide all of the files in the PERSONAL sub-directory. `ATTRIB -H C:\PERSONAL` will make them visible once again.

`BACKUP`	The `BACKUP` command may be used to copy one or more files present on a hard disk to a number of floppy disks for security purposes. It is important to note that the `BACKUP` command stores files in a compressed format (i.e. not in the same format as that used by the `COPY` command). The `BACKUP` command may be used selectively with various options including those which allow files to be archived by date. The `BACKUP` command usually requires that the target disks have been previously formatted; however, from MS-DOS 3.3 onwards, an option to format disks was included.
	Examples:
	`BACKUP C:*.* A:`
	backs up all of the files present on the hard disk. This command usually requires that a large number of (formatted) disks are available for use in drive A:. Disks should be numbered so that the data can later be restored in the correct sequence.
	`BACKUP C:\DEV\*.C A:`
	backs up all of the files with a C: extension present within the DEV sub-directory on drive C:.
	`BACKUP C:\PROCESS\*.BAS A:/D:01-01-99`
	backs up all of the files with a BAS extension present within the PROCESS sub-directory of drive C: that were created or altered on or after 1 January 1999.
	`BACKUP C:\COMMS\*.* A:/F`
	backs up all of the files present in the COMMS sub-directory of drive C: and formats each disk as it is used.

`CHKDSK`	The `CHKDSK` command reports on disk utilization and provides information on total disk space, hidden files, directories, and user files. `CHKDSK` also gives the total memory and free memory available. `CHKDSK` incorporates options which can be used to enable reporting and to repair damaged files. `CHKDSK` provides two useful switches: `/F` fixes errors on the disk and `/V` displays the name of each file in every directory as the disk is checked. Note that if you use the `/F` switch, `CHKDSK` will ask you to confirm that you actually wish to make changes to the disk's *file allocation table* (FAT).

(*continued*)

Command	Function

	Examples:
	`CHKDSK A:`
	Checks the disk placed in the A: drive and displays a status report on the screen.
	`CHKDSK C:\DEV\*.ASM/F/V`
	checks the specified disk and directory, examining all files with an ASM extension, reporting errors and attempting to correct them.
	If you make use of the `/F` switch, `CHKDSK` will ask you to confirm that you actually wish to correct the errors. If you do go ahead `CHKDSK` will usually change the disk's file allocation table (FAT). In some cases this may result in loss of data!

`COMP`	The `COMP` command may be used to compare two files on a line by line or character by character basis. The following options are available: `/A` use . . . to indicate differences `/B` perform comparison on a character basis `/C` do not report character differences `/L` perform line comparison for program files `/N` add line numbers `/T` leave tab characters `/W` ignore white space at beginning and end of lines *Example:* `COMP /B PROC1.ASM PROC2.ASM` carries out a comparison of the files PROC1.ASM and PROC2.ASM on a character by character basis.
`DISKCOMP`	The `DISKCOMP` command provides a means of comparing two (floppy) disks. `DISKCOMP` accepts drive names as parameters and the necessary prompts are generated when a single-drive disk comparison is made. *Example:* `DISKCOMP A: B:` compares the disk in drive A: with that placed in drive B:.
`EXE2BIN`	The `EXE2BIN` utility converts, where possible, an EXE program file to a COM program file (which loads faster and makes less demands on memory space). *Example:* `EXE2BIN PROCESS` will search for the program PROCESS.EXE and generate a program PROCESS.COM. `EXE2BIN` will not operate on EXE files that require more than 64 KB of memory (including space for the stack and data storage) and/or those that make reference to other memory segments (`CS, DS, ES,` and `SS` *must* all remain the same during program execution).

(*continued*)

Command	Function
FASTOPEN	The FASTOPEN command provides a means of rapidly accessing files. The command is only effective when a hard disk is fitted and should ideally be used when the system is initialized (e.g. from within the AUTOEXEC.BAT file). *Example:* `FASTOPEN C:32` enables fast opening of files and provides for the details of up to 32 files to be retained in RAM. FASTOPEN retains details of files within RAM and must not be used concurrently with ASSIGN, JOIN, and SUBST.

FDISK	The FDISK utility allows users to format a hard (fixed) disk. Since the command will render any existing data stored on the disk inaccessible, FDISK should be used with extreme caution. Furthermore, improved hard disk partitioning and formatting utilities are normally be supplied when a hard disk is purchased. These should be used in preference to FDISK whenever possible.
	To ensure that FDISK is not used in error, copy FDISK to a sub-directory that is not included in the PATH statement then erase the original version using the following commands:
	`CD\`
	`MD XDOS`
	`COPY C:\DOS\FDISK.COM C:\XDOS`
	`ERASE C:\DOS\FDISK.COM`
	Finally, create a batch file, FDISK.BAT, along the following lines and place it in the DOS directory:
	`ECHO OFF`
	`CLS`
	`ECHO ***** You are about to format the hard disk! *****`
	`ECHO All data will be lost - if you do wish to continue`
	`ECHO change to the XDOS directory and type FDISK again.`

FIND	The FIND command can be used to search for a character string within a file. Options include: /C display the line number(s) where the search string has been located /N number the lines to show the position within the file /V display all lines which do not contain the search string *Example:* `FIND/C "output" C:/DEV/PROCESS.C` searches the file PROCESS.C present in the DEV sub-directory for occurrences of the word 'output'. When the search string is located, the command displays the appropriate line number.

(*continued*)

Command	Function

FORMAT	The FORMAT command is used to initialize a floppy or hard disk. The command should be used with caution since it will generally not be possible to recover any data which was previously present. Various options are available including:
	/1 single-sided format
	/8 format with 8 sectors per track
	/B leave space for system tracks to be added (using the SYS command)
	/N:8 format with 8 sectors per track
	/S write system tracks during formatting (note that this must be the last option specified when more than one option is required)
	/T:80 format with 80 tracks
	/V format and then prompt for a volume label
	Examples:
	`FORMAT A:`
	formats the disk placed in drive A:.
	`FORMAT B:/S`
	formats the disk placed in drive B: as a system disk.

JOIN	The JOIN command provides a means of associating a drive with a particular directory path. The command must be used with care and must not be used with ASSIGN, BACKUP, DISKCOPY, FORMAT, etc.
KEYB	The KEYB command invokes the DOS keyboard driver. KEYB replaces earlier utilities (such as KEYBUK) which were provided with DOS versions prior to MS-DOS 3.3. The command is usually incorporated in an AUTOEXEC.BAT file and must specify the country letters required. *Example:* `KEYB UK` selects the UK keyboard layout.
LABEL	The LABEL command allows a volume label (maximum 11 characters) to be placed in the disk directory. *Example:* `LABEL A: TOOLS` will label the disk present in drive A: as TOOLS. This label will subsequently appear when the directory is displayed.
MODE	The MODE command can be used to select a range of screen and printer options. MODE is an extremely versatile command and offers a wide variety of options.

(*continued*)

Command	Function

	Examples:
	`MODE LPT1: 120,6`
	initializes the parallel printer LPT1 for printing 120 columns at 6 lines per inch.
	`MODE LPT2: 60,8`
	initializes the parallel printer LPT2 for printing 60 columns at 8 lines per inch.
	`MODE COM1: 1200,N,8,1`
	initializes the COM1 serial port for 1200 baud operation with no parity, 8 data bits and 1 stop bit.
	`MODE COM2: 9600,N,7,2`
	initializes the COM2 serial port for 9600 baud operation with no parity, 7 data bits and 2 stop bits.
	`MODE 40`
	sets the screen to 40-column text mode.
	`MODE 80`
	sets the screen to 80-column mode.
	`MODE BW80`
	sets the screen to monochrome 40-column text mode.
	`MODE CO80`
	sets the screen to colour 80-column mode.
	`MODE CON CODEPAGE PREPARE=((850)C:\DOS\EGA.CPI)`
	loads code page 850 into memory from the file EGA.CPI located within the DOS directory.
	The `MODE` command can be used to redirect printer output from the parallel port to the serial port using `MODE LPT1:=COM1:`. Normal operation can be restored using `MODE LPT1:`.

`PRINT`	The `PRINT` command sends the contents of an ASCII text file to the printer. Printing is carried out as a background operation and data is buffered in memory. The default buffer size is 512 bytes; however, the size of the buffer can be specified using `/B:` (followed by required buffer size in bytes). When the utility is first entered, the user is presented with the opportunity to redirect printing to the serial port (COM1:). A list of files (held in a queue) can also be specified.
	Examples:
	`PRINT README.DOC`
	prints the file README.DOC from the current directory.
	`PRINT /B:4096 HELP1.TXT HELP2.TXT HELP3.TXT`

(*continued*)

Command	Function
	establishes a print queue with the files HELP1.TXT, HELP2.TXT, and HELP3.TXT and also sets the print buffer to 4 KB. The files are sent to the printer in the specified sequence.

`RESTORE`	The `RESTORE` command is used to replace files on the hard disk which were previously saved on floppy disk(s) using the `BACKUP` command. Various options are provided (including restoration of files created before or after a specified date).
	Examples:
	`RESTORE C:\DEV\PROCESS.COM`
	restores the files PROCESS.COM in the sub-directory named DEV on the hard disk partition, C:. The user is prompted to insert the appropriate floppy disk (in drive A:).
	`RESTORE C:\BASIC /M`
	restores all modified (altered or deleted) files present in the sub-directory named BASIC on the hard disk partition, C:.

`SYS`	The `SYS` command creates a new boot disk by copying the hidden DOS system files. `SYS` is normally used to transfer system files to a disk which has been formatted with the `/S` or `/B` option. `SYS` cannot be used on a disk which has had data written to it after initial formatting.
`TREE`	The `TREE` command may be used to display a complete directory listing for a given drive. The listing starts with the root directory.
`XCOPY`	The `XCOPY` utility provides a means of selectively copying files. The utility creates a copy which has the same directory structure as the original. Various options are provided: `/A` only copy files which have their archive bit set (but do not reset the archive bits) `/D` only files which have been created (or that have been changed) after the specified date `/M` copy files which have their archive bit set but reset the archive bits (to avoid copying files unnecessarily at a later date) `/P` prompt for confirmation of each copy `/S` copy files from sub-directories `/V` verify each copy `/W` prompt for disk swaps when using a single-drive machine *Example:* `XCOPY C:\DOCS\*.* A:/M` copy all files present in the DOCS sub-directory of drive C:. Files will be copied to the disk in drive A:. Only those files which have been modified (i.e. had their archive bits set) will be copied. Always use `XCOPY` in preference to `COPY` when sub-directories exist. As an example, `XCOPY C:\DOS\*.* A:\ /S` will copy all files present in the DOS directory on drive C: together with all files present in any sub-directories, to the root directory of the disk in A:.

Using batch files

Batch files provide a means of avoiding the tedium of repeating a sequence of operating system commands many times over. Batch files are nothing more than straightforward ASCII text files which contain the commands which are to be executed when the name of the batch is entered. Execution of a batch file is automatic; the commands are executed just as if they had been types in at the keyboard. Batch files may also contain the names of executable program files (i.e. those with a COM or EXE extension), in which case the specified program is executed and, provided the program makes a conventional exit to DOS upon termination, execution of the batch file will resume upon termination.

Batch file commands

DOS provides a number of commands which are specifically intended for inclusion within batch files.

Command	Function

`ECHO`	The `ECHO` command may be used to control screen output during execution of a batch file. `ECHO` may be followed by `ON` or `OFF`, or by a text string which will be displayed when the command line is executed.
	Examples:
	`ECHO OFF`
	disables the echoing (to the screen) of commands contained within the batch file.
	`ECHO ON`
	re-enables the echoing (to the screen) of commands contained within the batch file. (Note that there is no need to use this command at the end of a batch file as the reinstatement of screen echo of keyboard generated commands is automatic.)
	`ECHO Sorting data - please wait!`
	displays the message:
	`Sorting data - please wait!`
	on the screen.
	You can use `@ECHO OFF` to disable printing of the `ECHO` command itself. You will normally want to use this command instead of `ECHO OFF`.

`FOR`	`FOR` is used with `IN` and `DO` to implement a series of repeated commands.
	Examples:
	`FOR %A IN (IN.DOC OUT.DOC MAIN.DOC) DO COPY %A LPT1:`
	copies the files IN.DOC, OUT.DOC, and MAIN.DOC in the current directory to the printer.
	`FOR %A IN (*.DOC) DO COPY %A LPT1:`

(*continued*)

Command	Function
	copies all the files having a DOC extension in the current directory to the printer. The command has the same effect as `COPY *.DOC LPT1:`.
`IF`	If is used with `GOTO` to provide a means of branching within a batch file. `GOTO` must be followed by a label (which must begin with :). *Example:* `IF NOT EXIST SYSTEM.INI GOTO :EXIT` transfers control to the label `:EXIT` if the file SYSTEM.INI cannot be found in the current directory.
`PAUSE`	the pause command suspends execution of a batch file until the user presses any key. The message: `Press any key when ready...` is displayed on the screen.
`REM`	The `REM` command is used to precede lines of text which will constitute remarks. *Example:* `REM Check that the file exists before copying`

Creating batch files

Batch files may be created using an ASCII text editor or a word processor (operating in ASCII mode). Alternatively, if the batch file comprises only a few lines, the file may be created using the DOS `COPY` command. As an example, let us suppose that we wish to create a batch file which will:

1 erase all of the files present on the disk placed in drive B:;
2 copy all of the files in drive A having a TXT extension to produce an identically named set of files on the disk placed in drive B:;
3 rename all of the files having a TXT extension in drive A: to so that they have a BAK extension.

The required operating system commands are thus:

```
ERASE B:\*.*
COPY A:\*.TXT B:\
RENAME A:\*.TXT A:\*.BAK
```

The following keystrokes may be used to create a batch file named ARCHIVE.BAT containing the above commands (note that `<ENTER>` is used to terminate each line of input):

```
COPY CON: ARCHIVE.BAT
ERASE B:\*.*
COPY A:\*.TXT B:\
RENAME A:\*.TXT A:\*.BAK
<CTRL-Z>
```

If you wish to view the batch file which you have just created simply enter the command:

```
TYPE ARCHIVE.BAT
```

Whenever you wish to execute the batch file simply type:

```
ARCHIVE
```

Note that, if necessary, the sequence of commands contained within a batch file may be interrupted by typing:

```
<CTRL-C>
```

(i.e. press and hold down the CTRL key and then press the C key).

The system will respond by asking you to confirm that you wish to terminate the batch job. Respond with Y to terminate the batch process or N if you wish to continue with it.

Additional commands can be easily appended to an existing batch file. As an example, assume that we wish to view the directory of the disk in drive A: after running the archive batch file. We can simply append the extra commands to the batch files by entering:

```
COPY ARCHIVE.BAT + CON:
```

The system displays the filename followed by the `CON` prompt. The extra line of text can now be entered using the following keystrokes (again with each line terminated by `<ENTER>`):

```
DIR A:\
<CTRL-Z>
```

Passing parameters

Parameters may be passed to batch files by including the `%` character to act as a place holder for each parameter passed. The parameters are numbered strictly in the sequence in which they appear after the name of the batch file. As an example, suppose that we have created a batch file called REBUILD, and this file requires two file specifications to be passed as parameters. Within the text of the batch file, these parameters will be represented by `%1` and `%2`. The first file specification following the name of the batch file will be `%1` and the second will be `%2`. Hence, if we enter the command:

```
REBUILD PROC1.DAT PROC2.DAT
```

During execution of the batch file, `%1` will be replaced by PROC1.DAT whilst `%2` will be replaced by PROC2.DAT.

It is also possible to implement simple testing and branching within a batch file. Labels used for branching should preferably be stated in lower case (to avoid confusion with operating systems commands) and should be preceded by a colon when they are the first (or only) statement in a line. The following example which produces a sorted list of directories illustrates these points:

```
@ECHO OFF
IF EXIST %1 GOTO valid
```

```
ECHO Missing or invalid parameter
GOTO end
:valid
ECHO Index of Directories in %1
DIR %1 | FIND "<DIR>" | SORT
:end
```

The first line disables the echoing of subsequent commands contained within the batch file. The second line determines whether, or not, a valid parameter has been entered. If the parameter is invalid (or missing) the `ECHO` command is used to print an error message on the screen.

Simple menus can be created with batch files. As an example, the following batch files make a simple 'front-end' for four separate DOS applications. In this example, three of these applications are located in the root directory whilst the fourth, EDIT, is located in the TOOLS sub-directory:

Batch file	Contents

MENU.BAT	`ECHO OFF`
	`CLS`
	`CD\`
	`ECHO ******** MENU ********`
	`ECHO [1] = CONFIGURE`
	`ECHO [2] = START PROCESS`
	`ECHO [3] = SHUT DOWN`
	`ECHO [4] = TEXT EDITOR`
	`ECHO *********************`

1.BAT	`CONFIG.EXE`
2.BAT	`START.EXE`
3.BAT	`CLOSE.EXE`
4.BAT	`CD TOOLS` `EDIT.EXE`

In order to display the menu automatically it is necessary to include `MENU.BAT` in the AUTOEXEC.BAT file. (see the example on page 135). Note that all four of the batch files must be present in the root directory and that, when an application terminates and returns control to DOS, it will be necessary to run the MENU.BAT file again by simply typing `MENU` at the command prompt.

Using CONFIG.SYS

When DOS starts, but before the commands within the AUTOEXEC.BAT file are executed, DOS searches the root directory of the boot disk for a file called CONFIG.SYS. If this file exists, DOS will attempt to carry out the commands in the file. As with any batch file, the configuration sequence can be abandoned by means of `<CTRL-C>` or `<CTRL-BREAK>`. CONFIG.SYS is a plain ASCII text file with commands on separate lines. The file can be created using any text editor or word processor operating in ASCII mode (CONFIG.SYS can also be created using `COPY CON:` as described earlier for the creation of batch files).

Only the following subset of DOS commands is valid within the CONFIG.SYS file:

Command	Function
`BREAK`	Determines the response to a <CTRL-BREAK> sequence. If you set `BREAK ON` in CONFIG.SYS, DOS checks to see whether you have requested a break whenever a DOS call is made. If you set `BREAK OFF`, DOS checks for a break only when it is working with the video display, keyboard, printer, or a serial port.
`BUFFERS`	Sets the number of file buffers which DOS uses. This command can be used to significantly improve disk performance with early versions of DOS and when a disk cache (accessed via IBMCACHE.SYS or SMARTDRV.SYS) is not available. The use of buffers can greatly reduce the number of disk accesses that DOS performs (DOS only reads and writes full sectors). Data is held within a buffer until it is full. Furthermore, by reusing the least-recently used buffers, DOS retains information more likely to be needed next. It is worth noting that each buffer occupies 512 bytes of RAM (plus 16 additional bytes overhead). Hence, the number of buffers may have to be traded-off against the amount of conventional RAM available (particularly in the case of machines with less than the standard 640 KB RAM). In general, `BUFFERS=20` will provide adequate for most applications. `BUFFERS=40` (or greater) may be necessary for database or other applications which make intensive use of disk files. DOS uses a default value for `BUFFERS` of between 2 and 15 (depending upon the disk and RAM configuration). Later versions of DOS (e.g. MS-DOS 4.1) provide a much improved `BUFFERS` command which includes support for expanded memory and look-ahead buffers which can store sectors ahead of those requested by a DOS read operation. The number of look-ahead buffers must be specified (in the range 0–8) and each buffer requires 512 bytes of memory and corresponds exactly to one disk sector. The use of expanded memory can be enabled by means of a `/X` switch. *Example:* `BUFFERS=100,8 /X` sets the number of buffers to 100 (requiring approximately 52 KB of expanded memory) and also enables 8 look-ahead buffers (requiring a further 4 KB of expanded memory).
`COUNTRY`	Sets the country-dependent information.
`DEVICE`	Sets the hardware device drivers to be used with DOS. *Examples:* `DEVICE=C:\MOUSE\MOUSE.SYS` enables the mouse driver (MOUSE.SYS) which contained in a sub-directory called `MOUSE`.

(continued)

Command	Function

	`DEVICE=C:\DOS\ANSI.SYS`
	selects the ANSI.SYS screen driver (the ANSI.SYS file must be present in the DOS directory).
	`DEVICE=C:\WINDOWS\HIMEM.SYS`
	selects the Windows extended memory manager HIMEM.SYS (the HIMEM.SYS file must be present in the WINDOWS directory).
	`DEVICE=C:\DOS\DISPLAY.SYS CON=(EGA,850,2)`
	selects the DOS display driver and switches it to multilingual EGA mode (code page 850) with up to two code pages.
	Drivers often provide a number of 'switches' which allow you to optimize them for a particular hardware configuration. Always consult the hardware supplier's documentation to ensure that you have the correct configuration for your system.
	You may find it handy to locate all of your drivers in a common directory called DRIVERS, DEVICE, or SYS. This will keep them separate from applications and help you to find them at some later date.
	Finally, note that you can load as many device drivers as you need, but you must use a separate DEVICE line for each driver.
	Example:
	`DEVICE = C:\DRIVERS\ANSI.SYS`
	`DEVICE = C:\DRIVERS\CDROM.SYS`

`FCBS`	Sets the number of file control blocks that DOS can have open at any time (note that this command is now obsolete).
`FILES`	Sets the maximum number of files that DOS can access at any time.
`INSTALL`	Installs memory-resident programs. *Example:* `INSTALL = C:\DOS\FASTOPEN.EXE C:=100` installs the DOS `FASTOPEN` utility and configures it to track the opening of up to 100 files and directories on drive C:. Slightly less memory is used when memory-resident programs are loaded with this command than with AUTOEXEC.BAT. Don't, however, use `INSTALL` to load programs that use environment variables or shortcut keys or that require COMMAND.COM to be present to handle critical errors.
`LASTDRIVE`	Specifies the highest disk drive on the computer.
`REM`	Treates a line as a comment/remark.
`SHELL`	Determines the DOS command processor (e.g. COMMAND. COM).
`STACKS`	Sets the number of stacks that DOS uses.
`SWITCHES`	Disables extended keyboard functions.

Using configuration files and device drivers

DOS provides a number of device drivers and utility programs which, in an earlier DOS/Windows environment could be installed from CONFIG.SYS. Typical of these drivers were:

Function	Device driver (generic name)
Disk caches	IBMCACHE.SYS, SMARTDRV.SYS
RAM drives	RAMDRIVE.SYS, VDISK.SYS
Additional disk drives	DRIVER.SYS
Memory management	XMAEM.SYS, EMM386.SYS, EMM386.EXE
Display adapter configuration	DISPLAY.SYS
Printer configuration	PRINTER.SYS

In a modern Windows NT or XP environment CONFIG.SYS is replaced by CONFIG.NT. Unless a different start-up file is specified in an application's Program Interchange File (PIF), CONFIG.NT is used to initialize the MS-DOS environment.

By default, no information is displayed when the DOS environment is initialized. If required you can display CONFIG.NT/AUTOEXEC.NT information by simply adding the command `echoconfig` to the CONFIG.NT (or other start-up file). When you return to the command prompt from a TSR or while running a DOS-based application, Windows runs COMMAND.COM. This allows the TSR to remain active. To run CMD.EXE, the Windows command prompt, rather than COMMAND.COM, simply add the command `ntcmdprompt` to CONFIG.NT (or other start-up file).

Also by default, it is possible to start any type of application when running COMMAND.COM. However, if an application other than an MS-DOS-based application is started, any TSR that is running may be disrupted. To ensure that only MS-DOS-based applications can be started, you can add the command `dosonly` to CONFIG.NT (or other start-up file).

As an example of the use of a DOS driver, the expanded memory manager (EMM) is configured by means of the following command syntax:

```
EMM = [A=AltRegSets] [B=BaseSegment] [RAM]
```

Where: `AltRegSets` specifies the total Alternative Mapping Register Sets you want the system to support (in the range 1–255). The default value of `AltRegSets` is 8.

`BaseSegment` specifies the starting segment address in the DOS conventional memory you want the system to allocate for EMM page frames. The value must be given in hexadecimal in the range 0×1000 to 0×4000. Note that the value of `BaseSegment` is rounded down to a 16 KB boundary and the default value is 0×4000.

`RAM` specifies that the system should only allocate 64 KB address space from the upper memory block (UMB) area for EMM page frames and leave the rest (if available) to be used by DOS to support the `loadhigh` and `devicehigh` commands. By default, the system will allocate all possible and available *upper memory block* (UMB) for page frames.

The amount of EMM is determined by the *Program Interchange File* (PIF). This may either be one that is associated with an application or, if unspecified, will be _default.pif. If the size specified in the PIF file is 0, EMM will be disabled and the EMM line will be ignored.

A typical CONFIG.NT file (created using a simple text editor) might be as follows:

`dos=high, umb`	(load DOS into high memory)
`device=%SystemRoot%\system32\himem.sys`	(use the himem memory manager)
`files=40`	(allow for 40 open files)

Using AUTOEXEC.BAT

The AUTOEXEC.BAT file allows you to automatically execute a series of programs and DOS utilities to add further functionality to a system when the system is initialized. AUTOEXEC.BAT normally contains a sequence of DOS commands but in addition it can also contain the name of an application or shell that will be launched automatically when the system is booted. This is a useful facility if you always use the same shell or application whenever you power-up your system, or if you wish to protect the end-user from the need to remember rudimentary DOS commands (such as `MD`, `CD`, `XCOPY`, etc.).

AUTOEXEC.BAT is typically used to:

1 set up the system prompt (see page 118);
2 define the path for directory searches (using SET PATH, see page 118);
3 execute certain DOS utilities (e.g. SHARE);
4 load a mouse driver (e.g. MOUSE.COM);
5 change directories (e.g. from the root directory to a 'working' directory);
6 launch an application or menu program (e.g. MENU.BAT).

It is important to note that Windows and some DOS programs have their own built-in mouse drivers and can thus communicate directly with the mouse. However, if you regularly use a mouse with DOS applications, you will probably wish to include reference to your mouse driver within the AUTOEXEC.BAT file.

If you are operating from within a DOS environment and you do decide to experiment with your CONFIG.SYS and AUTOEXEC.BAT files, it is essential to make sure that you keep back-up copies of your original files (CONFIG.BAK, CONFIG.OLD, etc.). If you are experiencing problems with memory limitations you can use the `MEM` command with the `PROGRAM`, `DEBUG`, or `/CLASSIFY` switches to see the effect of changes made to DOS drivers and memory managers.

A typical AUTOEXEC.BAT file (once again created using nothing more than a simple text editor) might be as follows:

`PROMPT $P$G`	(prompt with directory path)
`LOADHIGH=C:\DOS\SHARE.EXE`	(permits file sharing and locking)
`SET COMSPEC=C:\DOS\COMMAND.COM`	(specify the location of the command interpreter)
`SET PATH=C:\DOS;C:\UTILITY;C:\TOOLS`	(search DOS, UTILITY, and APPS directories)
`MENU.BAT`	(launch the menu batch file)

```
C:\WINDOWS\System32\command.com
1662:011A CD21          INT     21

C:\QBASIC>debug
-u 100
1662:0100 B405          MOV     AH,05
1662:0102 B20A          MOV     DL,0A
1662:0104 CD21          INT     21
1662:0106 B20D          MOV     DL,0D
1662:0108 CD21          INT     21
1662:010A B241          MOV     DL,41
1662:010C B13E          MOV     CL,3E
1662:010E B405          MOV     AH,05
1662:0110 CD21          INT     21
1662:0112 FEC2          INC     DL
1662:0114 E2FC          LOOP    0112
1662:0116 B000          MOV     AL,00
1662:0118 B44C          MOV     AH,4C
1662:011A CD21          INT     21
1662:011C 3400          XOR     AL,00
1662:011E 51            PUSH    CX
1662:011F 16            PUSH    SS
-a 114
1662:0114 loop 0110
1662:0116
-
```

Figure 3.4 *A typical Debug session*

Using DEBUG

One of the most powerful (but all too often neglected) tools available within the DOS environment is the debugger, DEBUG.COM or DEBUG.EXE. This program provides a variety of facilities including single stepping a program to permit examination of the processor's registers and the contents of memory after execution of each instruction. On most modern Windows installations, DEBUG.EXE can be found in the System32 folder of the Windows directory.

The Debug command line can accept several arguments. Its syntax is as follows:

```
DEBUG [filespec] [parm1] [parm2]
```

where `[filespec]` is the specification of the file to be loaded into memory, `[parm1]` and `[parm2]` are optional parameters for the specified file.

As an example, the following MS-DOS command will load debug along with the file MYPROG.COM (taken from the disk in drive A:) ready for debugging:

```
DEBUG A:\MYPROG.COM
```

When debug has been loaded, the familiar MS-DOS prompt is replaced by a hyphen (-). This indicates that DEBUG is awaiting a command from the user. Commands comprise single letter (in either upper or lower case). Delimiters are optional between commands and parameters. They must, however, be used to separate adjacent hexadecimal values.

<CTRL-BREAK> can be use to abort a DEBUG command whilst <CTRL-NUM.LOCK> can be used to pause the display (any other keystroke restarts the output). Commands may be edited using the keys available for normal MS-DOS command editing.

All Debug commands accept parameters (except the Q command). You can separate parameters with commas or spaces, but these separators are required only between two hexadecimal values. Therefore the following commands are equivalent:

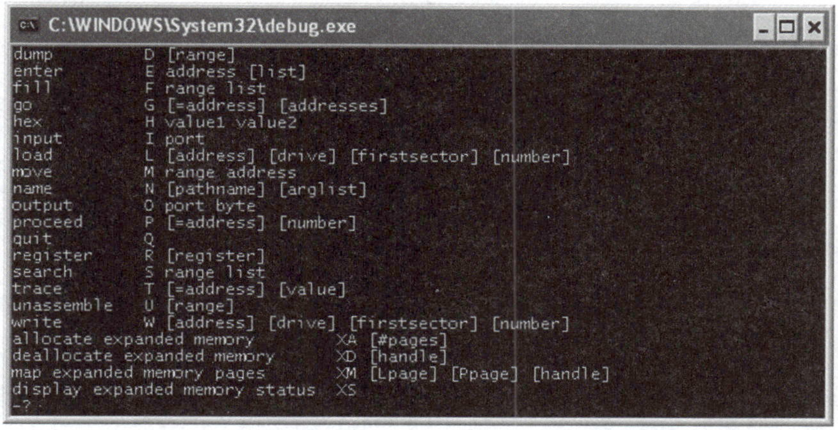

```
C:\WINDOWS\System32\debug.exe
dump          D [range]
enter         E address [list]
fill          F range list
go            G [=address] [addresses]
hex           H value1 value2
input         I port
load          L [address] [drive] [firstsector] [number]
move          M range address
name          N [pathname] [arglist]
output        O port byte
proceed       P [=address] [number]
quit          Q
register      R [register]
search        S range list
trace         T [=address] [value]
unassemble    U [range]
write         W [address] [drive] [firstsector] [number]
allocate expanded memory        XA [#pages]
deallocate expanded memory      XD [handle]
map expanded memory pages       XM [Lpage] [Ppage] [handle]
display expanded memory status  XS
-?
```

Figure 3.5 *Limited help information available from within Debug*

```
D CS:100 110
DCS:100 110
D,CS:100,110
```

Hard copy of Debug sessions can sometimes be very useful. If you need this facility, just type `<CTRL-P>` before the DEBUG command, and then all your screen output will be echoed to your printer. Press `<CTRL-P>` a second time in order to cancel the printer echo (Figures 3.4 and 3.5).

Debug commands

The following Debug commands are available:

Command	Meaning	Function

`A [addr]`	Assemble	Assemble mnemonics into memory from the specified address. If no address is specified, the code will be assembled into memory from address `CS:0100`. The `<ENTER>` key is used to terminate assembly and return to the Debug prompt.
		Examples:
		`A 200`
		starts assembly from address `CS:0200`.
		`A 4E0:100`
		starts assembly from address `04E0:0100` (equivalent to a physical address of 04F00).

`C range addr`	Compare	Compare memory in the specified range with memory starting at the specified address.
`D [addr]`	Dump	Dump (display) memory from the given starting address. If no Start address is specified, the dump will commence at DS:0100. *Examples:* `D 400`

(*continued*)

Command	Meaning	Function
		dumps memory from address DS:0400. `D CS:0` dumps memory from address CS:0000.
`D [range]`		Dump (display) memory within the specified range. *Example:* `D DS:200 20F` displays 16 bytes of memory from DS:0200 to DS:0210 inclusive.
`E addr [list]`	Enter	Enter (edit) bytes into memory starting at the given address. If no list of data bytes is specified, byte values are displayed and may be sequentially overwritten. <SPACE> may be used to advance, and <–> may be used to reverse the memory pointer. *Example:* `E 200,3C,FF,1A,FE` places byte values of 3C, FF, 1A, and FE into four consecutive memory locations commencing at DS:0200.

`F range list`	Fill	Fills memory in the given range with data in the list. The list is repeated until all memory locations have been filled.
		Examples:
		`F 100,10F,FF`
		fills 16 bytes of memory with FF commencing at address DS:0100.
		`F 0,FFFF,AA,FF`
		fills 65536 bytes of memory with alternate bytes of AA and FF.

`G [=addr]`	Go	Executes the code starting at the given address. If no address is specified, execution commences at address CS:IP. *Example:* `G =100` executes the code starting at address CS:0100.

`G [=addr]`
`[addr]`
`[addr]...`
		Executes the code starting at the given address with the specified breakpoints.
		Example:
		`G =100 104 10B`
		executes the code starting at address CS:0100 and with breakpoints at addresses CS:0104 and CS:010B.

`H value value`	Hexadecimal	Calculates the sum and difference of two hexadecimal values.
`I port`	Port input	Inputs a byte value from the specified I/O port address and display the value. *Example:* `I 302` inputs the byte value from I/O port address 302 and displays the value returned.

(*continued*)

Command	Meaning	Function
`L [addr]`	Load	Loads the file previously specified by the Name `(N)` command. The file specification is held at address CS:0080. If no load address is specified, the file is loaded from address CS:0100.
`M range addr`	Move	Moves (replicates) memory in the given range so that it is replicated starting at the specified address.
`N filespec`	Name	Names a file to be used for a subsequent Load `(L)` or Write `(W)` command. *Example:* `N A:\MYPROG.COM` names the file MYPROG.COM stored in the root directory of drive A: for a subsequent load or write command.
`O port byte`	Port output	Output a given byte value to the specified I/O port address. *Example:* `O 303 FE` outputs a byte value of FE from I/O port address 303.
`P [=addr] [instr]`	Proceed	Executes a subroutine, interrupt loop or string operation and resumes control at the next instruction. Execution starts at the specified address and continues for the specified number of instructions. If no address is specified, execution commences at the address given by CS:IP.
`Q`	Quit	Exits debug and return control to the current MS-DOS shell.
`R [regname]`	Register	Displays the contents of the specified register and allows the contents to be modified. If a name is not specified, the contents of all of the CPU registers (including flags) is displayed together with the next instruction to be executed (in hexadecimal and in mnemonic format).
`S range list`	Search	Search memory within the specified range for the listed data bytes. *Example:* `S 0100 0800 20,1B` searches memory between address DS:0100 and DS:0800 for consecutive data values of 20 and 1B.
`T [=addr] [instr]`	Trace	Traces the execution of a program from the specified address and executing the given number of instructions. If no address is specified, the execution starts at address CS:IP. If the number of instructions is not specified then only a single instruction is executed. A register dump (together with a disassembly of the next instruction to be executed) is displayed at each step. *Examples:* `T` traces the execution of the single instruction referenced by CS:IP. `T =200,4` traces the execution of four instructions commencing at address CS:0200.

(*continued*)

Command	Meaning	Function

`U [addr]`	Unassemble	Unassemble (disassemble) code into mnemonic instructions starting at the specified address. If no address is specified, disassembly starts from the address given by CS:IP.
		Examples:
		`U`
		disassembles code starting at address CS:IP.
		`U 200`
		disassembles code starting at address CS:0200.

`U [range]`		Unassemble (disassemble) code into mnemonic instructions within the specified range of addresses. *Example:* `U 200 400` disassembles the code starting at address CS:0200 and ending at address CS:0400.
`W [addr]`	Write	Writes data to disk from the specified address. The file specification is taken from a previous Name (`N`) command. If the address is not specified, the address defaults to that specified by CS:IP. The file specification is located at CS:0080.

Notes:

(a) Parameters enclosed in square brackets ([and]) are optional.

(b) The equal sign (=) must precede the start address used by the following commands: Go (`G`), Proceed (`P`), and Trace (`T`).

(c) Parameters have the following meanings:

Parameter	Meaning

`addr`	Address (which may be quoted as an offset or as the contents of a segment register or segment address followed by an offset). The following are examples of acceptable addresses:
	`CS:0100`
	`04C0:0100`
	`0200`

`byte`	A byte of data (i.e. a value in the range 0 to FF). The following are examples of acceptable data bytes:
	`0`
	`1F`
	`FE`

`filespec`	A file specification (which may include a drive letter and sub-directory, etc.). The following are examples of acceptable file specifications:
	`MYPROG.COM`
	`A:MYPROG.COM`
	`C:\PROGS\MYPROG.COM`

`instr`	The number of instructions to be executed within a Trace (`T`) or Proceed (`P`) command.

(*continued*)

Parameter	Meaning

`list`	A list of data bytes, ASCII characters (which must be enclosed in single quotes), or strings (which must be enclosed in double quotes). The following examples are all acceptable data lists:
	`C,2F,C2,00,10`
	`'A',':','/'`
	`"Insert disk and press ENTER"`

`port`	A port address. The following are acceptable examples of port addresses: `E` (the DMA controller) `30C` (within the prototype range) `378` (the parallel printer) (see page 31 for more information).
`range`	A range of addresses which may be expressed as an address and offset (e.g. CS:100,100) or as an address followed by a size (e.g. `DS:100 L 20`).

`regname`	A register name (see (d)). The following are acceptable examples of register names:
	`AX`
	`DS`
	`IP`

`value`	A hexadecimal value in the range 0 to FFFF.

(d) The following register and flag names are used within debug:

`AX, BX, CX, DX`	16-bit General-Purpose Registers
`CS, DS, ES, SS`	Code, Data, Extra, and Stack Segment Registers
`SP, BP, IP`	Stack, Base, and Instruction Pointers
`SI, DI`	Source and Destination Index Registers
`F`	Flag Register

(e) The following abbreviations are used to denote the state of the flags in conjunction with the Register (R) and Trace (T) commands:

Flag	Abbreviation	Meaning/status
Overflow	`OV`	Overflow
	`NV`	No overflow
Carry	`CY`	Carry
	`NC`	No carry
Zero	`ZR`	Zero
	`NZ`	Non-zero
Direction	`DN`	Down
	`UP`	Up
Interrupt	`EI`	Interrupts enabled
	`DI`	Interrupts disabled
Parity	`PE`	Parity even
	`PO`	Parity odd

(*continued*)

```
C:\WINDOWS\System32\debug.exe
-d 0040:0000
0040:0000  F8 03 F8 02 E8 03 E8 02-BC 03 78 03 78 02 C0 9F   ..........x.x...
0040:0010  22 C8 20 80 02 85 00 00-00 00 36 00 36 00 64 20   ".........6.6.d
0040:0020  20 39 30 0B 30 0B 34 05-30 0B 3A 27 30 0B 30 0B    90.0.4.0.:'0.0.
0040:0030  30 0B 30 0B 0D 1C 00 00-00 00 00 00 00 00 00 00   0.0.............
0040:0040  1E 00 C3 00 00 00 00 00-00 03 50 00 C0 0D 00 00   ..........P.....
0040:0050  00 06 00 00 00 00 00 00-00 00 00 00 00 00 00 00   ................
0040:0060  0F 0B 00 D4 03 29 30 00-00 00 00 00 DA 4C 0B 00   .....)0......L..
0040:0070  00 00 00 00 00 00 08 00-14 14 14 14 01 01 01 01   ................
-
```

Figure 3.6 *The Debug Dump* (D) *command being used to display the contents of 128 bytes of memory starting at address* 0040:0000 *(equivalent to a memory address of* 04000H)

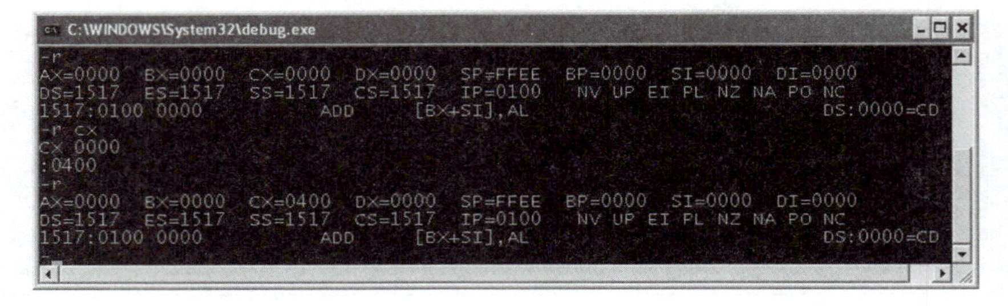

Figure 3.7 *The Debug Register* (r) *command being first used to display the contents of the CPU registers and then to change the contents of the CX register from* 0000 *to* 0400

Flag	Abbreviation	Meaning/status
Parity	PE	Parity even
	PO	Parity odd
Sign	NG	Negative
	PL	Positive
Auxiliary carry	AC	Auxiliary carry
	NC	No auxiliary carry

(f) All numerical values within Debug are in hexadecimal (Figures 3.6 and 3.7).

A Debug walkthrough

The following 'walkthrough' has been provided in order to give you an insight into the range of facilities offered by Debug. We shall assume that a short program TEST.EXE has been written to test a printer connected to the parallel port. The program is designed to generate a single line of upper- and lower-case characters but, since an error is present, the compiled program prints only a single character. The source code for the program (TEST.ASM) is shown in Figure 3.8.

The first stage in the debugging process is to invoke Debug from MS-DOS using the command:

```
DEBUG TEST.EXE
```

```
                ; Program name: ptest.com
                ;
                ; This program outputs a line of ASCII characters
                ; in the range 41H to 7FH to the parallel printer
                ; port.
                ;
                ; Registers used: AX, CL, DL
                ; Parameters passed: none
                ;
                TITLE ptest
_TEXT           SEGMENT             ; Define code segment
                ASSUME cs:_TEXT,ds:_TEXT,ss:_TEXT
                ORG  100H           ; Normal for a COM program
start:          MOV  AH,05H         ; Function code for printer output
                MOV  DL,0AH         ; First generate a
                INT  21H            ; line feed
                MOV  DL,0DH         ; Next generate a
                INT  21H            ; carriage return
                MOV  DL,41H         ; First character to print is 'A'
                MOV  CL,3EH         ; Number of characters to print
                MOV  AH,05H         ; Set up the function code
                INT  21H            ; and print the character
prch:           INC  DL             ; Get the next character
                LOOP prch           ; and go around again
                MOV  AL,00H         ; Set up the return code
                MOV  AH,4CH         ; and the function code
                INT  21H            ; for an exit to DOS
_TEXT           ENDS
                END  start
```

Figure 3.8 *Faulty source code used for the printer test program (ptest.com)*

```
1662:0100  B4 05 B2 0A CD 21 B2 0D-CD 21 B2 41 B1 3E B4 05   .....!...!.A.>..
1662:0110  CD 21 FE C2 E2 FC B0 00-B4 4C CD 21 34 00 51 16   .!.......L.!4.Q.
1662:0120  0A 3D 51 00 74 DF 3D 52-00 74 DA 8B 77 0A F7 C7   .=Q.t.=R.t..w...
1662:0130  40 00 74 09 AD F7 C7 10-00 74 02 AD AD 8B C7 25   @.t......t.....%
1662:0140  03 00 48 74 06 2E AD FF-D0 03 F0 2E AD 05 B4 2D   ..Ht...........-
1662:0150  89 46 FE F7 C7 08 00 74-0A 2E AD 89 46 FC 2E AD   .F.....t....F...
1662:0160  89 46 FA 80 3E 7B 2A 00-74 06 F7 C7 04 00 75 3D   .F..>{*.t.....u=
1662:0170  E8 3E E3 8B 46 FE A3 2A-14 E8 D1 E6 7E 1A 83 3E   .>..F..*....~..>
```

Figure 3.9 *Using Debug's Dump* (`d`) *command to view the printer test program in memory*

The command assumes that TEST.EXE is present in the current directory and that DEBUG.EXE is accessible either directly or via previous use of the SET PATH command.

After the Debug hyphen prompt appears, we can check that our code has loaded, we use the Dump (D) command. Entering the command `D100` at the Debug hyphen prompt produces the display shown in Figure 3.9.

The extreme left-hand column gives the address (in segment register:offset format). The next 16 columns comprise hexadecimal data showing the bytes stored at the 16 address locations starting at the address shown in the left-hand column. The first line in Figure 3.9 shows the hexadecimal contents of 16 bytes of memory starting at 1662:0100 (i.e. segment address = 1662, offset = 0100). The hexadecimal value of the first byte in the 16-byte block is B3 whilst the second is 05, and so on. The hexadecimal value of the last byte in the 16-byte block (address 1662:010F) is also 05.

An ASCII representation of the data is shown in the right-hand column of the screen dump. Byte values that do not correspond to printable ASCII characters are shown simply as a full-stop. Hence B4 and 05 (which are both non-printable characters) are shown by full-stops whilst 21 appears as !, and 41 as A.

In the context of executable code, the hexadecimal/ASCII dump shown earlier is not particularly useful and a more meaningful representation can be achieved by using the Unassemble (U) command. Entering the command U100 at the Debug hyphen prompt produces the display shown in Figure 3.10. The executable code starts at address 1662:0100 and ends at address 1662:011B. In total there are 28 (decimal) bytes of code.

The first instruction occupies 2 bytes of memory (addresses 1662:0100 and 1662:0101). The instruction comprises a move of 8 bits of immediate data (05) into the AH register. The last program instruction is at address 1662:011A and is a software interrupt relating to address 21 in the interrupt vector table.

At this point it is worth mentioning that the Unassemble command can sometimes produce some rather odd displays. This is simply because the command is unable to distinguish valid program code from data; Unassemble will quite happily attempt to disassemble something which is not actually a program!

Having disassembled the program code resident in memory we can check it against the original source code file. Normally, however, this will not be necessary unless the object code file has become changed or corrupted in some way.

The next stage is that of tracing program execution. The Debug Trace (T) command could be employed for this function; however, it is better to make use of the Proceed (P) command to avoid tracing execution of the DOS interrupt routines in order to keep the amount of traced code manageable.

The Proceed command expects its first parameter to be the address of the first instruction to be executed. This must then be followed by a second parameter

```
1662:0100 B405        MOV     AH,05
1662:0102 B20A        MOV     DL,0A
1662:0104 CD21        INT     21
1662:0106 B20D        MOV     DL,0D
1662:0108 CD21        INT     21
1662:010A B241        MOV     DL,41
1662:010C B13E        MOV     CL,3E
1662:010E B405        MOV     AH,05
1662:0110 CD21        INT     21
1662:0112 FEC2        INC     DL
1662:0114 E2FC        LOOP    0112
1662:0116 B000        MOV     AL,00
1662:0118 B44C        MOV     AH,4C
1662:011A CD21        INT     21
```

Figure 3.10 *Using Debug's Unassemble* (U) *command to disassemble the program code*

which gives the number of instructions to be traced. In this case, and since our program terminates normally, we can supply any sufficiently large number of instructions as the second parameter to the Proceed command. Hence the required command is `P=100,100` (note the use of the equals sign) and the resulting trace dump is shown in Figure 3.11.

```
AX=0500  BX=0000  CX=0000  DX=0000  SP=FFEE  BP=0000  SI=0000  DI=0000
DS=1662  ES=1662  SS=1662  CS=1662  IP=0102   NV UP EI PL NZ NA PO NC
1662:0102 B20A          MOV     DL,0A

AX=0500  BX=0000  CX=0000  DX=000A  SP=FFEE  BP=0000  SI=0000  DI=0000
DS=1662  ES=1662  SS=1662  CS=1662  IP=0104   NV UP EI PL NZ NA PO NC
1662:0104 CD21          INT     21


AX=050A  BX=0000  CX=0000  DX=000A  SP=FFEE  BP=0000  SI=0000  DI=0000
DS=1662  ES=1662  SS=1662  CS=1662  IP=0106   NV UP EI PL NZ NA PO NC
1662:0106 B20D          MOV     DL,0D

AX=050A  BX=0000  CX=0000  DX=000D  SP=FFEE  BP=0000  SI=0000  DI=0000
DS=1662  ES=1662  SS=1662  CS=1662  IP=0108   NV UP EI PL NZ NA PO NC
1662:0108 CD21          INT     21

AX=050D  BX=0000  CX=0000  DX=000D  SP=FFEE  BP=0000  SI=0000  DI=0000
DS=1662  ES=1662  SS=1662  CS=1662  IP=010A   NV UP EI PL NZ NA PO NC
1662:010A B241          MOV     DL,41

AX=050D  BX=0000  CX=0000  DX=0041  SP=FFEE  BP=0000  SI=0000  DI=0000
DS=1662  ES=1662  SS=1662  CS=1662  IP=010C   NV UP EI PL NZ NA PO NC
1662:010C B13E          MOV     CL,3E

AX=050D  BX=0000  CX=003E  DX=0041  SP=FFEE  BP=0000  SI=0000  DI=0000
DS=1662  ES=1662  SS=1662  CS=1662  IP=010E   NV UP EI PL NZ NA PO NC
1662:010E B405          MOV     AH,05

AX=050D  BX=0000  CX=003E  DX=0041  SP=FFEE  BP=0000  SI=0000  DI=0000
DS=1662  ES=1662  SS=1662  CS=1662  IP=0110   NV UP EI PL NZ NA PO NC
1662:0110 CD21          INT     21
A  <----
AX=0541  BX=0000  CX=003E  DX=0041  SP=FFEE  BP=0000  SI=0000  DI=0000
DS=1662  ES=1662  SS=1662  CS=1662  IP=0112   NV UP EI PL NZ NA PO NC
1662:0112 FEC2          INC     DL

AX=0541  BX=0000  CX=003E  DX=0042  SP=FFEE  BP=0000  SI=0000  DI=0000
DS=1662  ES=1662  SS=1662  CS=1662  IP=0114   NV UP EI PL NZ NA PE NC
1662:0114 E2FC          LOOP    0112

AX=0541  BX=0000  CX=0000  DX=007F  SP=FFEE  BP=0000  SI=0000  DI=0000
DS=1662  ES=1662  SS=1662  CS=1662  IP=0116   NV UP EI PL NZ NA PO NC
1662:0116 B000          MOV     AL,00

AX=0500  BX=0000  CX=0000  DX=007F  SP=FFEE  BP=0000  SI=0000  DI=0000
DS=1662  ES=1662  SS=1662  CS=1662  IP=0118   NV UP EI PL NZ NA PO NC
1662:0118 B44C          MOV     AH,4C

AX=4C00  BX=0000  CX=0000  DX=007F  SP=FFEE  BP=0000  SI=0000  DI=0000
DS=1662  ES=1662  SS=1662  CS=1662  IP=011A   NV UP EI PL NZ NA PO NC
1662:011A CD21          INT     21
```

Figure 3.11 *Program trace showing incorrect execution of the printer test program (note that only a single character, A, is sent to the printer)*

The state of the processor registers is displayed as each instruction is executed together with the *next* instruction in disassembled format. Taking the results of executing the first instruction (`MOV AH,05`) as an example, we see that 05 has appeared in the upper byte of `AX` (`AH`) and the Instruction Pointer (`IP`) has moved on to offset address `0102`. The next instruction to be executed (located at the address which `IP` is pointing to) is `MOV DL,0A`. The state of the processor flags is also shown within the register dump. In this particular case, none of the flags has been changed as a result of executing the instruction.

In order to obtain a hard copy of the program trace, a `<CTRL-P>` command can be issued immediately before issuing the Proceed (`P`) command. From that point onwards, screen output was echoed to the printer. Since the program directs is own output to the printer, this also appears amidst the traced output.

A single character, A, is printed after the eighth instruction (see arrow marked on Figure 3.11). Thereafter, the program executes the loop formed by the instructions at offset addresses `0112` and `0114`. However, no printing takes place within this loop even though the `DL` register is incremented through the required range of ASCII codes (`41` to `7F`). Clearly the loop is not returning to the `INT 21` instruction which actually makes the required calls into DOS.

Fortunately, we can easily overcome this problem from within the debugger without returning to the macro assembler. We simply need to modify the `LOOP` instruction at offset address `0114`. To do this we can make use of the Assemble (`A`) command to overwrite the existing instruction. The required command is:

```
A 114
```

The `CS:IP` prompt is then displayed (in this case it shows `1662:0114`) after which we simply enter:

```
LOOP 0110
```

However, the `CS:IP` prompt is incremented since we need to make no further changes to the code, we can simply escape from the Debug line assembler by simply pressing `<ENTER>`.

Having modified our code, we can again trace the program using the Proceed (`P`) command exactly as before. The traced output produced by the modified program is shown in Figure 3.12. Note that we have now succeeded in producing a line of printed output showing the full range of characters (see arrow marked on Figure 3.12).

Since no further errors have been found, we can exit from Debug, load the macro assembler, make the necessary changes to our source code, assemble and link to produce a modified EXE program file. The corrected source code is shown in Figure 3.13.

Using Debug's line assembler

Debug has an in-built line assembler which can be used to generate simple programs. The assembler is accessible from within Debug (as described in the previous section), but can also be accessed by means of a *script file* that can be generated by any word processor or text editor capable of producing an ASCII text file (or even by means of the DOS COPY command).

```
AX=0500  BX=0000  CX=0000  DX=0000  SP=FFEE  BP=0000  SI=0000  DI=0000
DS=1662  ES=1662  SS=1662  CS=1662  IP=0102   NV UP EI PL NZ NA PO NC
1662:0102 B20A          MOV     DL,0A

AX=0500  BX=0000  CX=0000  DX=000A  SP=FFEE  BP=0000  SI=0000  DI=0000
DS=1662  ES=1662  SS=1662  CS=1662  IP=0104   NV UP EI PL NZ NA PO NC
1662:0104 CD21          INT     21


AX=050A  BX=0000  CX=0000  DX=000A  SP=FFEE  BP=0000  SI=0000  DI=0000
DS=1662  ES=1662  SS=1662  CS=1662  IP=0106   NV UP EI PL NZ NA PO NC
1662:0106 B20D          MOV     DL,0D

AX=050A  BX=0000  CX=0000  DX=000D  SP=FFEE  BP=0000  SI=0000  DI=0000
DS=1662  ES=1662  SS=1662  CS=1662  IP=0108   NV UP EI PL NZ NA PO NC
1662:0108 CD21          INT     21

AX=050D  BX=0000  CX=0000  DX=000D  SP=FFEE  BP=0000  SI=0000  DI=0000
DS=1662  ES=1662  SS=1662  CS=1662  IP=010A   NV UP EI PL NZ NA PO NC
1662:010A B241          MOV     DL,41

AX=050D  BX=0000  CX=0000  DX=0041  SP=FFEE  BP=0000  SI=0000  DI=0000
DS=1662  ES=1662  SS=1662  CS=1662  IP=010C   NV UP EI PL NZ NA PO NC
1662:010C B13E          MOV     CL,3E

AX=050D  BX=0000  CX=003E  DX=0041  SP=FFEE  BP=0000  SI=0000  DI=0000
DS=1662  ES=1662  SS=1662  CS=1662  IP=010E   NV UP EI PL NZ NA PO NC
1662:010E B405          MOV     AH,05

AX=050D  BX=0000  CX=003E  DX=0041  SP=FFEE  BP=0000  SI=0000  DI=0000
DS=1662  ES=1662  SS=1662  CS=1662  IP=0110   NV UP EI PL NZ NA PO NC
1662:0110 CD21          INT     21
A
AX=0541  BX=0000  CX=003E  DX=0041  SP=FFEE  BP=0000  SI=0000  DI=0000
DS=1662  ES=1662  SS=1662  CS=1662  IP=0112   NV UP EI PL NZ NA PO NC
1662:0112 FEC2          INC     DL

AX=0541  BX=0000  CX=003E  DX=0042  SP=FFEE  BP=0000  SI=0000  DI=0000
DS=1662  ES=1662  SS=1662  CS=1662  IP=0114   NV UP EI PL NZ NA PE NC
1662:0114 E2FA          LOOP    0110
BCDEFGHIJKLMNOPQRSTUVWXYZ[\]^_`abcdefghijklmnopqrstuvwxyz{|}~   <----
AX=057E  BX=0000  CX=0000  DX=007F  SP=FFEE  BP=0000  SI=0000  DI=0000
DS=1662  ES=1662  SS=1662  CS=1662  IP=0116   NV UP EI PL NZ NA PO NC
1662:0116 B000          MOV     AL,00

AX=0500  BX=0000  CX=0000  DX=007F  SP=FFEE  BP=0000  SI=0000  DI=0000
DS=1662  ES=1662  SS=1662  CS=1662  IP=0118   NV UP EI PL NZ NA PO NC
1662:0118 B44C          MOV     AH,4C

AX=4C00  BX=0000  CX=0000  DX=007F  SP=FFEE  BP=0000  SI=0000  DI=0000
DS=1662  ES=1662  SS=1662  CS=1662  IP=011A   NV UP EI PL NZ NA PO NC
1662:011A CD21          INT     21
```

Figure 3.12 *Program trace showing the correct execution of the printer test routine (note that the full range of characters is now printed)*

During execution, Debug will take its input (redirected from the keyboard) from the script file. The script file will contain a sequence of Debug commands (which can include assembly language statements).

The two examples which follow show how Debug's assembler can be used to generate programs to, respectively, perform a 'warm' and 'cold' reboot.

```
             ; Program name: ptest.com
             ;
             ; This program outputs a line of ASCII characters
             ; in the range 41H to 7FH to the parallel printer
             ; port.
             ;
             ; Registers used: AX, CL, DL
             ; Parameters passed: none
             ;
             TITLE ptest
_TEXT        SEGMENT          ; Define code segment
             ASSUME cs:_TEXT,ds:_TEXT,ss:_TEXT
             ORG  100H        ; Normal for a COM program
start:       MOV  AH,05H      ; Function code for printer output
             MOV  DL,0AH      ; First generate a
             INT  21H         ; line feed
             MOV  DL,0DH      ; Next generate a
             INT  21H         ; carriage return
             MOV  DL,41H      ; First character to print is 'A'
             MOV  CL,3EH      ; Number of characters to print
             MOV  AH,05H      ; Set up the function code
prch:        INT  21H         ; and print the character
             INC  DL          ; Get the next character
             LOOP prch        ; and go around again
             MOV  AL,00H      ; Set up the return code
             MOV  AH,4CH      ; and the function code
             INT  21H         ; for an exit to DOS
_TEXT        ENDS
             END  start
```

Figure 3.13 *Corrected source code for the printer test program (ptest.com)*

Warm reboot

The following script file can be used with Debug to generate a program (WARM.COM). This program directs the program counter to the start of ROM BIOS but avoids the power-on memory check routine.

Assuming that the script file is to be produced by means of the DOS COPY command, the following keyboard entries will be required:

```
COPY CON WARM.DBG
A
XOR AX,AX
MOV ES,AX
MOV DI,0472
MOV AX,1234
STOSW
JMP FFFF:0000

NWARM.COM
RCX
10
W
Q
^Z
```

It is important to note that a newline, i.e. <ENTER>, should be used to terminate each line and the input should be terminated (after the newline which follows "Q") by means of <CTRL-Z> (shown as ^Z). The <CTRL-Z> should also be followed by a newline.

The keystrokes will generate a file (WARM.DBG) which can be used as input to Debug by means of the following command:

```
DEBUG < WARM.DBG
```

Debug will assemble the statements contained in the script file in order to generate an executable file, WARM.COM. This program can be executed directly from the DOS prompt by typing WARM followed by enter (Note: this will reboot your system!).

Cold reboot

If a cold reboot is required, the assembly code should be modified by changing the MOV AX,1234 to MOV AX,0. The following keyboard entries are required:

```
COPY CON COLD.DBG
A
XOR AX,AX
MOV ES,AX
MOV DI,0472
MOV AX,0
STOSW
JMP FFFF:0000

NCOLD.COM
RCX
10
W
Q
^Z
```

Again, note that the input should be terminated (after the Enter that follows "Q") by means of <CTRL-Z> (shown as ^Z) which is also followed by Enter.

The keystrokes will generate a file (COLD.DBG) which can be used as input to Debug by means of the following command:

```
DEBUG < COLD.DBG
```

Debug will assemble the statements contained in the script file in order to generate an executable file, COLD.COM. This program can be executed directly from the DOS prompt by typing COLD followed by enter. This should again reboot your system but this time the initial memory check routines will be performed.

4 Programming

Whilst many users of PC-based instrumentation and control systems will be able to make use of off-the-shelf software packages, others may have specific applications for which there is no existing software package available. This is often the case with dedicated process control systems where a particular operational configuration is unique to the system concerned or where an existing software package is limited in some way.

The control engineer should be perfectly capable of developing simple, robust, and efficient control programs without the assistance of a programmer or software engineer. However, where the software is complex, sophisticated, or requires a high degree of optimization, then the services of a software engineer/programmer will almost certainly be required.

At the outset, it should be stated that there is a great deal more to programming than simply entering code. Programming benefits from a disciplined approach and this is absolutely essential when developing software which must operate reliably and be easy to maintain.

Experience shows that electronic engineers (particularly those involved with control systems) generally make excellent software engineers. They have usually developed a high degree of familiarity with hardware (microprocessors and support devices) and will be only too well aware of the characteristics and constraints of such devices.

Software engineering should not be confused with programming. A programmer is not necessarily a software engineer neither is a software engineer necessarily a programmer. In fairness, a software engineer will normally be proficient in several computer languages; however, such proficiency will be relatively unimportant if the software he/she produces behaves erratically or is impossible to maintain.

This chapter introduces some of the basic concepts associated with the production of structured code which is both predictable and reliable and is easy to maintain. This information should be invaluable to the control or test engineer who may be increasingly involved with the development of programs to control PC-based systems. Please note, however, that the code fragments used as illustrations in this chapter are *not* complete programs and most will require additional code (such as appropriate C++ pre-processor directives) before they can be made into complete course code files from which executable programs can be built.

Choice of language

Sooner or later, the software developer must make some decisions concerning the choice of language used for software development. To some extent this decision will be crucial to the success of a project. The essential features to consider when selecting a language for software development in PC-based

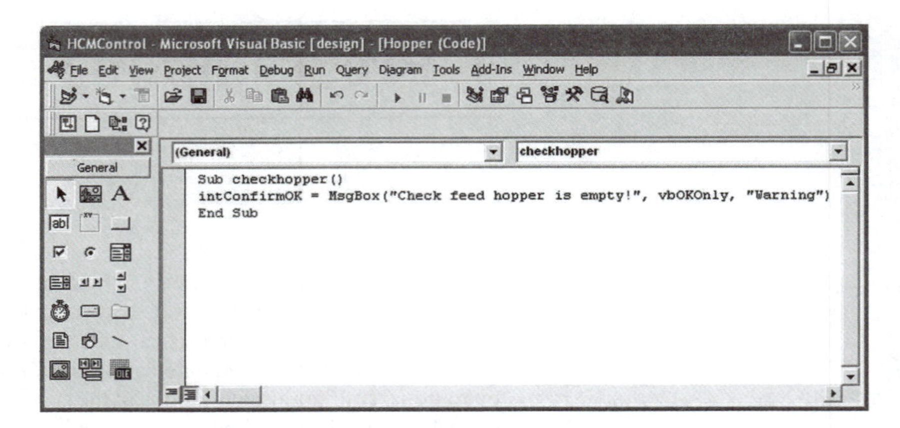

Figure 4.1 *It is possible to obtain the same results using different programming languages. This Visual Basic code produces the simple warning message shown in Figure 4.2*

Figure 4.2 *The warning message produced by the Visual Basic program shown in Figure 4.1*

instrumentation and control applications are as follows:

- *What control flow structures are provided to facilitate the development of structured code?*
  Such control structures may take several forms but should ideally include the ability to handle user-defined functions and procedures (with or without local variables) and such control structures as `IF ... THEN ... ELSE ... ENDIF`, `DO WHILE ... LOOP`, `SELECT ... CASE ... END SELECT`, and `WHILE ... WEND`.
- *What provision is there for handling I/O?*
  Most languages provide functions and statements (e.g. BASIC's `PEEK` and `POKE`) which facilitate direct access to memory. A language for PC-based instrumentation and control applications should have statements that allow reading from and writing to I/O port addresses. Taking BASIC as an example, functions such as:

  ```
  INP(port)
  ```

  and statements such as:

  ```
  OUT port, data
  ```

  make writing I/O routines extremely easy.
- *How easy is it to combine/interface modules written in the same or a different language?*
  A facility for combining/interfacing modules written in the same or a different language will be essential in any other than the simplest of applications (Figures 4.1 and 4.2). As an example, it may be convenient to develop an assembly language routine to handle some critical I/O process and then interface this to a high-level language program which deals with more mundane processes, such as keyboard input, display output, and disk filing. In such a case, it will generally be necessary to have some mechanism for

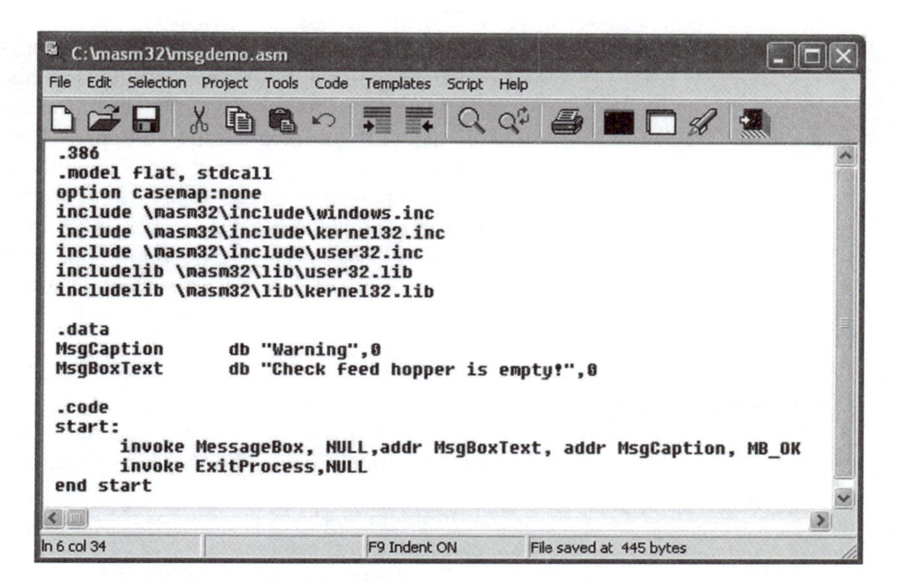

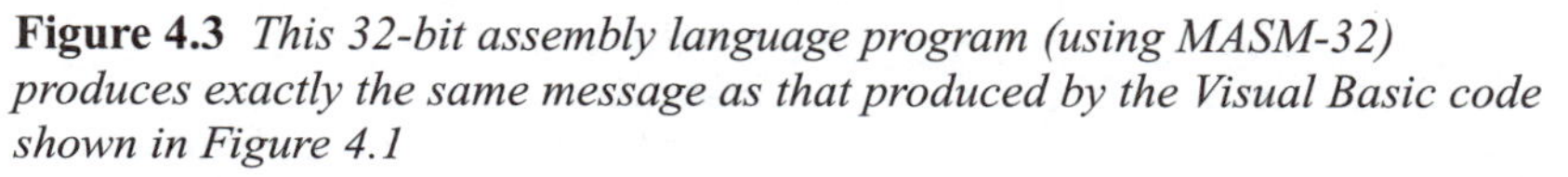

Figure 4.3 *This 32-bit assembly language program (using MASM-32) produces exactly the same message as that produced by the Visual Basic code shown in Figure 4.1*

Figure 4.4 *The warning message produced by the MASM-32 program shown in Figure 4.3*

passing parameters between the main program and the code generated by the assembly language module.

- *What, if any, provision is there for handling interrupts?*
  Some mechanism for allowing the user to incorporate his/her own interrupt handling routines will be essential in many real-time control applications.
- *What provision is there for event/error trapping?*
  The ability to include specific event/error trapping routines can be important in making the program robust and suitable for non-technical users. Error handling routines should permit meaningful error reporting as well as the ability to retain control of the program with an orderly shutdown when operation cannot continue.
- *Finally, will the language allow multi-tasking for use in event-driven processes?*
  In control applications, the ability to support multi-tasking is a highly desirable feature. In addition to the main process, the programmer will then be able to define one, or more, background tasks (sub-processes) to run concurrently with the main program. These tasks will be switched to repeatedly during program execution and thus effectively run in parallel with the main program.

Unfortunately, true multi-tasking can be a problem within an x86 DOS environment as the Real Mode provided by the x86 processor in the original PC, PC-XT, and PC-AT employed straightforward addressing with no inter-process protection. The limitation in available memory (640 KB under PC-DOS or MS-DOS) further mitigated against applications that were truly multi-tasking. Happily, with modern 32-bit operating systems and large memory environments this constraint no longer applies (Figure 4.3 and 4.4)!

If you can answer with an unqualified 'yes' to the majority of the foregoing questions, you can be assured that the language under consideration is an ideal candidate for software development in the control and instrumentation field. Coupled with an Integrated Development Environment (editor and debugger) it should be able to cope with almost anything!

With modern operating systems the *Protected Mode* environment provides an environment which can support true multi-tasking and this allows *event-driven programs* to be developed. Such programs allow a *main process* to exist along with a number of *sub-processes*, each of which shares some of the processor's time. We shall return to this important theme a little later in this chapter.

Software development

Software development should normally be a top-down process in which one moves from the general to the specific. The process can be divided into a number of identifiable phases which generally include:

1 Problem *analysis*, leading to
2 a *software specification*.
3 Development of an *algorithm* and
4 a *program definition*.
5 *Coding* and
6 *testing* (against the specification) and
7 *debugging*.
8 *Implementation* and
9 *evaluation*.
10 Finally, there will be a need for *ongoing maintenance*.

In practice, Steps (5), (6), and (7) will invariably be repeated a number of times in order to refine the software and eliminate errors made during the coding phase. At this stage, it is perhaps worth examining each of the phases in the *software development cycle* in a little more detail.

The first two stages (problem analysis and the production of a software specification) involve first determining the user's requirements, and then itemizing the functions and facilities expected of the software. The specification should, of course, be agreed with the user. Furthermore the initial stages will normally require a dialogue with the user in order to establish the parameters within which the system should operate. Very few users are able to give a precise definition of their requirements and, since it is important to consider all eventualities, it is important to explore with the user what should happen in abnormal circumstances as well as in routine situations.

As an example, consider the case of the operator of an aggregate processing plant that comprises several conveyor belts, processing drums, and a washing plant. The problem essentially involves delivering various grades of aggregate at rates which are sufficient to ensure that the capacity of the stockpile is not exceeded and that a certain minimum amount of each grade of material is always available. The software specification (agreed with the operator) will involve delivery rates and volumes. However, the plant operator may forget to mention that, in the event of an interruption of the water supply, part of the plant must shutdown with a consequent and drastic change in delivery rates.

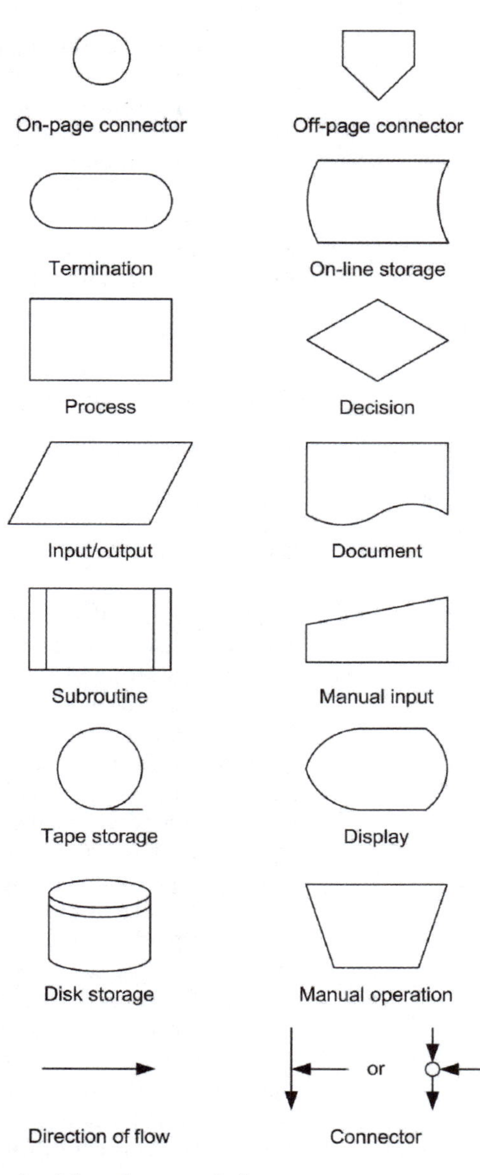

Figure 4.5 *Standard flowchart symbols*

In extreme cases, a problem of this type may only come to light when the system is commissioned. Clearly, this would not have happened if the initial stages of the development model had been rigorously followed.

Steps (3) and (4) can be considered to be the 'design' phases. The first of these (development of an algorithm) involves conceptualizing the means of solving the problem. This is often done with the aid of a flowchart or a data flow diagram and usually involves breaking down the problem into a number of smaller steps (processes). Figure 4.5 shows the set of standard symbols that are commonly used in flowcharts.

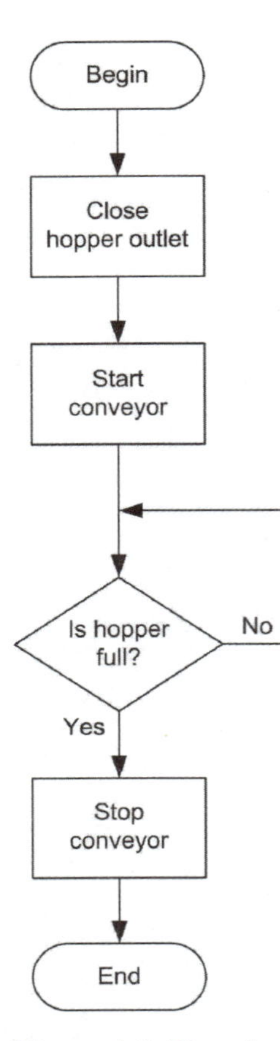

Figure 4.6 *Flowchart for the hopper filling process*

The second of the design phases involves defining the various program modules and procedures. These will often be associated with the individual stages of the flowchart model (or its equivalent) and may be separately documented. The action of each module can be summarized using structured English (or *pseudo code*). Each line of *pseudo code* will generally correspond to one, or more, lines of program code.

As an example, consider the case of a process employed within a grain drying plant which is responsible for filling a hopper from a conveyor. In structured English (*pseudo code*) the process can be summarized along the following lines:

```
Begin
Close hopper outlet
Start conveyor
While hopper not full
   Run conveyor
EndWhile
Stop conveyor
End
```

The equivalent flowchart for the hopper filling process is shown in Figure 4.6 (note the use of a conditional loop).

Steps (5), (6), and (7) of the software development cycle involve routine program entry, testing, and debugging. All but the simplest of programs should be developed on a modular basis making it possible to work on a single module (procedure) at a time. Modules can also be drawn from a standard library whenever one is available. Furthermore, whenever a module has been successfully developed and tested, it should be added to the appropriate library so that it is available for future use in other programs. A routine which will read a remote keypad and return its status to the system might, for example, be useful in a variety of applications.

Having produced a functional control program, the next stage is implementation. Since the software will almost certainly have been developed within a controlled environment removed from the environment in which it is to be finally imbedded, it will generally be necessary to install the software and carry out some rigorous testing with real (rather than simulated) inputs and/or outputs. This is often the most critical phase in the entire project cycle and it will sometimes reveal problems which were not foreseen during the earlier stages. Problems and difficulties are often associated with:

- *Speed of response*: the real-world system may be too fast or too slow in comparison with that of the simulated development environment.
- *Noise*: signals in the real-world environment are rarely ideal and often contain a significant amount of noise.

As an example, a system installed to monitor the flow of gas along a pipeline behaved erratically when an apparently functional (and fully debugged) program was installed within its industrial PC-based controller. Sixteen remote sensors (based on rotating vanes) were used to sample the flow rate at various points. Each sensor was connected, via an asynchronous serial data link, to the controller. Under certain conditions, the PC indicated that the flow rates were well outside the prescribed limits for the system. However, upon examination it was found that, not only did the sensors exhibit a reluctance to respond to very low flow rates but the signals from the furthermost sensors were regularly erroneous due to power-line induced switching transients and lack of RS-232 parity checking.

The moral, of course, is that one should attempt to anticipate problems at the earliest stages of hardware/software development. By planning for the unforeseen, it is possible to minimize the time taken to imbed the software into the target system to a bare minimum, reducing both costs and inconvenience.

Finally, it will usually be necessary to evaluate the performance of the system against the original specification. Such an evaluation will generally involve both qualitative and quantitative aspects. The qualitative evaluation will involve questions such as 'Does the user feel at ease with the system?' and 'Are the displays and prompts meaningful?' while the quantitative evaluation will be concerned with collecting data on response times, accuracy, repeatability, etc.

Control structures

In anything other than the simplest of applications, programs will involve some deviation from a straightforward linear sequence of processes. There may, for example, be a need for conditional forward branching (bypassing a particular process) depending upon some particular outcome, or for a certain process to be repeated a number of times until a particular result is obtained.

Several common control structures (available within the majority of today's programming languages) are illustrated in Figure 4.7. First, these (Figure 4.7(a)) involve a simple branch forwards depending on the outcome of the conditional test. A typical example of this control structure, expressed in pseudo code, is:

```
If tank empty
   Open valve
   Operate pump
EndIf
```

It should be noted that, if the test evaluates to 'false' (i.e. if the valve is open) none of the indented statements will be executed. Furthermore, the condition may take the form of a compound statement, as in the following example:

```
If temperature high and coolant off
   Display warning message
   Turn heat off
EndIf
```

The indented statements will only be executed if both the conditions evaluate true. If either condition is not satisfied (i.e. one or other evaluates false), the indented statements will not be executed.

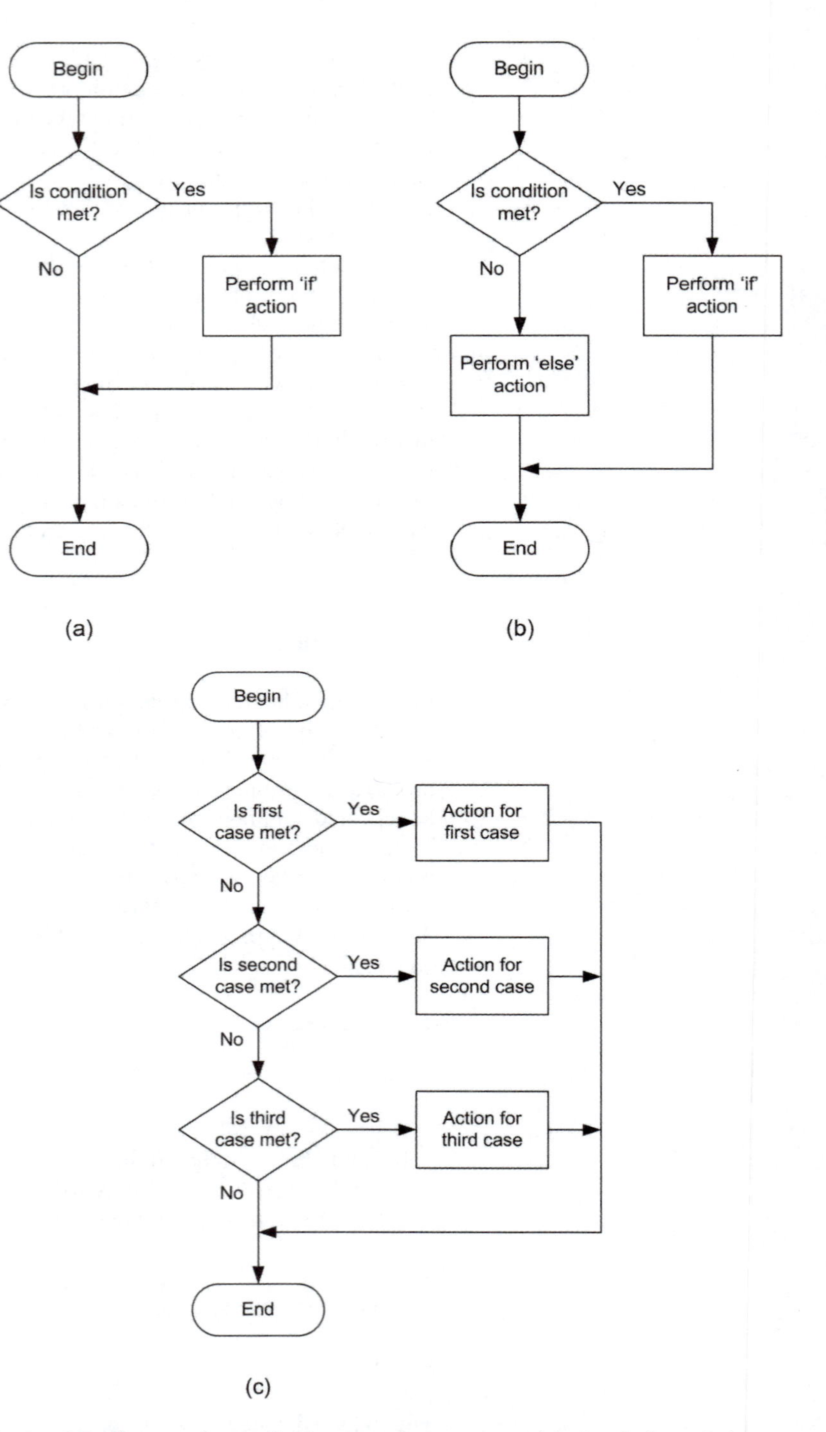

Figure 4.7 *Commonly available control structures: (a) simple branch* (`If ... EndIf`)*; (b) binary branch* (`If ... Else ... EndIf`)*; (c) multiple branch* (`Select ... Case ... EndSelect`)

A succession of `If ... EndIf` statements may be used where a number of outcomes need to be tested for. As an example, the following pseudo code describes part of a process employed in a flow soldering plant:

```
If temperature < 230°C
  Stop conveyor
  Turn on heater
EndIf

If temperature > 230°C and temperature < 270°C
  Start conveyor
  Turn on heater
EndIf

If temperature > 270°C
  Display warning message
  Stop conveyor
  Turn off heater
EndIf
```

Figure 4.7(b) shows a control structure which may be adopted where two outcomes are required. The pseudo code equivalent of this is known as `If ... Else ... EndIf`. A typical example of the use of this control structure is found in the following pseudo code:

```
If light level low
  Lights on
Else
  Lights off
EndIf
```

A further control structure provides for multiple branching (rather than binary branching, as in the case of `If ... Else ... EndIf`). This structure is illustrated in the flowchart of Figure 4.7(c) and a typical application might be in the selection of a main menu option, as described by the following representative pseudo code:

```
Select Case
   1, Input new data
   2, Get old data
   3, Sort data
   4, Print data
   5, Exit
Else warn user
EndSelect
```

This (apparently complex) pseudo code can quite easily be implemented in both BASIC and C. A typical BASIC routine to satisfy the pseudo code would take the form:

```
SELECT CASE R$
   CASE "1"
      CALL NewData
   CASE "2"
      CALL OldData
```

```
   CASE "3"
      CALL SortData
   CASE "4"
      CALL PrintData
   CASE "5"
      CALL UpdateFile
   CASE "6"
      EXIT DO
   CASE ELSE
      BEEP
   PRINT "Input not valid!"
END SELECT
```

while its equivalent in C would be:

```
switch(c)
{
   case '1':
      newdata();
      break;
   case '2':
      olddata();
      break;
   case '3':
      sortdata()
      break;
   case '4':
      printdata()
      break;
   case '5':
      exit ()
   default
      beep()
      printf("Input not valid!\n")
}
```

Loops

A loop structure (backwards branch) may be used in order to avoid the need to repeat blocks of code several times over whenever a process is to be repeated more than once. Various types of loops are possible (both conditional and unconditional) and these are supported by pseudo code statements such as `Do ... Loop While`, `Do ... Loop Until`, `Do While ... Loop`, and `Do Until ... Loop`.

As an example of a simple loop, the following C++ code fragment prints the numbers 1 to 10 separated by spaces:

```
for (count = 1; count <= 10; count++)
   cout << cout << "";
```

This routine uses the C++ increment operator, `count++`, however we could have obtained the same result using:

```
for (count = 1; count <= 10; count = count + 1)
   cout << cout << "";
```

The fragment of C++ code that follows is an example of the `Do ... Loop While` logical construct:

```
do
{
  print_label(code);
  cout << "Label printed\n";
  cout << "Enter Y to print another label, any other key to
      exit: ";
  cin = response;
} while(response = = 'y' || response = = 'Y')
cout << "Shutting down ... please wait!\n";
```

Loop structures are explained in detail in Chapter 6.

Error checking and input validation

Error checks and input validation routines should be incorporated whenever data is input and before the system accepts the data for processing. Error handling routines should be incorporated to warn the user that a fault has occurred and indicate from which source the error has arisen. This caveat also applies to operator input; an unacceptable input should be echoed to the user together with the range of acceptable responses. Care should be exercised when inputs are defaulted. The default response should result in inactivity rather than any form of positive action on the part of the system. Furthermore, the program should demand confirmation where a response or input condition will produce an irreversible outcome.

Event-driven programs

With an event-driven program the processor must share its time with the main process and any sub-processes responding, for example, to a user clicking on a button, an input from a sensor, or a timer signalling the end of a time period. Since these events may occur at any time and in any sequence this requires a somewhat different approach than that which would be appropriate for a strictly sequential process. The important thing to remember is that events can take different forms and they can occur virtually at any time. This makes it difficult, or even impossible, to describe the program by means of a conventional flowchart; the program must be able to make an effective response to an event *whenever* it occurs.

As an example of an event-driven process consider the work of a receptionist in a small but busy office. The main process associated with the job can simply be described as 'receptionist'. The sub-processes might then be 'telephone answering', 'greeting visitors', 'dealing with general enquiries', 'opening and sorting incoming mail', 'preparing outgoing mail', and so on. Within the main process events can occur in any order. For example, the telephone might ring, a visitor might arrive, the mail might be delivered, and so on. It might be difficult, or even impossible, to say when these events will occur in any particular working day. However, they *will* occur and each of the sub-processes must be handled correctly for the main process to be satisfied.

Testing

It is only possible to claim that a program has been validated after exhaustive testing in conjunction with the target hardware system. In many cases it may be possible to test individual code modules before they are linked into the final executable program. This may be instrumental in reducing debugging time at a later stage.

Testing the completed program requires simulating all conditions that can possibly arise and measuring the outcome in terms of the program's response. A common error is that of only presenting the system with a normal range of inputs. Comprehensive testing should also involve the simulation of each of the following:

- Unexpected or nonsensical responses from the operator or user.
- Failure of hardware components (including transducers, signal conditioning hoards, cables, and connectors).
- Out-of-tolerance supplies (including complete power failure).
- Noise and the effect of RF interference (RFI).
- Environmental changes (temperature, humidity, etc.).

Documentation

Programmers are usually woefully lacking where program documentation is concerned. Documentation, which is essential to make the program understandable, takes various forms, the most obvious of which is the comments included in the lines of source code text.

Comments

Comments should explain the action of the source code within the program as a whole and, since the function of the operation code and operand will usually be obvious (or can be found by referring to the instruction set) there is no point in expanding on it. Comments should be reasonably brief but not so brief that they become cryptic. Also, there is no need to attempt to confine a comment to a single statement line. Comments can be quite effective if they read clearly and are continued over several statements to which they refer.

Headers

Headers are extended comments which are included at the start of a program module, macro definition, or subroutine. Headers should include all relevant information concerning the section of code in question and should follow a standard format (see Figure 4.8). As a minimum, the following should be included:

- Name and purpose of the module or subroutine.
- Brief explanation of the action of the code (in terms of parameters passed, registers involved, etc.).

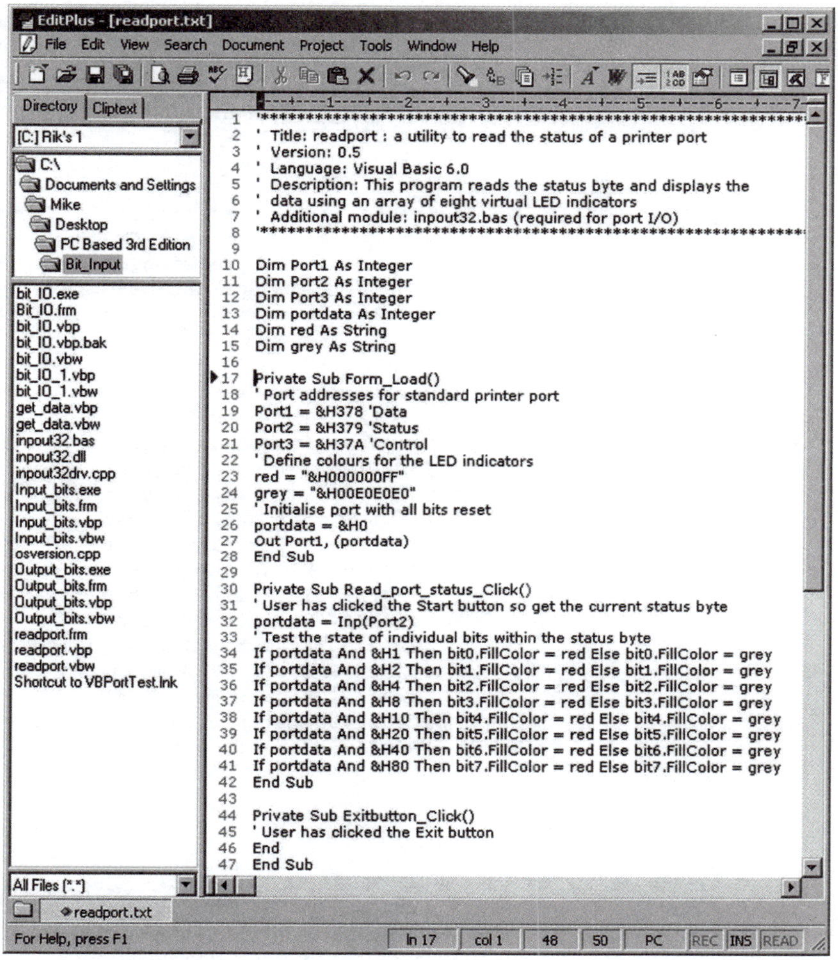
EditPlus - [readport.txt]
File Edit View Search Document Project Tools Window Help
Directory
Cliptext
[C:] Rik's 1
C:\
Documents and Settings
Mike
Desktop
PC Based 3rd Edition
Bit_Input
bit_IO.exe
Bit_IO.frm
bit_IO.vbp
bit_IO.vbp.bak
bit_IO.vbw
bit_IO_1.vbp
bit_IO_1.vbw
get_data.vbp
get_data.vbw
inpout32.bas
inpout32.dll
inpout32drv.cpp
Input_bits.exe
Input_bits.frm
Input_bits.vbp
Input_bits.vbw
osversion.cpp
Output_bits.exe
Output_bits.frm
Output_bits.vbp
Output_bits.vbw
readport.frm
readport.vbp
readport.vbw
Shortcut to VBPortTest.lnk
All Files (*.*)
readport.txt
For Help, press F1
ln 17
col 1
48
50
PC
REC
INS
READ
'***
' Title: readport : a utility to read the status of a printer port
' Version: 0.5
' Language: Visual Basic 6.0
' Description: This program reads the status byte and displays the
' data using an array of eight virtual LED indicators
' Additional module: inpout32.bas (required for port I/O)
'***
Dim Port1 As Integer
Dim Port2 As Integer
Dim Port3 As Integer
Dim portdata As Integer
Dim red As String
Dim grey As String
Private Sub Form_Load()
' Port addresses for standard printer port
Port1 = &H378 'Data
Port2 = &H379 'Status
Port3 = &H37A 'Control
' Define colours for the LED indicators
red = "&H000000FF"
grey = "&H00E0E0E0"
' Initialise port with all bits reset
portdata = &H0
Out Port1, (portdata)
End Sub
Private Sub Read_port_status_Click()
' User has clicked the Start button so get the current status byte
portdata = Inp(Port2)
' Test the state of individual bits within the status byte
If portdata And &H1 Then bit0.FillColor = red Else bit0.FillColor = grey
If portdata And &H2 Then bit1.FillColor = red Else bit1.FillColor = grey
If portdata And &H4 Then bit2.FillColor = red Else bit2.FillColor = grey
If portdata And &H8 Then bit3.FillColor = red Else bit3.FillColor = grey
If portdata And &H10 Then bit4.FillColor = red Else bit4.FillColor = grey
If portdata And &H20 Then bit5.FillColor = red Else bit5.FillColor = grey
If portdata And &H40 Then bit6.FillColor = red Else bit6.FillColor = grey
If portdata And &H80 Then bit7.FillColor = red Else bit7.FillColor = grey
End Sub
Private Sub Exitbutton_Click()
' User has clicked the Exit button
End
End Sub

Figure 4.8 *Example of commented source code showing program header information. The code is written in Visual Basic and is being edited using the EditPlus editor*

- Names of other modules, subroutines, or macros on which the module depends and, where applicable, names of relevant macro libraries in which definitions are held.
- Entry requirements (in terms of register and/or buffer contents before the module is executed).
- Exit conditions (in terms of register contents, buffers, and flags after the module has been executed).
- In the case of assembly language modules, a list of registers used during execution of the code (which may have their contents changed as a consequence).

When producing a program header, it is wise to include any information which may be required by another programmer who may subsequently need to debug or modify the code. Nothing should be taken for granted and all loose ends should be explained!

During the development phase, it is worth including a brief development history within the main program header, as shown in the following example:

```
' *******************************************
' *           Program: DSM3.BAS             *
' *              Version: 0.1               *
' *          Copyright USET 1999            *
' *******************************************
'
'
' Development history
' 12/01/99 Creation date
' 13/01/99 Structure defined
' 14/01/99 New sub-programs added
' 16/01/99 64K block save and load added
' 18/01/99 Viw sample added
' 20/01/99 Block size increased to 256K
' 28/01/99 Mixed language interface added
' 28/01/99 Interrupt enable/disable added
' 29/01/99 Various flags added
' 29/01/99 Assembly language modules added
' 04/02/99 Block size modified to 128K
' 05/02/99 View sample removed
' 10/02/99 Multiple blocks added
' 15/02/99 Save data file added
```

Names

Names used for variables, symbols, and labels should be meaningful and any abbreviations used should be as obvious as possible. In the case of the names used for constants, where standard abbreviations are in common use (e.g. ESC for Escape), they should be adopted. In a large program, there may be a large number of labels and/or constants and it will be necessary to distinguish between them.

As an example of the use of names, comments, and headers consider the following examples which, while functionally identical, illustrate the extremes of programming style:

Case converter subroutine

```
;
;       CASE CONVERTER
;
con:            CMP  AL,61H      ; Compare A with 61H
                JE   exit        ; Return if carry set
                CMP  AL,7BH      ; Compare A with7 BH
                JNE  exit        ; Return if carry reset
                SUB  AL,20H      ; Subtract 20H from A
exit:           RET              ; Return
```

Improved case converter subroutine (self-documenting)

```
; LOWER TO UPPER CASE CHARACTER CONVERSION
; PARAMETERS PASSED:
; ENTRY:      AL=ASCII character (upper or lower case)
; EXIT:       AL=ASCII character (upper case only)
; REGISTERS:        AL, F
;
upcase:       CMP   AL,'a'       ;     Is it already upper case?
              JL    exit         ;     If so, do nothing
              CMP   AL, 'z'      ;     Or is it punctuation?
              JG    exit         ;     If so, do nothing
              SUB   AL, 'a'-'A'  ;     Otherwise, change case
exit:         RET
```

The second example shows how a program module can be made largely self-documenting by the inclusion of effective comments and a meaningful header. Note that the name of the routine has been changed so that it is easier to remember and is less likely to be confused with others. Finally, the code itself has been modified so that its action is much easier to understand.

Documentation is particularly important where software development is being carried out by several members of a team. Each development phase will rely on the documentation prepared in earlier stages, hence documentation should be considered an ongoing task and a folder should be prepared to contain the following items:

- A detailed program specification (including any notes relevant to the particular hardware configuration required).
- Flowcharts or descriptions of the program written in structured English.
- Lists of all definitions and variable names.
- Details of macro or sub-routine libraries used.
- Details of memory usage (where appropriate).
- A fully commented listing of the program (latest version).
- A diary giving the dates at which noteworthy modifications are made together with details of the changes incorporated and the name of the programmer responsible.
- A test specification for the program with descriptions and results of diagnostic checks performed.

Presentation

Finally, attention should be given to the way in which the program interacts with the user and the aim should be that of making the software as 'user-friendly' as possible. Prompts and messages should always be meaningful and, whenever any doubt may exist, the user should be prompted with the range of acceptable values or valid responses and *help screens* should be provided wherever appropriate. Later chapters provide further examples of programs presented in different formats.

5 Assembly language programming

This chapter aims to provide readers with an overview of assembly language programming techniques, and explores the architecture and instruction set of the x86 microprocessor family used in the PC and compatible equipment. Rather than providing a complete guide to assembly language programming (which, in any event, would require a complete book in its own right!), the aim has been that of providing readers with sufficient information to decide whether assembly language is appropriate for a particular application, to outline the advantages and disadvantages of assembly language programming, and to introduce techniques used for the development of assembly language programs.

Readers wishing to develop their own assembly language programs will not only require complete documentation for the x86 family of processors (including a comprehensive explanation of the microprocessor's instruction set) but will also require development software comprising, as minimum, a macro assembler, a linker, and a debugger. Furthermore, despite the fact that one of the most powerful 32-bit assemblers, MASM32, is currently available as 'freeware'; readers should not underestimate the investment required (in terms of time) required to successfully follow this route.

Advantages of assembly language

Assembly language programs offer a number of advantages when compared with higher-level alternatives. The principal advantages are that the executable code produced by an assembler (and linker) will:

- invariably be more compact than an equivalent program written in a higher-level language;
- invariably run faster than an equivalent program written in a higher-level language;
- not require the services of a resident interpreter or a compiler run-time system;
- be able to offer the programmer unprecedented control over the hardware in the system.

It is this last advantage, in particular, that makes assembly language a prime contender for use in control applications. No other programming language can hope to compete with assembly language where control of hardware is concerned. Indeed, an important requirement of high-level languages used in control applications is that they can be interfaced with machine code modules designed to cope with problems arising from limitations of the language where input/output (I/O) control is concerned.

Disadvantages of assembly language

Unfortunately, when compared with higher-level languages, assembly language has a number of drawbacks; most notable of which are the following:

- Programs require considerably more development time (including writing, assembling, linking or loading, and debugging) than their equivalent written in a high-level language.
- Programs are not readily transportable between microprocessors from different families. Different microprocessors have different internal architectures and, in particular, the provision of registers accessible to the programmer will vary from one microprocessor to another. Differences in internal architecture is reflected in corresponding differences in the type and function of the software instructions provided for the programmer.
- The situation is further compounded by the fact that microprocessor manufacturers frequently adopt different terminology to refer to the same thing. The variety of names used to describe the register used to indicate the outcome of the last arithmetic logic unit (ALU) operation (and the internal status of the microprocessor) is a case in point. This is variously referred to as a Flag Register, Status Register, Condition Code Register, and Processor Status Word.
- In practice this means that the system designer is constrained to select one particular microprocessor type or family, and develop code exclusively for this particular device. This, of course, is not a particular problem in the case of the PC and compatible equipment which are all based on the standard x86 and Pentium families.
- Unless liberally commented, the action of an assembly language program is not obvious from merely reading the source text. Programs written in high-level language are usually easy to comprehend and their structure is usually self-evident.
- The production of efficient assembly language programs requires a relatively high degree of proficiency on the part of the programmer. Such expertise can usually only be acquired as a result of practical experience aided by appropriate training.

Developing assembly language programs

The process of developing an assembly language program depends on a number of factors including the hardware configuration available for software development and the range of software tools available to the developer. As a minimum, the task normally involves the following steps:

1. Analysing the problem and producing a specification for both hardware and software (see Chapter 4).
2. Developing the overall structure of the program, defining the individual elements and modules within it, and identifying those which already exist (or can be easily modified or extended) within the programmer's existing library.
3. Coding each new module required using assembly language mnemonics, entering the text using an editor, and saving each source code module to disk using an appropriate filename.
4. Assembling each source code module (using an assembler) to produce an intermediate relocatable object code file.

5 Linking modules (including those taken from the user's library) in order to produce a complete executable program.
6 Testing, debugging, and documenting the final program prior to evaluation and/or acceptance testing by the end-user (see Chapter 4).

In practice, the development process is largely iterative and there may also be some considerable overlap between phases. In order to ensure that the target specification is met within the constraints of time and budget, an ongoing appraisal is necessary in order to maximize resources in the areas for which there is much need.

Software tools

The following items of utility software (software tools) are normally required in the development process:

- an ASCII text editor (e.g. Microsoft's M);
- a macro assembler (e.g. Microsoft's MASM);
- a linker (e.g. Microsoft's LINK).

In addition, three further software tools may be found to be invaluable. These are:

- a cross-referencing utility (e.g. Microsoft's CREF);
- a library manager (e.g. Microsoft's LIB);
- a utility which can help automate the program development cycle (e.g. Microsoft's MAKE).

Note that, in order to assist the programmer and to help automate the production of executable code, an *Integrated Development Environment* (IDE) is often used. This acts as a 'shell' which launches the various software tools, passing any required parameters without requiring the user to be aware of the necessary command syntax. However, for the benefit of the newcomer to assembly language programming, we shall briefly explain the function of each of the basic tools and their role in the production of assembly language programs.

Editors

Editors allow users to create and manipulate text files. Such files can be thought of as a sequence of keystrokes saved to disk. An assembly language *source code* file is simply a text file written using assembly language mnemonics and containing appropriate *assembler directives.*

The *Microsoft Editor* (M) is invoked using a command line of the form:

```
M <options><file list>
```

The options include that of allowing the user to load a previously saved configuration file (TOOLS.INI). This file contains settings which will be used to initialize the editor and thus the user may easily customize the software to his/her own particular requirements. The file list is simply a list of files that will be loaded into the editor. The first file in the list will be the first to be edited. Then, when the user selects the exit option (F8), the next file in the list, ready for editing, is loaded.

The Microsoft editor is extremely powerful. It provides the usual *cut and paste*, and *search and replace* facilities together with macros which can be invoked from a single keystroke. Furthermore, to reduce the overall edit/assemble cycle time, it is possible to assemble a program from within the editor, view, and correct any errors that may have occurred, then re-assemble. Multiple source files can easily be handled and a split-screen windowing facility can be used to examine and edit different parts of the same file simultaneously.

When preparing source text using an editor, it is important to bear in mind the requirements of the assembler concerning the format of source code statements. In the case of most x86 assemblers (and Microsoft's MASM in particular), each line of source code is divided into four fields, as shown in the example below:

```
Symbol     Operation  Argument   Comments
maxcount   DB         16         ; initialize maximum count
```

The first entry in the line of code is known as a *symbol.* The symbols used in a program are subject to certain constraints imposed by the assembler but are chosen by the individual programmer. *Labels* are a particular form of symbol which are referred to by one, or more, statements within a program. Labels are used to mark the entry point to the start of a particular section of code or the point at which a branch or loop is to be directed. During the assembly process, labels (wherever they appear in the program) are replaced by addresses.

Entries in the *operation field* may comprise an *operation code (opcode)*, a *pseudo-operation code (pseudo-op)*, an *expression*, or the name of a *macro.* Operation codes are those recognized by the microprocessor as part of its instruction set (e.g. `MOV, ADD, JMP`, etc.) whereas pseudo-ops are *directives* which are recognized by the assembler and are used to control some aspect of the assembly process. Typical pseudo-ops are `DB` (define byte), `DW` (define word), `ORG` (origin or program start address), and `INCLUDE`. The last-named directive instructs the assembler to search a named macro library file and to expand macro definitions using this library.

The *argument field* may contain constants or expressions, such as `0DH, 42, 64*32, 512/16, 'A', 'z'-'A'`, or the operands required by microprocessor operation codes (represented by numbers, characters, and symbols, which are extended opcodes).

The *comment field* contains a line of text, added by the programmer, which is designed to clarify the action of the statement within the program as a whole (see Chapter 4).

In the example shown previously, the variable *maxcount* has been declared in the symbol column. The operation field contains a pseudo-op *(assembler directive)* which instructs the assembler to reserve a byte of storage and initialize its value to 16. Thereafter, any references to *maxcount* will take the value 16 (at least until the value is next modified by the program). The programmer has added a comment (following the obligatory semicolon) which reminds him or her that *maxcount* is the symbol used to hold the current maximum value of the counter.

Not all source code lines involve entries in all four fields, as in the next example:

```
Symbol  Operation  Argument      Comments
        Mov        AL,maxcount   ; get maximum count
```

Here, the symbol field is blank since the instruction does not form part of the start of a block of code. The operation, MOV, is an opcode that instructs the microprocessor to perform an operation which will move data from one location to another. The operand required by the instruction specifies the lower half of the 16-bit accumulator (AL) as the destination for the data and the contents of the variable storage location *maxcount* as the source of the data. The programmer has again added a brief comment to clarify the action of the line.

It should be noted that any line of source code starting with a semicolon is ignored by the assembler and treated as a comment. This allows the programmer to include longer comments as well as program or module headers which provide lengthy information on the action of the statements that follow. Furthermore, since the four fields in each source code statement are each separated by *white space* it is not essential that they are precisely aligned in columns. However, whereas the following lines of source text are perfectly legal:

```
start: MOV AH,05H ; Printer output function code
MOV DL,OAH ; First generate
INT 21H ; a line feed
MOV DL,00H ; Next generate
INT 21H ; a carriage return
MOV DL,41H ; First character to print is A
MOV CL,3EH ; Number of characters to print
MOV AH,05H ; Set up the function code
prch: INT 21H ; and print the character
INC DL ; Get the next character
LOOP prch ; and go round again
MOV AL,00H ; Set up the return code
MOV AH,4CH ; and the function code
INT 21H ; for an exit to DOS
```

they would be much more readable had they been entered in the strict format shown below:

```
start: MOV    AH,05H  ; Printer output function code
       NOV    DL,OAH  ; First generate a
       INT    21H     ; line feed
       MOV    DL,00H  ; Next generate a
       INT    21H     ; carriage return
       MOV    DL,41H  ; First character to print
       MOV    CL,3EH  ; Number of characters to print
       NOV    AH,05H  ; Set up the function code
prch:  INT    21H     ; and print the character
       INC    DL      ; Get the next character
       LOOP   prch    ; and go round again
       MOV    AL,00H  ; Set up the return code
       MOV    AH,4CH  ; and the function code
       INT    21H     ; for an exit to DOS
```

Macro assemblers

A macro facility allows the programmer to write blocks of often-used code and incorporate these in programs by referring to them by name. The blocks of code are each defined as a *macro.* Thereafter, the macro assembler expands the macro call by automatically assembling the block of instructions that it represents into the program. The macro call can also be used to pass parameters (e.g. symbols, constants, or registers) to the assembler for use during the macro expansion.

As an example of the use of macros, the macro defined in the following code can be used to exchange the contents of two registers passed to the macro as parameters `reg1` and `reg2`:

```
;       MACRO TO EXCHANGE 16-BIT REGISTER CONTENTS
;       PARAMETERS PASSED:     reg1, reg2
;       REGISTERS AFFECTED:    reg1, reg2

Swap    MACRO reg1,reg2   ;   Specify registers to swap
        PUSH      reg1    ;   Stack contents of reg1 first
        PUSH      reg2    ;   then stack contents of reg2
        POP       reg1    ;   reg1 receives reg2 contents
        POP       reg2    ;   reg2 receives reg1 contents
        ENDM
```

The following line of code shows how the macro call is made:

```
Swap     AX,CX    ;   Call the macro
```

The macro assembler expands the call, replacing it with the code given in its definition. The code generated by the macro assembler (i.e. the *macro expansion*) will thus be:

```
PUSH AX
PUSH CX
POP AX
POP CX
```

A macro facility can be instrumental in making significant reductions in the size of source code modules. Furthermore, macros can be nested such that a macro definition can itself contain references to other macros which, in turn, can contain references to others. A notable disadvantage of using macros is that the resulting object code may contain a large number of identical sections of code and will also occupy more memory space than if an equivalent subroutine had been used. In practice, therefore, programmers should use macros with care since there may be occasions where subroutines would be more efficient even though they may not be quite so easy to implement.

As well as macros, most assemblers also support *conditional assembly*. This allows the programmer to specify conditions under which portions of the program are either assembled or not assembled. Conditional assembly allows the programmer to test for specific conditions (using statements such as `IF ... ELSE ... ENDIF`) and use the outcome to control the assembly process.

Assemblers generally make two passes through a source file. During the first pass, *macro calls are expanded* and a *symbol table* is generated. On the second pass, *relocatable code is generated* which can be saved in a disk file. Such files are, however, not directly executable and require the services of a linker in order to function as self-contained programs.

The Microsoft macro assembler (MASM) provides logical programming syntax which supports the segmented architecture of x86 microprocessors. The assembler produces relocatable *object modules* which are linked together using the Microsoft overlay linker, LINK.

MASM is invoked by a command line of the form:

```
M <options><file list>
```

The options, which may be selected, include the generation of additional statistics, error information, and data which may be used by the Microsoft's CodeView debugger. The file list must contain the name of the assembly language source code file. This filename may be followed by the names of the object code file, the listing file, and the cross-reference file. Where these last three named files are not specified, MASM will prompt for them.

The default file extension for the object code filename is OBJ, while those for the source listing and cross-reference files are LST and CRF, respectively.

Linkers

The linker is used to combine one, or more, object code files into a single executable program file. The output file produced by the linker is not bound by specific memory addresses (i.e. it is *relocatable*), and the operating system is able to load and execute the file at any convenient address.

The linker must resolve address references between modules such that any module which directs program execution outside itself (by means of a CALL, an external symbol, or an *include directive*) will be linked to the module which contains the corresponding code.

The Microsoft linker (LINK) is invoked by a command of the form:

```
LINK <options><file list>
```

The options which may be selected include the display of linker process information, the packing of executable files, and the listing of *public symbols*. The file list must contain the name of each of the object code files to be linked. These may be followed by the names of the executable program file, the map file, and the names of the library files. Where these last three named files are not specified, LINK will prompt for them.

The default file extension for the executable program file is EXE whilst those for the map and library files are MPA and LIB, respectively.

Cross-reference utilities

Cross-reference utilities can be invaluable when debugging since they can greatly speed up the search for symbols within a source code file during a debugging session. A cross-reference utility can be used to produce a specially created listing of all of the symbols used in an assembly language program. The listing is invariably alphabetical and each symbol in the list is followed by one, or more, line numbers that indicate the lines in the source code file which contain a reference to the symbol.

The Microsoft cross-reference utility (CREF) is invoked by a command line of the form:

```
CREF <file list>
```

where the file list consists of the name of the cross-reference file generated by MASM followed by the name of the readable (ASCII format) cross-reference file.

Library managers

A library manager allows the programmer to gather a number of object code files (i.e. those with an OBJ extension) into a single library file (having an LIB extension). This file will generally be used in the production of several different programs and the object code modules collected by the library manager may be special modules created by the programmer or modules taken from an existing library. An optional library list usually can also be created by the library manager.

The value of building a library is that the routines needed within a program can be very easily linked into an executable object code file. Routines taken from the library can be used to construct further libraries or combined, as necessary, into executable programs by the linker.

The Microsoft library manager (LIB) is invoked by a command of the form:

```
LIB <library name><file list>
```

where the file list contains the names of the object code modules (each preceded by a '±' and separated by a comma) which are to be added to the library. As an example, the command line:

```
LIB graphics +fill, +shape
```

will add the object code modules fill.obj and shape.obj to the graphics library.

Symbolic debuggers

A symbolic debugger is an item of utility software that is designed to facilitate interactive testing and debugging of programs. As a minimum, a debugger should provide the user with commands which can be used to:

- Examine and modify the contents of memory.
- Examine and modify the content of the CPU registers.
- Run a program (starting at a given address) with breakpoints at which execution may be halted to permit examination of the CPU registers.
- Single-step a program (starting at a given address) with a register dump at the completion of each instruction.
- Disassemble a block of memory into assembly language mnemonics.
- Relocate a given block of memory.
- Initialize a given block of memory with specified data.
- Load or save blocks of memory from/to disk.

The debugger provided with the MS-DOS operating system (DEBUG) can be used for simple debugging (see Chapter 3); however, the Microsoft macro assembler provides a much enhanced debugger which is known as CodeView. This package is a powerful window-oriented software tool which allows the programmer to quickly locate logical errors in programs. The debugger can display source and object code simultaneously (indicating the line which is about to be executed), dynamically watch values (local or global), and switch screens to display program output. Use of the debugger is, to a large extent, intuitive and it greatly outperforms the DEBUG package supplied with DOS.

A full description of the facilities and use of CodeView (or an equivalent debugger) is beyond the scope of this book. Familiarity with the use of a debugger is, however, strongly recommended to all potential assembly language programmers as it can be instrumental in quickly and effectively dealing with the vast majority of bugs and defects in assembly language programs.

A MASM walkthrough

As an example of using assembly language development tools from the CLI the simple printer test routine that we met earlier in this chapter (see page 171) and also in Chapter 3 (see page 148) is edited, assembled, linked, and tested using the following sequence of commands and screen output. Later in this chapter we provide an example of using a modern 32-bit *Integrated Development Environment* (IDE) which automates the software development process (see page 183).

Firstly, the source code file is entered using the Microsoft Editor:

```
C>m ptest.asm<ENTER>
```

When the editor is left, the source file ptest.asm is written to the hard disk. The macro assembler is then invoked:

```
C>masm ptest<ENTER>
```

The following screen output is generated by the macro assembler:

```
Microsoft (R) Macro Assembler Version 5.10
Copyright (C) Microsoft Corp 1981, 1988. All rights reserved.

Object filename  [ptest.OBJ]: <ENTER>
Source listing   [NUL.LST]: <ENTER>
Cross-reference  [NUL.CRF]: <ENTER>

     50144 + 394061 Bytes symbol space free

     0 Warning Errors
     0 Severe Errors
```

The Linker is now used to produce an executable program:

```
C>link ptest<ENTER>
```

The following screen output is generated by the linker:

```
Microsoft (R) Overlay Linker Version 3.64
Copyright (C) Microsoft Corp 1983-1988. All rights reserved.

Hun File [PTEST.EXE]: <ENTER>
```

```
List File [NUL.MAP]: <ENTER>
Libraries [.LIB]: <ENTER>
LINK : warning L4021: no stack segment
```

The warning is ignored and the executable file is then tested:

```
C>ptest<RETURN>
```

The following printed output appears when the program is run:

```
ABCDEFGHIJKLMNOPQRSTUVWXYZ[\]^_'abcdefghijklmnop~rstuvwxvz{|}~
```

8086 assembly language

Before attempting to provide readers with an introduction to assembly language programming techniques and the use of a modern 32-bit macro assembler, it is important that readers have a basic understanding of the internal architecture and registers available within the 8086 microprocessor. We shall start, however, by briefly summarizing the 8086 instruction set.

8086 instruction set summary

The following is a brief summary of the 8086 instruction set.

Data transfer instructions

MOV	Move byte or word to register or memory
IN, OUT	Input byte or word from port, output word to port
LEA	Load-effective address
LDS, LES	Load pointer using Data Segment, Extra Segment
PUSH, POP	Push word onto stack, pop word off stack
XCHG	Exchange byte or word
XLAT	Translate byte using look-up table

Logical instructions

NOT	Logical NOT of byte or word (one's complement)
AND	Logical AND of byte or word
OR	Logical OR of byte or word
XOR	Logical exclusive-OR of byte or word
TEST	Test byte or word (AND without storing)

Shift and rotate instructions

SHL, SHR	Logical shift left, right byte or word by 1 or CL
SAL, SAR	Arithmetic shift left, right byte or word by 1 or CL
ROL, ROR	Rotate left, right byte or word by 1 or CL
RCL, RCR	Rotate left, right through carry byte or word by 1 or CL

Arithmetic instructions

ADD, SUB	Add, subtract byte or word
ADC, SBB	Add, subtract byte or word and carry (borrow)
INC, DEC	Increment, decrement byte or word
NEG	Negate byte or word (two's complement)

`CMP`	Compare byte or word (subtract without storing)
`MUL, DIV`	Multiply, divide byte or word (unsigned)
`IMUL, IDIV`	Integer multiply, divide byte or word (signed)
`CBW, CWD`	Convert byte to word, word to double word (useful before multiply or divide)
`AAA, AAS, AAM, AAD`	ASCII adjust for addition, subtraction, multiplication, division (ASCII codes 30–39)
`DAA, DAS`	Decimal adjust for addition, subtraction (binary-coded decimal numbers)

Transfer instructions

`JMP`	Unconditional jump
`JA (JNBE)`	Jump if above (not below nor equal)
`JAE (JNB)`	Jump if above or equal (not below)
`JB (JNAE)`	Jump if below (not above nor equal)
`JBE (JNA)`	Jump if below or equal (not above)
`JE (JZ)`	Jump if equal (zero)
`JG (JNLE)`	Jump if greater (not less nor equal)
`JGE (JNL)`	Jump if greater or equal (not less)
`JL (JNGE)`	Jump if less (not greater nor equal)
`JLE (JNG)`	Jump if less or equal (not greater)
`JC, JNC`	Jump if carry set, carry not set
`JO, JNO`	Jump if overflow, no overflow
`JS, JNS`	Jump if sign, no sign
`JNP (JPO)`	Jump if no parity (parity odd)
`JP (JPE)`	Jump if parity (parity even)
`LOOP`	Loop unconditional, count in CX
`LOOPE (LOOPZ)`	Loop if equal (zero), count in CX
`LOOPNE (LOOPNZ)`	Loop if not equal (not zero), count in CX
`JCXZ`	Jump if CX equals zero

Subroutine and interrupt instructions

`CALL, RET`	Call, return from procedure
`INT, INTO`	Software interrupt, interrupt if overflow
`IRET`	Return from interrupt

String instructions

`MOVS`	Move byte or word string
`MOVSB, MOVSW`	Move byte, word string
`CMPS`	Compare byte or word string
`SCAS`	Scan byte or word string
`LODS, STOS`	Load, store byte, or word string
`REP`	Repeat
`REPE, REPZ`	Repeat while equal, zero
`REPNE, REPNZ`	Repeat while not equal (zero)

Control instructions

`STC, CLC, CMC`	Set, clear, complement carry flag
`STD, CLD`	Set, clear direction flag
`STI, CLI`	Set, clear interrupt enable flag
`LAHF, SAHF`	Load AH from flags, store AH into flags
`PUSHF, POPF`	Push flags onto stack, pop flags off stack
`ESC`	Escape to external processor interface
`LOCK`	Lock bus during next instruction
`NOP`	No operation (do nothing)
`WAIT`	Wait for signal on TEST input
`HLT`	Halt processor

8086 register model

The *register model* of the 8086 is shown in Figure 5.1. Of the fourteen 16-bit registers available, four may be described as general purpose and can be divided into separate 8-bit registers. As an example, the 16-bit extended accumulator (AX) can be divided into two 8-bit registers, AH and AL. The high byte of a 16-bit word placed in AX is stored in AH whilst the low byte is stored in AL. Instructions can be made to refer to various parts of the accumulator so that operations can be carried out on the word stored in AX, or the individual bytes stored in AH or AL.

The four Segment Registers are Code Segment (CS), Data Segment (DS), Stack Segment (SS), and Extra Segment (ES). By making appropriate changes to the contents of these registers, the programmer can dynamically change the allocation of workspace.

As briefly mentioned in Chapter 1, the 8086 forms a 20-bit address from the contents of one of the Segment Registers (either CS, DS, SS, or ES) and

Pointer and Index Registers

15 — 0
SP
BP
SI
DI

General Purpose Registers

	15 — 8	7 — 0
AX	AH	AL
BX	BH	BL
CX	CH	CL
DX	DH	DL

Segment Registers

15 — 0
CS
DS
SS
ES

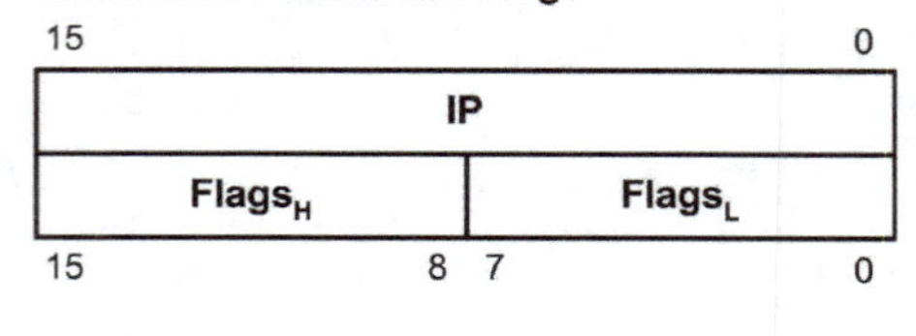

Figure 5.1 *8086 register model*

an offset taken from one (or more) of the other registers or from a memory reference within the program. The four Segment Registers (CS, DS, SS, and ES) effectively allow the programmer to set up individual 64KB workspace segments within the total 1MB address range.

We shall now briefly consider each of the 8086 registers in turn:

Accumulator, AX (AH and AL)

The accumulator is the primary source and destination for data used in a large number of 8086 instructions. The following data movement instructions give some idea of the range of options available:

`MOV AL,data`	Moves 8-bit immediate data into the least-significant byte of the accumulator, AL.
`MOV AH,data`	Moves 8-bit immediate data into the most-significant byte of the accumulator, AH.
`MOV AX,register`	Copies the contents of the specified 16-bit register to the 16-bit extended accumulator (AX).
`MOV AH,register`	Copies the byte present in the specified 8-bit register to the 8-bit register, AH.
`MOV AX,[address]`	Copies the 16-bit word at the specified address into the general-purpose base register, AX.

BX (BH and BL) register

The BX register is normally as a base register (address pointer). The following data movement instructions give some idea of the range of options that are available:

`MOV BX,data`	Moves 16-bit immediate data into the general-purpose base register (BX).
`MOV BX, [address]`	Copies the 16-bit word at the specified address into the general-purpose base register (BX).
`MOV BX,register`	Copies the contents of the specified register to the base register (BX).
`MOV [BX],AL`	Copies the contents of the AL register to the memory address specified by the BX register.

CX (CH and CL) register

The CX register is often employed as a *loop counter*. The 8086 LOOP instruction tests the contents of the CX register pair in order to determine whether the loop should be repeated, or not. This makes coding loops extremely simple as the following code fragment shows:

```
start:  MOV   CX,OCOOH   ; Number of times round the loop
delay:  LOOP  delay      ; Count down finished?
        RET
```

The CX and CL registers are also used to implement repeated string moves, shifts, and rotates. The following example shows how the contents of the accumulator can be rotated by the value placed in the CL register:

```
rote4:  MOV    CL,4         ; Number of bits to shift
        ROR    AX,CL        ; Rotate to the right
        RET
```

DX (DH and DL) register

The DX register is a general-purpose 16-bit register which can also be used as an extension of the AX register in 16-bit multiplication and division.

Stack Pointer (SP)

The SP register acts as a conventional Stack Pointer and points to the memory offset (relative to the paragraph address held in the Stack Segment register) of the current top of the stack. Adjustment of the SP register is automatic and programmers should avoid modifying the contents of this register if at all possible!

Base Pointer (BP), Destination Index (DI), Source Index (SI)

These registers are used in some of the more sophisticated of the 8086 addressing modes which permit the programmer to implement advanced data structures (such as two-dimensional arrays). All three registers are used to form addresses as shown in the following simple examples:

`MOV [BP+20],AX` Copies the word present in the AX register to an address offset by 20 bytes from the Base Pointer (BP).

`MOV [DI],space` Places 20H (previously defined by an equate of the form space EQU 20H) at the address pointed to by the Destination Index (DI).

`MOV SI,message` Move the start address of message into the Source Index (SI). Thereafter, SI can be used with an offset to point to a particular character within the string, message.

Instruction Pointer (IP)

The Instruction Pointer is a 16-bit register which points to the address of the next instruction to be executed. The Instruction Pointer is automatically updated by the CPU and the physical address of the instruction is found by adding the 16-bit value taken from the Instruction Pointer with the 16-bit value taken from the Code Segment Register shifted 4 bits to the left (see Chapter 1).

Flag Register (F)

The 8086 has a 16-bit Flag Register which contains nine status bits which may be set (0) or reset (1) depending upon the internal state of the CPU. Flags keep their status (either set or reset) until an instruction is executed which has an effect on them. The 8086 flags are shown in Figure 5.2.

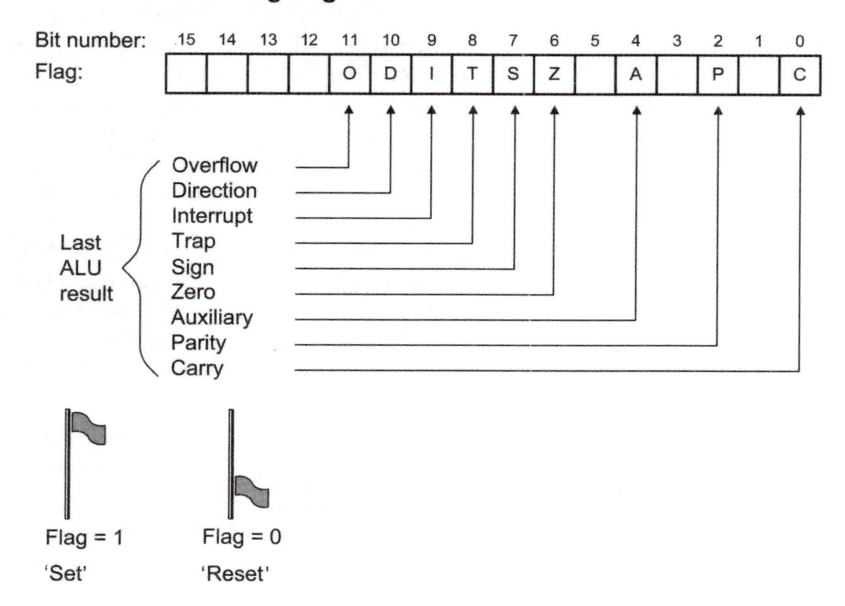

Figure 5.2 *8086 flags*

Segment Registers (CS, DS, SS, and ES)

We have already briefly mentioned the function of the four Segment Registers. Each register is associated with a separate workspace. The workspace defined by the Code Segment Register will contain program instructions, whilst the space defined by the Data and Extra Segments will generally contain data. In situations where RAM is limited there is no reason why the several Segment Registers should not have the same value (as in the case of a COM program). The code fragment:

```
MOV     AX,CS        ; Make Code and Data
MOV     DS,AX        ; Segments the same
```

can be used to make the Data Segment equal to the Code Segment (note that the instruction `MOV DS,CS` is not a valid 8086 instruction).

As a further example, the code fragment:

```
MOV     AX,vidram    ; Make Data Segment point
MOV     DS,AX        ; to video memory
```

can be used to make the Data Segment point to the start of a block of video RAM (vidram will previously have been the subject of an equate).

Interrupt handling

By comparison with earlier 8-bit microprocessors, the 8086 provides somewhat superior interrupt handling and uses a table of 256 4-byte pointers stored in the bottom 1 KB of memory (addresses 0000H to 03FFH). Each of the locations in the Interrupt Pointer Table can be loaded with a pointer to a different interrupt service routine. Each pointer contains 2 bytes for loading into the Code Segment (CS) Register and 2 bytes for loading into the Instruction Pointer (IP). This allows the programmer to place interrupt service routines in any appropriate place within the 1 MB physical address space.

Each of the 256 *Interrupt Pointers* is allocated a different type number. A Type 0 interrupt has its associated Interrupt Pointer in the lowest 4 bytes of memory (0000H to 0003H). A Type 1 interrupt will have its pointer located in the next 4 bytes of memory (0004H to 0007H), and so on.

The structure of the 8086 Interrupt Pointer Table is shown in Figure 5.3. Interrupt types 0 to 4 have dedicated functions while Types 5 to 31 are reserved. Hence there are 224 remaining locations in which Interrupt Pointers may be stored. The interrupting device places a byte on the data bus in response to an

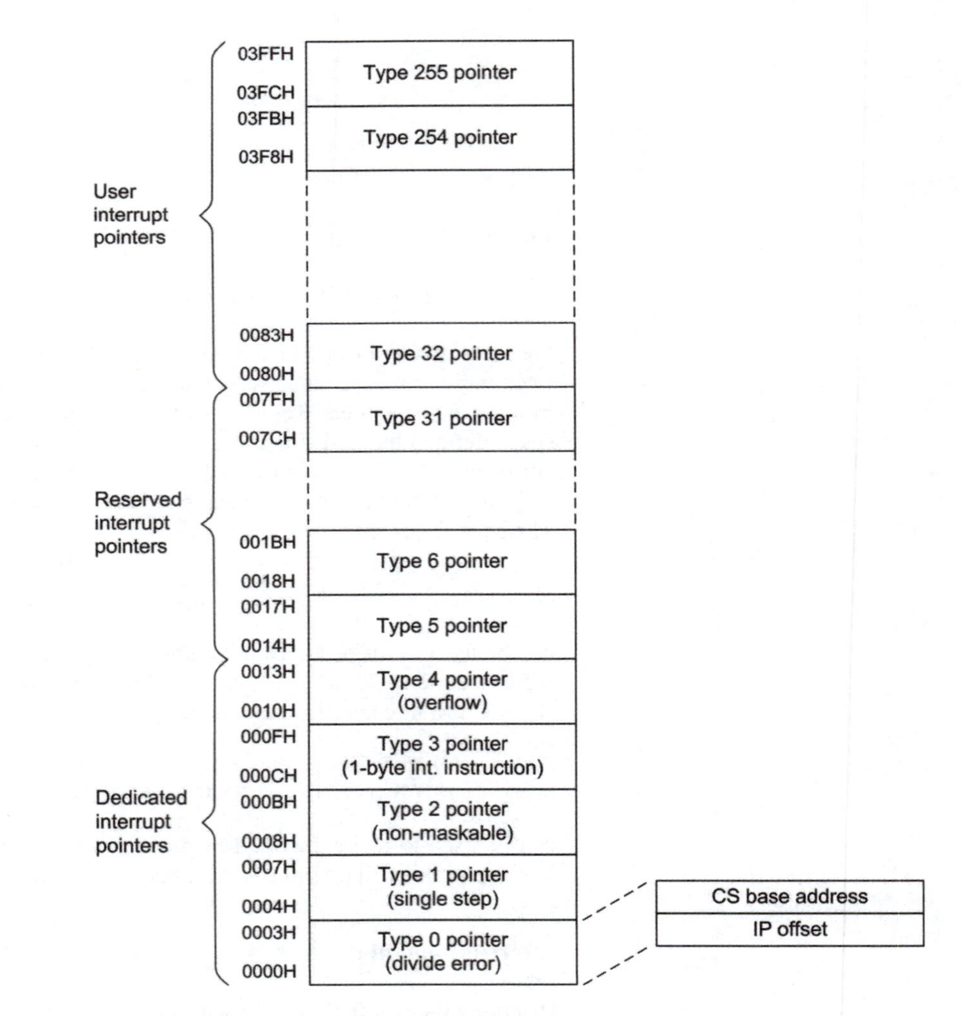

Figure 5.3 *8086 Interrupt Pointer table*

interrupt acknowledgement generated by the CPU. This byte gives the interrupt type and the 8086 loads its Code Segment and Instruction Pointer registers with the words stored at the appropriate locations in the Interrupt Pointer Table and then commences execution of the interrupt service routine.

The following code fragment shows how the Interrupt Pointer Table can be initialized to cope with three interrupt service routines:

```
; Initialise Interrupt Pointer Table
     MOV    AX,0          ; Point to start
     MOV    DS,AX         ; of memory.
     MOV    AX,CS         ; Get code segment.
; Type 32
     MOV    80H,dev1      ; Offset for device #1 ISR
     MOV    82H,AX        ; and segment address
; Type 33
     MOV    84H,dev2      ; Offset for device #2 ISR
     MOV    86H,AX        ; and segment address
; Type 255
     MOV    3FCH,dev3     ; Offset for device #3 ISR
     MOV    3FEH,AX       ; and segment address
```

MASM32

If you plan to make extensive use of assembly language programs it is worth moving to a modern 32-bit macro assembler and a full Integrated Development Environment (IDE). One of the most comprehensive packages is currently provided by MASM32 which is a complete freeware assembly language development environment. MASM32 has its roots in the original Microsoft MASM assembler but it is combined with elements of the Microsoft Windows DDK/SDK and it enjoys the support of an excellent Editor and IDE (see Figure 5.4).

For anyone contemplating using assembly language as the main vehicle for software development the current version of MASM32 assembler (together with its IDE and support tools) can be very highly recommended. Indeed, 32-bit assembler is both clearer and simpler than the DOS and 16-bit Windows code and is not cursed with the complexity of segment arithmetic. You no longer have to deal with using pairs of registers for long integers and there is no 64 KB boundary imposed by the segmented structure of 16-bit software.

The complexity of writing 32 bit Windows software is related to the structure of Windows and the sheer range of functions in the Attachment Packet Interface (API) set. It differs from DOS code only in so far as the parameters are passed on the stack rather than in registers as in the DOS interrupts. While the sheer range of functions in Window can be a bit intimidating, it also puts in the hands of the assembler language programmer, a massive set of capacities that were never available in DOS.

One of the advantages of writing in assembler is that it comfortably handles the C-code format of the Windows APIs with no difficulty. Zero-terminated strings, structures, pointers, data sizes, etc., are all part of writing assembler.

The following code fragment (see also page 153) shows how a simple message box can be produced using just a few lines of assembly code:

```
.386
.model flat, stdcall
option casemap:none
```

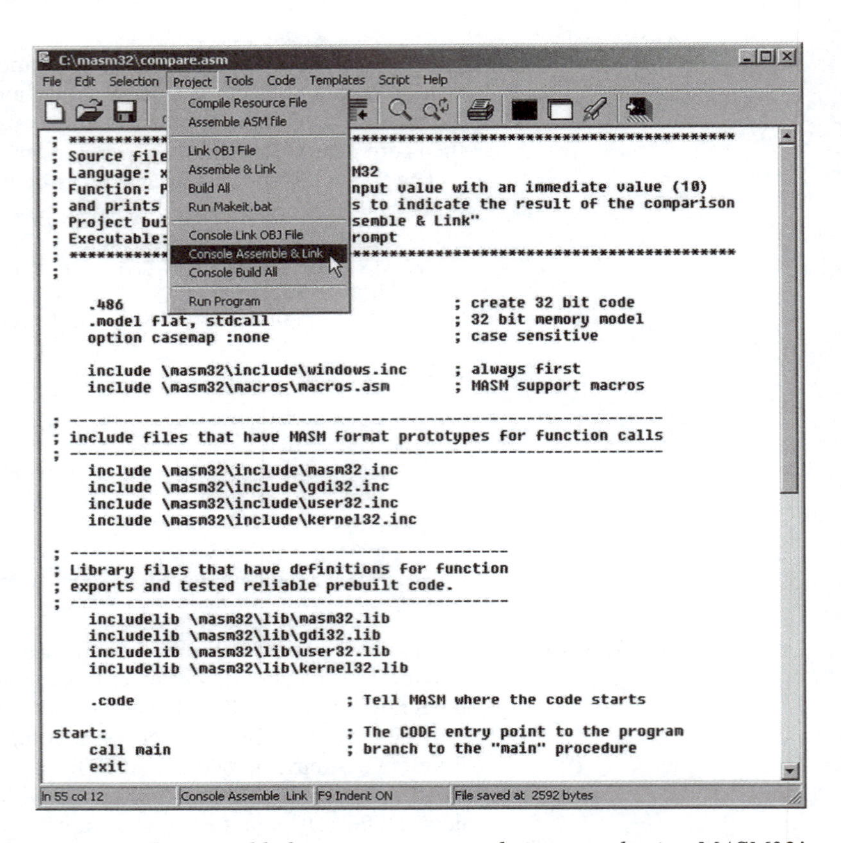

Figure 5.4 *The assembly language source code is entered using MASM32's integrated editor and Console Assemble and Link is used to automatically produce an executable program*

```
include \masm32\include\windows.inc
include \masm32\include\kernel32.inc
include \masm32\include\user32.inc
includelib \masm32\lib\user32.lib
includelib \masm32\lib\kernel32.lib

.data
MsgCaption      db "Warning",0
MsgBoxText      db "Check feed hopper is empty!",0

.code
start:
   invoke MessageBox, NULL,addr MsgBoxText, addr MsgCaption,
      MB_OK
   invoke ExitProcess,NULL
end start
```

The macro that handles the production of the message box in the previous example is as follows:

```
MsgBox MACRO handl, TxtMsg, TxtTitle, styl
        LOCAL Msg1
        LOCAL Titl
```

```
    If @InStr(1,<TxtMsg>,<ADDR>) eq 0
      If @InStr(1,<TxtTitle>,<ADDR>) eq 0
      .data
        Msg1 db TxtMsg,0
        Titl db TxtTitle,0
      .code
        invoke MessageBox,handl,ADDR Msg1,ADDR Titl,styl
        EXITM
      EndIf
    EndIf
    If @InStr(1,<TxtMsg>,<ADDR>) gt 0
      If @InStr(1,<TxtTitle>,<ADDR>) eq 0
      .data
        Titl db TxtTitle,0
      .code
        invoke MessageBox,handl,TxtMsg,ADDR Titl,styl
        EXITM
      EndIf
    EndIf
    If @InStr(1,<TxtMsg>,<ADDR>) eq 0
      If @InStr(1,<TxtTitle>,<ADDR>) gt 0
      .data
        Msg1 db TxtMsg,0
      .code
        invoke MessageBox,handl,ADDR Msg1,TxtTitle,styl
        EXITM
      EndIf
    EndIf
    If @InStr(1,<TxtMsg>,<ADDR>) gt 0
      If @InStr(1,<TxtTitle>,<ADDR>) gt 0
        invoke MessageBox,handl,TxtMsg,TxtTitle,styl
        EXITM
      EndIf
    EndIf
ENDM
```

Hopefully, this simple example should help to convince you of the advantages of using macros and the immense time saving that this can potentially offer the programmer!

The following data types are supported by MASM32.

Register	Data	Size
al	BYTE	8 bit
ax	WORD	16 bit
eax	DWORD	32 bit
mm(0)	QWORD	64 bit

It is worth comparing the above lost with the complexity of the higher-level programming languages, such as C++ which can have as many more data types! To assist with the conversion of data types a WINDOWS.INC include file is supplied with MASM32.

A MASM32 walkthrough

Finally, to illustrate how MASM32 assemble, disassemble and provide debugging and other information, the following assembly language code is for a simple number guessing game:

```
; ************************************************************************
; Source file: compare.asm
; Language: x86 assembler for MASM32
; Function: Program compares an input value with an immediate value (10)
; and prints one of three messages to indicate the result of the comparison
; Project build: use "Console Assemble & Link"
; Executable: runs from command prompt
; ************************************************************************
;
    .486                                    ; create 32 bit code
    .model flat, stdcall                    ; 32 bit memory model
    option casemap :none                    ; case sensitive

    include \masm32\include\windows.inc     ; always first
    include \masm32\macros\macros.asm       ; MASM support macros
; -----------------------------------------------------------------
; include files that have MASM format prototypes for function calls
; -----------------------------------------------------------------
    include \masm32\include\masm32.inc
    include \masm32\include\gdi32.inc
    include \masm32\include\user32.inc
    include \masm32\include\kernel32.inc
; ------------------------------------------------
; Library files that have definitions for function
; exports and tested reliable prebuilt code.
; ------------------------------------------------
    includelib \masm32\lib\masm32.lib
    includelib \masm32\lib\gdi32.lib
    includelib \masm32\lib\user32.lib
    includelib \masm32\lib\kernel32.lib
; ------------------------------------------------
    .code                         ; Tell MASM where the code starts
start:                            ; The CODE entry point to the program
    call main                     ; branch to the "main" procedure
    exit
main proc
    LOCAL var1:DWORD              ; space for a DWORD variable
    LOCAL str1:DWORD              ; a string handle for the input data
    mov var1, sval(input("Please enter a number between 1 and 20: "))
    cmp var1, 10                  ; compare the input number with 10
    je equal                      ; Is the input number equal to 10?
    jg greater                    ; Is the input number greater than 10?
    jl less                       ; Is the input number less than 10?
  equal:
    print chr$("The number you entered is 10",13,10)
    jmp over
  greater:
    print chr$("The number you entered is greater than 10",13,10)
    jmp over
```

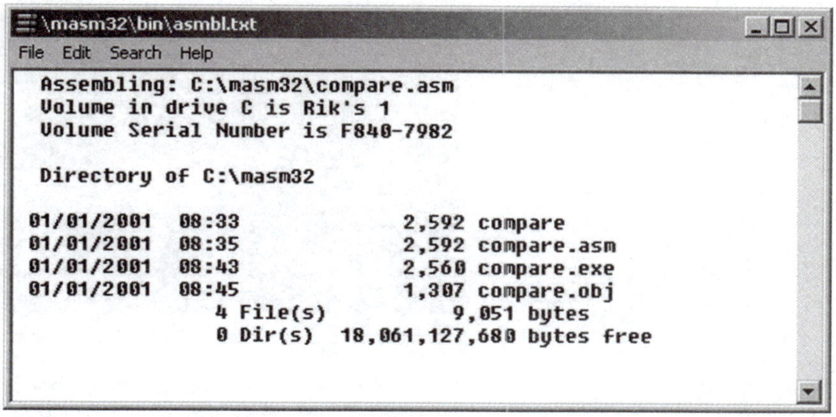

```
\masm32\bin\asmbl.txt
File Edit Search Help
Assembling: C:\masm32\compare.asm
Volume in drive C is Rik's 1
Volume Serial Number is F840-7982

Directory of C:\masm32

01/01/2001  08:33              2,592 compare
01/01/2001  08:35              2,592 compare.asm
01/01/2001  08:43              2,560 compare.exe
01/01/2001  08:45              1,307 compare.obj
               4 File(s)          9,051 bytes
               0 Dir(s)  18,061,127,680 bytes free
```

Figure 5.5 *Screen messages produced during the Console Assemble and Link process. The executable file is named compare.exe whilst the object code file is compare.obj.*

```
    less:
      print chr$("The number you entered is less than 10",13,10)
    over:
      ret
main endp
end start                      ; Tell MASM where the program ends
```

The code is entered using MASM32's editor (part of the IDE). After saving the source code file as compare.asm, Console, Assemble, and Link are used to automatically produce and execute program Link (see Figures 5.4 and 5.5).

The executable file (compare.exe) can then be tested by entering the name of the executable file at the command line (see Figure 5.6). Notice that the current directory has been changed to C:\MASM32 in which the executable file has been saved.

Finally, if necessary, MASM32's disassembler can be used to provide information about the executable code. The following is a short extract from the information generated by MASM32's disassembler:

```
C:\masm32\compare.exe                     (hex)           (dec)

.EXE size (bytes)                           490            1168
Minimum load size (bytes)                   450            1104
Overlay number                                0               0
Initial CS:IP                         0000:0000
Initial SS:SP                         0000:00B8             184
Minimum allocation (para)                     0               0
Maximum allocation (para)                  FFFF           65535
Header size (para)                            4               4
Relocation table offset                      40              64
Relocation entries                            0               0
Portable Executable starts at                        b0
Signature                                      00004550 (PE)
```

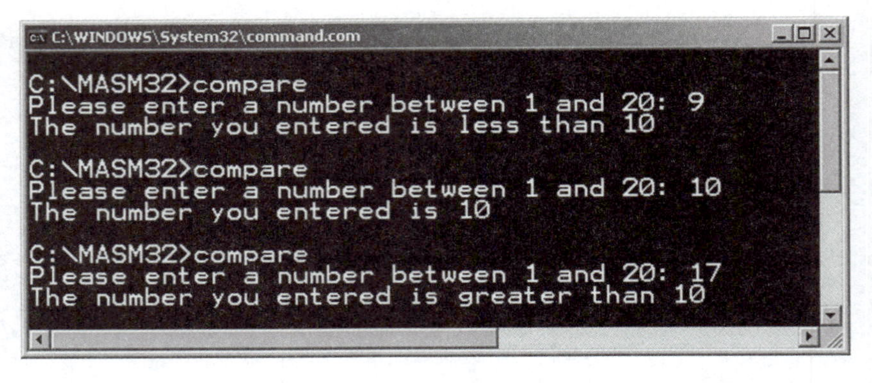

Figure 5.6 *Results of running compare.exe from the CLI. The program has been tested with three different input values.*

```
Machine                                   014C (Intel 386)
Sections                                  0003
Time Date Stamp                       3A50440A Mon Jan 1 08:47:06 2001
Symbol Table                          00000000
Number of Symbols                     00000000
Optional header size                      00E0
Characteristics                           010F
   Relocation information stripped
   Executable Image
   Line numbers stripped
   Local symbols stripped
   32 bit word machine
Magic                                     010B
Linker Version                            5.12
Size of Code                          00000200
Size of Initialized Data              00000400
Size of Uninitialized Data            00000000
Address of Entry Point                00001000
Base of Code                          00001000
Base of Data                          00002000
Image Base                            00400000
Section Alignment                     00001000
File Alignment                        00000200
Operating System Version                  4.00
Image Version                             0.00
Subsystem Version                         4.00
reserved                              00000000
Image Size                            00004000
Header Size                           00000400
Checksum                              00000000
Subsystem                                 0003 (Console)
DLL Characteristics                       0000
Size Of Stack Reserve                 00100000
Size Of Stack Commit                  00001000
Size Of Heap Reserve                  00100000
Size Of Heap Commit                   00001000
Loader Flags                          00000000
Number of Directories                 00000010
```

6 BASIC programming

Despite increasing competition from other languages such as Pascal/Delphi and C/C++, BASIC remains extremely popular in the field of instrumentation and process control; the language is relatively easy to learn and programs can be quickly developed by those with little previous programming experience. Furthermore, modern implementations of the language put it on a par with many of its more powerful competitors. Gone are the days when BASIC programs were constrained to show a lack of structure by the absence of control structures such as `DO ... LOOP` and `WHILE ... WEND`. Furthermore, BASIC procedures, subprograms, and user-defined functions all aid the programmer since they promote modularity and aid flexibility.

The availability of compilers adds a further dimension to the language since compiled BASIC programs can be indistinguishable from those written in other (supposedly superior) languages. Such programs are compact, execute at high speed, and are relatively straightforward to develop and maintain. Such factors conspire to make modern structured and compiled BASICs worthy contenders for most applications within the field of instrumentation and control.

Since the majority of readers will have at least a passing acquaintance with the BASIC programming language, we shall deal only with topics which are directly relevant to the development of efficient programs for instrumentation and control applications. Readers with no previous knowledge are advised to consult one of the many tutorial books aimed at newcomers to BASIC programming (see Appendix E). There is no shortage of material to choose from and most texts will provide a more than adequate introduction to the subject. We begin this chapter by introducing some of the most popular BASIC compilers including those that were developed strictly for a DOS environment as well as more modern variants designed to fully exploit the features offered by the 32-bit Windows operating system.

Microsoft BASIC for DOS

To simplify some of the terminology, I've coined the phrase 'Microsoft BASIC for DOS' to encompass a number of BASIC compilers and associated development tools produced by Microsoft over the last two decades. Such tools include QuickBASIC, BASIC 6.0, and the BASIC 7.1 Professional Development System. Later versions of QuickBASIC are also sometimes referred to as QuickBASIC Extended (QBX).

Microsoft takes pains to stress that their BASIC for DOS products are only certified for use with MS-DOS and PC-DOS systems and, because of this and the advent of Visual programming languages (such as Visual BASIC) their DOS-based products are no longer actively supported. Nevertheless, Microsoft BASIC for DOS *will* work within a Windows environment but with a few restrictions, the most notable of which is that applications that access

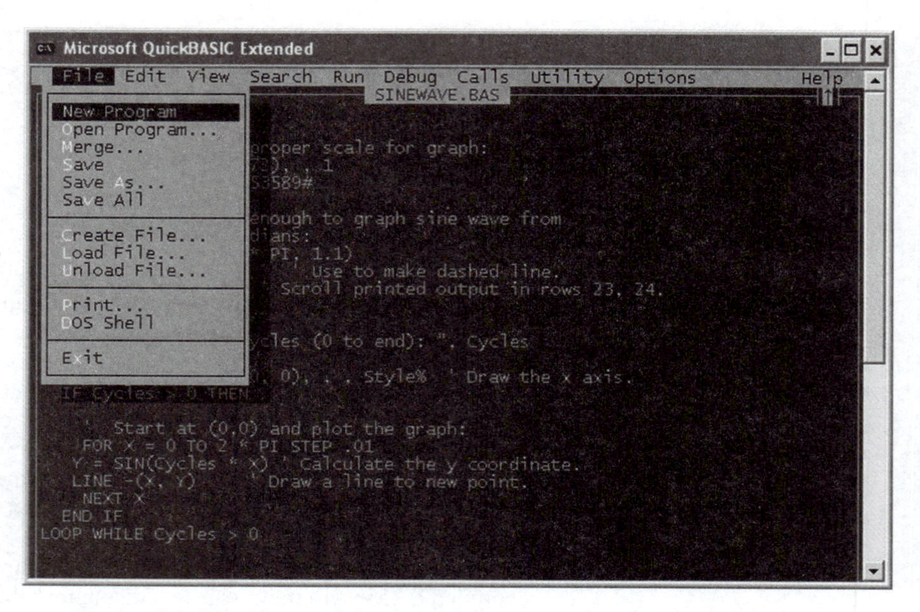

Figure 6.1 *The Microsoft BASIC Professional Development System IDE*

hardware directly are unsuitable for deployment in a Windows *Protected Mode environment*. In order to overcome the restrictions imposed by later versions of Windows (i.e. beyond Windows 9x), your applications may have to be prepared with a view to operation in a pure DOS environment (rather than in a DOS Window). We will return to this important point later in this chapter.

Many of the BASIC programs that appear in this book were originally developed using QuickBASIC and Microsoft BASIC 6.0, and most are compatible with modern forms of BASIC (such as QBX and PowerBASIC 3.5). The early Microsoft DOS BASIC packages provide the user with a simple 'no-frills' *Integrated Development Environment* (IDE) (see Figure 6.1). The IDE allows program entry, editing, running, and debugging without having to leave the IDE's shell. Programs can be tested during development with minimal fuss and then free-standing executable programs (EXE files) can be produced when the user is reasonably confident that the program is robust and bug-free.

Microsoft DOS BASIC also offers the user comprehensive context-sensitive online help. Using the resident BASIC text editor, syntax errors are reported immediately when the code is entered, and debugging is aided by the availability of *breakpoints* and *watchpoints* which can be freely imbedded within the code.

Modular programming is encouraged and current modules are co-resident in memory during program development. Multiple editing Windows allow the programmer to view the main code along with the code for a subprogram (procedure). The programmer can also exit to DOS, carry out a DOS operation (such as formatting a disk) and then return to the BASIC environment at the point at which it was left.

Despite its age, Microsoft DOS BASIC can still provide a useful environment for developing simple and compact DOS-based applications. So, as long as you don't need a full Windows environment for your application these compilers can make a useful starting point.

Developing Microsoft BASIC for DOS programs

Since it is relatively easy to write and enter Microsoft BASIC for DOS programs, it is unfortunately all too easy to develop bad habits. Furthermore, the end result produced by an unstructured program (i.e. 'quick and dirty code') can sometimes be indistinguishable from that produced by a program which is highly structured. The difference only becomes important when the time comes for extending, modifying, or maintaining the program. With structured code this is a relatively simple matter. An unstructured program, on the other hand, may be a tangled nest of haphazard code and a major modification to the program may well result in the need for a complete rewrite. This can hardly be described as efficient!

There are a number of techniques that can be used to assist in the production of efficient structured code. First and foremost, it is vitally important to get into the habit of being consistent in the layout of your programs and in the names used for variables. Failure to do this will make it extremely difficult to port sections of code from one program to another. This is a highly desirable feature which will save many hours of work. An efficient procedure for, say, accepting keyboard input and verifying that it is numeric, truncating it to integer, and confirming that it is within a given range, can be useful in a huge variety of control applications. There is absolutely no reason why an efficient code module that performs such a function should not be included in every program that you write. Once written, you will never do it again!

Variable types

Wherever possible, integer numeric variables should be used in order to minimize storage space and increase processing speed. Floating point variables, which have considerable processing and storage overhead, should be avoided. Integer variables are normally recognized by a trailing %. Thus `t` represents a floating point numeric variable while t% represents an integer numeric variable and `t$` represents a string variable.

Integer variables require 2 bytes for storage and values can be whole numbers (i.e. no decimal points) ranging from −32 768 to +32 767. Microsoft BASIC for DOS also supports long integers (in which each occupy 4 bytes of storage), and both single- and double-precision floating point numbers (see Table 6.1).

String variables comprise a sequence of characters (letters, numbers, and punctuation). Microsoft BASIC for DOS supports both fixed and variable length strings (the length of the former type must be declared). In either case, the maximum length permitted is 32 767 characters.

Variable names

In order to aid readability, it has become fashionable to use relatively long names for variables. Happily, where a BASIC program will eventually be compiled, the overhead associated with long variable names applies only to the source files. It is thus permissible to use more meaningful variable names in such applications. Whether or not one is using a compiled BASIC, it is essential to maintain

Table 6.1 *QuickBASIC variable types*

Class	Type	Number of bytes for storage	Range of values	Identifier	Examples	Notes
Numeric	Integer	2	−32 768 to +32 767	%	min%	Stored in 16-bit 2's complement format
Numeric	Long integer	4	−2 147 483 648 to +2 147 483 647	&	alt&	Stored in 16-bit 2's complement format
Numeric	Single-precision floating point	4	−3.4E+38 to +3.4E+38 (approx.)	!	val!	Stored in IEEE format (accurate to seven decimal places)
Numeric	Double-precision floating point	8	−1.797E+308 to +1.797E+308 (approx.)	#	max#	Stored in IEEE format (accurate to 15 or 16 digits)
String	Character	Fixed	n/a	$	input$	Length must be declared (max. 32 767)
String	Character	Variable	n/a	$	file$	Length can be variable (max. 32 767)

consistency with the choice of variable names and also to ensure that, as far as possible, they are descriptive. Examples of acceptable variable names are:

`chan%`	channel number (integer)
`col%`	column number (integer)
`date$`	date (string – typical format `mmddyyyy`)
`day$`	day of the week (string)
`error$`	error message (string)
`file$`	filename (string)
`inp%`	input port address (integer)
`lim%`	limit value (integer)
`outp%`	output port address (integer)
`prompt$`	user prompt (string)
`r$`	general user response (string – usually a single character)
`time$`	time (string – typical format hhmmss)
`vel&`	velocity (single precision floating point)
`ver$`	version number (string)
`x%`	x-axis displacement (integer)
`y%`	y-axis displacement (integer)

BASIC command summary

The following is a summary of BASIC commands that are found in most modern versions of BASIC. Note that Windows specific commands have been excluded from this list and not all of the listed commands may be available in any particular version of the language. In all cases, readers are advised to familiarize themselves with the command set and command syntax applicable to the version of BASIC that they will actually be using!

Command	Function/operation
ABS	Return the absolute value of a numeric expression
AND	Perform a logical bitwise AND operation
ASC	Return the ASCII code of the specified character in a string
ASM	Identify an assembly language statement
ATN	Return the arctangent of an argument
BEEP	Generate a beep using the PC's internal speaker
BIN$	Return a string that is the binary (base 2) representation of a value
BIT	Set/reset the value of a particular bit in an integer-class variable
CALL	Invoke a procedure, subroutine or function)
CBYT	Convert a value to a Byte data type
CDBL	Convert a value to a double-precision data type
CDWD	Convert a value to a double-word data type
CHR$	Convert one or more ASCII codes into ASCII character(s)
CINT	Convert a value to a integer data type
CLNG	Convert a value to a long-integer data type
CLS	Clear screen (erases current screen data)
COS	Return the cosine of an argument
CSNG	Convert a value to a single-precision data type
DATA	Declare an array of constants
DATE$	Set or retrieve the system date
DEFBYT	Declare the default variable type to be Byte
DEFDBL	Declare the default variable type to be double precision
DEFDWD	Declare the default variable type to be double word
DEFINT	Declare the default variable type to be integer
DEFLNG	Declare the default variable type to be long integer
DEFSNG	Declare the default variable type to be single precision
DEFSTR	Declare the default variable type to be string
DEFWRD	Declare the default variable type to be word
DIM	Declare and dimension arrays, scalar variables, and pointers
DIR$	Return a filename that matches the given mask
DISKFREE	Return the amount of available space on a disk, in bytes
DISKSIZE	Return the total amount of space on a disk, in bytes
DO/LOOP	Define a group of program statements that are executed repetitively
END SELECT	Closes a SELECT CASE block
EOF	Return the end-of-file status of a file
EXIT	Transfer program execution out of a block structure
EXP	Return a number raised to a power of e (inverse natural logarithm)
FIX	Truncate a floating point number to an integer
FOR/NEXT	Define a loop of program statements controlled by a counter
FORMAT$	Format numeric data according to a string mask expression
FUNCTION	Define the start of a function block
GET	Read a record from a random-access file
GET$	Read a string from a file opened in binary mode
GLOBAL	Declare global (shared) variables between subs and functions
GOSUB	Invoke a local subroutine

(*continued*)

Command	Function/operation
GOTO	Transfer program execution to the statement identified by a label
HEX$	Return a hexadecimal (base 16) string representation of an argument
IF	Test a condition and execute one or more program statements
IF/END IF	Create a IF/THEN/ELSE block with multiple lines and conditions
IN	Input byte data from a port
INPUT#	Load variables with data from a sequential file
INSTR	Search a string for the first occurrence of a character or string
INT	Convert a numeric expression to an integer-class value
KILL	Delete a disk file
LCASE$	Return a lowercase version of a string argument
LEFT$	Return the left-most n characters of a string
LEN	Return the logical length of a variable
LET	Assign a numeric variable or string
LINE INPUT#	Read line(s) from a sequential file into a string variable or string array
LOC	Determine the current seek position in an open disk file
LOCAL	Declare local variables in a sub or function
LOCATE	Define a point on the screen
LOF	Return the length of an open disk file
LOG	Return the natural (base e) logarithm of an argument
LSET	Left-align a string within the space of another string
LSET$	Return a string containing a left-justified (padded) string
LTRIM$	Return a string with leading characters or strings removed
MID$	Return a portion of a string
NOT	Logical bitwise NOT operation
OCT$	Return a string that is a octal (base 8) representation of a value
ON ERROR	Specify an error handling routine; enable/disable trapping
ON GOSUB	Call one of several subroutines based on a numeric expression
ON GOTO	Send program flow to one of several labels based on a value
ON KEY	Used for trapping key events
OPEN	Prepare a file or device for reading or writing
OR	Logical bitwise OR arithmetic operation
OUT	Output byte data to a port
PEEK	Return the byte at a specific memory location
PEEK$	Return a sequence of bytes starting at a specific memory location
POKE	Store a byte at a specific memory location
POKE$	Store a sequence of bytes starting at a specific memory location
PRINT#	Write a string or a complete array to a sequential file
PUT	Write a record to a random-access file or a variable to a binary file
PUT$	Write a string to a file opened in binary mode
RANDOMIZE	Seed the random number generator

(*continued*)

Command	Function/operation
READ$	Retrieve string data from a local DATA list
REDIM	Declare dynamic arrays, allocate, deallocate, or reallocate memory
REM	Indicate the remainder of a line of source code is a remark or comment
RESET	Set a variable, array element or an entire array to zero
RESUME	Continue execution after error handling with ON ERROR GOTO
RETURN	Return from a subroutine (GOSUB) to its caller
RIGHT$	Return the rightmost n characters of a string
RND	Return a random number
RSET	Right justify a string into the space of a string variable
RSET$	Return a string containing a right-justified (padded) string
RTRIM$	Return a copy of a string with trailing characters or strings removed
SEEK	Set the position in a file for the next input or output operation
SELECT CASE	Control program flow based on the value of an expression
SGN	Return the sign of a numeric expression
SHIFT	Shift the bits in an integer-class variable
SIN	Return the sine of an argument
SOUND	Generate a sound
SPACE$	Return a string consisting of a specified number of spaces
SQR	Return the square root of an argument
STATIC	Declare static variables inside of a sub or function
STR$	Return the string representation of a number in printable form
STRING$	Return a string with multiple copies of the specified character
SUB/END SUB	Define a sub (procedure) block
TAN	Return the tangent of an argument
TIME$	Read and/or set the system time
TIMER	Return the number of seconds that have elapsed since midnight
UCASE$	Return an all-uppercase (capitalized) version of a string
USING$	Format one or more string/numeric expressions using a mask string
VAL	Return the numeric equivalent of a string argument
VARPTR	Return the 32-bit address of a variable or string handle
WHILE/WEND	Define a block of program statements that are executed repeatedly
WRITE#	Output data to a sequential file in a delimited format
XOR	Perform a logical or a bitwise exclusive-OR operation

Subroutines

Subroutines can be instrumental in making very significant reductions in the size of BASIC programs and they should be employed whenever a section of code is to be executed more than once. Note, however, that if your version of BASIC supports the use of subprograms, procedures or user-defined functions, then these should normally be used instead! A typical example of the use of a subroutine might involve a delay routine which is required at various points in a program. Assuming that such a routine was to be used in an early version of BASIC which employs line numbers and that the subroutine starts at line 10100, it might take the following form:

```
10100 REM Delay subroutine
10110 FOR c% = 0 TO 10000: NEXT e%
10120 RETURN
```

The subroutine may be called from several points within the main program as follows:

```
340
350 GOSUB 10100
360
...
440
450 GOSUB 10100
460
...
710
720 GOSUB 10100
730
```

In each case, program execution resumes at the line immediately following the `GOSUB` statement. Also note that, on exit from the subroutine, `c%` will have the value 10001.

We could make the delay subroutine even more flexible (allowing for variable length delays) by altering the upper limit of the loop using a variable which is set immediately prior to the subroutine call. The modified subroutine would then become:

```
10100 REM Delay subroutine
10110 FOR c% = 0 TO lim%: NEXT c%
10120 RETURN
```

As before, the routine may be called from several points in the main program as follows:

```
340
350 lim% = 10000: GOSUB 10100
360
...
440
450 lim% = 20000: GOSUB 10100
460
...
710
720 lim% = 15000: GOSUB 10100
730
```

On exit from the subroutine, the value of c% will have been modified to 1 greater than the value of `lim%` immediately prior to the subroutine call.

Since line numbers are not required using QuickBASIC (in common with later versions of Microsoft BASIC for DOS as well as virtually all modern BASIC compilers) a *label* is used (instead of a line number) to mark the start of the delay subroutine, as shown below:

```
REM Start of main program
...
...
GOSUB Delay
...
...
GOSUB Delay
...
...
GOSUB Delay
...
...

END
REM Delay subroutine
Delay:
FOR c% = 0 TO 10000: NEXT c%
RETURN
```

It is important to note that the label (`Delay`) is immediately followed by a colon and that the main body of program code must be terminated by an `END` statement in order to prevent execution of the subroutine when the end of the code has been reached. If this should ever happen, an error condition will result as the `RETURN` statement does not have a matching `GOSUB`.

Procedures

A user-defined procedure can be thought of as a named subroutine. The procedure is simply `CALL`ed by name rather than by `GOSUB` followed by a label. This can be instrumental in not only making the resulting code more readable but it also ensures that the structure of the program can be easily understood. A further advantage of procedures is that parameters may be passed into procedures and values returned to the main program. Variables which are to be common with the main program may be declared at the start of the procedure using the `SHARED` statement (otherwise all variables internal to the procedure will be strictly local).

Procedures are defined using statement of the form `SUB` <name> and are terminated by `END SUB`. Procedures may also contain references to other procedures (i.e. procedures can be 'nested'). Procedure names should be chosen so they do not conflict with any variable names nor should they be BASIC reserved words.

The previous delay subroutine can be easily written as a procedure:

```
REM Delay procedure
SUB Delay(lim%)
FOR c% = 0 TO lim%: NEXT c%
END SUB
```

It should be noted that the QuickBASIC editor will recognize the `SUB` statement and will treat the procedure as a separate subprogram (with an automatic declaration inserted at the start of the main program code). Thereafter, the subprogram can be viewed and edited independently of the main program code.

The method of calling the delay procedure is more elegant than that used with the equivalent subroutine and takes the following format:

```
REM Start of main program
DECLARE SUB Delay(lim%)
...
...
CALL Delay(10000)
...
...
CALL Delay(20000)
...
...
CALL Delay(15000)
...
...
```

The values within parentheses are parameters passed into the procedure as `lim%`. Such values are local to the procedure and external references to `lim%` will remain unchanged by the action of the procedure. It should also be noted that an overflow error will occur if values passed into the subprogram should ever exceed 32 766. Longer delays can be produced using floating point variables, as follows:

```
REM Delay procedure
SUB Delay(lim)
FOR c = 0 TO lim: NEXT c
END SUB
```

while the relevant code in the main program should run along the following lines:

```
REM Start of main program
DECLARE SUB Delay(lim)
...
...
CALL Delay(10000)
...
...
CALL Delay(20000)
...
...
CALL Delay(15000)
...
...
```

Since we are using floating point variables, values passed into the subprogram can now exceed the 32 766 limit imposed on integers (see Table 6.1 for details).

User-defined functions

User-defined functions are similar to user-defined procedures but return values (integer, float, or string) to the main program. As with user-defined procedures, functions are called by name (or `FN name`). An example of a user-defined function (`FNConfirm%`) appears later in this chapter.

Logical constructs

Modern versions of BASIC provide us with a number of other useful constructs which can be instrumental in the production of efficient structured code. As an example, a somewhat more elegant delay procedure can be produced using the `WHILE ... WEND` construct. This routine uses a single variable rather than the two that were required in the `FOR ... NEXT` construct used earlier.

```
REM Delay procedure
SUB Delay(lim%)
WHILE lim% > 0
lim% = lim% - 1
WEND
END SUB
```

The condition in the `WHILE` stated is tested and, as long as it remains true (i.e. evaluates to non-zero), the code within the loop will be repeated. It should, perhaps, be stated that there is no particular advantage in using `WHILE ... WEND` in this simple delay subroutine and a straightforward `FOR ... NEXT` loop would, in practice, be perfectly adequate!

The `DO ... LOOP` construct offers an even more powerful alternative to `FOR ... NEXT` and `WHILE ... WEND`. Several forms of `DO ... LOOP` structure are available with tests for the loop condition at the start of the loop (`DO WHILE ...` and `DO UNTIL ... LOOP`) and tests at the end of the loop (`DO ... LOOP WHILE` and `DO ... LOOP UNTIL`). The logic of these constructs is contrasted in Figures 6.2 and 6.3,

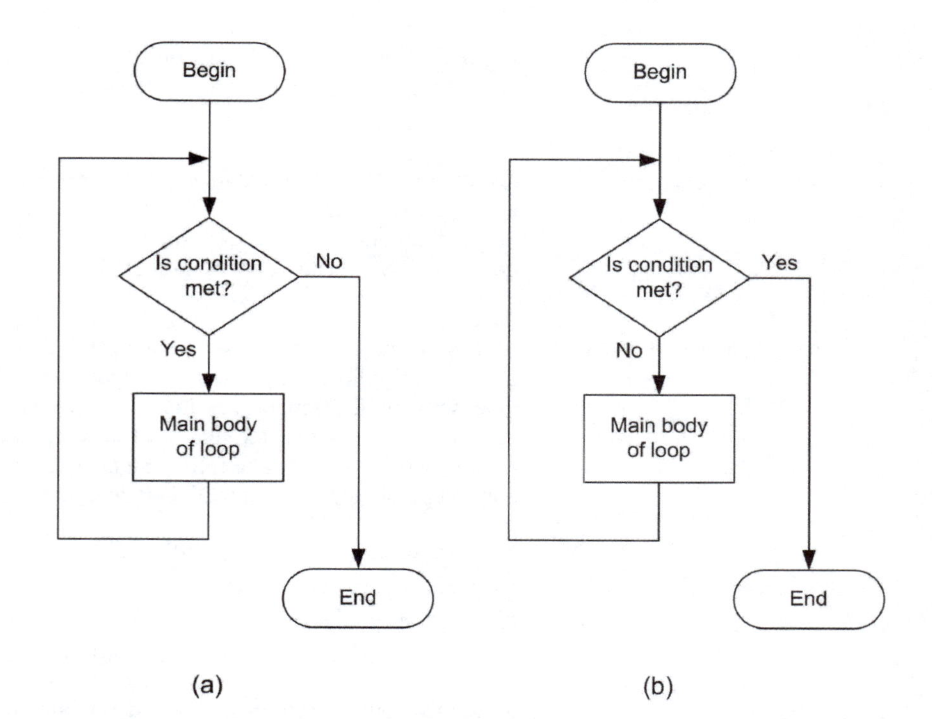

Figure 6.2 *Flowcharts illustrating the logic of* `DO WHILE ... LOOP` *and* `DO UNTIL ... LOOP` *structures. (a)* `DO WHILE ... LOOP`*; (b)* `DO UNTIL ... LOOP`

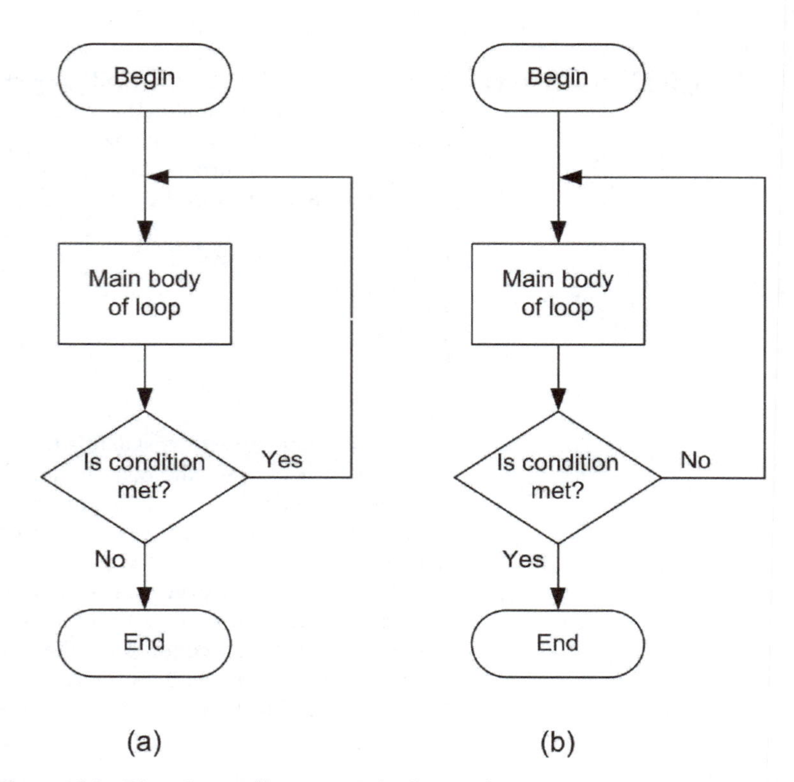

Figure 6.3 *Flowcharts illustrating the logic of* `DO ... LOOP WHILE` *and* `DO ... LOOP UNTIL` *structures. (a)* `DO ... LOOP WHILE`*; (b)* `DO LOOP UNTIL`

respectively. It is important to note that the main body of loop statements within a `DO ... LOOP WHILE` or `DO ... LOOP UNTIL` structure is executed *at least once* whilst the main body of loop statements within a `DO WHILE ... LOOP` or `DO UNTIL ... LOOP` need *never* be executed.

Prompts and messages

Any program to be used by a person other than the originator should incorporate meaningful prompts and messages to aid the user. Prompts should also give some indication of the input required from the user in terms of the acceptable keystrokes, the length of an input string, and the need to include a RETURN keystroke. The following are examples of typical text prompts that are acceptable for use in a DOS Window or a total DOS environment:

```
Do you wish to quit? (Y/N)
Press [SPACE] to continue.
Enter today's date (MM:DD:YY) followed by [RETURN]
Enter filename (max. 8 characters) followed by [RETURN]
```

Messages, unlike prompts, demand no immediate input from the user and should be included at any point in the program at which the user may require information concerning the state of the system. Messages should be written in plain English and should not assume any particular level of technical knowledge on the part of the user. The following are examples of acceptable messages:

```
Loading data file from disk . . . please wait!
Printer is not responding - please check paper supply.
Warning! Strain transducer on Channel 4 is not responding.
```

Keyboard entry

Keyboard input from the user will be required in a variety of applications. Such input may take one of three basic forms summarized below:

1. Single keystrokes. Keystrokes may either be a letter, number, or punctuation and will generally not require the use of the RETURN or ENTER key.
2. Numerical inputs (comprising one or more keystrokes terminated by RETURN or ENTER). Each keystroke must be a number (or decimal point in the case of floats) and the input will normally be assigned to a numeric variable (either integer or floating point).
3. String inputs (comprising one or more keystrokes terminated by RETURN). Each keystroke may be a number, letter, or punctuation. The string input by the user will normally be assigned to a string variable.

Single key inputs

Single key inputs will be required in a wide variety of applications. Such inputs can take various forms including menu selections or simple 'yes/no' confirmations. In either case, it is important to make the user aware of which keys are valid in each selection and, where the consequences of a user's input is irrevocable, a warning should be issued and further confirmation should be sought.

A simple typical 'yes/no' dialogue would take the following form:

```
INPUT "Are you sure (Y/N) "; r$
IF r$ = "Y" THEN ... ELSE ...
```

This piece of code has a number of shortcomings not the least of which is that it will accept any input from the user including a default (i.e. RETURN or ENTER used on its own). Other problems are listed below:

- The user may not realize that the input has to be terminated by ENTER or RETURN.
- A response of 'N' is not distinguished from a default (or any input other than 'Y').
- The routine does not allow a lower case input and the user may not realize that the SHIFT key has to be applied.
- If the user replies with 'YES' or 'yes', this would be equivalent to 'N'!
- Finally, since we would probably want to use the routine at several points within the program, it should be coded as a procedure or user-defined function.

A much better solution to the problem would take the following form:

```
REM Confirm function
DEF FNConfirm%
r$ = ""
f% = -1
```

```
PRINT "Are you sure? (Y/N)"
DO
  DO
    r$ = INKEY$
  LOOP WHILE r$ = ""
  IF r$ "y" OR r$ = "Y" THEN f% = 1
  IF r$ = "n" OR r$ = "N" THEN f% = 0
LOOP WHILE f% <> 1 AND f% <> 0
FNConfirm% = f%
END DEF
...
REM Main program starts here
...
...
IF FNConfirm% = 1 THEN ... ELSE ...
...
...
```

The function returns a flag, `f%`, which is true (non-zero) if the user presses Y or y and is false (zero) if the user presses N or n. The function waits until a character is available from the keyboard (via the inner `DO ... LOOP`) and then checks to ensure that it is one of four acceptable responses (i.e. upper and lower case Y and N). Any other keyboard input is invalid and the program continues to wait for further keyboard input until an acceptable value is returned (during this time the prompt message remains on the screen and does not scroll).

When a valid input is received, the function returns the appropriate flag in `f%`. It is important to note that `INKEY$`, unlike `INPUT`, does not require the use of the `RETURN` or `ENTER` key as a terminator and that a function definition must always precede the main body of code which calls it.

The problem could equally well have been solved by means of a procedure (rather than a user-defined function). In this case the code would have been as follows:

```
REM Main program starts here
DECLARE SUB Confirm(f%)
...
...
CALL Confirm(f%)
IF f% = 1 THEN ... ELSE.
...
...
REM Confirm procedure
SUB Confirm(f%)
r$ = ""
f% = -1
PRINT "Are you sure? (Y/N)"
DO
  DO
    r$ = INKEY$
  LOOP WHILE r$ = ""
  IF r$ = "y" OR r$ = "Y" THEN f% = 1
  IF r$ = "n" OR r$ = "N" THEN f% = 0
LOOP WHILE f% <> 1 AND f% <> 0
END SUB
```

Note that, as with all subprograms, the procedure is declared at the beginning of the main code and defined at the end.

Now, to take a more complex example, let's consider the case of a main menu selection. Suppose we are dealing with a control system which has four main functions (each of which is to be handled by a secondary menu) together with a function which closes down the system and exits from the program. The five main functions will be as follows:

```
1 Set parameters
2 Heater control
3 Pump control
4 Print report
5 Close down
```

The following code can be used for the main program loop:

```
DECLARE SUB Setparams()
DECLARE SUB Heatcontrol()
DECLARE SUB Pumpcontrol()
DECLARE SUB Printreport()
DECLARE SUB Closedown()
REM Main menu selection
WHILE 1
   CLS
   LOCATE 3, 36
   PRINT "MAIN MENU"
   LOCATE 6, 30
   PRINT "[1] Set parameters"
   LOCATE 8, 30
   PRINT "[2] Heater control"
   LOCATE 10, 30
   PRINT "[3] Pump control"
   LOCATE 12, 30
   PRINT "[4] Print report"
   LOCATE 14, 30
   PRINT "[5] Close down"
   LOCATE 16, 30
   PRINT "Option required (1-5)?"
   DO
     r$ = INKEY$
     k% = VAL(r$)
   LOOP UNTIL k% < 6 AND k% > 0
   IF k% = 1 THEN CALL Setparams
   IF k% = 2 THEN CALL Heatcontrol
   IF k% = 3 THEN CALL Pumpcontrol
   IF k% = 4 THEN CALL Printreport
   IF k% = 5 THEN CALL Closedown
WEND
```

Notice that the main program loop consists of an infinite `WHILE ... WEND` loop. The single character string returned by `INKEY` is converted to an integer and then tested to see whether it is within range of the valid keyboard responses (note that depressing the first key returns a value in `k%` of 1, and so on). The `DO ... LOOP` is only exited when a valid keystroke is detected. Having obtained a valid keystroke, the program checks the response to see which key was depressed using a series of IF ... THEN statements so that the desired procedure can be CALLed.

If, for example, the user had pressed the fifth key, the result of the IF ... THEN statement would have been found to be true (all others having been false) and program execution would be diverted to the procedure named Closedown. In this case, and since the result of the Closedown routine is irrevocable, the user should be given the option of returning to the main menu. Hence the Closedown procedure should take the following form:

```
REM Close down and exit
SUB Closedown
CLS
PRINT "You have selected the CLOSE DOWN option."
CALL Confirm(f%)
IF f% = 1 THEN END
END SUB
```

As before, the confirmation function returns a flag, f%, which is true (non-zero) if the user presses Y or y but is false (zero) if the user presses N or n. If the user decides not to continue with the Closedown procedure, END SUB ensures that the procedure is abandoned and execution resumes at the statement which follows the procedure CALL. The WEND statement then diverts the program back to the beginning of the main menu selection routine.

QuickBASIC offers a more powerful logical construct which is particularly useful when making menu selections. The construct is based on SELECT CASE and eliminates the multiple use of IF ... THEN. The equivalent SELECT CASE menu selection program is as follows:

```
DECLARE SUB Setparams()
DECLARE SUB Heatcontrol()
DECLARE SUB Pumpcontrol()
DECLARE SUB Printreport()
DECLARE SUB Closedown()
REM Main menu selection
WHILE 1
  CLS
  LOCATE 3, 36
  PRINT "MAIN MENU"
  LOCATE 6,30
  PRINT "[1] Set parameters"
  LOCATE 8, 30
  PRINT "[2] Heater control"
  LOCATE 10, 30
  PRINT "[3] Pump control"
  LOCATE 12, 30
  PRINT "[4] Print report"
  LOCATE 14, 30
  PRINT "[5] Close down"
  LOCATE 16, 30
  PRINT "Option required (1-5)?"
  DO
    r$ = INKEY$
  LOOP UNTIL r$<>""
  SELECT CASE r$
```

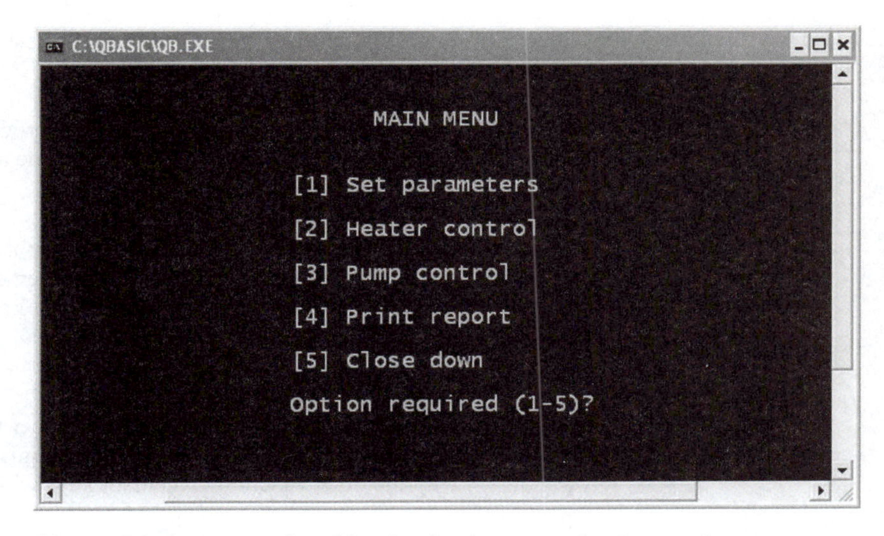

Figure 6.4 *Output produced by the simple menu selection routine*

```
    CASE "1"
      CALL Setparams
    CASE "2"
      CALL Heatcontrol
    CASE "3"
      CALL Pumpcontrol
    CASE "4"
      CALL Printreport
    CASE "5"
      CALL Closedown
    CASE ELSE
      SOUND 60, 2
  END SELECT
WEND
```

Since we are no longer testing for a valid key entry (in the form of a figure in the range 1 to 5) we have included a 'catchall' in the form of the `CASE ELSE` statement. We have also made the program a little more 'user-friendly' by providing the user with an audible warning if an input keystroke is unacceptable. The output produced by the program is shown in Figure 6.4.

Whilst on the topic of 'user-friendly' programs, it is perhaps worth mentioning that good use can be made of QuickBASIC's ability to trap key events. As an example, let us assume that the user is to be provided with an online help facility available from any point in the program when the F1 key is depressed. The following steps are required:

1 Code the subroutine (in this example we shall name it Help) along the lines described earlier in this chapter.
2 Inform QuickBASIC that the Help subroutine is to be associated with the F1 key. The required statement is:

```
ONKEY(1) GOSUB Help
```

Table 6.2 *Function and cursor key numbers for use in conjunction with the QuickBASIC KEY(n) statement*

Key to be trapped	Value of n
F1	1
F2	2
F3	3
F4	4
F5	5
F6	6
F7	7
F8	8
F9	9
F10	10
F11	30
F12	31
Cursor up	11
Cursor left	12
Cursor right	13
Cursor down	14

3 Enable trapping of the F1 key using the statement:

```
KEY(1) ON
```

4 If, at any time (e.g. during some critical process) it is subsequently necessary to disable F1 key trapping, simply include a statement of the form:

```
KEY(1) OFF
```

5 Finally, to temporarily inhibit F1 key trapping but, at the same time remembering whether or not the F1 key has been depressed (so that the event trap can later be executed when a subsequent `KEY ON` statement is encountered) the following statement can be used:

```
KEY(1) STOP
```

For readers who may wish to make further use of QuickBASIC's key event trapping facility, Table 6.2 gives the requisite numerical values associated with the other function and cursor keys within the `KEY(n)` statement.

Numerical inputs

The simple method of dealing with numerical input involves using a BASIC statement of the form:

```
INPUT "Value required"; n%
```

Sadly, this line of code will only work properly if the user realizes that a numeric value is required. Since BASIC cannot assign a letter to a numeric variable, the program will either crash or assign a value of zero if the user inadvertently presses a letter rather than a number. Furthermore, it would be useful to be able to impose a range of acceptable values on the user. The program should reject input values outside this range, warn the user that his input is invalid, and prompt again for further input. Again, such a routine would be ideally coded as a procedure.

The procedure call could typically take the form:

```
prompt$ = "Temperature required"
CALLNumberin(prompt$, 60, 90, num%)
```

while the procedure itself would be coded along the following lines:

```
REM General purpose integer numerical input
SUB Numberin (prompt$, min%, max%, num%)
DO
  PRINT prompt$;
  INPUT num$
  num% = VAL (num$)
  IF n% <= max% AND n% >= min% THEN EXIT SUB
  PRINT "Value outside permissible range!"
LOOP
END SUB
```

The procedure prints the prompt string (`prompt$`) and assigns the user's input to a string variable in order to avoid the program crashing if a letter is inadvertently pressed. The string is subsequently converted to an equivalent numeric variable using the VAL function. The resulting integer is then tested to see whether it lies within the acceptable range. If the integer is within range, the procedure is exited (via EXIT SUB) with num% containing a valid integer input. If the integer is not within range, the user is warned and prompted for further input. A similar routine can be produced for floating point input and, if desired, the prompt string can be included in the list of parameters to be passed into the function.

String inputs

The simple method of dealing with string input involves using a BASIC statement of the form:

```
INPUT "Filename"; n$
```

This line of code is fortunately not quite so prone to problems as its equivalent for numeric input. It is, however, worth considering what action we should take if the user should default the input (i.e. just presses RETURN or ENTER) or proceeds to input an unacceptably long string the latter is an important consideration when dealing with filenames). Hence our general-purpose string input routine should allow for the substitution of a default string and should also truncate the user's input to a specified length. The procedure call might take the following form:

```
prompt$ = "Filename"
CALL Stringin(prompt$, 8, "MYSAMPLE", inputstr$)
```

while the procedure itself would be coded along the following lines:

```
REM General purpose string input
SUB Stringin(prompt$, length%, default$, inputstr$)
  PRINT prompt$;"? ";
  LINE INPUT r$
  IF r$ = "" THEN r$ = default$
  inputstr$ = LEFT$(r$, length%)
END SUB
```

As before, the procedure prints the prompt string (prompt$) and assigns the user's input to a string variable. The use of LINE INPUT (rather than just INPUT) ensures that the user can include punctuation. The user's response (r$) is then checked to determine whether it is a null string (i.e. the user has defaulted) and, ifso, the specified default string is substituted. Lastly, the string is truncated to the specified length using the LEFT$ string function.

The following gives typical user entries and resulting values returned to the main program (in inputstr$) by the foregoing code when length% takes the value 8:

User input	Value returned
OLD_DATA	OLD_DATA
NEW_SAMPLE	NEW_SAMP
CONTROL_DATA	CONTROL_
(default)	MYSAMPLE

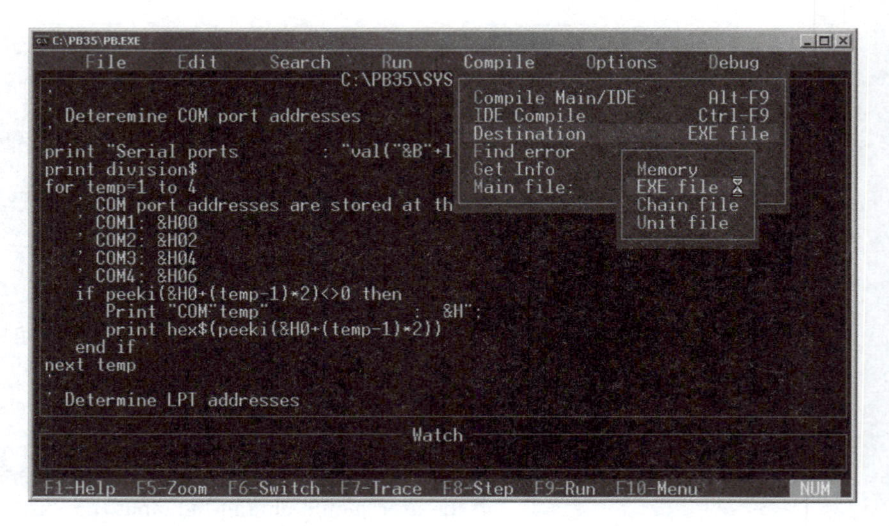

Figure 6.5 *The PowerBASIC 3.5 for DOS IDE*

Table 6.3 *Summary of PowerBASIC 3.5 for DOS variable types*

Variable type	Indicator	Element size (bytes)	DEF type (see Note 1)	Type keyword
Pointer	@	4		PTR
Integer	%	2	DEFINT	INTEGER
Long integer	&	4	DEFLNG	LONG
Quad integer	&&	8	DEFQUD	QUAD
Byte	?	1	DEFBYT	BYTE
Word	??	2	DEFWRD	WORD
Double Word	???	4	DEFDWD	DWORD
Single precision	!	4	DEFSNG	SINGLE
Double precision	#	8	DEFDBL	DOUBLE
Extended precision	##	10	DEFEXT	EXT
BCD fixed point	@	8	DEFFIX	FIX
BCD floating point	@@	10	DEFBCD	BCD
String (see Note 2)	$	2	DEFSTR	STRING
Flex string (see Note 2)	$$	2	DEFFLX	FLEX
Fixed-length string	n/a	n/a		STRING * *x*
ASCIIZ string	n/a	n/a		ASCIIZ * *x*

Notes: 1 DEF type refers to all 13 variable type declaration statements.
2 Only the string handle number is contained in a string array element. The string data itself is stored elsewhere in memory and it occupies as many bytes as the string has characters.

PowerBASIC for DOS

The PowerBASIC package includes two compilers, an Integrated Development Environment (IDE) (see Figure 6.5), and a command-line compiler. The integrated environment provides a text editor, a compiler, a debugger, pull-down menus, Windows, input boxes, and context-sensitive help.

PowerBASIC represents a significant enhancement to earlier DOS BASIC compilers. Power BASIC 2.0 was released in May 1990 in the same year

Table 6.4 *Summary of PowerBASIC 3.5 for DOS data types*

Data type	Size	Range
Integer	16 bits (2 bytes) signed	−32 768 to 32 767
Long integer	32 bits (4 bytes) signed	−2 147 483 648 to 2 147 483 647
Quad integer	64 bits (8 bytes) signed	$\pm 9.22 \times 10^{18}$
Byte	8 bits (1 byte) unsigned	0 to 255
Word	16 bits (2 bytes) unsigned	0 to 65 535
Double Word	32 bits (4 bytes) unsigned	0 to 4 294 967
Single precision	32 bits (4 bytes)	$\pm 8.43 \times 10^{-37}$ to $\pm 3.37 \times 10^{38}$
Double precision	64 bits (8 bytes)	$\pm 4.19 \times 10^{-307}$ to $\pm 1.67 \times 10^{308}$
Extended precision	80 bits (10 bytes)	$\pm 3.4 \times 10^{-4932}$ to $\pm 1.2 \times 10^{4932}$
BCD fixed point	64 bits (8 bytes)	$\pm 9.99 \times 10^{-63}$ to $\pm 9.99 \times 10^{63}$
BCD floating point	80 bits (10 bytes)	$\pm 9.99 \times 10^{-63}$ to $\pm 9.99 \times 10^{63}$

that Microsoft's BASIC Professional Development Systems Version 7.1 was released. However, unlike the latter product, Power BASIC continued to be developed and supported and the current version (Version 3.5 of Power BASIC for DOS) represents the ultimate 'state of the art' in tools for the development of BASIC programs in a DOS environment.

PowerBASIC supports direct generation of 80286 and 80386 processor code and 80287/80387 math coprocessor code. A fast procedure-based math package performs IEEE standard floating point operations and there is support for a full complement of variables and data types (see Tables 6.3 and 6.4).

Accessing assembly language from within BASIC programs

The ability to include assembly language routines within a BASIC program can be invaluable when developing code for instrumentation, data acquisition and control applications. This can be done in two basic ways: the use of inline assembly language statements or the ability to link to an external assembly language module. For those who are already developing assembly language code the latter might be an attractive option but for most of us the ability to introduce assembly language code within a set of BASIC program statements will satisfy most, if not all, requirements.

Here is a simple example of using assembly language within PowerBASIC 3.5 to access the PC's internal speaker:

```
' Title: assfunc.bas      Version: 0.2       Modified: 24/08/04
' Language:  PowerBASIC 3.5
' Function: Demonstrates function calls to inline assembly
' language routines. Generates sounds using the speaker
'
' Main loop to obtain input from user
'
division$ = String$(40, Chr$(205))
blank$ = String$(48, Chr$(32))
color 15, 1
cls
print division$
print "Assembly language function call demo."
print division$
```

```
print "Enter frequency in the range 100Hz to 5000HZ"
do
   locate 6, 1
   print blank$
   locate 6, 1
   print "Frequency (Hz) or <return> to quit: ";
   input freq$
   if freq$ = "" then call stopsound: end
   freq% = val(freq$)
   if freq% > 99 and freq% < 5001 then
     call speaker(freq%)
     locate 5, 1
     print blank$
     locate 5, 1
     print "Current frequency = ";freq%;"Hz"
   end if
loop
'
' Assembly language routines
'
function speaker(BYVAL freq%)
count% = 1190000/freq%
asm    in al, &H61
asm    or al, &H03
asm    out &H61, al
asm    mov al, &Hb6
asm    out &H43, al
asm    mov ax, count%
asm    out &H42, al
asm    mov al, ah
asm    out &H42, al
end function
'
sub stopsound
asm    in al, &H61
asm    and al, &Hfc
asm    out &H61, al
end sub
```

There are two important things to note from this example. Firstly, each line of assembly language code is preceded by the `asm` keyword. Secondly, the assembly language code has direct access to BASIC variables (such as `count%`).

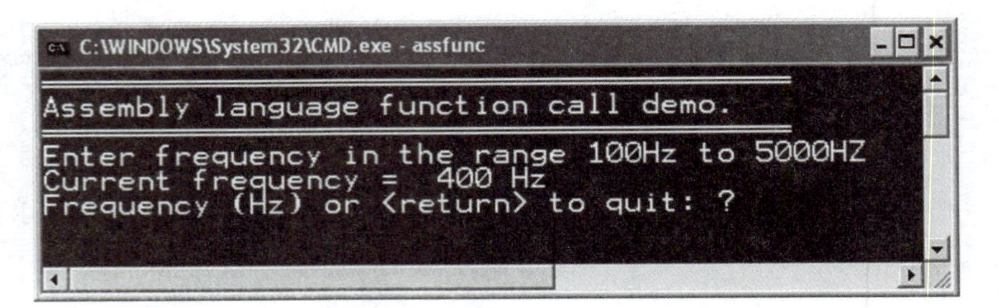

Figure 6.6 *Output produced by the PowerBASIC 3.5 speaker demonstration program*

This is an extremely powerful feature and one that helps to make mixed language programming very straightforward. The output produced by the program is shown in Figure 6.6.

Accessing the I/O ports in DOS or Windows 9x environments

Direct port access from BASIC or assembly language is eminently feasible from a true DOS environment. It is also possible using the Windows 95 or Windows 98 operating systems. Unfortunately, with modern 32-bit operating systems this no longer applies (this topic is further developed later in this chapter when we discuss methods of accessing I/O ports from within the Windows Protected Mode environment on page 215).

The following PowerBASIC 3.5 program demonstrates alternative methods of directly accessing the PC's parallel printer port. This program will operate from a true DOS prompt or from a DOS Window in a Windows 9x operating system:

```
' Title: portio.bas Version: 0.5 Modified: 25/08/04
' Language: PowerBASIC 3.5
' Function: Demonstrates methods of writing to and
' reading from I/O ports. Please see text for information
' about operation in a Windows XP environment
' For test purposes port addresses are currently set to LPT1
' Data data port = &H378 (output) : Status byte = &H379
    (input)
'
color 15, 1
division$ = String$(40, Chr$(205))
'
' Display menu and get option from user
'
do
  cls
  print division$
  print "I/O port read/write demonstration"
  print division$
  print "Select an option..."
  print " [A] for in-line assembly language"
  print " [B] for BASIC"
  print " [Q] to quit"
  do
    r$ = Ucase$(Inkey$)
  loop until r$ <> "" and instr("ABQ", r$)
  print division$
  if r$ = "Q" then goto shutdown
  if r$ = "A" then gosub assem
  if r$ = "B" then gosub basic
loop
'
' Demonstration of BASIC access to I/O ports
'
basic:
status% = inp(&H379) ' read value of status byte
print "Using BASIC: status byte (hex) = "; hex$(status%)
out &H378,&HAA        ' write 10101010 to data port
print "Using BASIC: port data = 10101010"
delay 2
```

```
out &H378, &H55       ' write 01010101 to data port
print "Using BASIC: port data = 01010101"
delay 2
return
'
' Demonstration of in-line assembler access to I/O ports
'
assem:
' read status byte
asm mov dx, &H379
asm in al, dx
asm mov status%, al
print "Using inline assembler: status byte = "; hex$(status%)
' write 10101010 to data port
asm mov al, &HAA
asm mov dx, &H378
asm out dx, al
print "Using inline assembler: port data = 10101010"
delay 2
' write 01010101 to data port
asm mov al, &H55
asm mov dx, &H378
asm out dx, al
print "Using inline assembler: port data = 01010101"
delay 2
return
'
' Reset all data port bits and shutdown
'
shutdown:
asm mov al, &H00
asm mov dx, &H378
asm out dx, al
print "Shutting down ..."
delay 2
cls
end
```

The output produced by the program is shown in Figure 6.7.

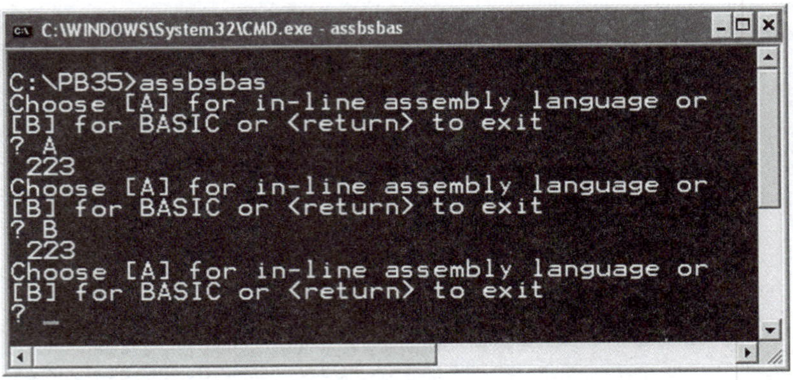

Figure 6.7 *Output produced by the PowerBASIC 3.5 port read/write demonstration program*

Microsoft Visual Basic

Microsoft Visual Basic is an event-driven (see page 161), object-based structured programming language that has become extremely popular for use in the development of data acquisition, control, and measurement applications. Visual Basic supports a wide range of objects starting with the form that contains an application into which further objects (known as *controls*) are placed. Command buttons, labels, text boxes, shapes, and timers are all examples of controls that might be needed in a typical application.

As with virtually all modern programming applications, Visual Basic provides its own Integrated Development Environment (see Figure 6.8). This is a multiple Windows-based environment which provides access to a Toolbox which contains a selection of controls (more specialized controls can be added to the Toolbox where needed), a Form window onto which the controls are placed and modified (e.g. re-sized), a Code window where Visual Basic code is inspected, entered and edited, and a Properties window which provides a means of viewing and editing the properties of and currently selected object. Typical object properties are caption, colour, font, and whether the control is to appear enabled or disabled. The Project window lists all of the forms and modules contained within a project and provides easy access to any that might need editing whilst an Immediate window is provided for debugging purposes.

Visual Basic supports a wide variety of data types (including byte, Boolean, integer, long integer, single- and double-precision floating point, currency, decimal, date, fixed and variable length strings, and variant (this latter type supports both numerical and string data).

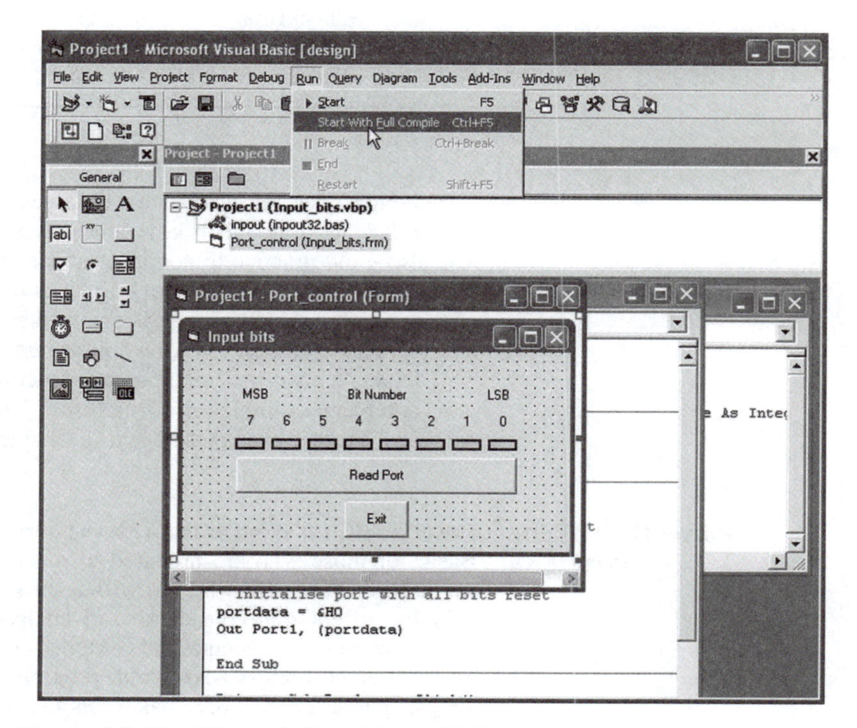

Figure 6.8 *The Microsoft Visual Basic IDE*

Table 6.5 *Recommended Visual Basic variable naming convention*

Data type	Identifier	Example
Byte	b (prefix)	bPortData
Boolean	f (suffix)	ValidDataf
Integer	i (prefix)	iSetPoint
Long	l (prefix)	lResult
Single (floating point)	f (prefix)	fVelocity
Double (floating point)	df (prefix)	fFieldStrength
String	s (prefix)	sStatusText
Object	o (prefix)	oHeaderTank
Variant	v (prefix)	vMessage1

Table 6.6 *Recommended Visual Basic control naming convention*

Control	Prefix	Example
CommandButton	cmd	cmdStart
Label	lbl	vStatus
OptionButton	opt	optFast
TextBox	tb	tbCurrentlResult
CheckBox	cb	cbHeatOn
Shape	sh	shLed
Timer	tmr	tmrOnTime

Although Visual Basic does not enforce the explicit declaration of variables before they are used, this is now considered to be good practice. Furthermore, it is also good practice to adopt a standard naming convention for variables, objects, and code modules. Not only does this help to improve the readability of the code but it will also assists considerably with maintenance and future development. Table 6.5 is a recommended convention for naming variables whilst Table 6.6 shows a recommended method of naming controls.

A complete example of a Microsoft Visual Basic program appears on page 217. Further examples appear in Chapter 13.

PowerBASIC for Windows

PowerBASIC for Windows provides a powerful alternative to Microsoft Visual Basic for those who are prepared to move out of a Microsoft development framework. For those who are willing to make the jump, Power BASIC for Windows offers some considerable advantages not least of which is a significant improvement in the execution of compiled code coupled with the ability to generate extremely compact executable programs. This, coupled with a deceptively simple IDE (see Figure 6.9) makes this a very attractive alternative for software development.

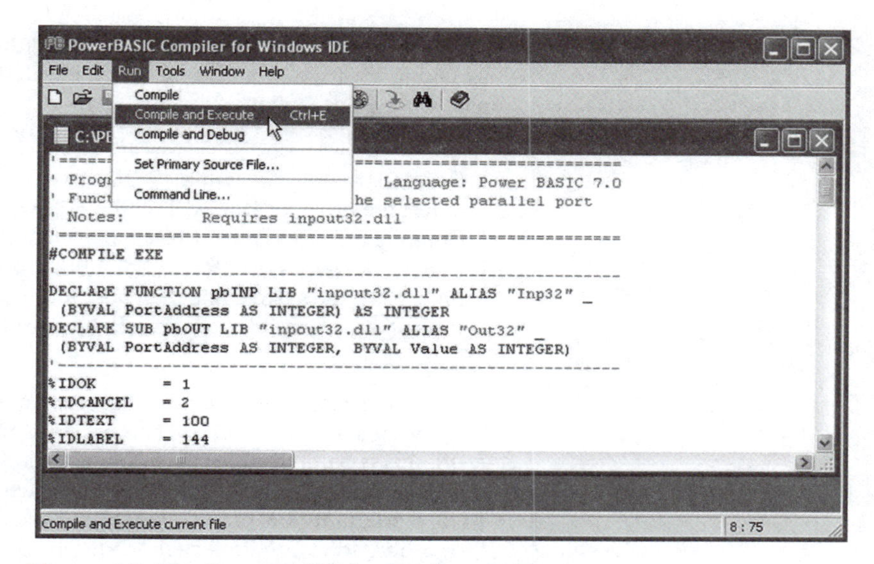

Figure 6.9 *The PowerBASIC for Windows IDE*

A complete example of a PowerBASIC for Windows program appears on page 218.

Using dynamic link library (DLL) files

A dynamic link library (DLL) is a Windows executable library module containing one or more Subs or Functions that can be called by executables (or by other DLLs). DLLs allow you to re-use a common set of procedures without having to include them in each application that needs them. This can significantly reduce the size of executables.

DLLs make efficient use of resources because only one copy of the DLL needs to be present in memory at any particular time in order to offer its services (i.e. access to its Subs and Functions) to any program (or any other DLL) that may need them. DLLs have multiple entry points, one for each exported Sub or Function. The next section shows how DLLs can be used to overcome the problem of accessing ports within the Windows Protected Mode environment.

Accessing the I/O ports from the Windows Protected Mode environment

Modern versions of Windows (Windows NT, Windows 2000, and Windows XP) allow programs to be executed in Protected Mode using a 32-bit flat memory model. The rationale for this was to make Windows more robust by preventing poorly constructed or faulty software from attempting to directly access the system hardware. Unfortunately, this is precisely what we are attempting to do with BASIC code such as:

```
out &H378,&HAA
```

and assembly language code such as:

```
mov al, &HAA
mov dx, &H378
out dx, al
```

It is, however, possible to overcome the limitations of Windows Protected Mode by taking the following steps that will ensure that our program is run within a pure DOS environment:

1 Develop the program within the IDE in the normal way (but note that attempts to test the program from within the IDE will have unpredictable results)
2 Compile the program and save a copy to a floppy disk
3 Use the Windows disk formatting utility to format a DOS boot disk
4 Shut down the system and then boot it directly into DOS using the boot disk
5 Insert the program disk and run the program from the DOS prompt.

Despite the fact that a program that accesses low-level hardware will run quite happily on a Windows XP system provided that it is first booted directly into DOS, this is hardly an elegant solution for long-term software development. It therefore becomes necessary to find a way that will allow us to access hardware from a program that runs under Windows NT/2000/XP without falling foul of the limitations of Protected Mode. To do this we need to make use of a *kernel mode* driver that runs at the highest privileged level and that *does* allow I/O instructions. Writing a kernel mode is not an easy task but fortunately a number of such drivers are currently available for downloading from the Web. They include:

- inpout32.dll from logix4u (http://www.logix4u.net)
- io.dll from Fred Bulback (http://www.geekhideout.com/iodll.shtml)
- NTPort Library 2.5 from Zeal SoftStudio (http://www.zealsoft.com)
- WinIo v2.0 from Yariv Kaplan (http://www.internals.com)
- DriverLINX Port I/O Driver (DLPortIO.DLL) from Scientific Software Tools, Inc. (http://www.sstnet.com).

Inpout32.dll

In order to provide readers with an example of using a kernel mode driver we shall describe the use of inpout.dll from logix4u. The functions in the inpout.dll kernel mode driver are defined in two source files, osversion.cpp and inpout32drv.cpp. The first routine checks the version of operating system (it is highly desirable for a kernel mode driver to be able to operate with all Windows versions) whilst the second routine installs the kernel mode driver (where required) and then performs the required port I/O routines.

The two functions available from inpout32.dll are:

`Inp32` which reads data from the specified parallel port register,

and

`Out32` which writes data to the specified parallel port register.

Various other functions are implemented within Inpout32.dll including those that check the operating system, load and unload the hardware interface, and create the service.

The following is an example of how Inpout32.dll can be used in conjunction with a Visual Basic program that displays the status of each bit of the PC's standard parallel port:

Inpout32.bas contains the following declarations:

```
Public Declare Function Inp Lib "inpout32.dll" _
Alias "Inp32" (ByVal PortAddress As Integer) As Integer
Public Declare Sub Out Lib "inpout32.dll" _
Alias "Out32" (ByVal PortAddress As Integer, ByVal Value As
    Integer)
```

and the main project file is as follows:

```
Dim Port1 As Integer
Dim Port2 As Integer
Dim Port3 As Integer
Dim portdata As Integer
'
Private Sub Exitbutton_Click()
End
End Sub
'
Private Sub Form_Load()
' Port addresses for standard printer port
Port1 = &H378 'Data
Port2 = &H379 'Status
Port3 = &H37A 'Control
' Initialise port with all bits reset
portdata = &H0
Out Port1, (portdata)
End Sub
'
Private Sub Read_port_Click()
portdata = Inp(Port2)
If portdata And 1 Then bit0_led.FillColor = "&H000000FF" _
Else bit0_led.FillColor = "&H00E0E0E0"
If portdata And &H2 Then bit1_led.FillColor = "&H000000FF" _
Else bit1_led.FillColor = "&H00E0E0E0"
If portdata And &H4 Then bit2_led.FillColor = "&H000000FF" _
Else bit2_led.FillColor = "&H00E0E0E0"
If portdata And &H8 Then bit3_led.FillColor = "&H000000FF" _
Else bit3_led.FillColor = "&H00E0E0E0"
If portdata And &H10 Then bit4_led.FillColor = "&H000000FF" _
Else bit4_led.FillColor = "&H00E0E0E0"
If portdata And &H20 Then bit5_led.FillColor = "&H000000FF" _
Else bit5_led.FillColor = "&H00E0E0E0"
If portdata And &H40 Then bit6_led.FillColor = "&H000000FF" _
Else bit6_led.FillColor = "&H00E0E0E0"
If portdata And &H80 Then bit7_led.FillColor = "&H000000FF" _
Else bit7_led.FillColor = "&H00E0E0E0"
End Sub
```

The output produced by the inpout.dll Visual Basic demonstration program is shown in Figure 6.10.

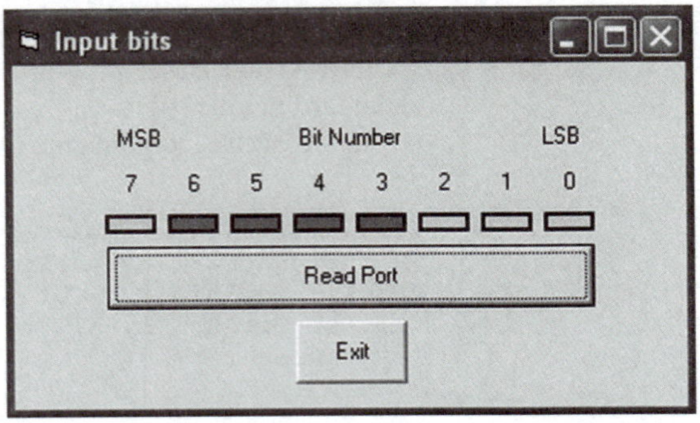

Figure 6.10 *Output produced by the inpout32.dll Visual Basic demonstration program*

As a further example, this PowerBASIC for Windows program is a complete port test utility routine that uses inpout.dll as a means of accessing the parallel ports:

```
'============================================================
' Program Name: porttest.bas Language: Power BASIC 7.0
' Function: Writes data to the selected parallel port
' Notes: Requires inpout32.dll
'============================================================
#COMPILE EXE
'------------------------------------------------------------
DECLARE FUNCTION pbINP LIB "inpout32.dll" ALIAS "Inp32" _
 (BYVAL PortAddress AS INTEGER) AS INTEGER
DECLARE SUB pbOUT LIB "inpout32.dll" ALIAS "Out32" _
 (BYVAL PortAddress AS INTEGER, BYVAL Value AS INTEGER)
'------------------------------------------------------------
%IDOK = 1
%IDCANCEL = 2
%IDTEXT = 100
%IDLABEL = 144
%IDSTATUS = 145
%BN_CLICKED = 0
%BS_DEFAULT = 1
%MF_ENABLED = 0
%MF_CHECKED = 8
%MF_UNCHECKED = 0
%WM_COMMAND = &H111
%ID_LPT1 = 401
%ID_LPT2 = 402
%ID_LPT3 = 403
%ID_HELP = 404
%ID_ABOUT = 405
'------------------------------------------------------------
GLOBAL gsUserInput AS STRING
GLOBAL current_port AS INTEGER
GLOBAL hMenu AS DWORD
GLOBAL hDlg AS DWORD
GLOBAL lResult AS LONG
'------------------------------------------------------------
CALLBACK FUNCTION OkButton()
```

```
  IF CBMSG = %WM_COMMAND AND CBCTLMSG = %BN_CLICKED THEN
    CONTROL GET TEXT CBHNDL, %IDTEXT TO gsUserInput
    DIALOG END CBHNDL, 1
    FUNCTION = 1
  END IF
END FUNCTION
'------------------------------------------------------------
CALLBACK FUNCTION CancelButton()
  IF CBMSG = %WM_COMMAND AND CBCTLMSG = %BN_CLICKED THEN
    DIALOG END CBHNDL, 0
    FUNCTION = 1
  END IF
END FUNCTION
'------------------------------------------------------------
CALLBACK FUNCTION DlgProc()
  IF CBMSG = %WM_COMMAND THEN
    IF CBCTL = %ID_LPT1 THEN
      MSGBOX "Port &H378 selected", &H00002000&
      current_port = &H378
      MENU SET STATE hMenu, BYCMD %ID_LPT1, %MF_CHECKED
      MENU SET STATE hMenu, BYCMD %ID_LPT2, %MF_UNCHECKED
      MENU SET STATE hMenu, BYCMD %ID_LPT3, %MF_UNCHECKED
      FUNCTION = 1
    END IF
    IF CBCTL = %ID_LPT2 THEN
      MSGBOX "Port &H278 selected", &H00002000&
      current_port = &H278
      MENU SET STATE hMenu, BYCMD %ID_LPT1, %MF_UNCHECKED
      MENU SET STATE hMenu, BYCMD %ID_LPT2, %MF_CHECKED
      MENU SET STATE hMenu, BYCMD %ID_LPT3, %MF_UNCHECKED
      FUNCTION = 1
    END IF
    IF CBCTL = %ID_LPT3 THEN
      MSGBOX "Port &H3BC selected", &H00002000&
      current_port = &H3BC
      MENU SET STATE hMenu, BYCMD %ID_LPT1, %MF_UNCHECKED
      MENU SET STATE hMenu, BYCMD %ID_LPT2, %MF_UNCHECKED
      MENU SET STATE hMenu, BYCMD %ID_LPT3, %MF_CHECKED
      FUNCTION = 1
    END IF
    IF CBCTL = %ID_HELP THEN
      MSGBOX "Click on Port from the menu bar to select the required port address" + _
      $CRLF + "(if unselected the routine will default to &H378 - the conventional" + _
      $CRLF + "address for LPT1)." + $CRLF + _
      $CRLF + "Please note that port data must be entered in binary format and" + _
      $CRLF + "all eight binary digits should be entered!", &H00002000&
      FUNCTION = 1
    END IF
    IF CBCTL = %ID_ABOUT THEN
      MSGBOX "Port Test Routine written in Power BASIC. For further information" + _
      $CRLF + "please see 'PC-Based Instrumentation and Control' by Mike Tooley", _
      &H00002000&
      FUNCTION = 1
    END IF
  END IF
END FUNCTION
'------------------------------------------------------------
FUNCTION PBMAIN () AS LONG
```

```
  LOCAL hPopup1 AS DWORD
  portdata% = 0
  current_port = &H378
  ' Top-level menu
  MENU NEW BAR TO hMenu
  ' Pop-up menu for Port selection
  MENU NEW POPUP TO hPopup1
  MENU ADD POPUP, hMenu, "&Port", hPopup1, %MF_ENABLED
  MENU ADD STRING, hPopup1, "&&H378", %ID_LPT1, %MF_CHECKED
  MENU ADD STRING, hPopup1, "&&H278", %ID_LPT2, %MF_ENABLED
  MENU ADD STRING, hpopup1, "&&H3BC", %ID_LPT3, %MF_ENABLED
  ' Pop-up menu for Help and About
  MENU NEW POPUP TO hPopup1
  MENU ADD POPUP, hMenu, "&Help", hPopup1, %MF_ENABLED
  MENU ADD STRING, hPopup1, "&Help",  %ID_HELP, %MF_ENABLED
  MENU ADD STRING, hPopup1, "-",      0, 0
  MENU ADD STRING, hPopup1, "&About", %ID_ABOUT, %MF_ENABLED
  ' Create the dialog and add controls to it
  DIALOG NEW 0, "Port Test Routine", ,, 160, 84, 0, 0 TO hDlg
  CONTROL ADD TEXTBOX, hDlg, %IDTEXT, "00000000", 100, 20, 48, 12, 0
  CONTROL ADD BUTTON, hDlg, %IDOK, "Write to port", 14, 44, 80, 14, %BS_DEFAULT _
  CALL OkButton
  CONTROL ADD BUTTON, hDlg, %IDCANCEL, "Cancel", 104, 44, 40, 14, 0 _
  CALL CancelButton
  CONTROL ADD LABEL, hDlg, %IDLABEL, "Enter binary data to write:", 14, 20, 80, 14
  MENU ATTACH hMenu, hDlg
  ' Display the dialog and check the returned result
  DIALOG SHOW MODAL hDlg, CALL DlgProc TO lResult
  IF lResult THEN
      x& = VERIFY(gsUserInput, "01")
      IF x& = 0 AND LEN(gsUserInput) = 8 THEN
         portdata% = VAL("&B" + gsUserInput)
         pbOut current_port, portdata%
         MSGBOX HEX$(portdata%) & "H written to " & HEX$(current_port) & "H"
      ELSE
         MSGBOX "Data must comprise eight binary digits!"
    END IF
  END IF
END FUNCTION
'------------------------------------------------------------
```

The output produced by the input32.dll PowerBASIC for Windows demonstration program is shown in Figure 6.11. This program contains many aspects of good programming practice, including input verification, user help, and a clear source code layout which includes appropriate comments.

Data files

The ability to store data acquired by a control or instrumentation system is important where a detailed analysis of data or system performance is required. Data may be stored in one, or more, disk files in a disk-based system. Such files can readily be manipulated from BASIC.

The stages required for saving data in a disk file are as follows:

1 Open the file for output (using `OPEN ... FOR OUTPUT`) and include a filename or complete file specification and an associated channel number which will be used to buffer operations on the disk file.

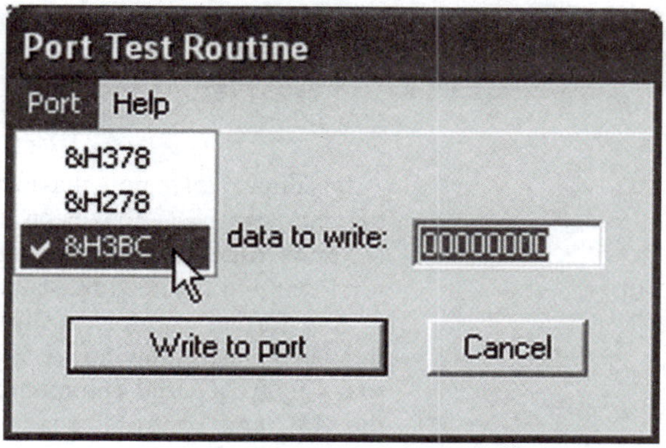

Figure 6.11 *Output produced by the inpout32.dll PowerBASIC for Windows demonstration program*

2 Send data to the file (using `PRINT#`).
3 Close the file (using `CLOSE#`).

As an example, let's assume that we have an integer array of 32 floating values, a(), to be stored in a disk file. If the data file is to be called `'TEMP.DAT'` and is to be stored on a floppy disk inserted into drive A:, the following code can be used in simple DOS BASIC:

```
REM Save data in disk file
SUB Savedata
  SHARED a()
  PRINT "Saving data on disk - please wait!"
  OPEN "A:TEMP.DAT" FOR OUTPUT AS #1
    FOR N% = 0 TO 31
  PRINT #1, a(n%)
  NEXT n%
  CLOSE #1
END SUB
```

The stages required for loading data from a disk file are as follows:

1 Open the file for input (using `OPEN ... FOR INPUT`) and include a filename or complete file specification and an associated channel number for the buffer which will be used for subsequent operations on the file.
2 Retrieve data from the file (using `INPUT#`).
3 Close the file (using `CLOSE#`).

The following code can be used to retrieve the data stored by the previous example, loading it back into array `a()`:

```
REM Load data from disk file
SUB Loaddata
  PRINT "Loading data from disk - please wait!"
  OPEN "A:TEMP.DAT" FOR INPUT AS #1
  FOR n% = 0 TO 31
```

```
    INPUT #1, a(n%)
  NEXT N%
    CLOSE #1
END SUB
```

It is important to note that in both the `Loaddata` and `Savedata` subprograms, the data array, `a()`, has been declared as `SHARED` between the procedure and the main code. The subprograms therefore have access to the data held in the array without the need for values to be passed in the form of a parameter list.

As a further example of simple file handling, the following routines written in Visual Basic show how it is possible to read and write parameters required to configure a serial communications port (note that this application requires the MSComm control from the Visual Basic Toolbox). In order to write the five string data values required to configure a serial port we could use:

```
Public Sub WriteDataParamFile()
Dim FILENUM As Byte
FILENUM = FreeFile ' get ID for the first available free file
Open "SetCom.txt" For Output As #FILENUM
Print #FILENUM, sComPort
Print #FILENUM, sBaudRate
Print #FILENUM, sParity
Print #FILENUM, sDataBits
Print #FILENUM, sStopBits
Close #FILENUM
End Sub
```

Typical default values for the string data might be:

```
sComPort = "1"
sBaudRate = "9600"
sParity = "N"
sDataBits = "8"
sStopBits = "1"
```

The stored data file will thus comprise the following data:

```
1,9600,N,8,1
```

In order to read the data file we will need a routine like the following:

```
Public Sub ReadDataParamFile()
Dim FILENUM As Byte
FILENUM = FreeFile ' get ID for the first available free file
Open "SetCom.txt" For Input As #FILENUM
Input #FILENUM, sComPort
Input #FILENUM, sBaudRate
Input #FILENUM, sParity
Input #FILENUM, sDataBits
Input #FILENUM, sStopBits
Close #FILENUM
End Sub
```

Note that, in practice the foregoing routine would require a simple error handler (using, for example, `On Error GoTo`) to cope with the eventuality that the file was not found.

7 C and C++ programming

The C programming language was the brainchild of Dennis Ritchie and the language was originally implemented on a DEC PDP-11 running under the UNIX operating system. Despite its origins and close association with UNIX, C, and its latest incarnation, C++, is now available in a variety of implementations. These include the immensely popular Borland C++ and Microsoft Visual C.

The C language is comparatively small but it employs a powerful range of control flow and data structures. It is, therefore, not surprising that it has become increasingly popular amongst programmers and software engineers. The language is well suited to the development of effective real-time applications aid is ideally suited to the world of control and instrumentation (C is an excellent choice for small, tight, and fast applications).

The relatively small core of the language has been instrumental in ensuring a high degree of portability from one hardware configuration to another. C offers some significant advantages in the development of software for real-time applications. The language, which promotes the use of structure, is highly portable and it yields code that is relatively compact. Furthermore, when compiled, it can offer execution speeds which are far in excess of those which can be obtained with comparable interpreted languages.

To the newcomer, C and C++ source code can appear somewhat cryptic. Indeed, programmers experienced in other (less structured) languages may have difficulty when making the transition to C. Indeed, it is often said that it is easier to learn C if one has not had the misfortune of acquiring preconceptions developed as a result of a familiarity with BASIC:. Whilst this may be demonstrably true, the fact is that most of today's learners of C and C++ will already be proficient in one or more other languages and these will invariably include BASIC.

Those wishing to develop proficiency with C/C++ programming should not underestimate the amount of time required. As always, the best way to learn is to test out each new concept as it is introduced. Furthermore, it is best not to dwell on comparisons between C and other languages (such as BASIC). It is first necessary to understand something of the structure of C programs before progressing to such topics as data types, pointers, functions, and control structures. The rewards for perseverance are considerable!

The code fragments and complete examples given in this chapter have been written using several different C/C++ compilers but those written using the Borland C++ 4.5 are based on ANSI C and will work with any C compiler. Provided that an effective *Integrated Development Environment* (IDE) is available, program development in C/C++ is simple and straightforward and the process of entering code, compiling, and linking is fully automated.

Figures 7.1–7.4 show examples of four different IDEs for use with C/C++ compilers.

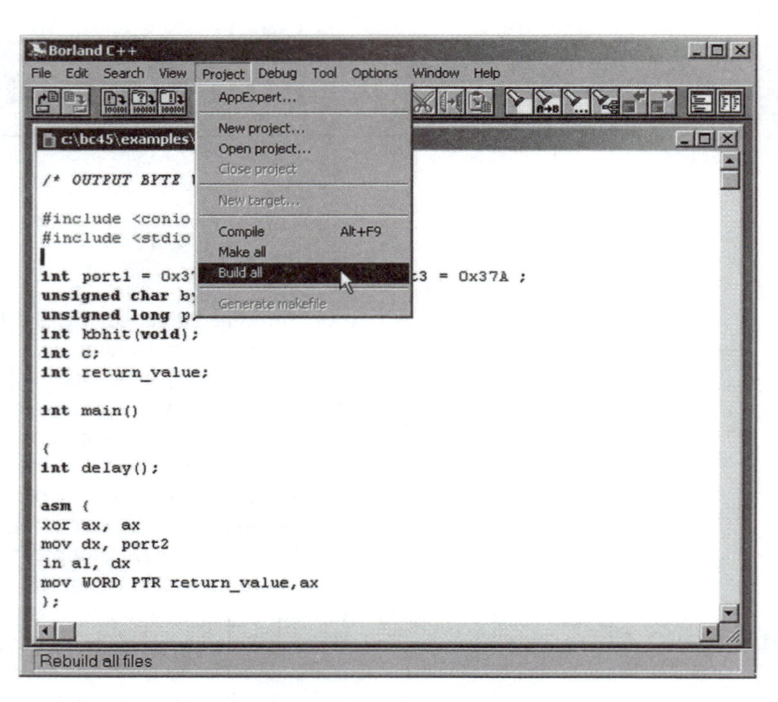

Figure 7.1 *The Borland C++ 4.5 IDE*

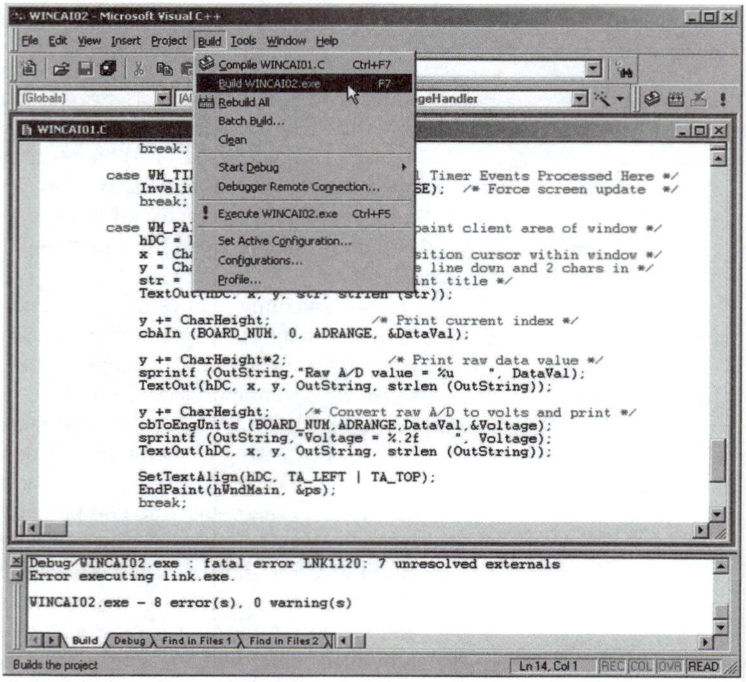

Figure 7.2 *Microsoft Visual C++ IDE*

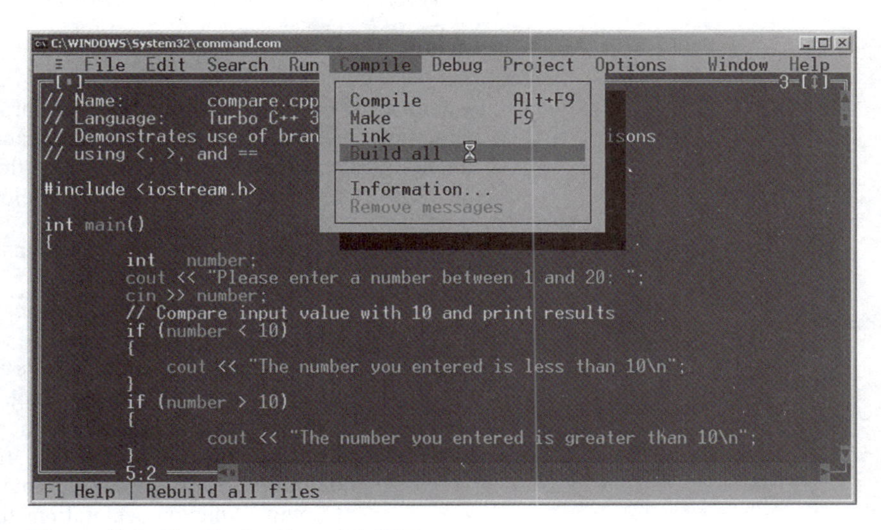

Figure 7.3 *The Turbo C++ 3.0 IDE*

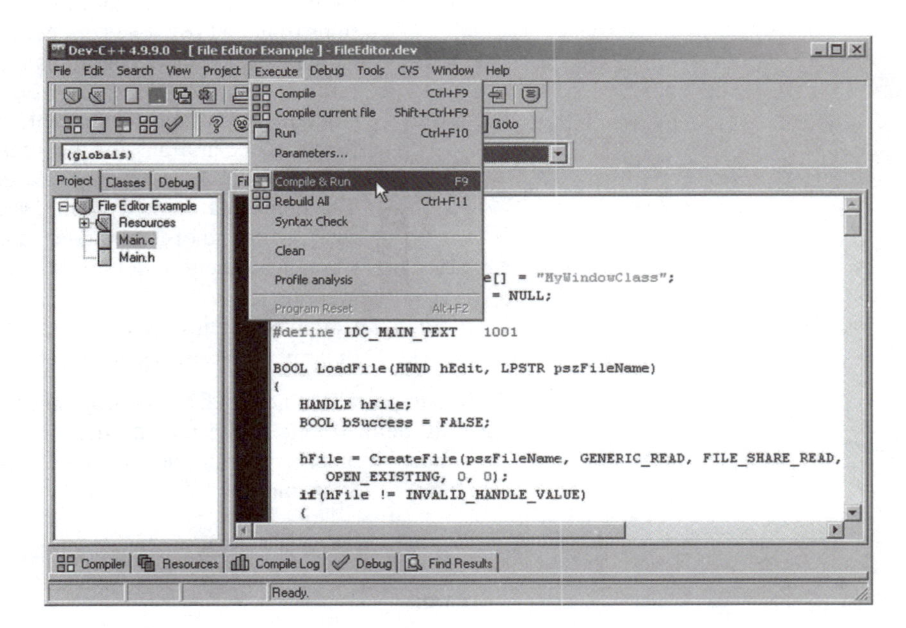

Figure 7.4 *The Dev-C++ IDE*

C programming techniques

There are numerous texts devoted to C/C++ programming. Hence, rather than devote space in this chapter to introducing readers to the basic concepts associated with C programming, we shall adopt the same approach to that used in Chapter 6 by providing a tutorial aimed specifically at showing how C can be used in control applications. Topics have therefore been included that have particular relevance to control, instrumentation and data acquisition. Newcomers to the language are advised to refer to one or more of the recommended texts in Appendix L prior to, or concurrently with, reading this chapter.

Include files

A number of *include* (or *header*) *files* are provided within a set of C run-time library. These files contain macro and constant definitions, type definitions and function declarations. Such files are given the file extension, `.h` and some of the more common include files are listed below:

`bios.h` Contains functions, declarations, and structure definitions for the BIOS service routines.

`conio.h` Contains function declarations for the console and port I/O routines (e.g. `cgets`, `cputs`, `getch`, `inp`, `inpw`, `outp`, and `outpw`).

`ctype.h` Defines unacros and constants audi declares global arrays used in character classification (e.g. `isalnum`, `isalpha`, `islower`, `isupper`, `toascii`, `tolower`, `toupper`, etc.).

`dos.h` Contains macro definitions, function declarations, and type definitions for the MS-DOS interface.

`io.h` Contains function declarations for file handling and low-level I/O functions, such as `open`, `close`, `read`, and `write`.

`malloc.h` Contains function declarations for the memory allocation functions (e.g. `malloc`, `calloc`, `free`, etc.).

`math.h` Contains function declarations for all floating-point mathematics routines (e.g. `abs`, `sin`, `cos`, `log`, `log10`, `exp`, etc.).

`stdio.h` Contains definitions of constant, macros, and types. Also contains function declarations for the stream I/O functions. The function definitions include `fopen`, `fclose`, `fread`, `printf`, and `scanf`. The constants defined within `stdio.h` include `BUFSIZ` (buffer size), `EOF` (end of file marker), and `NULL`.

`stdlib.h` Contains function definitions which include abort, exit, and system.

`string.h` Contains definitions for the string manipulation functions (e.g. `strcpy`, `strlen`, and `strcat`).

It is important to note that many programs use macros, constants, and types that are defined in separate include files. Each file containing such a definition must be specified within the source file (using the pre-processor directive `#include`), for example:

```
#include <stdio.h>
```

Streams

Streams are an abstraction used in C and C++ for input and output operations through a system of I/O based on *characters*. Streams operate with files, keyboard, printer, screen, and I/O ports. When a program that includes `stdio.h` begin its execution, three predefined streams are opened:

`stdin` This is the *standard input* stream. By default stdin corresponds to the keyboard, but this can be redirected by the operating system.

`stdout` This is the *standard output* stream. By default stdout is directed to the screen, but the operating system can redirect it to a file or any other output device.

`stderr` This is the *standard error* stream and is an output stream specifically intended to receive error messages. By default is directed to the standard output (like `stdout`), but it can be redirected to a log file or any other output device.

A `stdio.h` stream is represented by a pointer to a `FILE` structure that contains internal info about properties and indicators of a file. Normally data contained in these structures is not referenced directly. When using `stdio.h`, pointers to `FILE` structures are used to pass parameters to I/O functions.

Stdio.h function summary

`clearerr`	Reset error indicators
`fclose`	Close a stream
`feof`	Check if End Of File (`EOF`) has been reached
`ferror`	Check for errors
`fflush`	Flush a stream
`fgetc`	Get next character from a stream
`fgetpos`	Get position in a stream
`fgets`	Get string from a stream
`fopen`	Open a file
`fprintf`	Print formatted data to a stream
`fputc`	Write character to a stream
`fputchar`	Write character to `stdout`
`fputs`	Write string to a stream
`fread`	Read block of data from a stream
`freopen`	Reopen a file using a different file mode
`fscanf`	Read formatted data from a stream
`fseek`	Reposition stream's position indicator
`fsetpos`	Reposition file pointer to a saved location
`ftell`	Return the current position of the file pointer
`fwrite`	Write block of data to a stream
`getc`	Get the next character
`getchar`	Get the next character from `stdin`
`gets`	Get a string from `stdin`
`getw`	Get the next `int` value from a stream
`perror`	Print error message
`printf`	Print formatted data to `stdout`
`putc`	Write character to a stream
`putchar`	Write character to `stdout`
`puts`	Write a string to `stdout`
`putw`	Write an integer to a stream
`remove`	Delete a file
`rename`	Rename a file or directory
`rewind`	Reposition file pointer to the beginning of a stream
`scanf`	Read formatted data from `stdin`
`setbuf`	Change stream buffering
`setvbuf`	Change stream buffering
`sprintf`	Format data to a string

`sscanf`	Read formatted data from a string
`tmpfile`	Open a temporary file
`tmpnam`	Generate a unique temporary filename
`ungetc`	Push a character back into stream

Stdlib.h function summary

The standard C library functions (`stdlib.h`) can be divided into several groups according to application which can include:

- Conversion (`atof`, `atoi`, `atol`, `ecvt`, `fcvt`, `itoa`, `ltoa`, `strtod`, `strtol`, `strtoul`, `ultoa`).
- Dynamic memory allocation/deallocation (`calloc`, `free`, `malloc`, `realloc`).
- Program control and environment variables (`abort`, `atexit`, `exit`, `getenv`, `putenv`, `system`).
- Sorting and searching (`bsearch`, `lfind`, `lsearch`, `qsort`, `swab`).
- Mathematical operations (`abs`, `div`, `labs`, `ldiv`).

The available functions are as follows:

`abort`	Abort current process returning error code
`abs`	Return absolute value of integer parameter
`atexit`	Specifies a function to be executed at exit
`atof`	Convert string to double
`atoi`	Convert string to integer
`atol`	Convert string to long
`bsearch`	Binary search
`calloc`	Allocate array in memory
`div`	Divide two integer values
`ecvt`	Convert floating-point value to string
`exit`	Terminate calling process
`fcvt`	Convert floating-point value to string
`free`	Deallocate dynamically allocated memory
`gcvt`	Convert floating-point value to string
`getenv`	Get string from environment
`itoa`	Convert integer to string
`labs`	Return absolute value of a long integer
`ldiv`	Divide two long-integer values
`lfind`	Linear search
`lsearch`	Linear search
`ltoa`	Convert long-integer value to string
`malloc`	Allocate memory block
`max`	Return the greater of two parameters
`min`	Return the smaller of two parameters
`putenv`	Create or modify environment variable
`qsort`	Sort using quicksort algorithm
`rand`	Generate random number
`realloc`	Reallocate memory block

`srand`	Initialize random number generator
`strtod`	Convert string to double-precision floating-point value
`strtol`	Convert string to long integer
`strtoul`	Convert string to unsigned long integer
`swab`	Swap bytes
`system`	Execute command
`ultoa`	Convert unsigned long integer to string

Note that a number of the functions listed here are *not* part of the ANSI-C standard but nevertheless they are commonly supported by compilers.

String.h function summary

The `string.h` standard C library to manipulate C strings:

`memchr`	Search buffer for a character
`memcmp`	Compare two buffers
`memcpy`	Copy bytes to buffer from buffer
`memmove`	Copy bytes to buffer from buffer
`memset`	Fill buffer with specified character
`strcat`	Append string
`strchr`	Find character in string
`strcmp`	Compare two strings
`strcoll`	Compare two strings using locale settings
`strcpy`	Copy string
`strcspn`	Search string for occurrence of character set
`strerror`	Get pointer to error message string
`strlen`	Return string length
`strncat`	Append substring to string
`strncmp`	Compare some characters of two strings
`strncpy`	Copy characters from one string to another
`strpbrk`	Scan string for specified characters
`strrchr`	Find last occurrence of character in string
`strspn`	Get length of substring composed of given characters
`strstr`	Find substring
`strtok`	Sequentially truncate string if delimiter is found
`strxfrm`	Transform string using locale settings

Time.h

The `time.h` library provides access to time and date related functions:

`asctime`	Convert `tm` structure to string
`clock`	Return number of clock ticks since process start
`ctime`	Convert `time_t` value to string
`difftime`	Return difference between two times
`gmtime`	Convert `time_t` value to `tm` structure as UTC time
`localtime`	Convert `time_t` value to `tm` structure as local time
`mktime`	Convert `tm` structure to `time_t` value
`time`	Get current time

Note that `clock_t` and `time_t` are long-data types returned by clock and time functions, respectively, whilst `tm` is a structure returned (or used by) the `asctime`, `gmtime`, `localtime`, and `mktime` functions.

Math.h

The `math.h` library provides access to the following maths functions:

`abs`	Return absolute value of integer parameter
`acos`	Calculate arccosine
`asin`	Calculate arcsine
`atan`	Calculate arctangent
`atan2`	Calculate arctangent with two parameters
`atof`	Convert string to double
`ceil`	Return the smallest integer that is greater or equal to `x`
`cos`	Calculate cosine
`cosh`	Calculate hyperbolic cosine
`exp`	Calculate exponential
`fabs`	Return absolute value of floating point
`floor`	Round down value
`fmod`	Return remainder of floating-point division
`frexp`	Get mantissa and exponent of floating-point value
`labs`	Return absolute value of long-integer parameter
`ldexp`	Get floating-point value from mantissa and exponent
`log`	Calculate natural logarithm
`log10`	Calculate logarithm base-10
`modf`	Separate floating-point value into fractional and integer parts
`pow`	Calculate numeric power
`sin`	Calculate sine
`sinh`	Calculate hyperbolic sine
`sqrt`	Calculate square root
`tan`	Calculate tangent
`tanh`	Calculate hyperbolic tangent

Using C functions

The fundamental building blocks of C programs are called *functions*. Once written, functions (like BASIC procedures) may be incorporated in a variety of programs whenever the need arises. For example, the following function definition provides a delay:

```
delay()
{
 long x;
 for (x = 1; x<200000; ++x);
}
```

The delay function is called from a main program by a statement of the form:

```
delay()
```

A complete program to produce a delay would take the form:

```
/* delay1.c */
main()
{
 delay();
}
delay()
{
 long x;
 for (x = 1; x<200000; ++x);
}
```

It is important to note that no semicolon follows the closing bracket of a function definition, whereas when the function is called the program statement is terminated by a semicolon. The main body of the function is enclosed between curly braces ({and}). Since C/C++ is essentially a 'free-form' language (i.e. the compiler ignores white space within the source text) the programmer is able to adopt his/her own style of layout within the source text. The C functions and programs presented in this chapter will, however, follow the convention adopted by the author summarized below:

- Matching opening and closing braces, {and}, are vertically aligned with one another.
- Statements within the body of a function are indented by three columns with respect to their opening and closing braces.
- Expressions (enclosed in brackets) used in conjunction with `for` and `while` statements are placed on the same line as the matching `for` or `while`.
- Where readability needs to be improved, blank lines are used to separate function definitions.
- The first function defined in a program is `main()`.

Returning to the previous example, readers will probably have spotted a fundamental weakness in the simple delay function arising from the fact that it is only capable of providing a fixed delay. The function can be made more versatile by passing a *parameter* into it. The following modified delay function achieves this aim:

```
delay(limit)
long limit;
{
 long x;
 for(x = 1; x < limit; ++x);
}
```

The *argument* (contained in parentheses after the function name) is defined as a `long` type before the function body. The function is then called using a statement of the form:

```
delay(200000);
```

Thereafter, the value 20 000 is passed to the function and is used as the value for *limit.* A simple delay program would then take the form:

```
/* delay2.c */
main ()
```

```
{
 delay(200000);
}
delay(limit)
long limit;
{
 long x;
 for(x = 1; x < limit; ++x);
}
```

Where more than one argument is to be passed to a function, they are simply listed and separated by commas. The data type for each argument must then be defined before the opening brace of the function body. A function definition dealing with port output, for example, might be declared with statements of the form:

```
out (port, byte)
int port, byte;
{
 ...
 ...
 ...
}
```

The corresponding function call would require a statement of the form:

```
out(255,128);
```

In this case, the value 255 would be passed into `port` whilst the value 128 would be passed into `byte`.

I/O functions

The following types of I/O function are available within C:

Stream I/O	In which a data file or data item is treated as a stream of individual characters. Examples of stream I/O functions include `fopen`, `fgetc`, `fgets`, and `fclose`.
Low-level I/O	Routines which do not perform buffering and formating but which, instead, directly invoke the I/O capabilities of the operating system. Examples of low-level I/O functions include `open`, `close`, `read`, and `write`.
Console and port I/O	An extension of stream I/O which permits reading and writing to a console/terminal or sending/receiving bytes of data via an I/O port. Examples of console and port I/O functions include `getch`, `cgets`, `cputs`, `inp`, and `outp` (the latter are used in many control applications).

Messages

Messages in C can be printed using statements of the form:

```
printf(message string goes here)
```

The standard C `printf` statement is, however, more versatile than its equivalent in BASIC as it allows a wide variety of formatting variations. They include:

`\h`	for backspace
`\f`	for form feed
`\n`	for new line
`\t`	for tab

The following example prints the message 'Warning!' immediately preceded and immediately followed by two blank lines:

```
printf("\n\nWarning!\n\n");
```

Variables can be included within the formatted print statement, as the following example shows:

```
printf("Tank number %d temperature %d\n", tankno, temp);
```

The current values of `tankno` and `temp` are printed as integer decimal numbers within the string. Thus if `tankno` and `temp` currently had the values 4 and 56, the resulting output generated would be:

```
Tank number 4 temperature 56
```

C allows a wide range of conversion characters to be included within formatted print strings. These usually include:

`%c`	for single character
`%d`	for signed decimal
`%o`	for unsigned octal
`%s`	for string
`%u`	for unsigned decimal
`%x`	for lower-case hexadecimal
`%X`	for upper-case hexadecimal.

The following example shows how *conversion specifiers* can be used to print the decimal, hexadecimal, and octal value of the same number:

```
printf("Decimal %d, hexadecimal %X, octal %o", num, num, num);
```

It is important to note that each conversion specifier must correspond to an argument within the list. The following code fragment prints the hexadecimal and octal equivalents of the decimal number, 191:

```
/* bconv1.c */
main()
{
 int num;
 num = 191;
 printf("Decimal %d, Hex %X, octal %o", num, num, num);
}
```

Loops

Loops can be easily implemented in C programs. The program that follows (written in Borland C++ 4.5 – see Figure 7.5) prints the ASCII character set and uses the `%d`, `%X`, and `%c` *conversion specifiers* to provide the decimal, hexadecimal, and ASCII representation of the loop index (`byte`). The loop is executed for byte values in the range 32 to 127 and the variable `byte` is

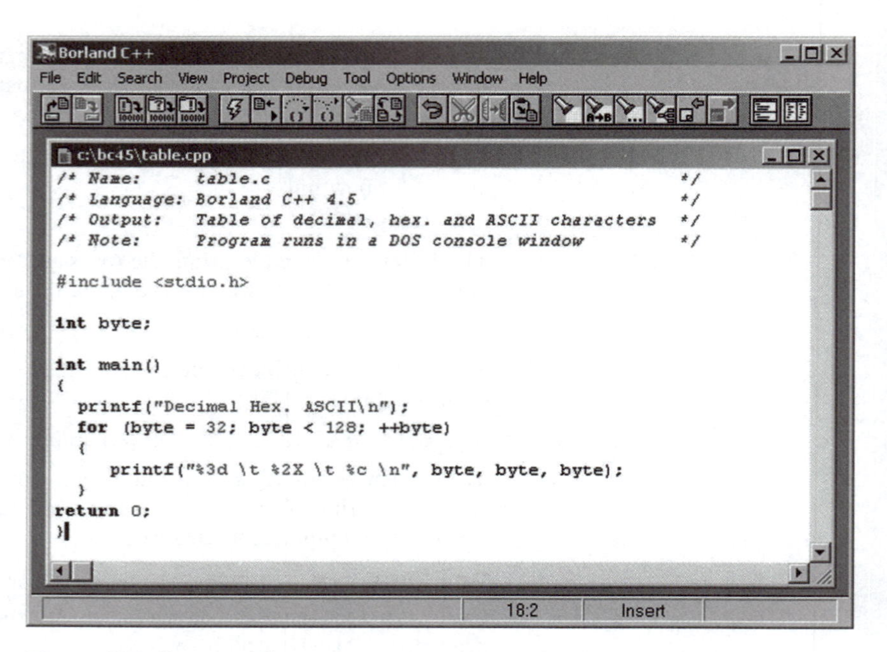

Figure 7.5 *Source code as it appears in the Borland C++ 4.5 editor*

increased by 1 on each pass round the loop. Formatted output is achieved by printing column headings before the loop is entered and including *field-width specifiers* within the format string.

```
/* Name:       table.c                                      */
/* Language:   Borland C++ 4.5                              */
/* Output:     Table of decimal, hex. and ASCII characters  */
/* Note:       Program runs in a DOS console window         */

#include <stdio.h>

int byte;

int main()
{
    printf("Decimal Hex. ASCII \n");
    for (byte = 32; byte < 128; ++byte)
    {
      printf("%3d \t %2X \t %c \n", byte, byte, byte);
    }
return 0;
}
```

The output produced by the program is shown in Figure 7.6.

Loops can also be nested to any required depth. The following Borland C++ 4.5 program provides a simple example based on the use of `while` rather than `for`:

```
/* Name:       loop1.c                                      */
/* Language:   Borland C++ 4.5                              */
/* Output:     Table of decimal, hex. and ASCII characters  */
/* Note:       Program runs in a DOS console window         */
```

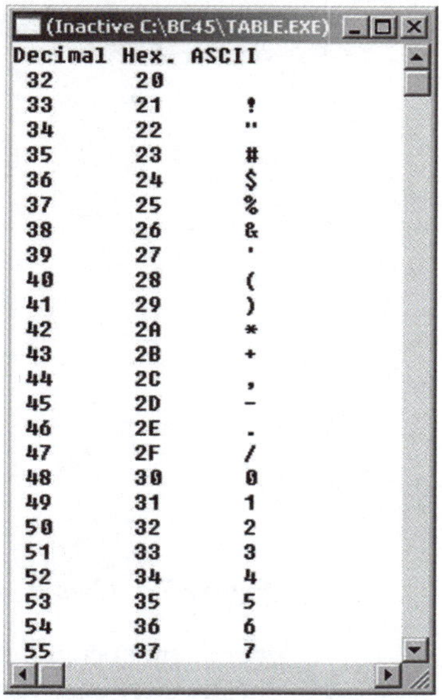

```
(Inactive C:\BC45\TABLE.EXE)
Decimal Hex. ASCII
 32      20
 33      21      !
 34      22      "
 35      23      #
 36      24      $
 37      25      %
 38      26      &
 39      27      '
 40      28      (
 41      29      )
 42      2A      *
 43      2B      +
 44      2C      ,
 45      2D      -
 46      2E      .
 47      2F      /
 48      30      0
 49      31      1
 50      32      2
 51      33      3
 52      34      4
 53      35      5
 54      36      6
 55      37      7
```

Figure 7.6 *Screen output as it appears in the console window when the program in Figure 7.5 is executed*

```
#include <stdio.h>
int main()
{
  int s;
  s = 0;
  while (s < 4)
  {
  ++s;
  printf("Outer loop count = %d\n", s);
  inner();
  }
return 0;
}
int inner()
{
  int t;
  t = 0;
  while (t < 4)
  {
  ++t;
  printf("\tInner loop count = %d\n", t);
  }
}
```

The outer loop is executed four times (with `s` taking the values 0 to 3 in the expression following `while`). The inner loop is executed four times (with `t` taking the values 0 to 3 in the expression following `while`) on each pass through the outer loop. The output produced by the program is shown in Figure 7.7.

There are a few things worth noting about the loop demonstration program. Firstly, the inner loop is defined as a separate function (a sub-process). Locally defined integer variables `s` and `t` are used as loop counters, and the ++

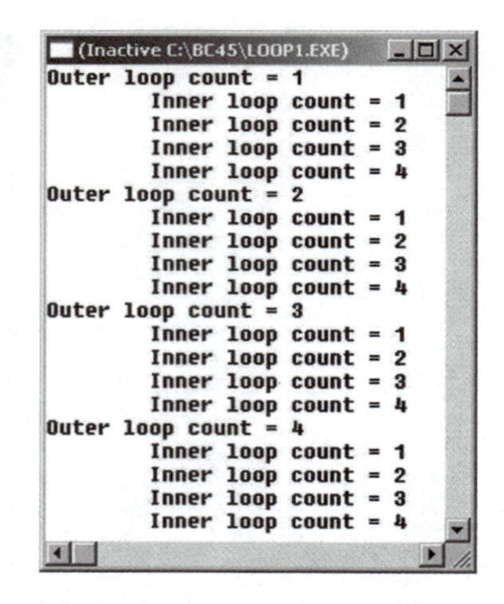

Figure 7.7 *Output produced by the nested loop example program*

increment operator is used to increment the count. Because `s` and `t` are defined locally there is, in fact, no need to use different variables and the following code would produce *exactly* the same result:

```
/* Name:        loop2.c                                        */
/* Language:    Borland C++ 4.5                                */
/* Output:      Table of decimal, hex. and ASCII characters */
/* Note:        Program runs in a DOS console window           */

#include <stdio.h>

int main()
{
  int s;
  s = 0;
  while (s < 4)
  {
      ++s;
      printf("Outer loop count = %d\n", s);
      inner();
  }
return 0;
}

int inner()
{
  int s;
  s = 0;
  while (s < 4)
  {
      ++s;
      printf("\tInner loop count = %d\n", s);
  }
}
```

Inputs and prompts

A single character can be returned from the standard input (usually the keyboard) by means of the `getchar()` function. The following routine shows how a single character can be returned from the keyboard:

```
/* Name:       getin1.c                                       */
/* Language:   Borland C++ 4.5                                */
/* Output:     Table of decimal, hex. and ASCII characters    */
/* Note:       Program runs in a DOS console window           */

#include <stdio.h>

int main()
{
  char c;
  c = inchar("Enter option required... ");
  printf("\n\nOption selected = %c\n", c);
  return 0;
}
int inchar(prompt)
char *prompt;
{
  printf("\n%s",prompt);
  return(getchar());
}
```

Here we have defined a function, `inchar`, which returns an integer to main. This is automatically converted to a character and assigned to the variable, `c`. It is important to note that the return key is used to terminate user input and, where the user provides more than one input character before pressing the return key, only the first character is returned by `getchar()`.

Where a multiple (rather than single) character string is required, the `scanf()` function can be used, as shown in the following code fragment:

```
int getcode()
{
  char code[16];
  printf("Enter operator code... ");
  scanf("%s", code);
}
```

The following Borland C++ 4.5 program shows how a string of characters can be accepted from the user and then printed on the screen:

```
/* Name:       getin2.c                                       */
/* Language:   Borland C++ 4.5                                */
/* Output:     Table of decimal, hex. and ASCII characters    */
/* Note:       Program runs in a DOS console window           */

#include <stdio.h>

int main()
{
  char code[16];
  printf("Enter operator code... ");
  scanf("%s", code);
  printf("\nCode entered: %s", code);
  return 0;
}
```

The `scanf()` function allows a similar set of conversion characters to that available for use within `printf()`. It is important to note that `scanf()` terminates input when a return, space or tab character is detected. Furthermore, the array must be sufficiently large to accommodate the longest string likely to be input. Since the string is automatically terminated by a null character, the array must be dimensioned so that its number of elements is one greater than the maximum string length.

Multiple arguments may be included within `scanf()`. It is important to note that, unlike `printf()`, the arguments to `scanf()` are pointers (not variables themselves). This point regularly causes confusion. Finally, since `scanf()` involves considerable overhead, simpler functions may be preferred where the space for code is strictly limited.

The following example illustrates the combined use of `getchar()`, `printf()`, and `scanf()` in a simple decimal to hexadecimal conversion utility:

```
/* Name:      hexdec.c                                         */
/* Language:  Borland C++ 4.5                                  */
/* Output:    Hexadecimal equivalent of decimal input          */
/* Note:      Program runs in a DOS console window             */

#include <stdio.h>
#include <stdlib.h>

int main()
{
  char number[16];
  int num, c;
  num = 1;
  printf("DECIMAL TO HEXADECIMAL CONVERSION\n");
  printf("=================================");
  while(num != 0)
  {
   printf("\n\nEnter decimal number (max. 65535) or 0 to
       quit: ");
   scanf("%s", number);
   num = atoi(number);
   printf("Decimal %u = %X hexadecimal", num, num);
  }
  return 0;
}
```

The expression following while evaluates true if the current value of `num` is non-zero. In such cases, the code following while is executed and the ASCII character string is converted to an integer by means of the `atoi()` function. If the user responds with 0 (or with a non-numeric character string) and expression evaluates false, the code following while is not executed and the program terminates. The output produced by the program is shown in Figure 7.8.

Menu selection

It is often necessary to provide users with a choice of several options at some point in a control program. Fortunately, C offers the switch case statement which is ready made for this particular purpose. Complex menu selections can be very easily implemented using the `switch case` logical construct. The following example shows how:

```
/* Name:      menu1.c                                          */
/* Language:  Borland C++ 4.5                                  */
/* Output:    Menu routine                                     */
/* Note:      Program runs in a DOS console window             */
```

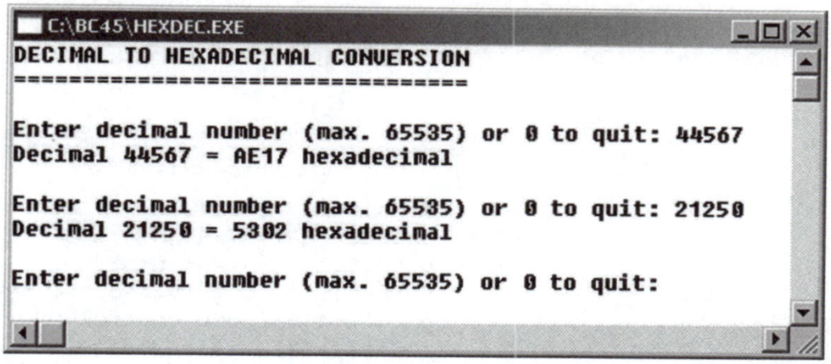

```
C:\BC45\HEXDEC.EXE
DECIMAL TO HEXADECIMAL CONVERSION
=================================

Enter decimal number (max. 65535) or 0 to quit: 44567
Decimal 44567 = AE17 hexadecimal

Enter decimal number (max. 65535) or 0 to quit: 21250
Decimal 21250 = 5302 hexadecimal

Enter decimal number (max. 65535) or 0 to quit:
```

Figure 7.8 *Output produced by the decimal to hexadecimal conversion program*

```
#include <stdio.h>
#define FOREVER 1

int main()
{
  char c;
  while (FOREVER)
  {
  menu();
  c = getchar();
  switch(c)
  {
  case '1':
    init();
    break;
   case '2':
    pump();
    break;
   case '3':
    mix();
    break;
   case '4':
    deliver();
    break;
   case '5':
    exit();
   default:
    printf("Invalid input!\n");
    printf("Please enter a number in the range [1] to [5]\n");
   }
   c = getchar();
  }
  return 0;
}

scroll(lines)
int lines;
{
  int x;
  for(x = 0 ; x < lines; ++x)
```

```
    {
        printf("\n");
    }
}

menu()
{
  scroll(1);
  printf("MAIN MENU\n");
  printf("=========\n");
  printf("[1] Initialise the system\n");
  printf("[2] Pump control\n");
  printf("[3] Mixer control\n");
  printf("[4] Delivery control\n");
  printf("[5] Close down and exit\n");
  scroll(1);
  printf("Enter option required... ");
}

init()
{
  scroll(2);
  printf("INITIALISING SYSTEM - PLEASE WAIT!\n");
  scroll(2);
  /* More code goes here */
}

pump()
{
  scroll(2);
  printf("PUMP CONTROL\n");
  scroll(2);
  /* More code goes here */
}

mix()
{
  scroll(2);
  printf("MIXER CONTROL\n");
  scroll(2);
  /* More code goes here */
}

deliver()
{
  scroll(2);
  printf("DELIVERY CONTROL\n");
  scroll(2);
  /* More code goes here */
}
```

The output produced by the menu program is shown in Figure 7.9.

Passing arguments into main

A useful facility available within C running under MS-DOS is that of passing arguments into programs from which the command line input by the user when the program is first loaded. The `main()` function allows two arguments: `argc`

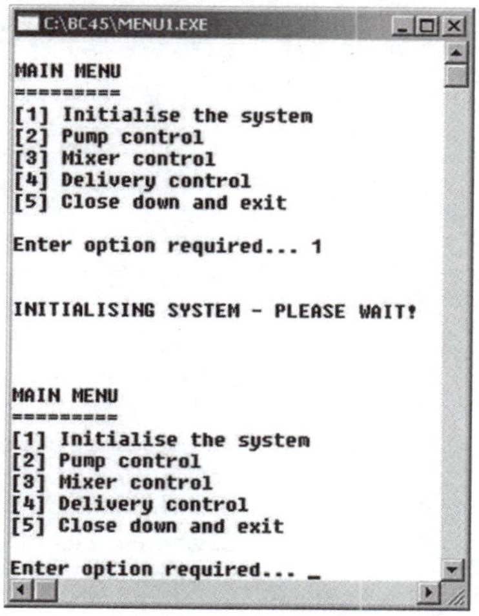

```
C:\BC45\MENU1.EXE
MAIN MENU
=========
[1] Initialise the system
[2] Pump control
[3] Mixer control
[4] Delivery control
[5] Close down and exit

Enter option required... 1

INITIALISING SYSTEM - PLEASE WAIT!

MAIN MENU
=========
[1] Initialise the system
[2] Pump control
[3] Mixer control
[4] Delivery control
[5] Close down and exit

Enter option required... _
```

Figure 7.9 *Output produced by the example menu program*

and `argv`. When main is called, `argc` is the number of elements in `argv`, and `argv` is an array of pointers to the strings which appear in the command line.

The following demonstration illustrates the method of passing parameters:

```
/* argdemo.c */

#include<stdio.h>

main(argc, argv)
char *argv[];
int argc;
{
  int i
  printf("argc: %d\n", argc);
  for(i = 0;i < argc; i++)
  printf("argv[%d]: %s\n", i ,arqv[i]);
}
```

After compilation into an executable program the routine is invoked from the MS-DOS in the following manner:

```
ARGDEMO ONE TWO THREE
```

The parameters to be passed are, in this case, the strings: one, two, and three.

The program generates the following output:

```
argc: 4
argv[0]: C\MC\BIN\ARGDEMO.EXE
argv[1]: one
argv[2]: two
argv[3]: three
```

The total number of parameters passed is given in `argc`. In this case, four parameters have been passed (including the current directory and program name which appears as `argv[0]`). The three strings (one, two, and three) appear as `argv[1]`, `argv[2]`, and `argv[3]`.

The following code fragment shows how the technique of parameter passing can be used to display the contents of a named file.

```
#include<stdio.h>

FILE *stream;

int main(argc, argv)
char *argv[];
int argc;
{
  int c;
  c = 1;
  stream = fopen(argv[1], "rb");
  while(c != EOF)
  {
    c = getc(stream);
    printf("%c", c);
  }
  fclose(stream);
   return 0;
}
```

Assuming that the compiled program is named SPRINT.EXE, the command line entered after the operating system prompt, would take the form:

```
SPRINT filename
```

To print a file called HELLO.DOC, the command would be:

```
SPRINT HELLO.DOC
```

More than one argument can be passed into main. The following Borland C++ 4.5 program is an example of a simple utility program for renaming files. This program passes two arguments (`argc` and `argv`) into main:

```
/* Name:       rname.c                                       */
/* Language:   Borland C++ 4.5                               */
/* Output:     Menu routine                                  */
/* Note:       Program runs in a DOS console window          */

#include<io.h>
#include<stdio.h>

int main(argc, argv)
char *argv[];
int argc;
{
  int result;
  result = rename(argv[1], argv[2]);
  if(result != 0)
    printf("\nUnable to rename file!\n");
  else
    printf("\Rename successful!\n");
return 0;
}
```

Assuming that the program is called RNAME, the command line entered after the operating system prompt, would take the form:

```
RNAME oldname newname
```

To change the name of a file called HELLO.DOC to GOODBYE.DOC the command would be:

```
RNAME HELLO.DOC GOODBYE.DOC
```

Disk files

File handling is quite straightforward using streams in C. Files must be opened before they can be used using a statement of the form:

```
stream = fopen(filename, mode);
```

The filename can be a file specification or the name of a logical device. The mode can be 'r' for read, 'w' for write, and 'u' for update. If the file cannot be opened (e.g. it is not present on the disk) `fopen()` returns 0 otherwise `fopen` returns the stream number to be used in conjunction with subsequent read or write operations.

After use, files must be closed using statement of the form:

```
fclose(stream);
```

where `stream` is the channel number returned by a previous `fopen` statement.

As a further example of file handling in C, the following Borland C++ 4.5 program converts the case of an ASCII file to upper case:

```
/* Name:      ucase.c                                              */
/* Language:  Borland C++ 4.5                                      */
/* Output:    Menu routine                                         */
/* Note:      Program runs in a DOS console window                 */

#include<stdio.h>
#include<io.h>

FILE *fp, *fq;
int pc;

int main(argc, argv)
int argc;
char *argv[];
{
printf("CONVERTING FILE TO UPPER CASE n");
fp = fopen(argv[1], "r");
if (fp == 0)
  {
  printf("Unable to open input file: %s \n" , argv[1]);
  return(0);
  }
fq = fopen(argv[2], "w");
if (fq == 0)
  {
  printf("Unable to create output file: %s n", argv[2]);
  return(0);
  }
printf("\nInput file: %s \n", argv[1]);
printf("\nOutput file: %s \n", argv[2]);
while((pc = getc(fp)) != EOF)
  {
  pc = toupper(pc);
  putc(pc,fq);
```

```
  putchar(pc);
  }
  fclose(fq);
  fclose(fp);
  exit();
return(0);
}
```

Difference between C and C++

As well as enhancements to the basic language, C++ provides *object-oriented* extensions to C required for modern 'visual' environments. In effect, C++ is a superset of ANSI C and C programs will normally compile without any problems in a C++ environment. C++ also attempts to resolve some of the problems with the basic C language.

Differences between C and C++ are usually apparent from a quick examination of the code. For example, a comment in C would look like this:

```
/* A comment in C */
```

or, spanning several lines:

```
/* A comment
in C
*/
```

Comments in C++ can use either of the following formats (the second example is an in-line comment and the // designates the entire line as a comment):

```
/* A comment in C++ */
```

or, spanning several lines:

```
/* A comment
in C++
*/
```

and:

```
// An in-line comment in C++
```

C++ enforces a higher level of type checking than does C. Furthermore, all functions must be prototyped before use. The next two code fragments show two ways of doing this.

Firstly, the function can be defined before it is used (in this way the definition itself forms the prototype):

```
/* Cube function */
int cube(int x)
{
  return x*x*x;
}
/* Main function */
int main()
{
  int y;
  y = cube(8);
}
```

Alternatively, the prototype can be declared before the function is used:

```
/* Prototype declaration */
int cube(int x);

/* Main function */
void main()
{
  int y;
  y = cube(8);
}

/* Cube function */
int cube(int x)
{
  return x*x*x;
}
```

A simple 'Hello, World!' program written in ANSI C would appear as follows:

```
/* Name:       hello.c                   */
/* Language:   ANSI C                    */
/* Output:     Prints "Hello, World!"    */

#include <stdio.h>

void main()
{
  printf("Hello, World!"\n);
}
```

Whilst its equivalent in C++ would be:

```
/* Name:      hello.cpp                  */
/* Language: C++                         */
/* Output:    Prints "Hello, World!"     */

#include <iostream.h>

using namespace std;
int main()
{
  cout << "Hello, World!\n");
  return 0;
}
```

Note that the `namespace` is a group of definitions and the `cout` object is defined within the standard, `std`, namespace in the `iostream.h` header file (`cout` works in much the same way as `printf` which is defined in `stdio.h`). There are, of course, many more differences between C and C++ but a detailed explanation is beyond the scope of this book.

As a further example of the use of C++ the following program written in Turbo C++ for DOS performs the same function as the MASM32 program (`compare.asm`) shown on page 186. The output produced by the program is shown in Figure 7.10.

```
// Name:        compare.cpp
// Language:    Turbo C++ 3.0
// Output:      Compares input value with 10
// Note:        Demonstrates branching based on
//              <, >, and == comparisons in C++
```

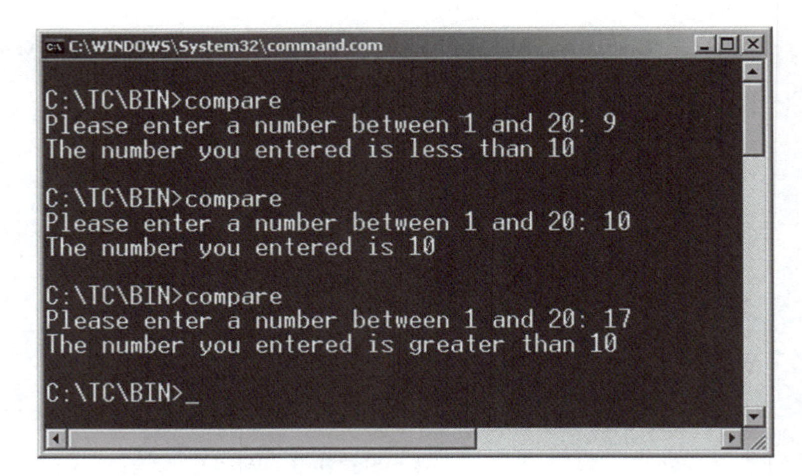

Figure 7.10 *Output produced by the example comparison program*

```
#include <iostream.h>

int main()
{
  int number;
  cout << "Please enter a number between 1 and 20: ");
  cin >> number;
  // Compare input number with 10 and display results
  if (number < 10)
  {
    cout << "The number you entered is less than 10\n";
  }
  if (number > 10)
  {
    cout << "The number you entered is greater than 10\n";
  }
  if (number == 10)
  {
    cout << "The number you entered is 10\n";
  }
return 0;
}
```

Port I/O in C and C++

Finally, the example programs that follow show how easy it is to read and write data to/from I/O ports in C. Note, however, that the restrictions apply to these programs when run from within a Protected Mode environment (see page 215).

The first example reads the state of three ports and displays the input data in decimal whilst the second example output bit-field data to a port. The third example shows how it is possible to include assembly language code within a C/C++ program. All three examples are written using Borland C++ 4.5.

Example 1:

```
/* Name:       readin1.c                                        */
/* Language:   Borland C++ 4.5                                  */
/* Output:     Reads input ports and displays in decimal        */
```

```
#include <conio.h>
#include <stdio.h>

int c, byte_val1, byte_val2, byte_val3;
int port1 = 0x378, port2 = 0x379, port3 = 0x37A ;
int kbhit(void);

int main()
{
gotoxy(0,6);
printf("Press any key to stop ....");
do {
 byte_val1 = inp(port1);
 byte_val2 = inp(port2);
 byte_val3 = inp(port3);
 gotoxy(0,0);
 printf("The value from port1 (%d decimal) is %d \n", port1,
byte_val1);
 printf("The value from port2 (%d decimal) is %d \n", port2,
byte_val2);
 printf("The value from port3 (%d decimal) is %d \n", port3,
byte_val3);
 } while (!kbhit());
return 0;
}
```

Example 2:

```
/* Name:       bitfld1.c                     */
/* Language:   Borland C++ 4.5               */
/* Output:     Sends output data to a port   */

#include <stdio.h>
#include <conio.h>
#include <math.h>

int byte_value[8];
int i;
int bit_no;
int port1 = 0x378;
int out_byte;
char c;

void main()
{
byte_value[0]=0;
byte_value[1]=1;
byte_value[2]=1;
byte_value[3]=0;
byte_value[4]=1;
byte_value[5]=0;
byte_value[6]=1;
byte_value[7]=0;

/* Display bitfield data in binary */
bit_no = 0;
while(bit_no <=8)
```

```
 {
 i = byte_value[bit_no];
 printf("%i",i);
 bit_no++;
 };
printf("\n");

/* Output bitfield data to port */
bit_no = 0;
out_byte = 0;
while(bit_no <=8)
 {
 if (byte_value[bit_no]==1)
   {
   out_byte = out_byte + pow(2, bit_no);
   };
 bit_no++;
 };

outp(port1, out_byte);
printf("Done!\n");
}
```

Example 3:

```
/* Name:       portin1.c                                       */
/* Language:   Borland C++ 4.5                                 */
/* Input:      Reads a port using assembly language            */

#include <conio.h>
#include <stdio.h>

int c, byte_val;
int port = 889;

int main()
{
  byte_val = inp(port);

  asm
  {
  mov ax, 0x0e07
  xor bx, bx
  int 0x10
  }

  printf("The value from port %d is %d \n", port, byte_val);
  return 0;
}
```

8 The IEEE-488 bus

The IEEE-488 bus, also known as the *Hewlett Packard Instrument Bus (HPIB)* and the *General-Purpose Instrument Bus (GPIB)*, provides a means of interconnecting a PC controller with a vast range of test and measuring instruments. The bus is ideally suited to the implementation of *automatic test equipment (ATE)* and it has become increasingly popular over the last two decades with a myriad of applications that range from routine production test to the solution of highly complex and specialized measurement problems. This chapter introduces the IEEE-488 standard and describes the programming of a typical IEEE-488 interface.

In the past, IEEE-488 facilities have tended to be available within only on the more expensive test equipment. The necessary interface is, however, becoming increasingly commonplace in medium- and low-priced instruments. This trend reflects not only an increased demand from the test equipment user but also the availability of low-cost dedicated IEEE-488 controller chips.

Nowadays, most items of modern electronic test equipment (such as digital voltmeters and signal generators) are either fitted with the necessary IEEE-488 interface as standard or can be upgraded with optional IEEE-488 interface cards. This provision allows them to be connected to a PC controller via the IEEE-488 bus such that the controller can be used both to supervise their operation and process the data that they collect.

Automated measurement is important in many applications, not just within the production test environment. Advantages of IEEE-488-based measurement systems incorporating PC-based controllers include:

- Elimination of repetitive manual operation (freeing the test engineer for more demanding tasks).
- Equipment settings are highly repeatable thus ensuring consistency of measurement.
- Increased measurement throughput (measurement rates are typically between 10 and 100 times faster than those which can be achieved by conventional manual methods).
- Reduction of errors caused by maladjustment or incorrect readings.
- Consistency of measurement (important in applications where many identical measurements are made).
- Added functionality (stored data may be analysed and processed in a variety of ways).
- Reduction in skill level of operators (despite the complexity of equipment, user-friendly software can guide operators through the process of connection and adjustment).

The original IEEE-488 standard is often referred to as IEEE-488.1 whilst the most recent developments of the standard are known as IEEE-488.2.

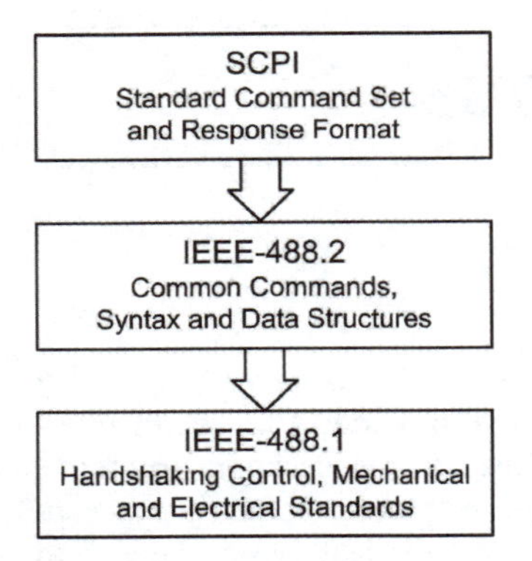

Figure 8.1 *Relationship between the SCPI, IEEE-488.2, and IEEE-488.1 standards*

This enhanced specification more precisely defines the ways in which controllers and instruments communicate with one another. A further improvement in the IEEE-488 standard is the *Standard Commands for Programmable Instruments* (SCPI) specification which provides the IEEE-488.2 specification with a comprehensive command set suitable for all instruments (see Figure 8.1).

The most recent version of the IEEE-488 standard (IEEE.488.1-2003) introduces the high-speed protocol, HS488, for data transfers. This specification supports transfers data rates of up to 8 MB/s although actual rates will depend upon host architecture and system configuration. Furthermore, because HS488 is a superset of the IEEE-488.1 standard, it is possible to mix non-HS488 GPIB devices with devices that are high-speed compatible without having to modify software applications.

IEEE-488 devices

The IEEE-488 standard provides for the following categories of device: listeners, talkers, talkers and listeners, and controllers. We shall briefly examine the role of each type of device.

Listeners

Listeners can receive data and control signals from other devices connected to the bus but are not capable of generating data. An obvious example of a listener is a signal generator.

Talkers

Talkers are only capable of placing data on the bus and cannot receive data. Typical examples of talkers are magnetic tape, magnetic stripe, and bar code readers. Note that, whilst only one talker can be active (i.e. presenting data to the bus) at a given time, it is possible for a number of listeners to be simultaneously active (i.e. receiving and/or processing the data).

Talkers and listeners

The function of a talker and listener can be combined in a single instrument. Such instruments can both send data to and receive from the bus. A digital multimeter is a typical example of a talker and listener. Data is sent to it in order to change ranges and returned to the bus in the form of digitized readings of voltage, current, and resistance.

Controllers

Controllers are used to supervise the flow of data on the bus and provide processing facilities. The controller within an IEEE-488 system is

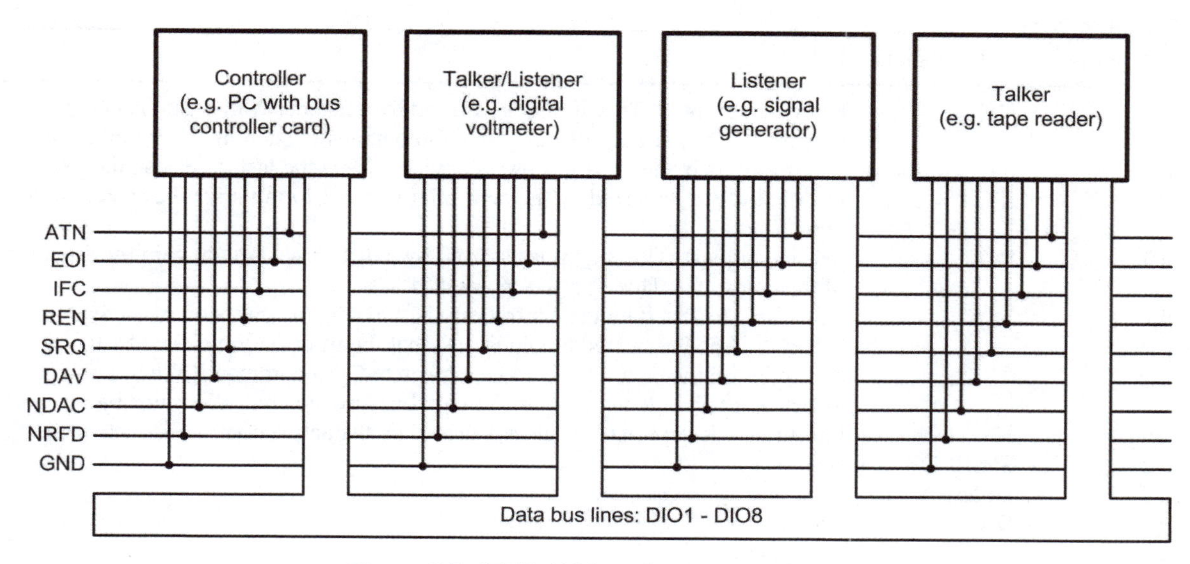

Figure 8.2 *IEEE-488 bus showing signals and devices*

invariably a microcomputer and, whilst some manufacturers provide dedicated microprocessor-based IEEE-488 controllers, this function is often provided by means of a PC or PC-compatible microcomputer.

IEEE-488 bus signals

The IEEE-488 bus uses eight multi-purpose bi-directional parallel data lines (see Figure 8.2). These are used to transfer data, addresses, commands and status bytes. In addition, five bus managements and three handshake lines are provided.

The connector used for the IEEE-488 bus is invariably a 24-pin type (as shown in Figure 8.3) having the following pin assignment:

Pin number	Abbreviation	Function
1	DIO1	Data line 1
2	DIO2	Data line 2
3	DIO3	Data line 3
4	DIO4	Data line 4
5	EOI	End or identify. This signal is generated by a talker to indicate the last byte of data in a multi-byte data transfer. EOI is also issued by the active controller to perform a parallel poll by simultaneously asserting EOI and ATN.
6	DAV	Data valid. Thus signal is asserted by a talker to indicate that valid data has been placed on the bus.
7	NRFD	Not ready for data. This signal is asserted by a listener to indicate that it is not yet ready to accept data.

(*continued*)

Pin number	Abbreviation	Function
8	NDAC	Not data accepted. This signal is asserted by a listener whilst data is being accepted. When several devices are simultaneously listening, each device releases this line at its own rate (the slowest device will be the last to release the line).
9	IFC	Interface clear. Asserted by the controller in order to initialize the system in a known state.
10	SRQ	Service request. This signal is asserted by a device wishing to gain the attention of the controller. This line is wire – OR'd.
11	ATN	Attention. Asserted by the controller when placing a command on to the bus. When the line is asserted this indicates that the information placed by the controller on the data lines is to be interpreted as a command. When it is not asserted, information placed on the data lines by the controller must be interpreted as data. ATN is always driven by the active controller.
12	SHIELD	Shield
13	DIO5	Data line 5
14	DIO6	Data line 6
15	DIO7	Data line 7
16	DIO8	Data line 8
17	REN	Remote enable. This line is used to enable or disable bus control (thus permitting an instrument to be controlled from its own front panel rather than from the bus).
18–24	GND	Ground/common signal return

Notes:

1 Handshake signals (DAV, NRFD, and NDAC) employ active low open–collector outputs which may be used in a wired-CR configuration.

2 All remaining signals are fully TTL compatible and are active low (asserted low).

3 Pins 18 to 23 are intended for use with twisted pair grounds for the control signals (DAV, NRFD, etc.) that appear on pins 6 to 11 on the other side of the connector).

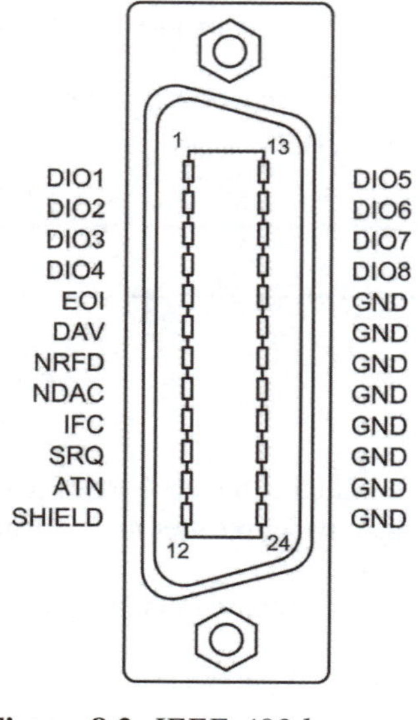

Figure 8.3 *IEEE-488 bus connector*

Commands

Bus commands are signalled by taking the ATN line low. Commands are then placed on the bus by the controller and directed to individual devices by placing a unique address on the lower five data bus lines. Alternatively, universal commands may be issued to all of the participating devices (see page 255).

Handshaking

The IEEE-488 bus uses three handshake lines (DAV, NRFD, and NDAC). The handshake protocol adopted ensures that reliable data transfer occurs at a rate determined by the slowest listener.

A talker wishing to place data on the bus first ensures that NDAC is in a released state. This indicates that all of the listeners have accepted the previous data byte. The talker than places the byte on the bus and waits until NRFD is released. This indicates that all of the addressed listeners are ready to accept

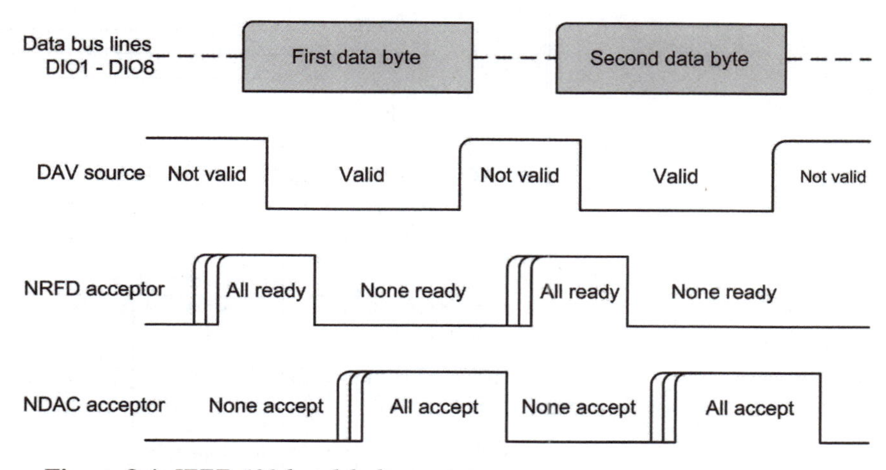

Figure 8.4 *IEEE-488 handshake sequence*

the data. Finally, the talker asserts DAV to indicate that the data on the bus is valid. Figure 8.4 illustrates this sequence of events.

Service requests

The service request (SRQ) line is asserted whenever a device wishes to attract the attention of the active controller. SRQ essentially behaves as a shared interrupt line since all devices have common access to it. In order to determine which device has generated a service request, it is necessary for the controller to carry out a poll of the devices present. The polling process may be carried out either serially or in parallel.

In the case of serial polling, each device will respond to the controller by placing a status byte on the bus. DIO7 will be set if the device in question is requesting service, otherwise this data bit will be reset. The active controller continues to poll each device present in order to determine which one has generated the service request. The remaining bits within the status byte are used to indicate the status of a device and, once the controller has located the device that requires service, it is a fairly simple matter to determine its status and instigate the appropriate action.

In the case of parallel polling, each device asserts an individual data line. The controller can thus very quickly determine which device requires attention. The controller cannot, however, at the same time ascertain the status of the device that has generated the service request. In some cases it will therefore be necessary to carry out a subsequent serial poll of the same device in order to determine its status.

Multi-line commands

The controller sends multi-line commands over the bus as data bytes with ATN asserted. Multi-line commands are divided into five groups, as in the table below. Figure 8.5 summarizes the IEEE-488 command codes.

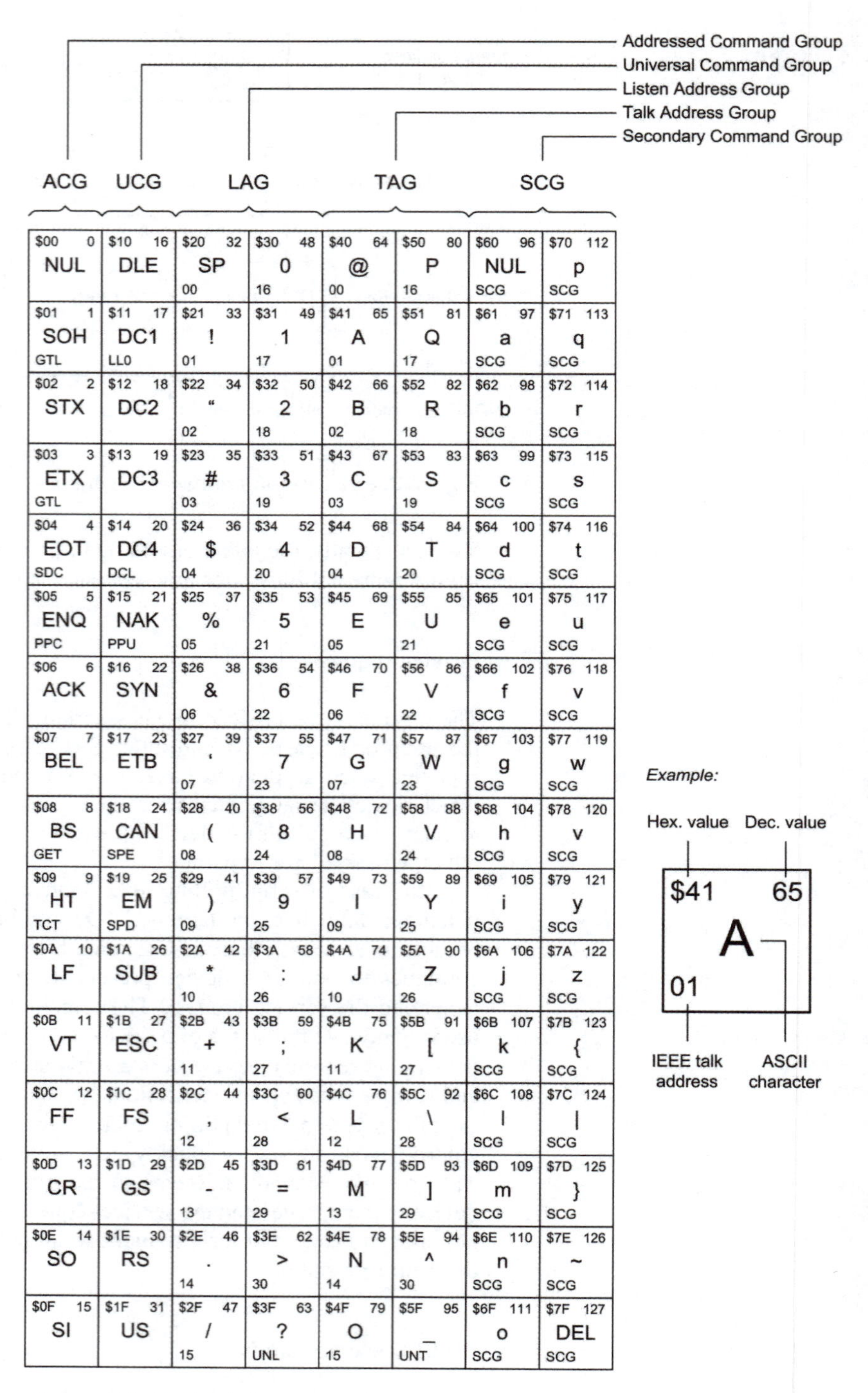

ACG	UCG	LAG		TAG		SCG	
$00 0 NUL	$10 16 DLE	$20 32 SP 00	$30 48 0 16	$40 64 @ 00	$50 80 P 16	$60 96 NUL SCG	$70 112 p SCG
$01 1 SOH GTL	$11 17 DC1 LLO	$21 33 ! 01	$31 49 1 17	$41 65 A 01	$51 81 Q 17	$61 97 a SCG	$71 113 q SCG
$02 2 STX	$12 18 DC2	$22 34 “ 02	$32 50 2 18	$42 66 B 02	$52 82 R 18	$62 98 b SCG	$72 114 r SCG
$03 3 ETX GTL	$13 19 DC3	$23 35 # 03	$33 51 3 19	$43 67 C 03	$53 83 S 19	$63 99 c SCG	$73 115 s SCG
$04 4 EOT SDC	$14 20 DC4 DCL	$24 36 $ 04	$34 52 4 20	$44 68 D 04	$54 84 T 20	$64 100 d SCG	$74 116 t SCG
$05 5 ENQ PPC	$15 21 NAK PPU	$25 37 % 05	$35 53 5 21	$45 69 E 05	$55 85 U 21	$65 101 e SCG	$75 117 u SCG
$06 6 ACK	$16 22 SYN	$26 38 & 06	$36 54 6 22	$46 70 F 06	$56 86 V 22	$66 102 f SCG	$76 118 v SCG
$07 7 BEL	$17 23 ETB	$27 39 ‘ 07	$37 55 7 23	$47 71 G 07	$57 87 W 23	$67 103 g SCG	$77 119 w SCG
$08 8 BS GET	$18 24 CAN SPE	$28 40 (08	$38 56 8 24	$48 72 H 08	$58 88 V 24	$68 104 h SCG	$78 120 v SCG
$09 9 HT TCT	$19 25 EM SPD	$29 41) 09	$39 57 9 25	$49 73 I 09	$59 89 Y 25	$69 105 i SCG	$79 121 y SCG
$0A 10 LF	$1A 26 SUB	$2A 42 * 10	$3A 58 : 26	$4A 74 J 10	$5A 90 Z 26	$6A 106 j SCG	$7A 122 z SCG
$0B 11 VT	$1B 27 ESC	$2B 43 + 11	$3B 59 ; 27	$4B 75 K 11	$5B 91 [27	$6B 107 k SCG	$7B 123 { SCG
$0C 12 FF	$1C 28 FS	$2C 44 , 12	$3C 60 < 28	$4C 76 L 12	$5C 92 \ 28	$6C 108 l SCG	$7C 124 | SCG
$0D 13 CR	$1D 29 GS	$2D 45 - 13	$3D 61 = 29	$4D 77 M 13	$5D 93] 29	$6D 109 m SCG	$7D 125 } SCG
$0E 14 SO	$1E 30 RS	$2E 46 . 14	$3E 62 > 30	$4E 78 N 14	$5E 94 ^ 30	$6E 110 n SCG	$7E 126 ~ SCG
$0F 15 SI	$1F 31 US	$2F 47 / 15	$3F 63 ? UNL	$4F 79 O 15	$5F 95 _ UNT	$6F 111 o SCG	$7F 127 DEL SCG

Figure 8.5 *IEEE-488 command codes*

Command group	Abbreviation	Function	Command byte
Addressed command	ACG	Used to select bus function affecting listeners (e.g. GTL which restores local front panel control of an instrument).	00-0F
Universal command	UCG	Used to select bus functions which apply to all devices (e.g. SPE which instructs all devices to output their serial poll status byte when they become the active talker).	10-1F
Listen address	LAG	Sets a specified device to listen.	20-3E
	UNL	Sets all devices to unlisten status.	3F
Talk address	TAG	Sets a specified device to talk.	40-5E
	UNL	Sets all devices to untalk status.	5F
Secondary address	SCG	Used to specify a device sub-address or sub-function (also used in a parallel poll configure sequence).	60-7F

Bus configurations

Since the physical distance between devices is usually quite small (less than 20 m), data rates can be relatively fast. In fact, data rates of between 50 and 250 KB/s are typical, however, to cater for variations in speed of response, the slowest listener governs the speed at which data transfer takes place. In order to achieve the highest data rates (up to 1 MB/s) it is advisable to restrict the overall length of the bus and to ensure that the maximum separation between devices is about 2 m. Furthermore, no more than 15 devices should be present on the bus and at least two-thirds of those present should be in the powered on state.

Figures 8.6 and 8.7 show two possible arrangements. The first of these (Figure 8.6) shows a basic *daisy chain bus* arrangement where each device is linked to the next device in the chain whilst the second arrangement (Figure 8.7) shows

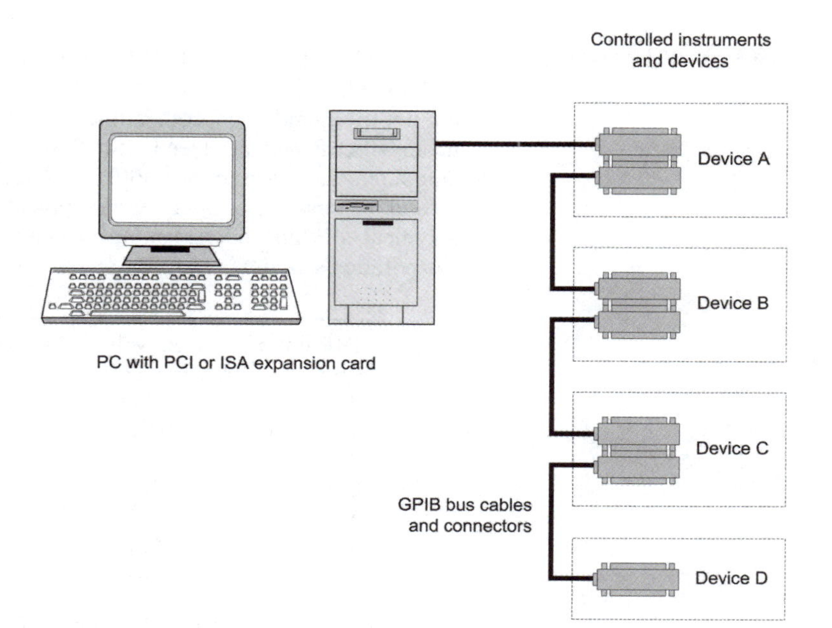

Figure 8.6 *Typical IEEE-488 bus configuration (daisy chain configuration)*

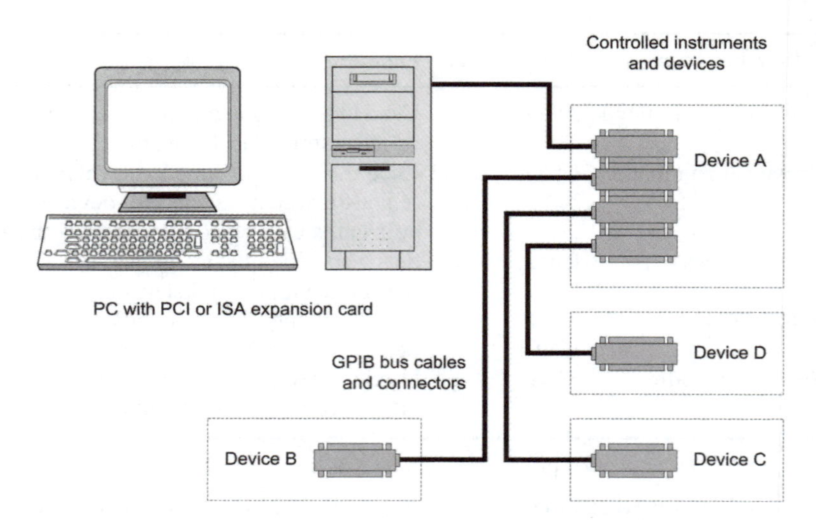

Figure 8.7 *Typical IEEE-488 bus configuration (star configuration)*

a *star bus* arrangement in which one device is connected to three others. Any combination of these two methods is possible provided that the restrictions mentioned in the previous paragraph are obeyed.

In all cases it is advisable to ensure that no instrument cable exceeds 4 m in length and that good quality double shielded (foil and braid plus earth) and twisted pair conductors are used. Connectors can be single or piggyback types in order to permit daisy chaining of devices. Standard cable lengths of 1, 2, and 4 m are available from various suppliers (see Appendix J).

IEEE-488 controllers

IEEE-488 controllers (or *GPIB controllers*) are available in a variety of forms including ISA, PCI, PC/104, and PCMCIA types. ISA and PC/104 cards (which are not 'plug-and-play' compatible) usually require base address selection by means of DIP switches (see Figure 8.8) or by means of PCB links. Since most computers have base address 300H (768 decimal free), this is usually the default setting for this type of board. Other typical addresses include 310H and 330H. A typical DIP/link address setting convention (as used on Metrabyte Computing Corporation's cards) is as follows:

DIP/link no.	Hex. value	Decimal value	Default setting
9	200H	512	ON
8	100H	256	ON
7	80H	128	OFF
6	40H	64	OFF
5	20H	32	OFF
4	10H	16	OFF

The base address is found by simply adding the values of the switches/links that are in the ON position. Thus the default address in the table shown above is

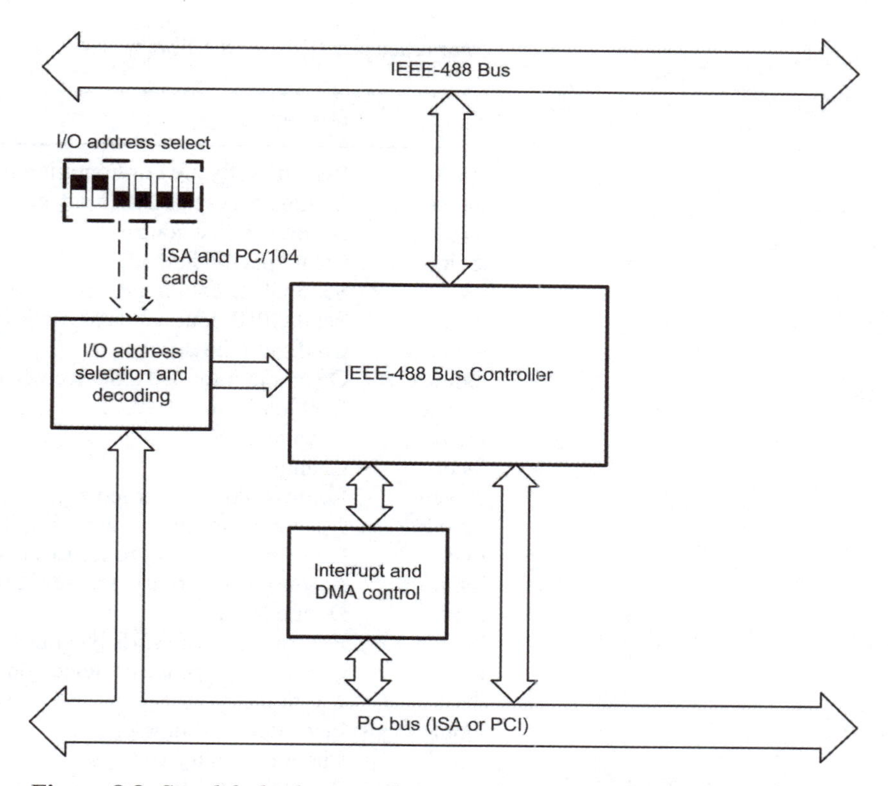

Figure 8.8 *Simplified schematic diagram of an IEEE-488/GPIB controller*

(200H + 100H) = 300H (or 768 decimal). Happily, PCI and PCMCIA cards are 'plug-and-play' compatible and therefore hardware base address selection is not required. The simplified schematic diagram of an IEEE-488/GPIB controller is shown in Figure 8.8.

IEEE-488 software

In order to make use of an IEEE-488 bus interface, it is necessary to have a resident driver to simplify the task of interfacing with control software. The requisite driver is invariably supplied with the interface hardware (i.e. the IEEE-488 expansion card) and is installed when the hardware is fitted. Thereafter, software will be able to communicate with the card using calls to a standard library. The user and/or programmer is then able to access the facilities offered by the IEEE-488 bus using appropriate commands and function calls. For example, the following commands might be used with a multimeter (e.g. Fluke 45):

```
*RST 'Reset the meter
VDC  'Select Volts DC Range
VAL? 'Take a measurement and send it over the GPIB bus
```

Standard libraries are provided in order to simplify the process of programming IEEE-488 bus instruments. The following is an example of a comprehensive IEEE-488.1 library:

Name	Description
ibask	Returns software configuration information
ibbna	Change access board of device
ibcac	Become active controller
ibclr	Clear specified device
ibcmd	Send GPIB commands from a string
ibcmda	Send GPIB commands asynchronously from a string
ibconfig	Configure the driver
ibdev	Open and initialize a device when the device name is unknown
ibdma	Enable/disable DMA
ibeos	Change EOS
ibeot	Change EOI
ibevent	Returns oldest recorded event
ibfind	Open a device and return its unit descriptor
ibgts	Go from active controller to standby
ibinit	Re-initializes library, reloads software configuration
ibist	Define IST bit
iblines	Return status of GPIB bus lines
ibln	Check for presence of device on bus
ibloc	Got to local
ibonl	Place device online/offline
ibpad	Change primary address
ibpct	Pass control
ibppc	Parallel poll configure
ibrd	Read data to a string
ibrda	Read data asynchronously
ibrdf	Read data to file
ibrdi	Read data to integer array
ibrdia	Read data asynchronously to integer array
ibrpp	Conduct parallel poll
ibrsc	Request/release system control
ibrsp	Return serial poll byte
ibrsv	Request service
ibsad	Define secondary address
ibsic	Send IFC
ibsre	Set/clear REN line
irsrq	Install an SRQ interrupt routine
ibstop	Stop asynchronous I/O operation
ibtmo	Define time limit
ibtrg	Trigger selected device
ibwait	Wait for event
ibwrt	Write data from a string
ibwrta	Write data asynchronously from a string
ibwrtf	Write data from file
ibwrti	Write data from integer array
ibwrtia	Write data asynchronously

Commands can be sent to instruments using simple single lines of code. As an example, the following BASIC call is used to send a command string (`cmd$`) to a specified board (`board%`):

```
CALL ibcmd (board%, cmnd$)
```

The alternative in C/C++ would be:

```
ibcmd (int board, char cmnd[], long bytecount)
```

where `board` is an integer containing the board handle, `cmnd` is the command string to be sent, and `bytecount` is the number of command bytes to be transferred.

The following code fragment shows how a standard IEEE-488.2 library can be used for *board level I/O* from BASIC:

```
const METER_ADR = 3
const BOARD_NUM = 0
buffer$ = space$(100)                  ' Reserve space for returned data
DevClear (BOARD_NUM, METER_ADR)        ' Clear the device
Send (BOARD_NUM, METER_ADR, "*RST", DABend)
Send (BOARD_NUM, METER_ADR, "VDC", DABend)
Send (BOARD_NUM, METER_ADR, "VAL?", DABend)
Receive (BOARD_NUM, METER_ADR, buffer$, 100, STOPend)
PRINT "Voltage = "; buffer$            ' Display the returned reading
```

The equivalent in C/C++ would be:

```
#define METER_ADR 3
#define BOARD_NUM 0
int board;
char buffer[100];                      // Reserve space for returned data
DevClear (BOARD_NUM, METER_ADR);       // Clear the device
Send (BOARD_NUM, METER_ADR, "*RST", DABend);
Send (BOARD_NUM, METER_ADR, "VDC", DABend);
Send (BOARD_NUM, METER_ADR, "VAL?", DABend);
Receive (BOARD_NUM, METER_ADR, buffer, STOPend);
printf ("Voltage = %s\n", buffer);     // Display the returned reading
```

Alternatively, the same results can be obtained using *device level I/O* and the IEEE-488.1 from BASIC:

```
buffer$ = space$(100)                  ' Reserve space for returned data
CALL ibfind ("VoltMeter", device%)     ' First open the voltmeter device
CALL ibclr (device%)                   ' and then clear the device
CALL ibwrt (device%, "*RST")           ' Send the command to reset the meter
CALL ibwrt (device%, "VDC")            ' and then select the DC voltage range
CALL ibwrt (device%, "VAL?")           ' Request the current voltage reading
CALL ibrd (device%, buffer$)           ' and read the value into the buffer
PRINT "Voltage = "; buffer$;           ' Now display the returned reading
```

The equivalent in C/C++ would be:

```
int device;                               // Reserve space for returned data
char buffer[100];                         // Reserve space for returned data
device = ibfind ("Volt Meter");           // First open the voltmeter device
ibclr (device);                           // and then clear the device
ibwrt (device, "*RST");                   // Send the command to reset the meter
ibwrt (device, "VDC");                    // and then select the DC voltage range
ibwrt (device, "VAL?");                   // Request the current voltage reading
ibrd (device, buffer,100);                // and read the value into the buffer
printf ("Voltage = %s\n" buffer);         // Now display the returned reading
```

Note that board I/O requires a more detailed understanding of the IEEE-488 bus and programs must explicitly send all of the command codes to set up each bus operation. For example, the following code fragments (the first in BASIC and the second in C/C++) will send the commands UNL, UNT, MTA13, MLA0 (see Figure 8.5) to a board:

```
command$ = "?" + "_" + "M" + " "          ' Create the command string
CALL ibcmd (board%, command$)             ' and send it to the board

command = "\0x3f\0x5f\0x4d\0x20";         // Create the command string
ibcmd (board%, command, 4);               // and send it to the board
```

Troubleshooting the IEEE-488 bus

The IEEE-488 bus is generally well tempered and easy use. Despite this, occasions do arise when the would-be system integrator is confounded by recalcitrant hardware and software which just will not behave as expected. Fortunately, fault finding on the IEEE-488 bus is usually very much simpler than when performing a similar task on an asynchronous serially based system (e.g. an RS-422-based network). There are two main reasons for this: firstly, the IEEE-488 bus standard is open to much less variation in implementation and secondly, all signals use standard TTL voltage levels. This latter fact permits the use of conventional digital instruments (such as logic probes and pulsers – see page 428). Furthermore, the controlling software often contains its own diagnostic routines and will warn the user if, for example, an external device is not responding to commands placed on the bus.

Where necessary, simple routines can be generated to exercise the bus (reading and displaying status codes for each device and transaction). It should be a relatively easy matter to isolate the fault by this means. Alternatively, remote instruments can be checked by interfacing in a different (perhaps simpler) bus configuration and checking that they perform correctly.

Finally, before delving into hardware, it is always worth checking the configuration of the software and the assignment of addresses to the various devices employed within the system at an early stage. If it is necessary to check the state of the various control signal lines (including EOI, SRQ NRFD, NDAC, etc.), a common-or-garden logic probe can be used to check for activity (remember that lines are active low).

9 Interfacing

This chapter aims to introduce readers to the general principles of interfacing sensors and transducers to PC bus I/O cards. We shall describe a variety of common sensors and transducers and, for those who do not wish to make use of 'off-the-shelf' signal conditioning modules, details of the circuitry necessary to interface such devices to several commonly available I/O cards has been provided. Before embarking on this task, it is perhaps worth mentioning some of the more important characteristics and limitations of conventional digital and analogue digital I/O ports.

Characteristics of digital I/O ports

The digital I/O ports provided by most PC expansion cards are invariably byte wide (i.e. each port comprises eight individual I/O lines). Such ports are usually implemented with the aid of one, or more, programmable parallel I/O devices (e.g. the 8255 described on page 29).

Where expansion card parallel I/O devices are connected directly to the outside world via a rear panel-mounted I/O connector, care should be taken to ensure that no output line is excessively loaded nor that any input level exceeds the manufacturer's recommended limits.

As far as outputs are concerned, the Port B lines of a programmable parallel I/O device are usually able to source sufficient current to permit the direct connection of the base of a high current gain (preferably Darlington) NPN transistor. To minimize loading on the remaining I/O lines it will generally be necessary to employ the services of one, or more, octal TTL buffers. In any event, it is important to note that, when sourcing appreciable current, the high-level output voltage present on a port line may fall to below 1.5 V. This will be acceptable when driving a conventional or Darlington transistor but represents an illegal voltage level as far as TTL devices are concerned.

Some digital I/O expansion cards incorporate buffers between the parallel I/O device and the rear panel-mounted expansion I/O connector. Others make use of octal tri-state buffers and transceivers (e.g. 74LS245) rather than a VLSI parallel I/O device. Such devices can often source and sink as much as 15 and 24 mA, respectively.

Where a much higher output current capacity is required, external circuitry will generally be required in order to boost the output current. Alternatively (and provided that switching speed is unimportant) an interface card fitted with medium/high relays may be used. Such a card may also be employed when a high degree of isolation is required between an output load and a PC-based controller.

An expansion card which uses programmable I/O devices (rather than conventional buffers and latches) will require software configuration. A typical configuration routine for an interface based on two 8255 Programmable Peripheral Interace (PPI) devices (providing 48 digital I/O lines in six groups of eight lines) would involve initializing Ports A, B, and C of both devices as either inputs or outputs, as required. This is carried out by simply writing appropriate control words to the *control register* of each device.

Having configured the I/O port, it is then relatively easy matter to send data to it or read data from it. Each port will appear as a unique address within the PC I/O map and data can be read from or written to the port using appropriate IN and OUT statements (or equivalent). Where the digital I/O lines within a port group have individual functions, appropriate *bit masks* can be included in the software so that only the state of the line in question is affected during execution of an OUT command.

Characteristics of analogue I/O ports

PC bus expansion cards for analogue I/O generally provide up to 16 analogue input lines and several analogue output lines. Analogue I/O ports are often based on one or more of the following devices:

Device	Resolution	Function	Package	Notes
AD573JN	10-bit	ADC	20-pin	
AD557JN	8-bit	DAC	16-pin	
AD574	12-bit	ADC	28-pin	
AD667JN	12-bit	DAC	28-pin	
AD1674JN	12-bit	ADC	28-pin	
AD7226KN	8-bit	DAC	20-pin	4-channel
AD7528JN	8-bit	DAC	20-pin	2-channel
AD7528JN	8-bit	DAC	20-pin	2-channel
AD7542KN	12-bit	DAC	16-pin	
AD7545KN	12-bit	DAC	20-pin	
AD7547JN	12-bit	DAC	24-pin	2-channel
AD7569JN	8-bit	ADC/DAC	24-pin	I/O port
AD7579KN	12-bit	ADC	24-pin	
AD7578KN	12-bit	aDC	24-pin	CMOS
AD7580JN	10-bit	ADC	24-pin	
AD7681JN	8-bit	ADC	28-pin	8-channel
AD7672KN	12-bit	ADC	24-pin	high-speed
AD7824KN	8-bit	ADC	24-pin	4-channel
AD7846AD	16-bit	DAC	28-pin	
AD7870JN	12-bit	ADC	24-pin	high-speed
AD7853N	12-bit	ADC	24-pin	high-speed
AD7893BN	12-bit	ADC	8-pin	serial interface
ADC0804LCN	8-bit	ADC	20-pin	
ADC0809CCN	8-bit	ADC	28-pin	
DAC0800LCN	8-bit	DAC	16-pin	
TLC7524CN	8-bit	DAC	20-pin	
ZN425E	8-bit	DAC	16-pin	
ZN427E	8-bit	ADC	18-pin	
ZN428E	8-bit	DAC	16-pin	
ZN435E	8-bit	DAC	18-pin	
ZN439E	8-bit	ADC	22-pin	
ZN448E	8-bit	ADC	18-pin	
ZN502E	10-bit	ADC	28-pin	

Analogue inputs generally exhibit a high resistance (50 kΩ or more) and operational amplifier buffers are usually fitted to provide voltage gain adjustment and additional buffering between the analogue input and the input of the ADC chip.

Analogue outputs are usually available at a relatively low-output impedance (100 Ω or less) and are invariably buffered from the DAC by means of operational amplifier stages. Typical output voltages produced by an analogue output port utilizing an 8-bit DAC range from 0 to 5.1 V (20 mV/bit) when configured for *unipolar operation* or −5.1 to +5.1 V (40 mV/bit) when *bipolar operation* is selected.

The procedure for reading values returned by an analogue input port will vary depending upon the type of ADC used. A typical sequence of operations for use with a multi-channel analogue input card with 8-bit resolution based on the ZN448E ADC would take the following form:

1 Select the desired input channel and start conversion. Send the appropriate byte to the status latch in order to select the required channel and input multiplexer. Conversion starts automatically when data is written to the status latch address.
2 Either
   (i) Wait 10 μs (this is just greater than the 'worst-case' conversion time) using an appropriate software delay.
   or
   (ii) Continuously poll the ADC to sense the state of the end-of-conversion (EOC) line. This signal appears as a single bit in the status byte and, when low, it indicates that conversion is complete and valid data is available from the ADC.
3 Read the data. Having ensured that conversion is complete, the valid data byte can be read from the appropriate ADC address.

The byte read from the port will take a value between 00H and FFH. If the ADC has been configured for unipolar operation, a value of 00H will correspond to 0 V while a value of FFH will correspond to full-scale positive input (typically 5.1 V). When bipolar operation is used, a data byte of 00H will indicate the most negative voltage (typically −5.1 V) whilst FFH will indicate the most positive voltage (typically +5.1 V).

It is important to note that the values returned by conventional successive approximation ADCs will not be accurate unless the input voltage has remained substantially constant during the conversion process. Furthermore, where some variation is inevitable, several samples should be taken and averaged.

Analogue output ports are generally much easier to use than their analogue input counterparts. It is usually merely sufficient to output a byte to the appropriate port address. In most cases, analogue output ports will be configured for unipolar operation and, in the case of an 8-bit DAC, a byte value of 00H will result in an output of 0 V whilst a byte value of FFH will result in a full-scale positive output (typically 5.1 V).

Sensors

Sensors provide a means of inputting information to a process control system. This information relates to external physical conditions such as temperature,

Photo 9.1 *Liquid flow sensor (digital output)*

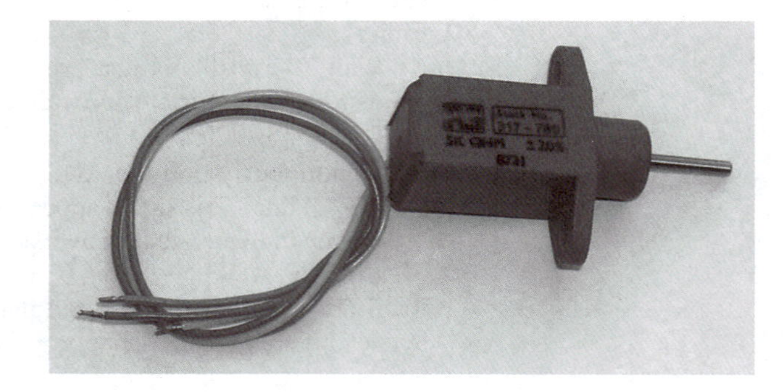

Photo 9.2 *Linear position sensor (analogue output)*

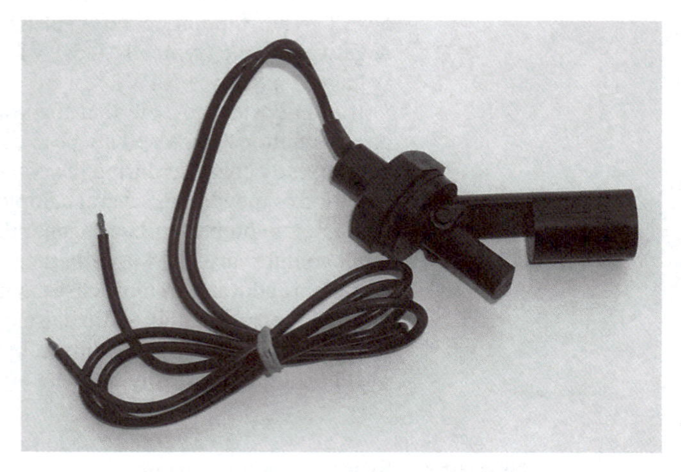

Photo 9.3 *Liquid level float switch*

position, and pressure. The data returned from the sensors together with control inputs from the operator (where appropriate) will subsequently be used to determine the behaviour of the system.

Any practical industrial process control system will involve the use of a number of devices for sensing a variety of physical parameters. The choice of sensor will be governed by a number of factors including accuracy, resolution, cost, and physical size. The following table covers the range of sensors and inputs most commonly encountered in industrial process control systems. The list is not exhaustive and details of other types of sensor can be found in most texts devoted to measurement, instrumentation, and control systems.

Physical parameter	Type of sensor	Notes
Angular position	Resistive rotary position sensor*	Rotary track potentiometer with linear law produces analogue voltage proportional to angular position. Limited angular range. Analogue input port required.
	Optical shaft encoder*	Encoded disk interposed between optical transmitter and receiver (infra-red LED and photodiode or phototransistor). Usually requires signal conditioning based on operational amplifiers. Digital input port required.
	Differential transformer	Transformer with fixed E-laminations and pivoted I-laminations acting as a moving armature. AC source, rectifier, and filter required. Analogue input port required.
Angular velocity	Tachogenerator	Small DC generator with linear output characteristics. Analogue output voltage proportional to shaft speed. Requires an analogue input port.
	Toothed rotor tachometer	Magnetic pick-up responds to the movement of a toothed ferrous disk. May require signal conditioning (typically an operational amplifier and a Schmitt input logic gate). Some sensors contain circuitry to provide TTL-compatible outputs. The pulse repetition frequency of the output is proportional to the angular velocity. Digital input port required.
	Optical shaft encoder*	Encoded disk interposed between optical transmitter and receiver (infra-red LED and photodiode or phototransistor). Usually requires signal conditioning based on operational amplifiers. Digital input port required.
Flow	Rotating vane flow sensor*	Turbine rotor driven by fluid. Turbine interrupts infra-red beam. Pulse repetition frequency of output is proportional to flow rate. A counter/timer chip can be used to minimize software requirements. Digital input port required.
Linear position	Resistive linear position sensor*	Linear track potentiometer with linear law produces analogue voltage proportional to linear position. Limited linear range. Analogue input port required.

(*continued*)

Physical parameter	*Type of sensor*	*Notes*
	Linear variable differential transformer (LVDT)	Miniature transformer with split secondary windings and moving core attached to a plunger. Requires AC excitation and phase-sensitive detector. Analogue input port required.
	Magnetic linear position sensor	Magnetic pick-up responds to movement of a toothed ferrous track. Pulses are counted as the sensor moves along the track (typically using an operational amplifier and Schmitt input logic gates) but some sensors contain circuitry in order to produce TTL-compatible outputs. The pulse repetition frequency of the output is proportional to the linear velocity. Digital input port required.
Light level	Photocell	Voltage-generating device. The analogue output voltage produced is proportional to light level. Analogue input port required.
	Light dependent resistor (LDR)*	An analogue output voltage results from a change of resistance within a cadmium sulphide (CdS) sensing element. Usually connected as part of a potential divider or bridge. Analogue input port required (alternatively a comparator arrangement can be used for threshold switching). Maximum sensitivity falls within the visible spectrum.
	Photodiode*	Two-terminal device connected as a current source. An analogue output voltage is developed across a series resistor of appropriate value. Analogue input port required (alternatively a comparator arrangement can be used for threshold switching). Maximum sensitivity usually occurs within the infra-red range (i.e. outside the visible spectrum).
	Phototransistor*	Three-terminal device connected as a current source. An analogue output voltage is developed across a series resistor of appropriate value. Analogue input port required. Phototransistors are also available in the form of light-activated switches which provide TTL-compatible outputs (in which case a digital input port must be used). Maximum sensitivity usually falls within the infra-red spectrum.
Liquid level	Float switch*	Simple switch element that operates when a particular level is detected. Digital input port required.
	Capacitive proximity switch*	Switching device that operates when a particular level is detected. Ineffective with some liquids. Digital input port required.
	Diffuse scan proximity switch*	Switching device that operates when a particular level is detected. Ineffective with some liquids. Digital input port required.
Pressure	Microswitch pressure sensor	Microswitch fitted with actuator mechanism and range setting springs. Suitable for high-pressure applications. Digital input port required.
	Differential pressure vacuum switch	Microswitch with actuator driven by a diaphragm. May be used to sense differential pressure. Alternatively, one chamber may be evacuated and the sensed pressure applied to a second input. Digital input port required.

(*continued*)

Physical parameter	*Type of sensor*	*Notes*
	Piezo-resistive pressure sensor	Pressure exerted on diaphragm causes changes of resistance in attached piezo-resistive transducers. Transducers are usually arranged in the form of a four active element bridge which produces an analogue output voltage. Analogue input port required.
Operator	Switch or push-button*	Suitable for providing basic manual on/off control. Available in various formats including conventional toggle, rotary, slide, button, keyswitch, and foot-operated types. Digital input port required.
	Dual in-line (DIL) switch*	Printed circuit board mounting switch available in single of multiple forms. Normally only used for selecting options or setting parameters and unsuitable for frequent use due to small size. Digital input port required.
	Keypad*	More cost-effective than using a number of individual push-button switches. Also suitable for data entry. Keypads fitted with encoders require digital input ports. Unencoded keypads are usually configured as a matrix of rows and columns (e.g. 4×4 in the case of a 16-key keypad) and will require at least one digital I/O port.
	Keyboard*	Provides the ultimate in data entry (including generation of the full set of ASCII characters). Encoded keyboards are generally easier to use than unencoded types which are more suitable for memory mapped I/O. Digital input port required.
	Joystick*	Available in both digital and analogue forms. The former type is generally based on four microswitches (two for each axis) whilst the latter is based on conventional resistive potentiometers. Either form is suitable for providing accurate position control but 'contactless' types are more reliable. Analogue or digital input port required, as appropriate.
	Touch screen	LCD display (requiring a digital output port) with touch sensitive using finger or stylus and typically requiring a force of 40 g for operation. Contacts have a typical 'on' resistance of between 150 Ω and 1.3 kΩ and an 'off' resistance of up to 20 MΩ. Requires a digital input port.
Proximity	Microswitch*	Microswitch fitted with actuator mechanism. Requires physical contact with the target object and small operating force. Also functions as a limit switch. Digital input port required.
	Reed switch*	Reed switch and permanent magnet actuator. Only effective over short distances. Digital input port required.
	Inductive proximity switch*	Target object modifies magnetic field generated by the sensor. Only suitable for metals (non-ferrous metals with reduced sensitivity). Digital input port required.
	Capacitive proximity* switch	Target object modifies electric field generated by the sensor. Suitable for metals, plastics, wood, and some liquids and powders. Digital input port required.

(*continued*)

Physical parameter	*Type of sensor*	*Notes*
	Optical proximity switch*	Available in diffuse and through scan types. Diffuse scan types require reflective targets. Both types employ optical transmitters and receivers (usually infra-red emitting LEDs and photodiodes or phototransistors). Digital input port required.
Strain	Resistive strain gauge	Foil type resistive element with polyester backing for attachment to body under stress. Normally connected in full bridge configuration with temperature-compensating gauges to provide an analogue output voltage. Analogue input port required.
	Semiconductor strain gauge	Piezo-resistive elements provide greater outputs than comparable resistive foil types. More prone to temperature changes and also inherently non-linear. Analogue input port required.
Temperature	Thermocouple*	Small e.m.f. generated by a junction between two dissimilar metals. For accurate measurement, requires compensated connecting cables and specialized interface. Analogue input port required.
	Thermistor	Usually connected as part of a potential divider or bridge. An analogue output voltage results from resistance changes within the sensing element. Analogue input port required.
	Semiconductor temperature sensor*	Two-terminal device connected as a current source. An analogue output voltage is developed across a series resistor of appropriate value. Analogue input port required.
Vibration	Electromagnetic vibration sensor	Permanent magnet seismic mass suspended by springs within a cylindrical coil. The frequency and amplitude of the analogue output voltage are respectively proportional to the frequency and amplitude of vibration. Requires an analogue input port.
Weight	Load cell	Usually comprises four strain gauges attached to a metal frame. This assembly is then loaded and the analogue output voltage produced is proportional to the weight of the load. Requires an analogue input port.

* Further details, interface circuits or photographs can be found later in this chapter.

Interfacing switches and sensors

Sensors can be divided into two main groups according to whether they are *active* (generating) or *passive*. Another, arguably more important distinction in the case of PC-based process control systems, is whether they provide digital or analogue outputs. In the former case, one or more digital I/O boards will be required whereas, in the latter case one or more analogue input ports must be provided.

We shall deal first with techniques of interfacing switches and sensors which provide *digital outputs* (such as switches and proximity detectors) before examining methods used by interfacing sensors which provide *analogue outputs*. It should be noted that the majority of sensors (of either type) will require some form of signal conditioning circuitry in order to make their outputs acceptable to conventional PC expansion cards.

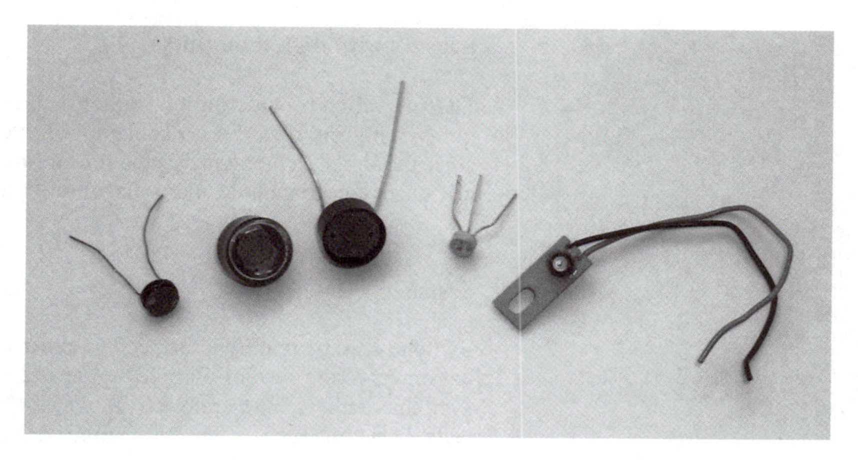

Photo 9.4 *Various optical and light-level sensors*

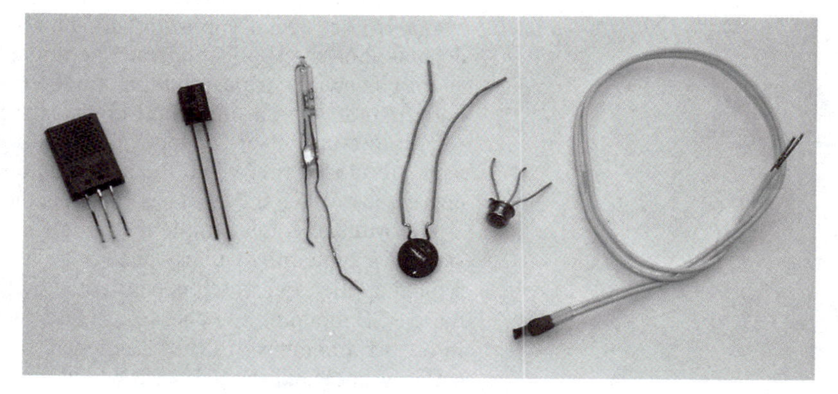

Photo 9.5 *Various temperature and gas sensors*

Photo 9.6 *Contactless joystick*

Sensors with digital outputs

Sensors which provide digital (rather than analogue) outputs can generally be quite easily interfaced with conventional PC bus expansion cards. However, since the signals generated by such sensors are seldom TTL compatible, it is usually necessary to include additional circuitry between the sensor and input port.

Switches

Switches can be readily interfaced to expansion cards in order to provide manual inputs to the system. Simple toggle and push-button switches are generally available with *normally open* (NO), *normally closed* (NC), or *changeover* contacts. In the latter case, the switch may be configured as either an NO or an NC type, depending upon the connections used.

Toggle, lever, rocker, rotary, slide, and push-button types are all commonly available in a variety of styles. Illuminated switches and key switches are also available for special applications. The choice of switch type will obviously depend upon the application and operational environment.

An NO switch or push-button may be interfaced to a digital I/O card using nothing more than a single pull-up resistor as shown in Figure 9.1.

The relevant bit of the input port will then return 0 when the switch contacts are closed (i.e. when the switch is operated or where the pushbutton is depressed). When the switch is inactive, the relevant port bit will return 1.

Unfortunately, this simple method of interfacing has a limitation when the state of a switch is regularly changing during program execution. However, a typical application which is unaffected by this problem is that of using one or more PCB mounted switches (e.g. a DIL switch package) to configure a system in one of a number of different modes. In such cases, the switches would be set only once and the software would read the state of the switches, and use the values returned to configure the system upon reset. Thereafter, the state of the switches would then only be changed in order to modify the operational parameters of the system (e.g. when adding additional I/O facilities). A typical DIL switch input interface to a digital input port is shown in Figure 9.2.

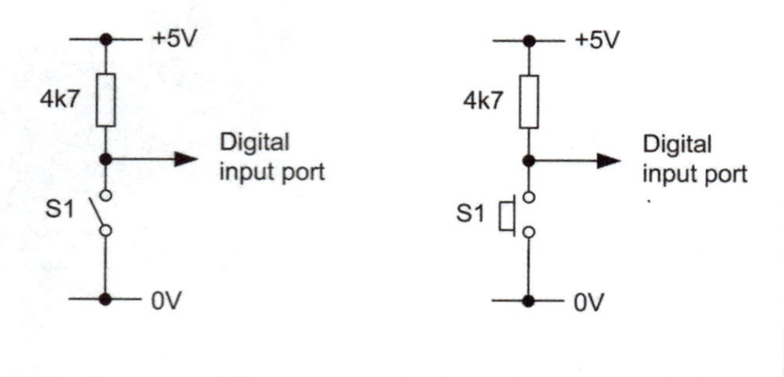

Figure 9.1 *Interfacing a normally open switch or push-button to a digital input port*

Switch debouncing

As mentioned earlier, the simple circuit of Figure 9.1 is unsuitable for use when the state of the switch is regularly changing. The reason for this is that the switching action of most switches is far from 'clean' (i.e. the switch contacts make and break several times whenever the switch is operated). This may not be a problem when the state of a switch remains static during program execution but it can give rise to serious problems when dealing with, for example, an operator switch bank or keypad.

The contact 'bounce' that occurs when a switch is operated results in rapid making and breaking of the switch until it settles into its new state. Figure 9.3

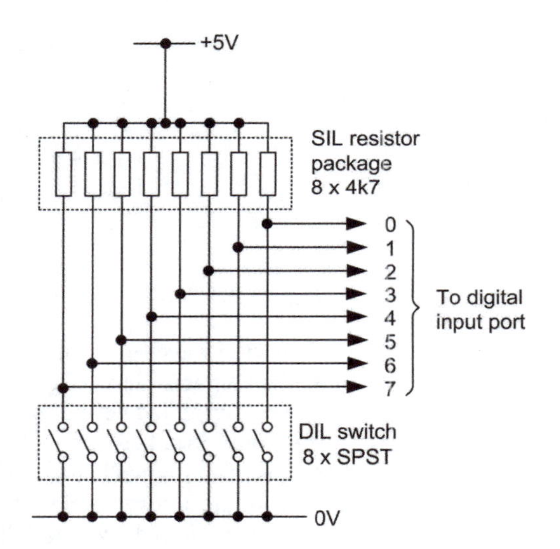

Figure 9.2 *Interfacing a DIL switch input to a digital input port*

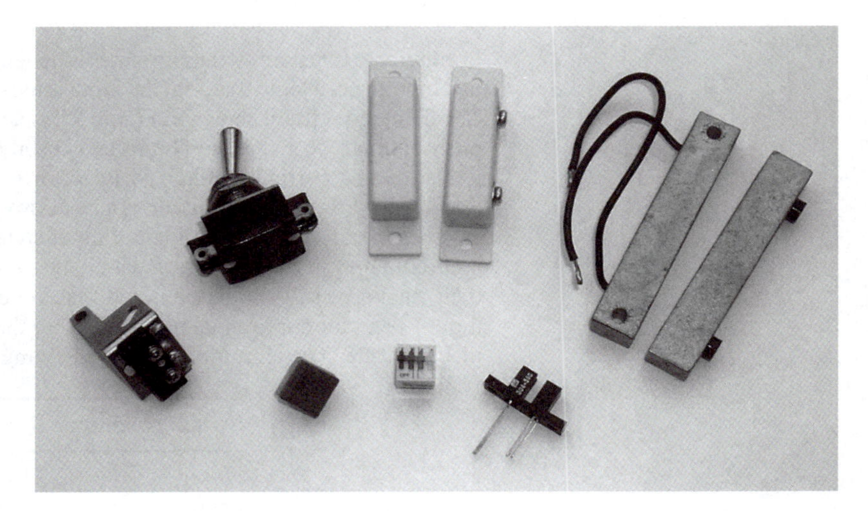

Photo 9.7 *Various switches and contacts*

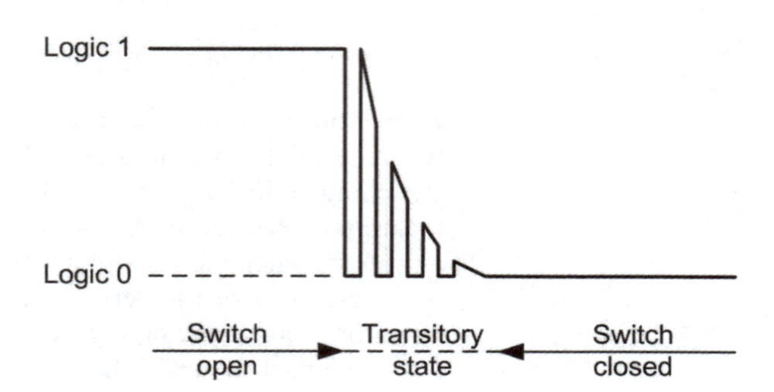

Figure 9.3 *Typical waveform produced by a switch closure*

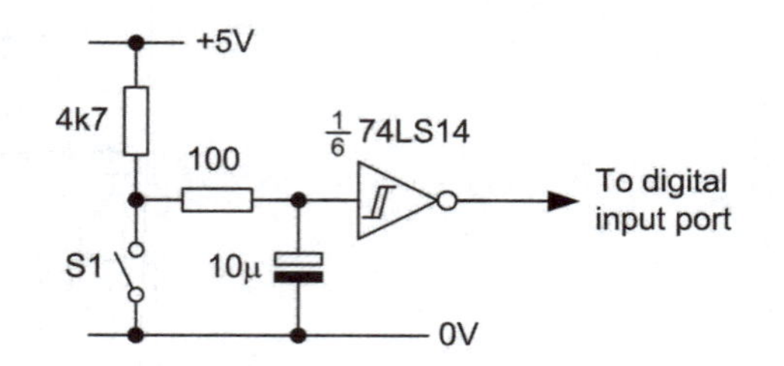

Figure 9.4 *Simple debounce circuit*

shows the waveform generated by the simple switch input circuit of Figure 9.1 as the contacts close. The spurious states can cause problems if the switch is sensed during the period in which the switch contacts are in motion, and hence steps must be taken to minimize the effects of bounce. This may be achieved by means of additional hardware in the form of a 'debounce' circuit or by including appropriate software delays (of typically 4 to 20 ms) so that spurious switching states are ignored. We shall discuss these two techniques separately.

Hardware debouncing

Immunity to transient switching states is generally enhanced by the use of active-low inputs (i.e. a logic 0 state at the input is used to assert the condition required). The debounce circuit shown in Figure 9.4 is adequate for most toggle, slide, and push-button type switches. The value chosen for R2 must take into account the low-state sink current required by IC1 (normally 1.6 mA for standard TTL and 400 μA for LS-TTL). R2 should not be allowed to exceed approximately 470 Ω in order to maintain a valid logic 0 input state. The values quoted generate an approximate 1 ms delay (during which the switch contacts will have settled into their final state). It should be noted that, on power-up, this circuit generates a logic 1 level for approximately 1 ms before the output reverts to a logic 0 in the inactive state. The circuit obeys the following state table:

Switch condition	Logic output
closed	1
open	0

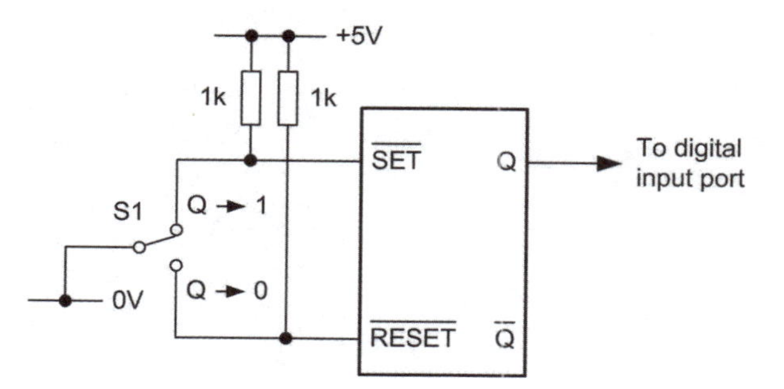

Figure 9.5 *Debounce circuit based on an RS bistable*

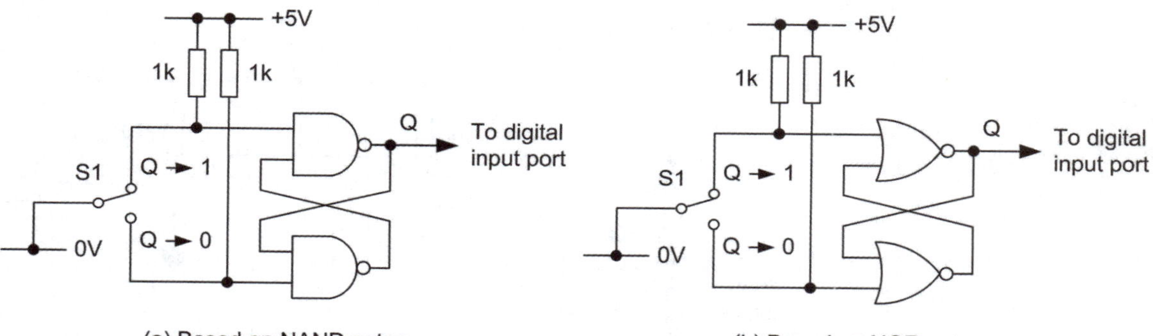

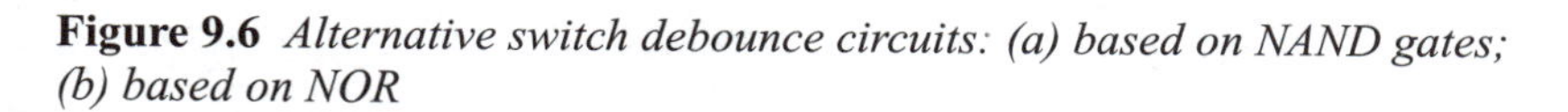

Figure 9.6 *Alternative switch debounce circuits: (a) based on NAND gates; (b) based on NOR*

An alternative, but somewhat more complex, switch de-bouncing arrangement is shown in Figure 9.5. Here a single-pole double-throw (SPDT) changeover switch is employed. This arrangement has the advantage of providing complementary outputs (Q and /Q) and it obeys the following state table:

Switch condition	Logic output	
	Q	/Q
Q→1	1	0
Q→0	0	1

Rather than use an integrated circuit RS bistable in the configuration of Figure 9.5 it is often expedient to make use of 'spare' two-input NAND or NOR gates arranged to form bistables using the circuits shown in Figures 9.6(a) and (b), respectively. Figure 9.7 shows a rather neat extension of this theme in the form of a touch-operated switch. This arrangement is based on a 4011 CMOS quad two-input NAND gate (though only two gates of the package are actually used in this particular configuration).

Finally, it is sometimes necessary to generate a latching action from an NO push-button switch. Figure 9.8 shows an arrangement in which a 74LS73 JK bistable is clocked from the output of a debounced switch.

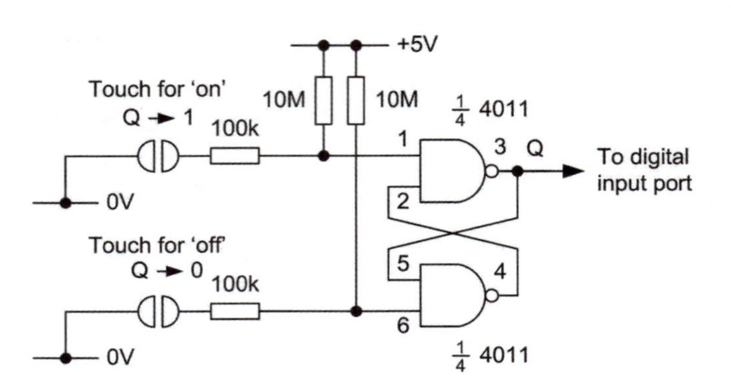

Figure 9.7 *Touch-operated switch*

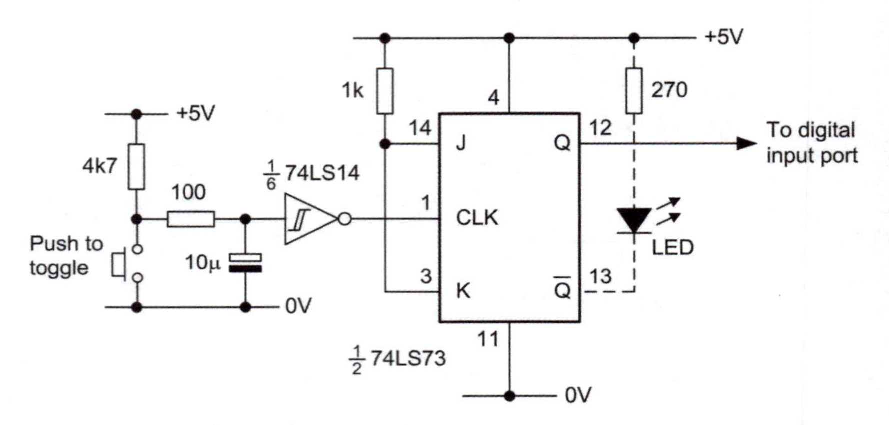

Figure 9.8 *Latching action switch*

Pressing the switch causes the bistable to change state. The bistable then remains in that state until the switch is depressed a second time. If desired, the complementary outputs provided by the bistable may be used to good effect by allowing the /Q output to drive an LED. This will become illuminated whenever the Q output is high.

Software debouncing

Software debouncing involves the execution of a delay routine whenever the state of a switch is read. The state of the switch at the start of the delay routine is compared with that at the end. If the same value is returned in both cases, the last value returned is assumed to represent the state of the switch. If the value has changed, the switch is read again. The period of the delay routine is chosen so that it is just greater than the maximum period of contact bounce expected (typically 4 to 10 ms).

A typical software debounce routine is given below:

```
readsw:  CALL  switch     ; Read the switch and
         MOV   BL, AL     ; store the value.
         CALL  swdelay    ; Wait and then
         CALL  switch     ; read it again.
```

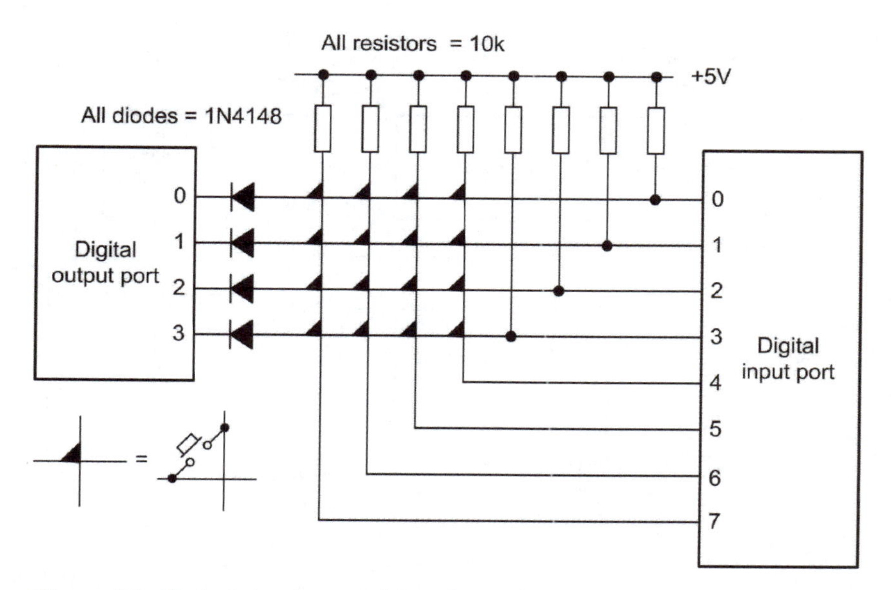

Figure 9.9 *Typical 4 × 4 matrix keypad interface*

```
          CMP    AL, BL      ; Has it changed?
          JNE    NZ, readsw  ; Yes, so try again.
          RET                ; No, so return with bit set in AL
switch:   IN     AL, porta   ; Get value from Port A
          AND    AL, mask    ; and check appropriate bit.
          RET                ; Go back ...
swdelay:  PUSH   AX          ; preserve the set bit
          MOV    CX, 0800H   ; and delay for a while.
sloop     LOOP   sloop
          POP    AX
          RET
```

Keypads

Keypads in process control applications vary from simple arrangements of dedicated push-button switches to arrangements of 16-keys (either coded or unencoded) in a standard 4 × 4 matrix. Keycaps may be engraved or fitted with suitable legends. Keypads sealed to 1P65 are available as similar units with individually illuminated keys.

Unencoded keypads are invariably interfaced using row and column lines to enable scanning of the keyboard. This arrangement is less demanding in terms of I/O lines than would be the case if the keypad contacts were treated as individual switches. A typical 16-key keypad arranged on a 4 × 4 matrix would make use of 12 digital I/O lines though it is possible to use just eight lines of a single port by alternately configuring the port for input and output. A representative arrangement is shown in Figure 9.9.

Unencoded keypads are generally preferred in high-volume applications where the cost of interfacing hardware has to be balanced at the expense of the extra overhead required by the software involved with scanning the keyboard, detecting, and decoding a keypress. In low-volume applications, and

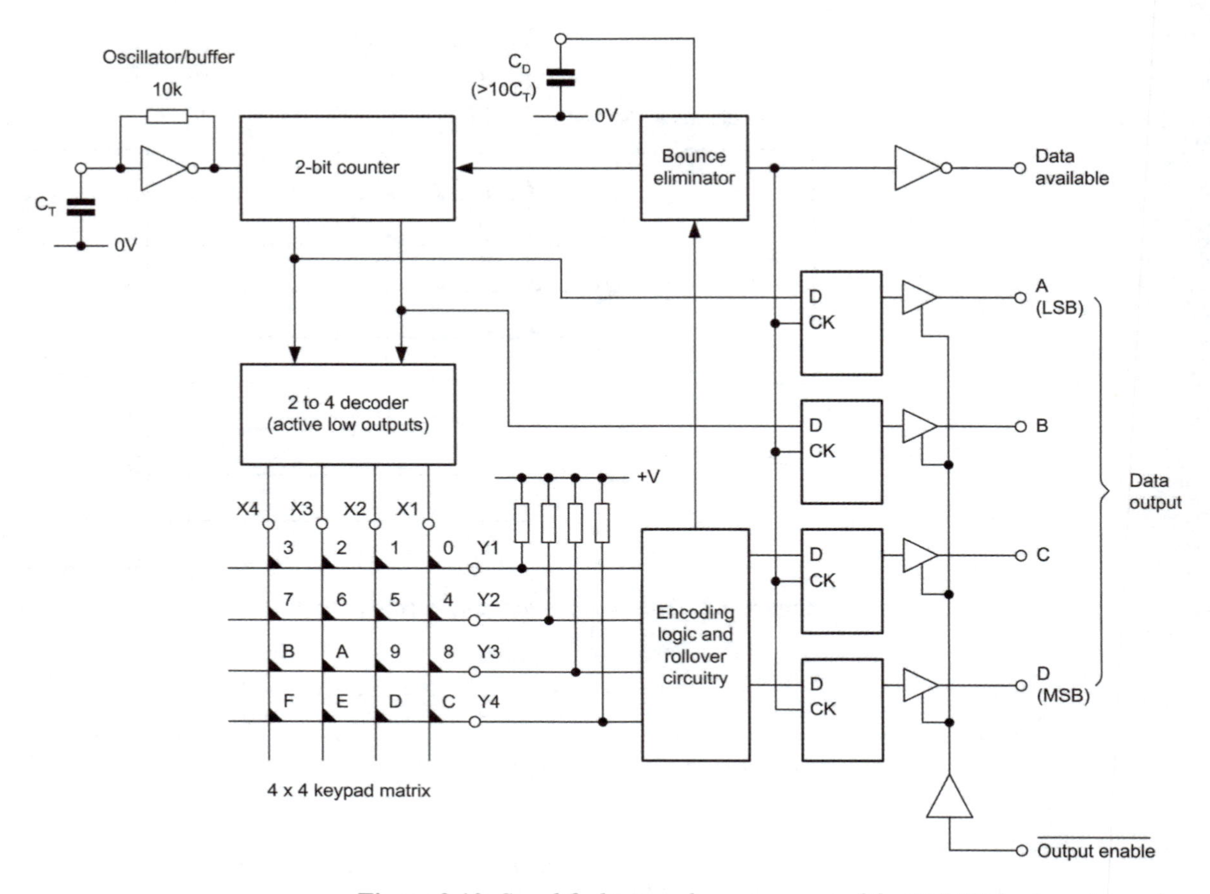

Figure 9.10 *Simplified internal arrangement of the 74C922 keyboard encoder*

where software overheads have to be minimized, the use of a fully encoded keyboard is obviously much to be preferred.

Encoded keypads employ dedicated encoder chips such as the 74C922. This device contains all the necessary logic to interface a 4 × 4 keypad matrix to four lines of a data bus or digital input port. The output is presented in binary coded decimal (BCD) form and an additional signal is provided to indicate that data is available from the keyboard. This active-high Data Available (DA) output can be used to drive an interrupt line when the keyboard is used in conjunction with a bus processor or may be connected via an open-collector inverter to one of the interrupt request (IRQ) lines of the PC expansion bus.

The simplified internal arrangement of the 74C922 is shown in Figure 9.10. The keypad scan may be implemented by the internal clock using an external timing capacitor (C_T) or may be over-driven by an external clock. On-chip pull-up resistors permit keypad switches with contact resistance of up to 50 kΩ. Internal debouncing is provided, the time constant of which is determined by an external capacitor (C_D).

The Data Available output goes high when a key is depressed and returns to low when a key is released even if another key is depressed. The Data Available output will return to high to indicate acceptance of the new key after a normal debounce period; this two key rollover is provided between any two switches. An internal register stores the last key pressed even after the key is released.

It should be noted that the LS-TTL-compatible outputs of the keypad encoder chip are tri-state, thus permitting direct connection to a data bus. Furthermore, the active-low Output Enable (/OE) input to the device can be used in a variety of configurations which permit asynchronous data entry as well as synchronous data entry and synchronous handshaking. Figure 9.11 shows how this can be achieved.

Proximity detectors

Proximity detectors are required in a wide variety of applications – from sensing the presence of an object on a conveyor to detecting whether a machine guard is in place. Simple proximity detectors need consist of nothing more than a microswitch and suitable actuator whereas more complex applications may require the use of inductive or capacitive sensors, or even optical techniques.

Microswitches

A microswitch is a simple electromechanical switch element which requires minimal operating and release force and which exhibits minimal differential travel. Microswitches are normally available in single-pole double-throw (SPDT) configurations and can thus be configured as either normally open (NO) or normally closed (NC).

The principal disadvantage of the humble microswitch is that it not only requires physical contact with the object sensed but also requires a force of typically 40 to 200 g for successful operation. Most common microswitch types (including the popular V3 and V4 types) can be fitted with a variety of actuator mechanisms. These include lever, roller, and standard button types. Metal-housed and environmentally-sealed micro-switches are available for more demanding environments.

Reed switches

Reed switches use an encapsulated reed switch that operates when in the proximity of a permanent magnetic field produced by an actuator magnet. Reed switches are generally available as either normally open (NO) or changeover types. The later may, of course, be readily configured for either NO or NC operation. Distances for successful operation (pull-in) of a reed switch are generally within 8–15 mm (measured between opposite surfaces of the actuator magnet and reed switch assembly). The release range, on the other hand, is generally between 10 and 20 mm.

Inductive proximity detectors

Inductive proximity detectors may be used for sensing the presence of metal objects without the need for any physical contact between the object and the sensor. Inductive proximity switches can be used to detect both ferrous and

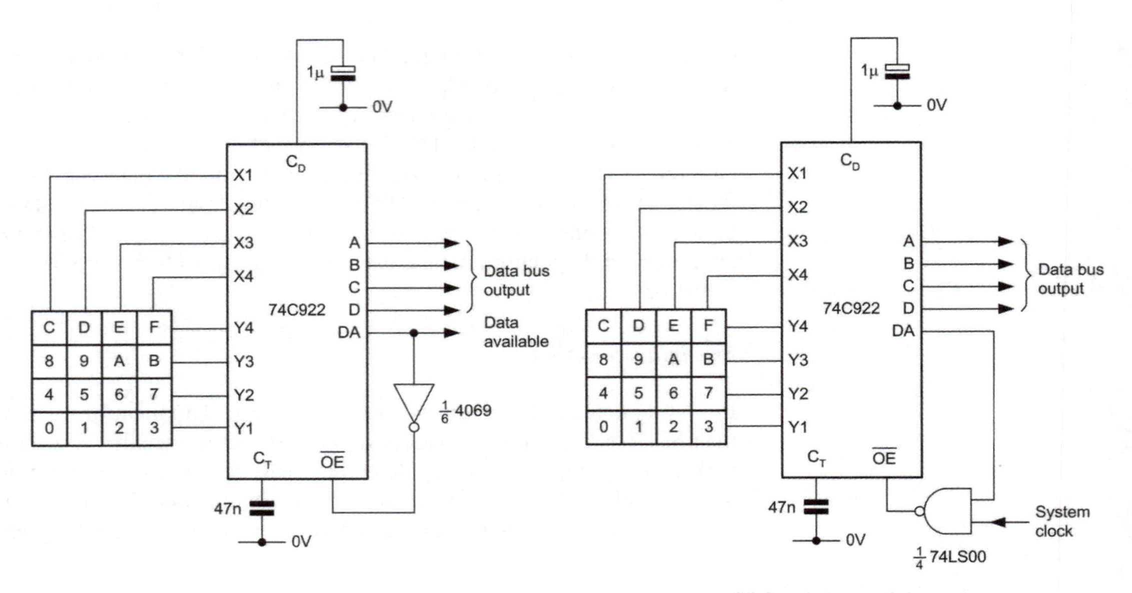

(a) Asynchronous data entry

(b) Synchronous data entry

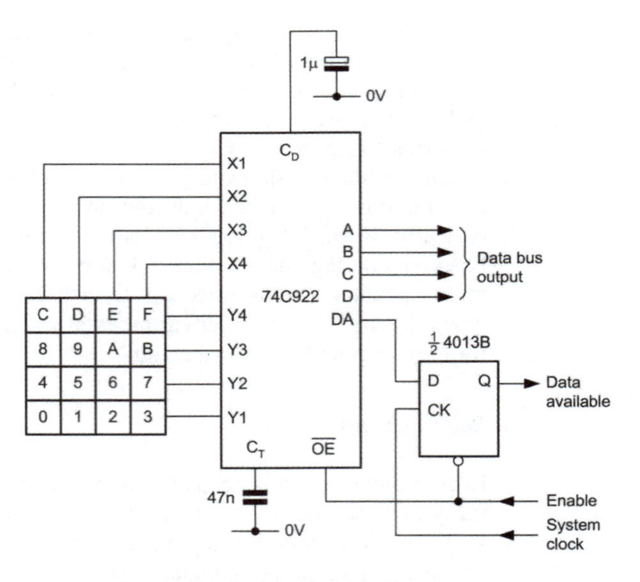

(c) Synchronous handshaking

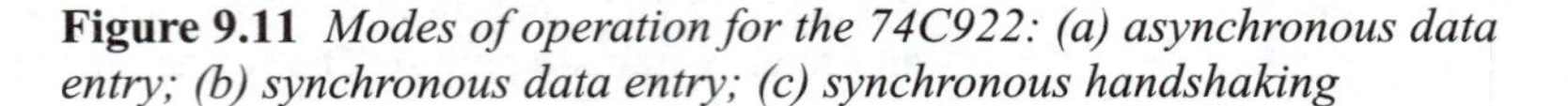

Figure 9.11 *Modes of operation for the 74C922: (a) asynchronous data entry; (b) synchronous data entry; (c) synchronous handshaking*

non-ferrous metals (the latter with reduced sensitivity). Hence metals such as aluminium, copper, brass, and steel can all be detected. Typical sensing distances for mild steel targets range from 1 mm for an object having dimensions $4 \times 4 \times 1$ mm to 15 mm for an object measuring $45 \times 45 \times 1$ mm. Note that sensitivity is reduced to typically 35% of the above for non-ferrous metals such as aluminium, brass, and copper.

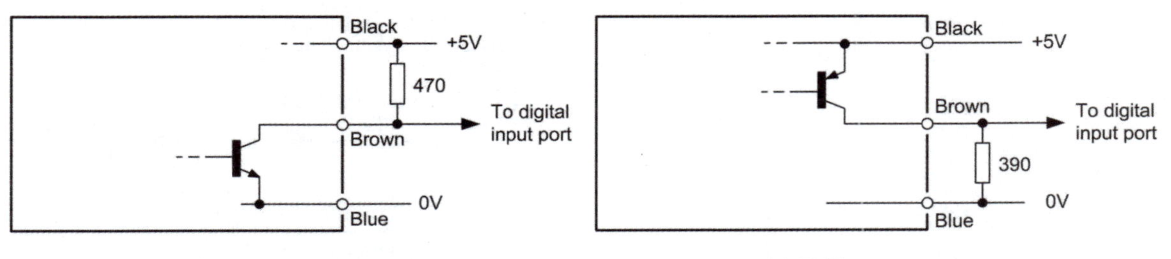

Figure 9.12 *Interfacing inductive proximity sensors: (a) NPN output types; (b) PNP output types*

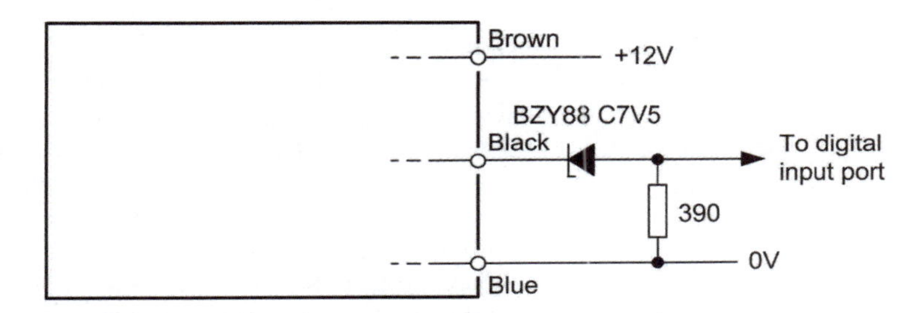

Figure 9.13 *Interface circuit for a typical capacitive proximity sensor*

Inductive proximity detectors are available with either NPN or PNP outputs (as shown in Figures 9.12(a) and (b)). An NPN type will return a logic 0 (low) when a target is detected whilst a PNP type will return logic 1 (high) in similar circumstances. When selecting a transducer for use with conventional I/O cards, it is advisable to choose a device which operates from a +5 V supply as this obviates the need for level shifting within the interface. A further consideration with such devices is the maximum speed at which they can operate. This is typically 2 kHz (i.e. 2000 pulses per second) but not that some devices are very much slower.

Capacitive proximity detectors

Capacitive proximity detectors provide an alternative solution to the use of inductive sensors. Unfortunately, such devices are also limited in their speed of response (typically 250 Hz maximum) and often require supply voltages in excess of the conventional +5 V associated with TTL signals. Capacitive proximity sensors will, however, detect the presence of materials such as cardboard, wood and plastics, as well as certain powders and liquids. Typical sensing distances range from 20 mm for metals to 4 mm for cardboard. As with inductive proximity sensors, the sensitivity of the detector is proportional to target size. A typical interface circuit for a DC-powered capacitive proximity detector is shown in Figure 9.13. This circuit provides a logic 0 (low) when a target is detected.

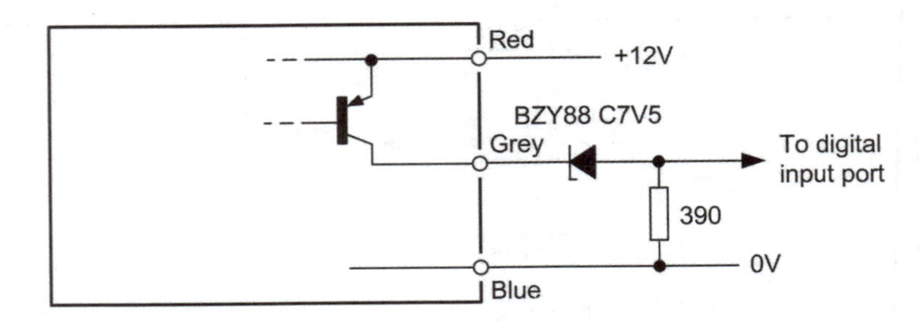

Figure 9.14 *Interface circuit for an optical proximity sensor*

Optical proximity detectors

Optical proximity detectors generally offer increased sensing ranges in comparison with both capacitive and inductive types. Optical proximity sensors are available in two basic forms: *diffuse scan* and *through scan* types. The former types rely on the target surface returning a proportion of the modulated light emitted by an optical transmitter which is mounted in the same enclosure as the receiver. In such an arrangement, a reflective target may be detected by the presence of a received signal. Through scan types, on the other hand, employ a separate transmitter and receiver and operate on the principle of the interrupted light beam (i.e. the target is detected by the absence of received light). Typical ranges vary from about 100 mm to 300 mm for diffuse scan sensors with plane white surfaces to up to 15 in (380 mm) for through scan sensors with opaque targets.

Proprietary sensor units are generally rather slow in operation and, for applications which involve rapid motion (such as counting shaft speeds) faster sensors should be employed. Here, a simple optical sensor (comprising an unmodulated infra-red emitting LED and photodiode) may be employed. Such devices are readily available in a variety of packages including miniature diffuse scan types and slotted through scan units. Figure 9.14 shows the circuitry required to interface such a device to a typical digital input port.

Position transducers

Position transducers can be used to provide an accurate indication of the position of an object and are available in a variety of forms (including linear and rotary types). Linear position sensors use linear law potentiometer elements (of typically 5 kΩ) and offer strokes of typically 10 or 100 mm. Rotary position sensors are also available. These provide indications over typically 105° and use linear law potentiometer elements similar to those found in conventional rotary potentiometer controls. A typical value for the resistive element is again 5 kΩ.

The output of linear and rotary position sensors is usually made available as an analogue voltage and a typical arrangement is shown in Figure 9.15. Note that the analogue input port should have a high impedance (say 500 kΩ or more) in order to avoid non-linearity caused by loading of the sensing potentiometer.

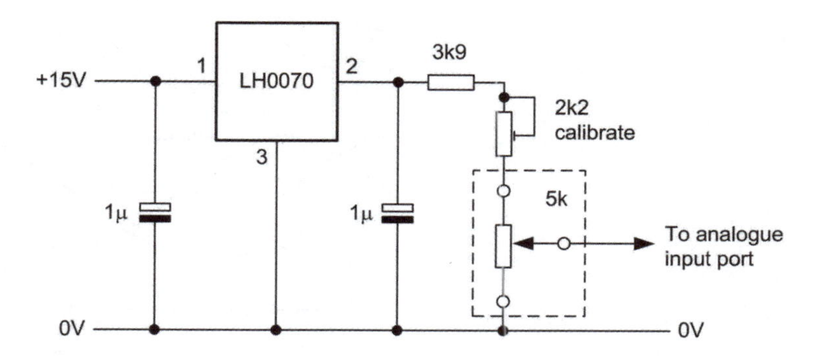

Figure 9.15 *Interface circuit for a resistive position transducer (either linear or rotary type)*

Shaft encoders

Shaft encoders can be used for sensing both rotary position and shaft speed. A typical shaft encoder produces 100 pulses per revolution and can thus provide a resolution of better than 1°. Such a device generally produces two phase-shifted outputs (to enable detection of direction of rotation) plus a third synchronizing pulse output (one pulse per revolution).

Shaft encoders are generally supplied in kit form comprising an encoder module, slotted disc, and hub. The encoder module usually contains three infra-red emitting LEDs and three matching photodetectors. The slotted disc is bonded to the hub ring which is, in turn, fitted to the rotating shaft. The encoder module is then mounted so that the disc is interposed between the LEDs and photodetectors.

The outputs of the encoder module are sinusoidal (as shown in Figure 9.16) and these must be converted to TTL-compatible input pulses in order to interface with a standard digital input port. For simple speed-sensing applications, a typical input stage based on an operational comparator and low-pass filter is shown in Figure 9.17.

Unfortunately, the simple circuit of Figure 9.17 is ineffective at very low frequencies and for stationary position indication. In such cases, the circuit shown in Figure 9.18 may be employed. Here, the potentiometer (RV1) must be adjusted so that the potential at the inverting input of the comparator is equal to that present at the non-inverting input. In this condition, the comparator produces a near 50% duty cycle.

A further refinement is that of providing an output which indicates the sense of rotation (i.e. clockwise or anticlockwise). This may be achieved with the aid of some additional logic and a single JK bistable element as shown in Figure 9.19. The Q output of the bistable goes high (logic 1) for clockwise rotation and low (logic 0) for anticlockwise rotation. Figure 9.20 shows typical waveforms for the logic shown in Figure 9.19.

Fluid sensors

A number of specialized sensors are available for use with fluids. These sensors include float switches (both horizontal and vertical types) and flow sensors.

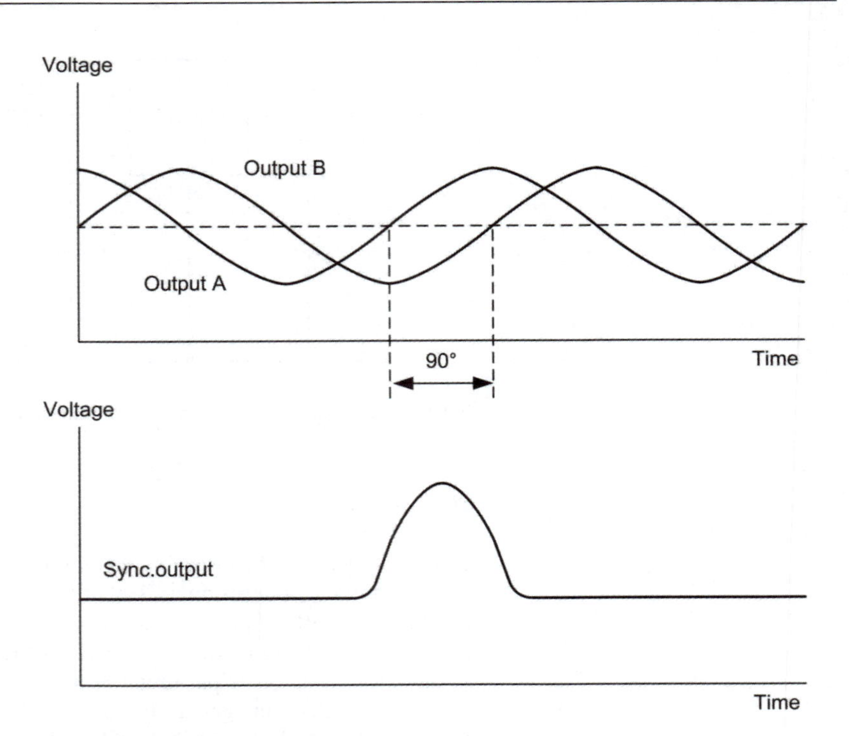

Figure 9.16 *Output waveforms produced by a typical shaft encoder*

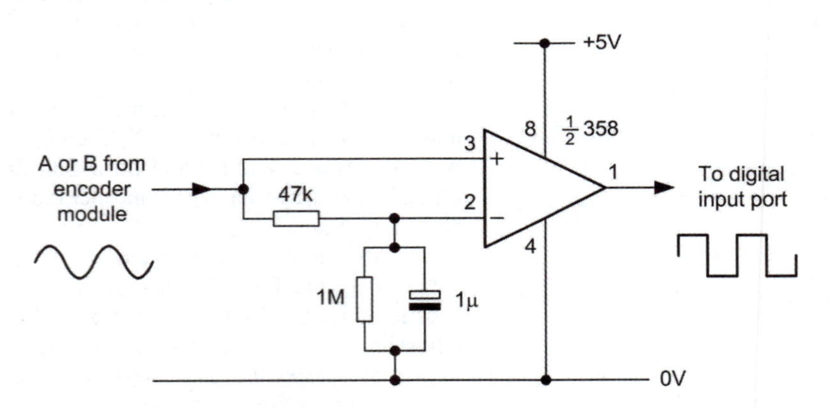

Figure 9.17 *Shaft encoder signal conditioning for measurement of rotational speed*

These latter devices incorporate a rotating vane and are suitable for use with flow rates ranging from 3 l/h to 500 l/h. Typical outputs range from 24 Hz at 10 l/h to 52 Hz at 20 l/h.

Optically isolated inputs

In a number of applications, it may be necessary to provide a high degree of electrical isolation between the source of a digital signal and its eventual

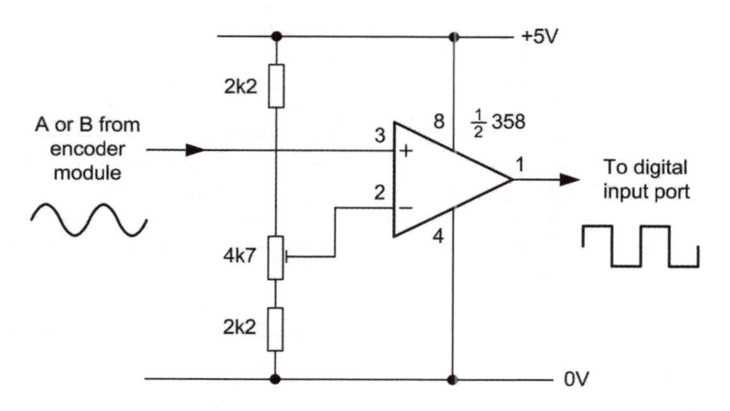

Figure 9.18 *Shaft encoder signal conditioning for low-speed applications and position sensing*

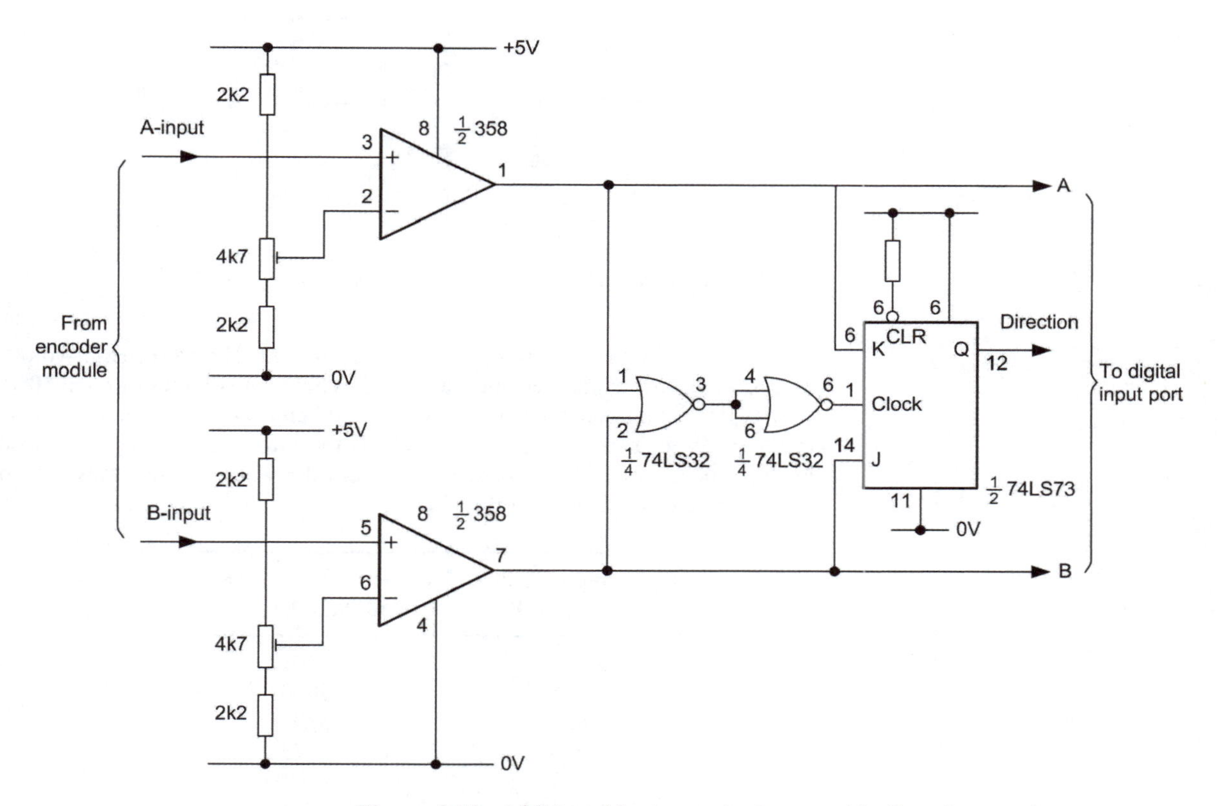

Figure 9.19 *Additional logic required to provide direction sensing*

connection to a digital input port. Such isolation can be achieved with the aid of an opto-isolator. These units comprise an optically coupled infra-red emitting LED and photodetector encapsulated in DIL package. The photodetector may take various forms including a *photodiode, phototransistor*, and *photo-Darlington*. Typical isolation voltages provided by such devices range from

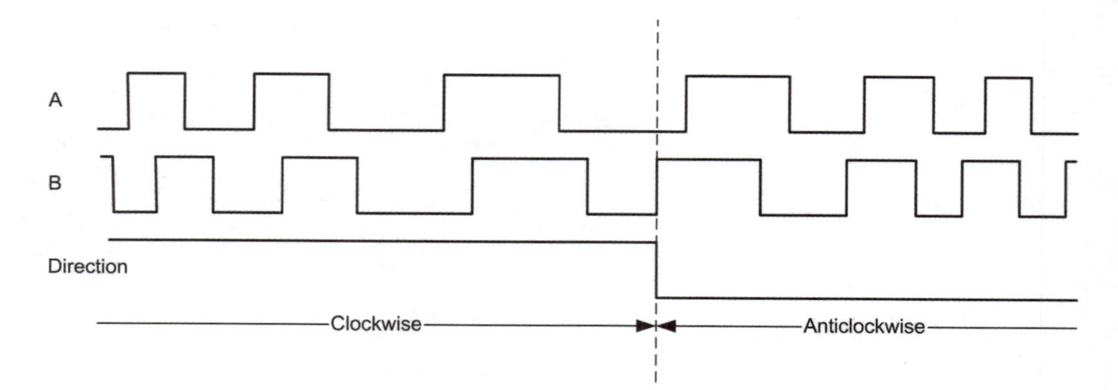

Figure 9.20 *Typical waveforms produced by the circuit of Figure 9.19*

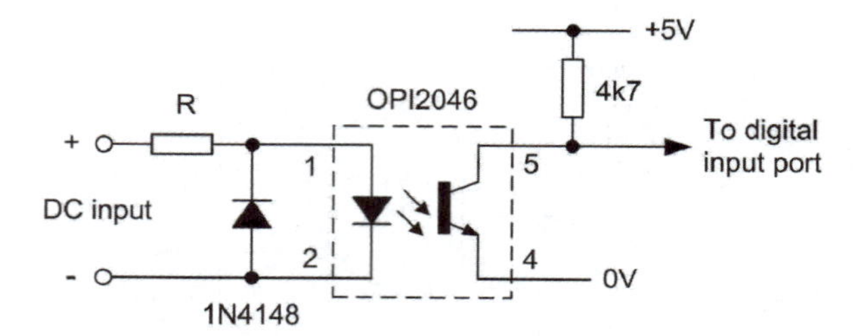

Figure 9.21 *Optically isolated digital input for DC signals*

500 V to 3 kV and switching rates may be up to 300 kHz, or so. High-voltage opto-isolators are available which will work reliably at voltages of up to 10 kV.

A typical single-channel optically isolated input arrangement is depicted in Figure 9.21. The external diode protects the infra-red emitting LED from inadvertent reversal of the input polarity and the value of the series resistor should be selected from the following table:

Input voltage (V) range (DC)	Series resistor (R) (all 0.25 W)
3 to 4	330 Ω
4 to 5	560 Ω
5 to 6	680 Ω
6 to 8	1 kΩ
8 to 11	1.5 kΩ
11 to 15	2.2 kΩ
15 to 30	3.9 kΩ

Bipolar optoisolators are available that will operate from inputs of either polarity. Such devices are useful when the input polarity is unknown but they can be unsuitable for general AC applications as the output will go high momentarily whenever an applied AC signal passes through zero volts. Figure 9.22 shows a

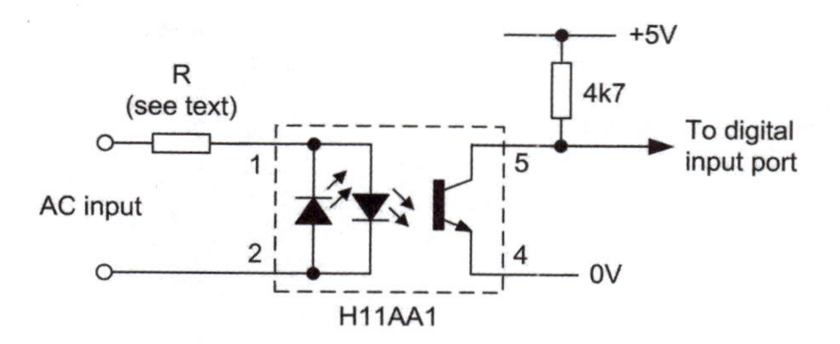

Figure 9.22 *Optically isolated digital input for bipolar DC signals*

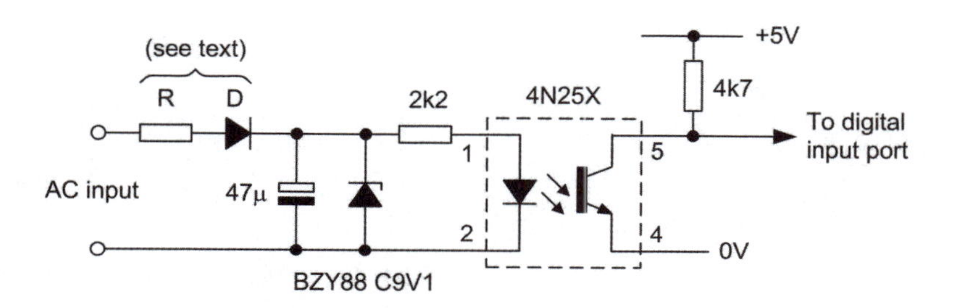

Figure 9.23 *Optically isolated digital input for AC signals*

typical arrangement (the value of series resistance is the same as that required for the circuit of Figure 9.21).

The optically-isolated input stage can be extended for monitoring AC voltages as shown in Figure 9.23. This arrangement is suitable for AC inputs of up to 240 V 50 Hz and may be used to sense the presence or absence of a mains supply.

Input voltage (V) range (RMS AC)	Series resistor (R) (0.25 W unless stated)	Diode (D)
9 to 12	1 kΩ	1N4001
12 to 15	1.5 kΩ	1N4001
15 to 24	2.7 kΩ	1N4001
24 to 35	3.9 kΩ	1N4002
50	6.8 kΩ 0.5 W	1N4003
110	18 kΩ 1 W	1N4004
220	39 kΩ 2.5 W	1N4007

Sensors with analogue outputs

Having dealt with a number of common sensors which provide digital outputs, we shall now turn our attention to a range of transducers which provide analogue outputs. These outputs may manifest themselves as changes in e.m.f, resistance, or current and, in the latter cases it will usually be necessary to incorporate additional signal conditioning circuitry so that an analogue input voltage can be provided for use with a standard PC analogue I/O card.

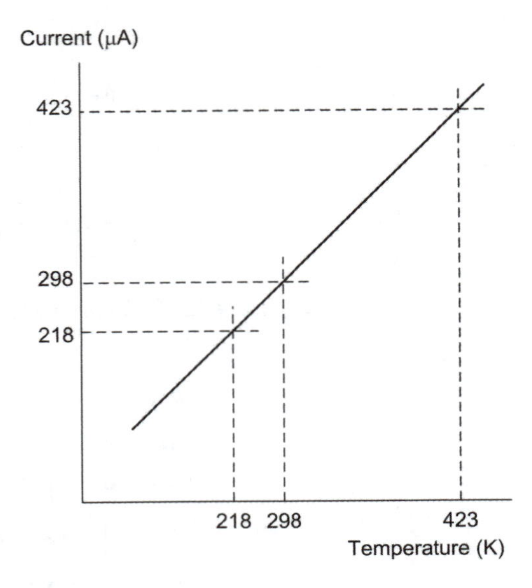

Figure 9.24 *Characteristic of the AD590 semiconductor temperature sensor*

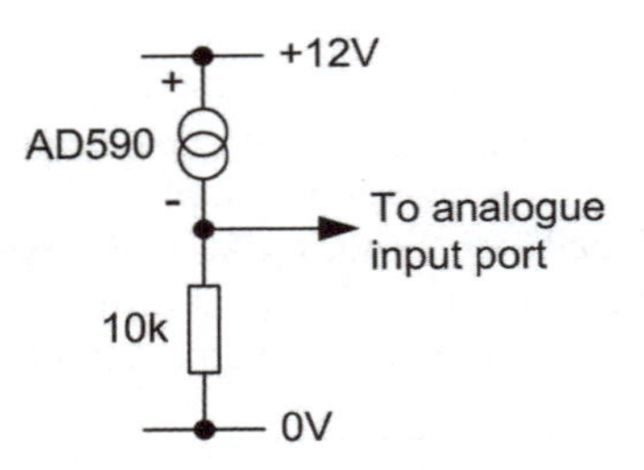

Figure 9.25 *Typical input interface for the AD590 semiconductor temperature sensor*

Semiconductor temperature sensors

Semiconductor temperature sensors are ideal for a wide range of temperature-sensing applications. The popular AD590 semiconductor temperature sensor, for example, produces an output current which is proportional to absolute temperature and which increases at the rate of 1 μA/K. The characteristic of the device is illustrated in Figure 9.24.

The AD590 is laser trimmed to produce a current of 298.2 μA (±2.5 μA) at a temperature of 298.2°C (i.e. 25°C). A typical interface between the AD590 and an analogue port is shown in Figure 9.25.

Thermocouples

Thermocouples comprise a junction of dissimilar metals which generate an e.m.f. proportional to the temperature differential which exists between the measuring junction and a reference junction. Since the measuring junction is usually at a greater temperature than that of the reference junction, it is sometimes referred to as the *hot junction.* Furthermore, the reference junction (i.e. the *cold junction)* is often omitted in which case the sensing junction is simply terminated at the signal conditioning board. This board is usually maintained at, or near, normal room temperatures.

Thermocouples are suitable for use over a very wide range of temperatures (from −100°C to +1100°C). Industry standard 'type K' thermocouples comprise a positive arm (conventionally coloured brown) manufactured from nickel/chromium alloy whilst the negative arm (conventionally coloured blue) is manufactured from nickel/aluminium.

The characteristic of a type K thermocouple is defined in BS 4937 Part 4 of 1973 (International Thermocouple Reference Tables) and this standard gives

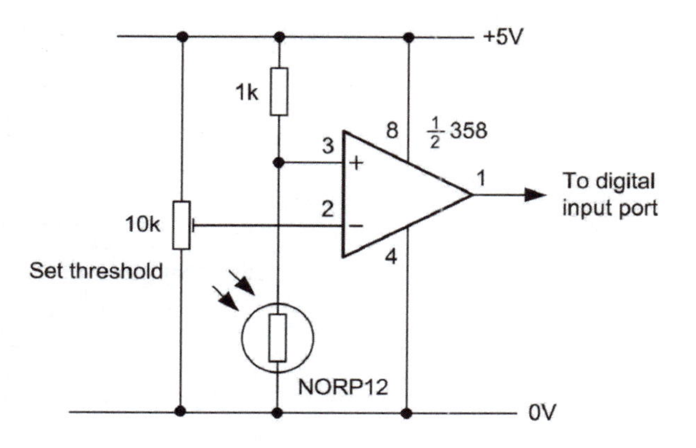

Figure 9.26 *Light-level threshold detector based on a light-dependent resistor (LDR)*

tables of e.m.f. versus temperature over the range 0°C to +1100°C. In order to minimize errors, it is usually necessary to connect thermocouples to appropriate signal conditioning using compensated cables and matching connectors. Such cables and connectors are available from a variety of suppliers and are usually specified for use with type K thermocouples.

Where thermocouples are to be used as sensors in conjunction with PC-based instrumentation systems, proprietary signal conditioning cards are available. These cards incorporate cable terminators and provide cold junction compensation as well as low-pass filtering to reduce the effects of 50 Hz noise induced in the thermocouple cables. The signal conditioning boards are then used in conjunction with one, or more, multi-channel analogue input ports.

Threshold detection with analogue output transducers

Analogue sensors are sometimes used in situations where it is only necessary to respond to a pre-determined threshold value. In effect, a two-state digital output is required. In such cases a simple one-bit analogue-to-digital converter based on a comparator can be used. Such an arrangement is, of course, very much simpler and more cost-effective than making use of a conventional analogue input port!

Simple threshold detectors for light level and temperature are shown in Figures 9.25–9.27. These circuits produce TTL-compatible outputs suitable for direct connection to a digital input port.

Figure 9.26 shows a light-level threshold detector based on a comparator and light-dependent resistor (LDR). This arrangement generates a logic 0 input whenever the light level exceeds the threshold setting, and vice versa. Figure 9.27 shows how light level can be sensed using a photodiode. This circuit behaves in the same manner as the LDR equivalent but it is important to be aware that circuit achieves peak sensitivity in the near infra-red region. Figure 9.28 shows how the spectral response of a typical light-dependent resistor (NORP12) compares with that of a conventional photodiode (BPX48). Note that the BPX48 can also be supplied with an integral daylight filter (BPX48F).

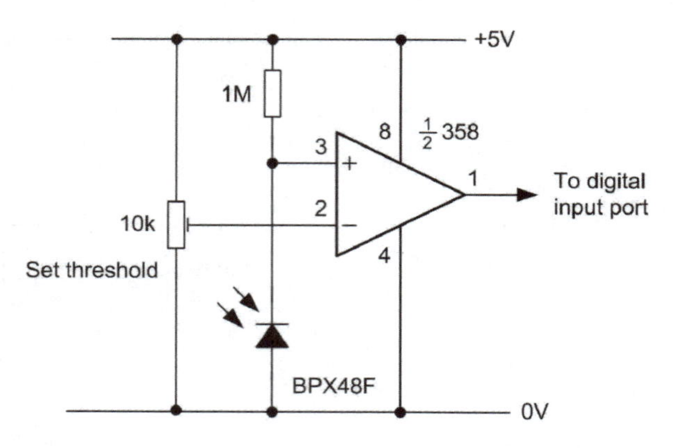

Figure 9.27 *Light-level threshold detector based on a photodiode*

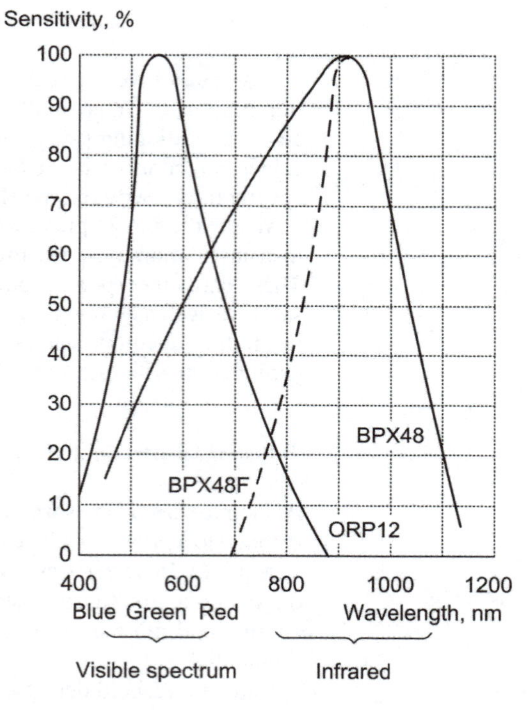

Figure 9.28 *Comparsion of typical spectral response of LDR and photodiodes*

Figure 9.29 shows how temperature thresholds can be sensed using the AD590 sensor described earlier. This arrangement generates a logic 0 input whenever the temperature level exceeds the threshold setting, and vice versa.

AC sensing

Finally, Figure 9.30 shows how an external AC source can be coupled to an input port. This arrangement produces TTL-compatible input pulses having 50% duty

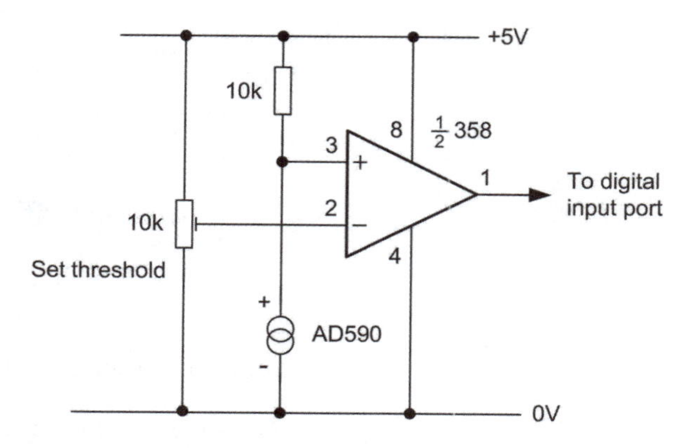

Figure 9.29 *Temperature threshold detector based on an AD590 semiconductor temperature sensor*

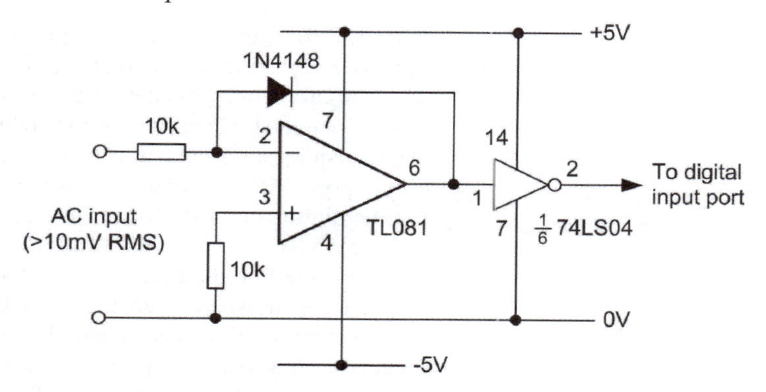

Figure 9.30 *Interface circuit to permit AC sensing*

cycle. The circuit requires an input of greater than 10 mV for frequencies up to 10 kHz and greater than 100 mV for frequencies up to 100 kHz.

The obvious application the arrangement shown in Figure 9.28 is the detection of audio frequency signals but, with its input derived from the low voltage secondary of a mains transformer (via a 10:1 potential divider), it can also function as a mains failure detector.

Output devices

Having dealt at some length with input sensors, we shall now focus our attention on output devices and the methods used for interfacing them. PC-based systems can readily be configured to work with a variety of different output transducers including actuators, alarms, heaters, lamps, motors, and relays. Ready-built output drivers are available for several types of load including relays and stepper motors. Many applications will, however, require custom-built circuitry in order to interface the necessary output devices.

Status and warning indications

Indicators based on light-emitting diodes (LEDs) are inherently more reliable than small filament lamps and also consume considerably less power. They are

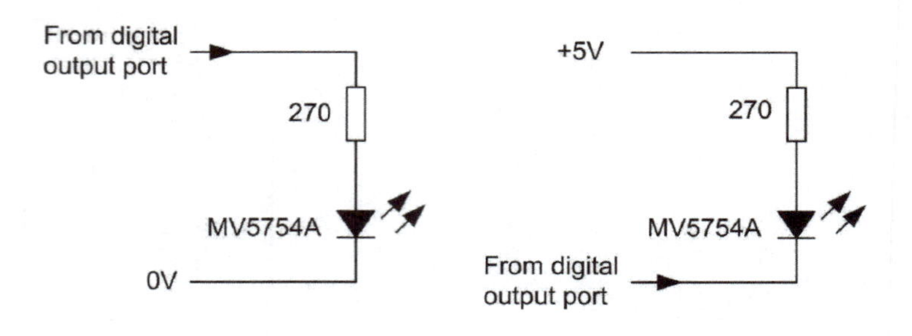

Figure 9.31 *Driving an LED from a buffered digital I/O port: (a) logic 1 to illuminate the LED; (b) logic 0 to illuminate the LED*

thus ideal for providing visual status and warning displays. LEDs are available in a variety of styles and colours, and 'high brightness' types can be employed where high-intensity displays are required.

A typical red LED requires a current of around 10 mA to provide a reasonably bright display and such a device may be directly driven from a buffered digital output port. Different connections are used depending upon whether the LED is to be illuminated for a logic 0 or logic 1 state. Several possibilities are shown in Figure 9.31.

Where a buffered output port is not available, an auxiliary transistor may be employed as shown in Figure 9.32. The LED will operate when the output from a PC expansion card is taken to logic 1 and the operating current should be approximately 15 mA (thereby providing a brighter display than the arrangements previously described). The value of LED series resistance will be dependent upon the supply voltage and should be selected from the table shown below:

Voltage (V)	Series resistance (all 0.25 W)
3 to 4	100 Ω
4 to 5	150 Ω
5 to 8	220 Ω
8 to 12	470 Ω
12 to 15	820 Ω
15 to 20	1.2 kΩ
20 to 28	1.5 kΩ

Driving LCD displays

A number of process-control applications require the generation of status messages and operator prompts. These can be easily produced using a conventional alphanumeric dot-matrix LCD display. Such displays are commonly available in a variety of formats ranging from 16 characters × 1 line to 40 characters × 4 lines and can usually display the full ASCII character set as well as user-defined

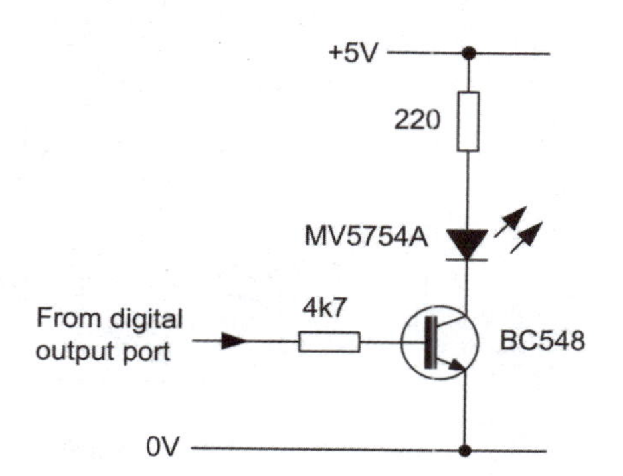

Figure 9.32 *Using an auxiliary transistor to drive an LED*

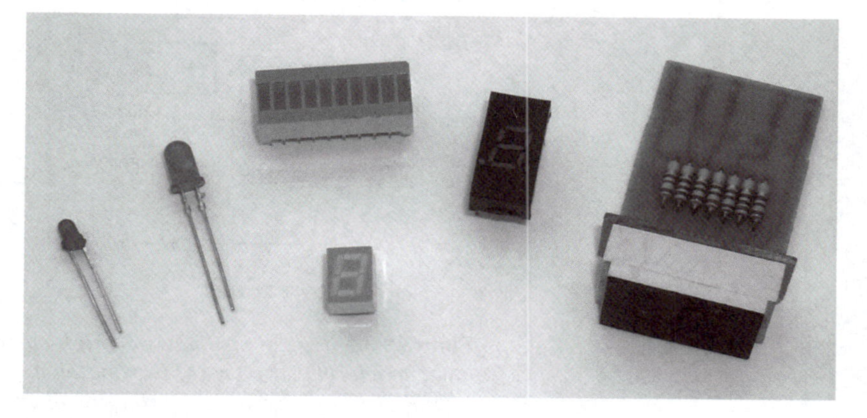

Photo 9.8 *Various light emitting diodes and indicators*

symbols. LCD displays are invariably fitted with the necessary hardware drivers and logic (sometimes in the form of a CMOS microprocessor) to interface directly with a digital I/O port.

Driving medium- and high-current loads

Due to the limited output current and voltage capability of most standard digital I/O expansion cards, external circuitry will normally be required to drive anything other than the most modest of loads. Figure 9.33 shows some typical arrangements for operating various types of medium- and high-current load. Figure 9.33(a) shows how an NPN transistor can be used to operate a low-power relay. Where the relay requires an appreciable operating current (say, 150 mA, or more) a plastic encapsulated Darlington power transistor should be used as shown in Figure 9.33(b). Alternatively, a power MOSFET may be preferred, as shown in Figure 9.33(c). Such devices offer very low values of 'on' resistance coupled with a very high 'off' resistance. Furthermore, unlike conventional bipolar transistors, a power FET will impose a negligible load on an I/O port. Figure 9.33(d) shows a filament lamp driver based on a plastic Darlington power transistor. This circuit will drive lamps rated at up to 24 V, 500 mA.

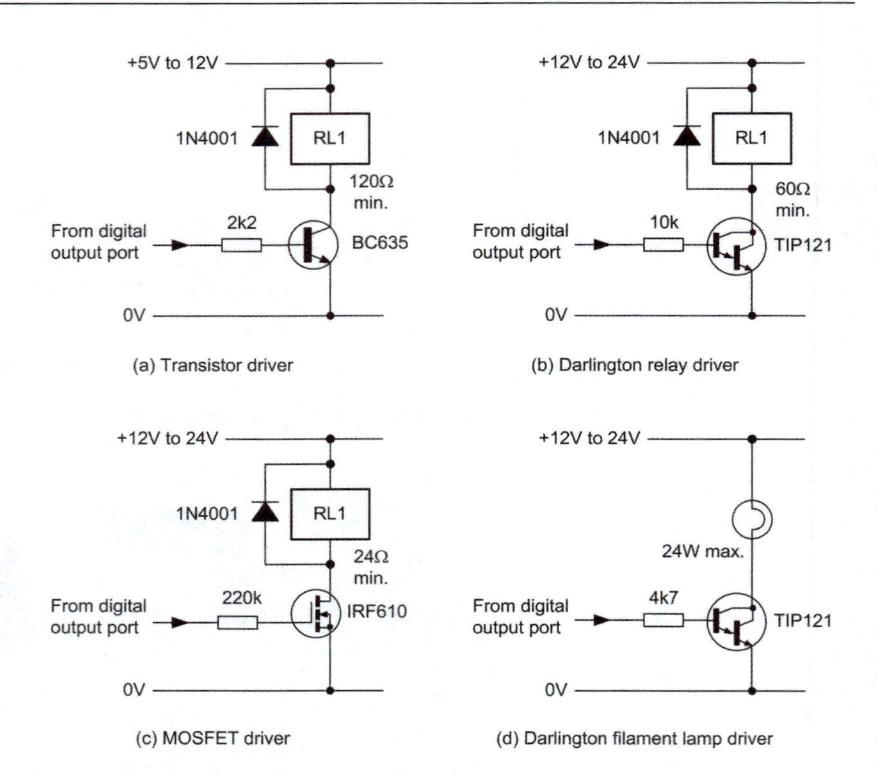

Figure 9.33 *Typical medium- and high-current driver circuits: (a) transistor low-current relay driver; (b) Darlington medium/high-current relay driver; (c) MOSFET relay driver; (d) Darlington filament lamp driver*

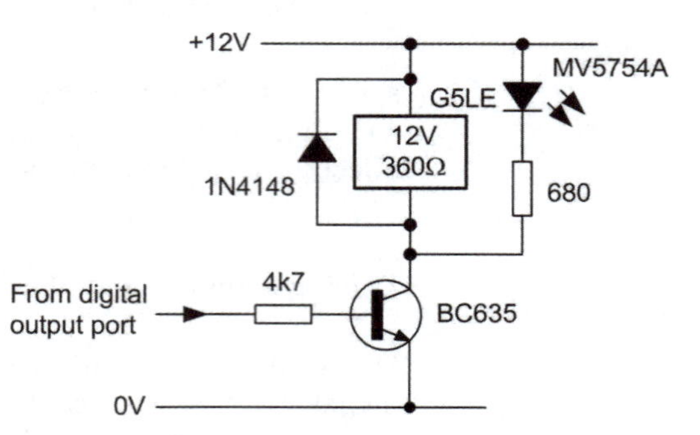

Figure 9.34 *An LED indicator can be easily added to a relay driver*

Finally, where visual indication of the state of a relay is desirable it is a simple matter to add an LED indicator to the driver stage, as shown in Figure 9.34.

Audible outputs

Where simple audible warnings are required, miniature piezo-electric transducers may be used. Such devices operate at low voltages (typically in the range

Photo 9.9 *Various types of relay*

3–15 V) and can be interfaced with the aid of a buffer, open-collector logic gate, or transistor. Figures 9.35(a)–(c) show typical interface circuits which produce an audible output when the port output line is at logic 1.

Where a pulsed rather than continuous audible alarm is required, a circuit of the type shown in Figure 9.36 can be employed. This circuit is based on a standard 555 timer operating in astable mode and operates at approximately 1 Hz. A logic 1 from the port output enables the 555 and activates the pulsed audio output.

Finally, the circuit shown in Figure 9.37 can be used where a conventional moving-coil loudspeaker is to be used in preference to a piezo-electric transducer. This circuit is again based on the 555 timer and provides a continuous output at approximately 1 kHz whenever the port output is at logic 1.

DC motors

Circuit arrangements used for driving DC motors generally follow the same lines as those described earlier for use with relays. As an example, the circuit shown in Figure 9.38 uses a power MOSFET to drive a low-voltage DC motor. This circuit is suitable for use with DC motors rated at up to 12 V with stalled currents of up to 3 A. In both cases, a logic 1 from the output port will operate the motor.

Output drivers

Where a number of output loads are to be driven from the same port, it is often expedient to make use of a dedicated octal driver chip rather than use eight individual driver circuits based on discrete components. Fortunately, a number of octal drivers are available and these invariably have TTL-compatible inputs which makes them suitable for direct connection to an output port.

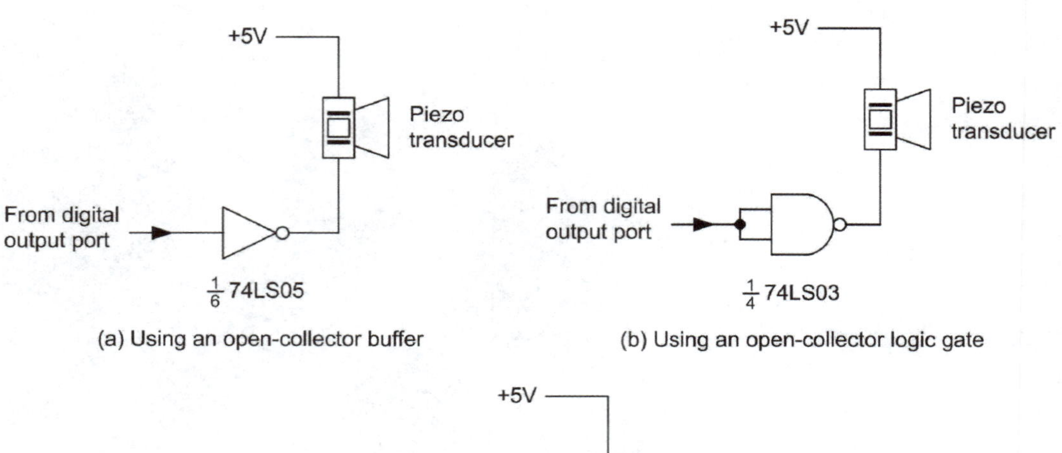

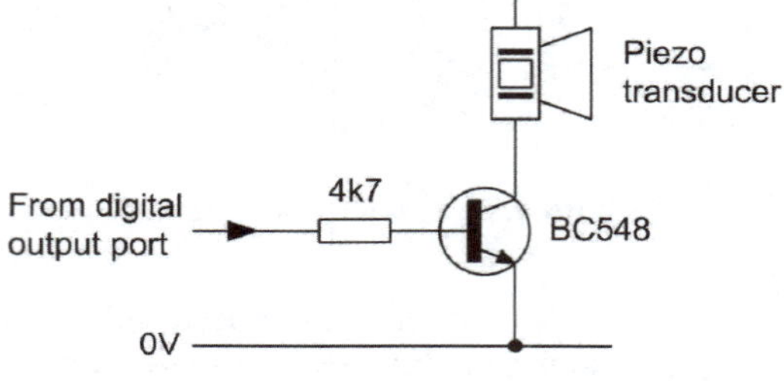

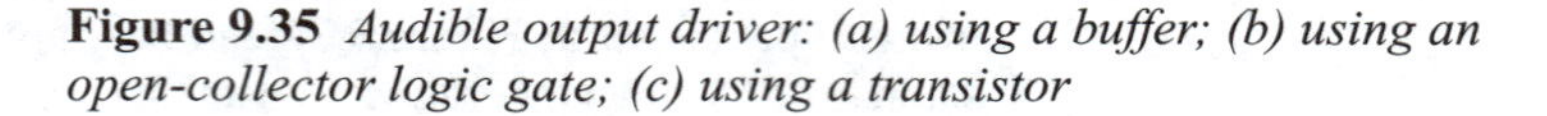

Figure 9.35 *Audible output driver: (a) using a buffer; (b) using an open-collector logic gate; (c) using a transistor*

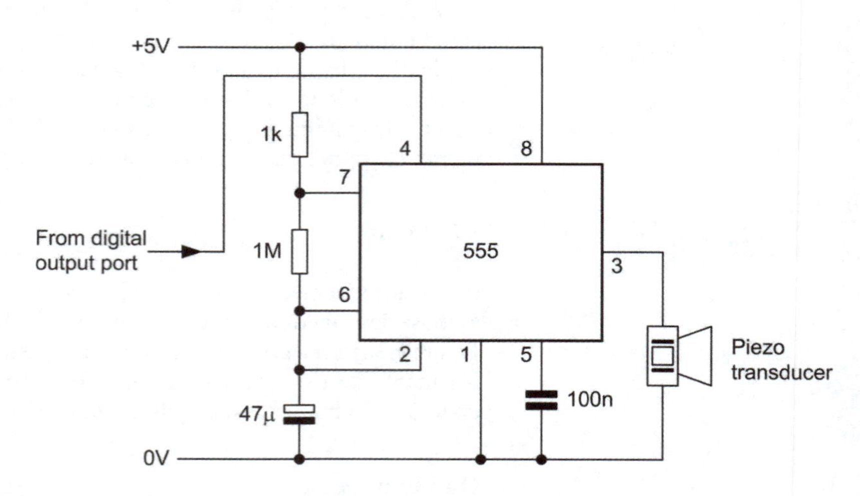

Figure 9.36 *Pulsed audible alarm based on a 555 astable*

A simple, general-purpose byte-wide output driver can be based around a dedicated octal latch/driver of which the UCN5801 is a typical example. This device is directly bus compatible but may also be used in conjunction with a conventional parallel I/O port. The UCN5801 has separate CLEAR, STROBE,

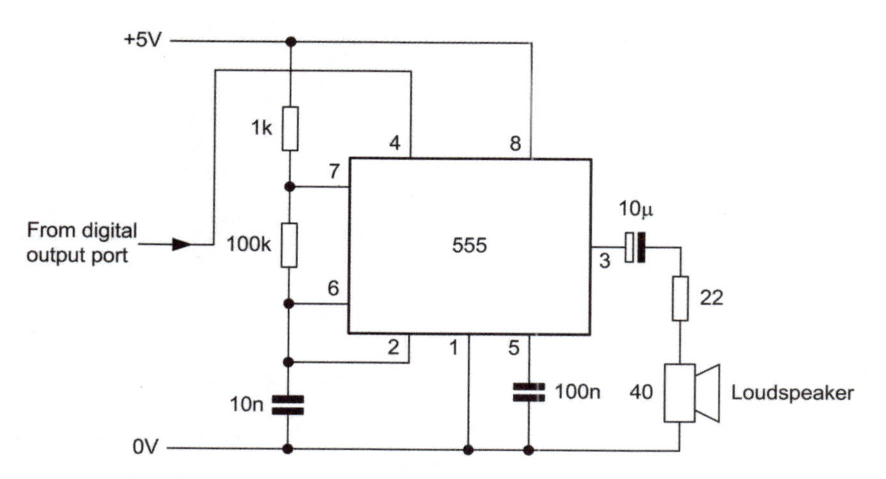

Figure 9.37 *Audible alarm with output to a moving coil loudspeaker*

Photo 9.10 *Various audible and ultrasonic transducers*

and output ENABLE control lines coupled to eight bipolar Darlington driver transistors. This configuration provides an extremely low-power latch with a very high-output current capability.

The eight outputs of the UCN5801 are all open-collector, the positive supply voltage for which may be anything up to 50 V. Each Darlington output device is rated at 500 mA maximum however; if that should prove insufficient for a particular application then several output lines may be paralleled together subject, of course, to the limits imposed by the rated load current of the high-voltage supply.

Figure 9.39 shows a typical arrangement of the UCN5801 in which the load voltage supply is +12 V. The state of the bus is latched to the output of ICI whenever the STROBE input is taken high; however, when used in conjunction

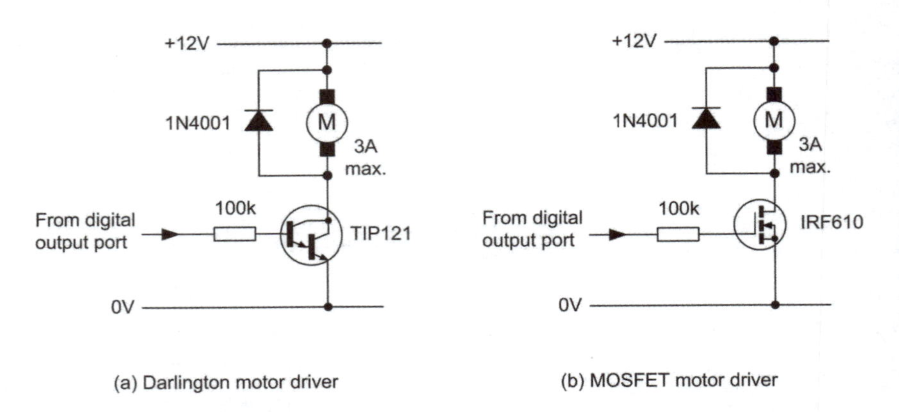

Figure 9.38 *Motor driver circuits: (a) Darlington motor driver; (b) MOSFET motor driver*

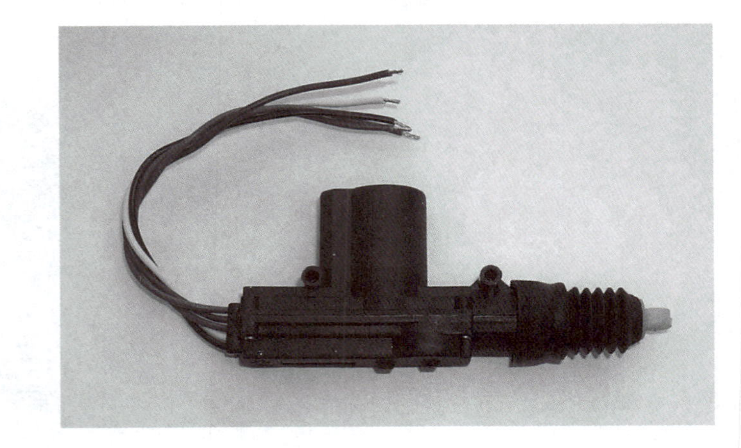

Photo 9.11 *Motorized linear actuator*

with a conventional I/O device, this line can be tied high. A logic 0 present on a particular data line will turn the corresponding Darlington output device 'off' whilst a logic 1 will turn it 'on'. It should also be noted that the output stages are protected against the effects of an inductive load by means of internal diodes. These are commoned at pin-12 and this point should thus be returned to the positive supply.

Driving mains connected loads

Control systems are often used in conjunction with mains connected loads. Modern solid-state relays (SSRs) offer superior performance and reliability when compared with conventional relays in such applications. SSRs are available in a variety of encapsulations (including DIL, SIL, flat-pack, and plug-in octal) and may be rated for RMS currents between l and 40 A.

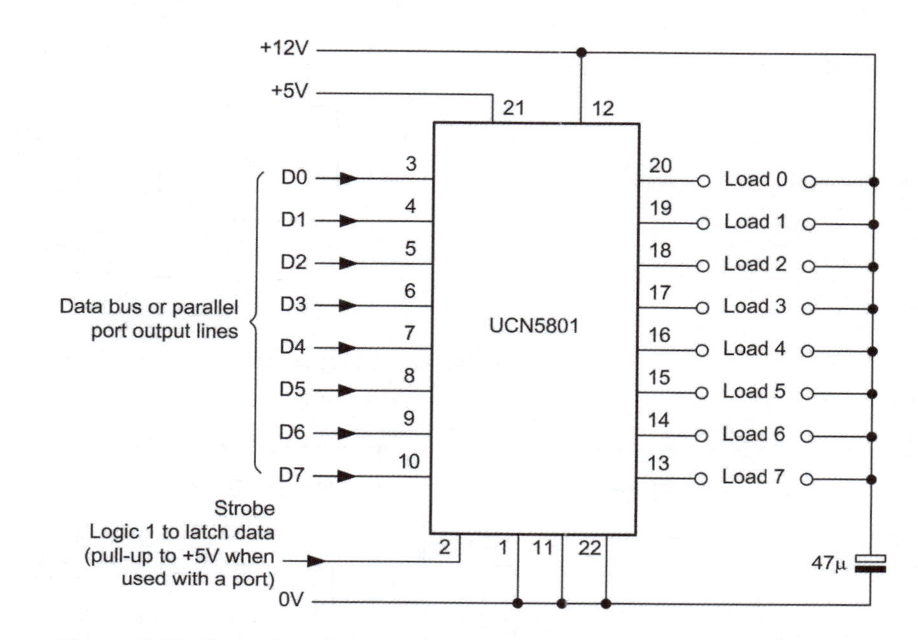

Figure 9.39 *Typical output driver arrangement based on the UCN5801*

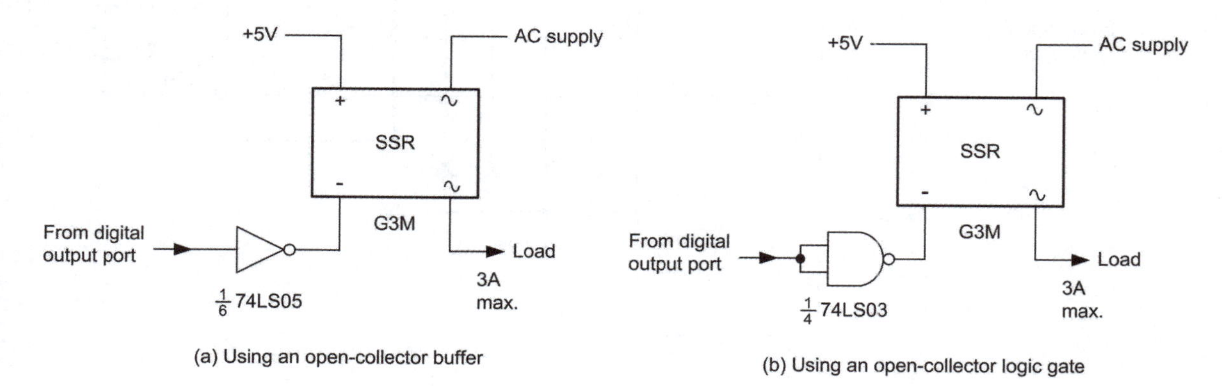

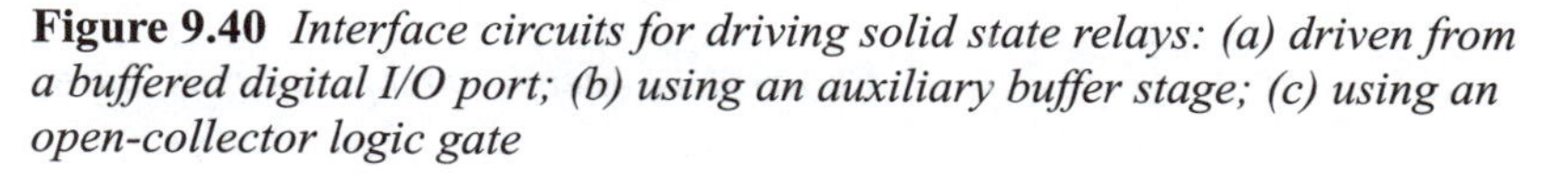

Figure 9.40 *Interface circuits for driving solid state relays: (a) driven from a buffered digital I/O port; (b) using an auxiliary buffer stage; (c) using an open-collector logic gate*

In order to provide a high degree of isolation between input and output, SSRs are optically coupled. Such devices require minimal input currents (typically 5 mA, or so, when driven from 5 V) and they can thus be readily interfaced with an I/O port that offers sufficient drive current. In other cases, it may be necessary to drive the SSR from an unbuffered I/O port using an open-collector logic gate. Typical arrangements are shown in Figure 9.40. Finally, it is important to note that, when an inductive load is to be controlled, a *snubber network* should be fitted, as shown in Figure 9.41.

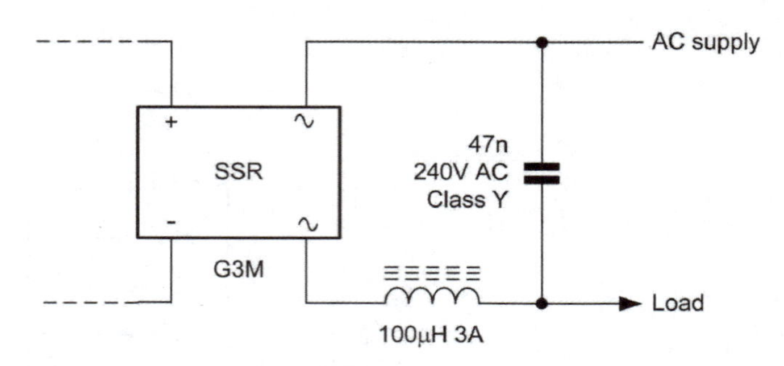

Figure 9.41 *Using a 'snubber' network with an inductive load*

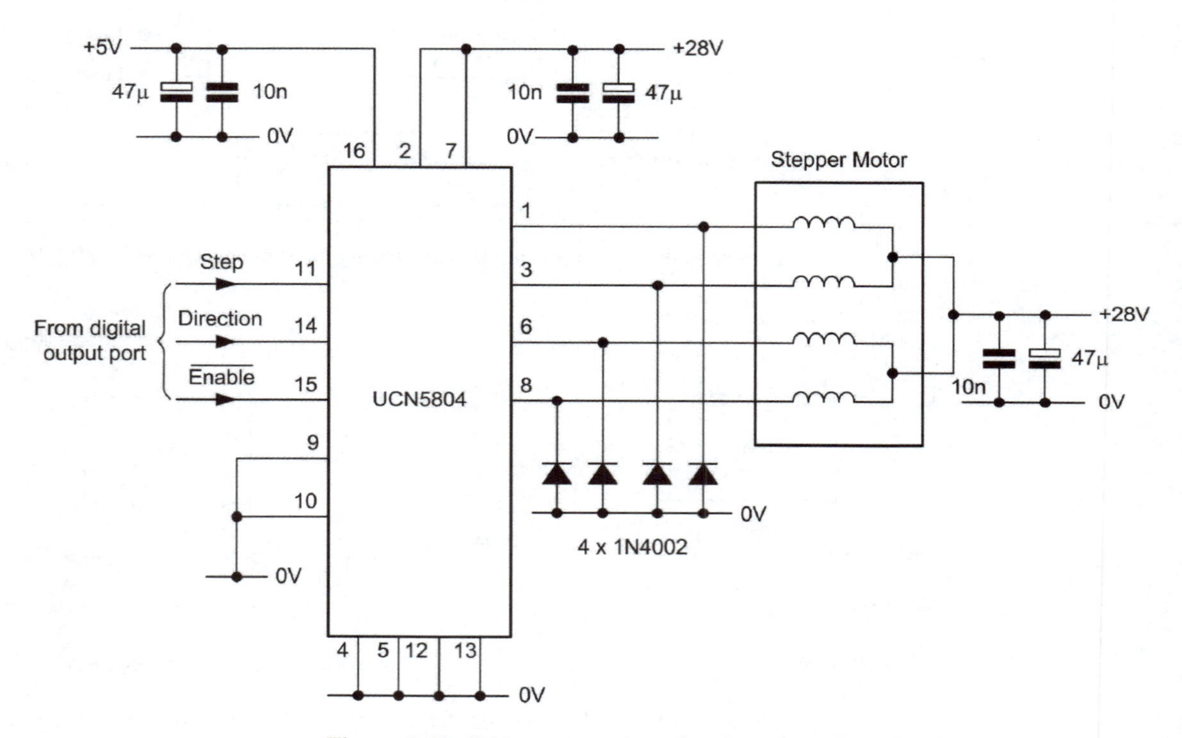

Figure 9.42 *Stepper motor interface based on the UCN5804*

Driving solenoids and solenoid-operated valves

Solenoids and solenoid-operated valves are generally available with coils rated for 110 V/240 V AC or 12 V/24 V DC operation. The circuitry for interfacing solenoids will thus depend on whether the unit is rated for AC or DC operation. In the case of AC-operated units, a suitably rated SSR should be employed (see Figure 9.41) while, in the case of DC-operated solenoids, interface circuitry should be identical to that employed with medium/high current relays (see Figure 9.33).

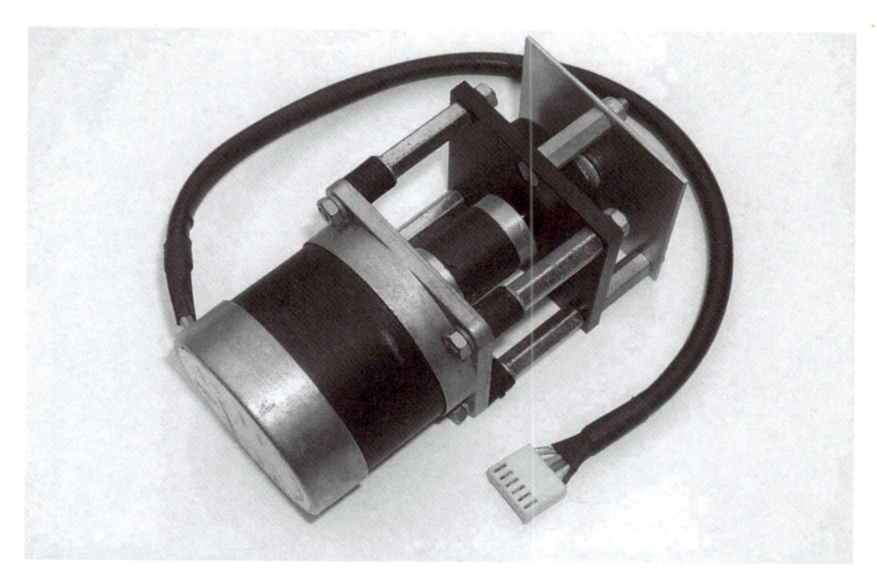

Photo 9.12 *Stepper motor*

Driving stepper motors

The complex task of interfacing a stepper motor to a PC-based system can be greatly simplified by using a dedicated stepper motor driver card. Alternatively, in many light-duty applications, a simple interface can be constructed based on a specialized stepper motor driver chip (such as the UCN5804). This device includes all necessary logic to drive a stepper motor as well as output drivers for each of the four phases. The chip can provide drive for a four-phase unipolar stepper motor with continuous rating of up to 1.25 A per phase (1.5 A startup) and 35 V. The inputs are compatible with standard PMOS, NMOS, and CMOS circuits (note that TTL or LS-TTL may require the use of pull-up resistors in order to ensure a proper logic 1 input high state).

Figure 9.42 shows a typical stepper motor interface based on the UCN5804. The stepper motor interface requires three port output lines that operate on the following basis:

- The STEP input is pulsed low to produce a single step rotation.
- The DIRECTION input determines the sense of rotation. A low on the DIRECTION input selects clockwise rotation. Conversely, a high on the DIRECTION input selects anticlockwise rotation.
- The /ENABLE input must be taken low to operate the motor.

The software routines for driving the stepper motor are quite straightforward and can be simply based on sequences of OUT (or equivalent) instructions which may be contained within loops where continuous rotation in a clockwise or anticlockwise direction is required. For further details (including the UCN5804's half-step mode) readers should refer to the relevant data sheets.

10 Software packages

It should be self-evident that the effectiveness of any PC-based data acquisition, instrumentation, or control system will be dependent not only upon the hardware employed but also upon the software which controls the system. Software has a vital role to play in that it provides an interface and acts as an intermediary between the user and the physical components of the system. Furthermore, the degree of control, flexibility, and ease of use will largely be dependent upon this software interface.

The newcomer to PC-based instrumentation and control systems can be forgiven for being baffled by the variety and complexity of software packages designed to assist him in his task. This chapter categorizes software packages on a variety of grounds and provides details of several of the most popular software products. The aim has been that of providing a yardstick by which the control and instrumentation engineer can judge his or her current and future software requirements.

As an example, a stand-alone process controller may, for example, require relatively unsophisticated software in the form of a simple 'turnkey' program developed in a high-level programming language. A complex distributed factory instrumentation system, on the other hand, which requires frequent re-configuration and which may necessitate interfacing with several other applications programs, may require the services of an applications programming environment.

Selecting a software package

Several factors need to be considered when selecting any software package. These are:

- Ease of use what level of expertise is required in order to make use of the package?
- Flexibility – can the package be easily adapted to differing requirements and can it be readily interfaced with other software?
- Performance – what performance criteria and specifications must be met?
- Functionality – does the package offer a suitable range of functions and will it interface correctly with the chosen hardware configuration?

To some extent, this last factor is paramount since, if the package cannot offer support for the particular hardware configuration in question, it may be of little use. Readers should be aware that a hardware specification will often be fixed before a full software specification has been developed.

Ease of use

The question of ease of use will largely depend upon the person (or persons) who will be using the system. In any event, a software package should be reasonably intuitive and the user should not be presented with outcomes which he or she did not expect. As an example, it should not be possible to quit from a package without being presented with a clear warning of the consequences (e.g. data loss). Similarly, on-screen controls and displays should, as far as possible, mimic those which the user will already be familiar with. All this may sound very obvious but programmers and software engineers often fail to identify with the level of expertise of the operator, and he or she may be left to 'muddle through' by trial and error.

Fortunately, there is a trend towards making software 'user-friendly' and this has been greatly aided by the availability of graphics-oriented operating systems such as Microsoft's Windows with programs that are designed to take full advantage of the visual environment. Such packages can be highly intuitive and provide an excellent user interface based on windows, icons, and pull-down menus (WIMP).

The WIMP interface is also available to any application that is co-resident with the operating system. Because of this, an application that needs to make use of the WIMP environment can simply make calls to appropriate system routines obviating the need to create its own stand-alone routines for manipulating windows and other graphical components.

Flexibility

Some software packages (particularly those which may have been written for a particular application) tend to be somewhat rigid since they are generally based on a pre-defined model. In many cases, such programs will not provide the operator with an opportunity to configure the system or select from a range of choices (e.g. via a menu screen). If a change does become necessary, the software has to be modified at the source code level. Sometimes this task can be tackled by a keen control engineer but, more often than not, it will require the services of a programmer or software engineer.

More flexible software packages will allow the user to configure the software for a particular hardware system (e.g. by specifying the system components and expansion capability) or, alternatively, will use plug-and-play capability when the operating system supports this. In all cases it is important to ensure that the installation and configuration process is made transparent to the user.

As users become fully aware of the advantages of a PC-based instrumentation and control system, a further consideration which will often become increasingly important is that of interfacing with other software packages so that data can be exported and imported in a fashion which is largely transparent to the user. This is often advantageous where measurements must be made within one program before the captured data (stored in a disk file) is imported into another program for statistical and/or graphical analysis. This opens up a completely new scenario in which data acquired by one program can be made to yield all manner of new information when analysed by another package. The availability of Dynamic Data Exchange (DDE) has made it possible for data to be exchanged on a transparent basis between suitably equipped programs. For example, DDE makes it possible to data sourced by a custom-written application to appear in an Excel spreadsheet without the need to first export the data and then import it into Excel.

Performance

Performance of a software package is often somewhat difficult to gauge since determining factors may differ from one application to another. Processing speed, for example, will be all important in some applications (e.g., in an oscilloscope) but largely irrelevant in others.

To a large extent, processing speed will depend upon simultaneous demands placed on the processor. Where a system has to carry out many functions at the same time (such as regularly updating a graphics display, servicing interrupt requests generated by several expansion cards, and responding to user input from a keyboard) it is hardly likely to run at what may be considered an acceptably high speed.

If speed is a paramount consideration and a great deal of 'number crunching' is expected, then a system with a high clock rate and plenty of memory will normally be essential. If such a system still fails to deliver the speed which is a pre-requisite of a particular real-time application then it is probably questionable as to whether a PC-based system should have been adopted in the first place!

Provided that processing speed is identified as a premium requirement, custom software can usually be made to offer significant speed advantages over 'off-the-shelf' packages. The programmer or software engineer can elect to optimize his or her code for speed rather than data integrity. As an example, input range checking could be abandoned in favour of faster throughput of data or opt for post-acquisition rather than real-time display of data.

Functionality

Functionality is becoming increasingly important in software selection and it relates to the fitness and suitability of a program for a particular application and hardware configuration. Most of today's applications packages offer additional functionality which may not appear as a part of an initial software specification. The ability to import and export data files is a prime example of this as the availability of a built-in IEEE-488 command language or custom language extensions such as user-written functions.

Software classification

In order to provide a frame of reference, Table below shows the continuum which exists between custom-written 'turnkey' software at one extreme and operating system utilities at the other.

Ease of use	Flexibility	Performance	Functionality	Examples
Very complex (requires a high level of programming expertise)	Designed to meet a particular set of requirements. Higly customizable.	Can be very fast and compact but may require substantial development time.	As required to meet a given specification.	Custom-written software using programming languages such as MASM32, Turbo C++, PowerBASIC, Visual Basic, or Visual C++.

(*continued*)

Ease of use	Flexibility	Performance	Functionality	Examples
Moderately complex (usually requires moderate programming expertise)	Flexible and adaptable. Offers significant customization.	Fast and efficient. Requires significant development time.	Offers a high level of functionality.	Custom-written software using programming languages in conjunction with extensions, external libraries, or Active X controls.
Average level of complexity (may require some programming expertise)	Moderately flexible and adaptable to most situations. Moderately customizable.	Reasonably fast and moderately efficient. Application development can be reasonably fast.	Offers a high degree of functionality but may require configuration for a particular application.	Programmable applications and data analysis tools (such as LABView), DASYLab, etc.).
Simple and relatively easy to use (requires no programming expertise)	Somewhat limited inflexibility and some packages may be dedicated to particular functions (such as strip recorders). Limited customization.	Usually fast and efficient. No development time required.	Levels of functionality vary but are often limited to a particular function.	Dedicated applications (such as TracerDAQ, WINDAQ, etc.).
Very simple to use (but may require interpretation)	Restricted to particular applications (such as port testing). Not customizable.	Usually fast and efficient. No development time required.	Provides a limited range of functions (such as reporting).	General tools and utilities (such as Norton Utilities, VB Port Test, etc.).
Very simple to use and generally highly intuitive	Not customizable.	Usually fast and efficient. No development time required.	Usually only provides reporting facilities.	Operating system utilities (such as Windows System Tools).

Custom-written software

Custom-written can be developed using a variety of programming languages such as assembly language, BASIC, and C/C++. The techniques for developing custom-written software have already been described at some length in earlier chapters and readers should refer to these for information as well as sample code written in MASM32, Microsoft BASIC, PowerBASIC, C/C++, and Visual Basic. You should not underestimate the task of developing your

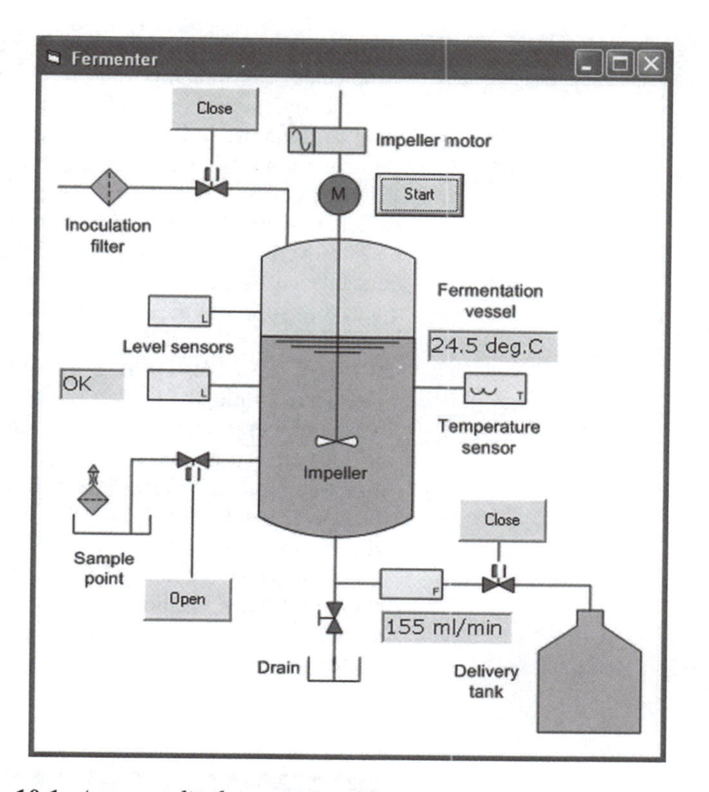

Figure 10.1 *A screen display generated by a program created using Visual Basic. The display shows a running fermentation process with fully interactive controls*

own applications (particularly if starting from scratch); however, in some cases, this may be the only effective solution to a particular problem.

With modern Visual programming languages (e.g. Visual Basic) it is eminently possible to produce attractive and meaningful graphical displays that show the inputs and outputs of a control or instrumentation system. Figure 10.1 shows a typical example of a screen display showing a graphical representation of a fermentation process. The window background was produced using Microsoft Visio and then imported into Visual Basic. The various controls were then placed at appropriate points within the schematic diagram. The controls and displays are fully interactive and the user is able to use the screen display to monitor the process.

Programming language extensions

Most modern high-level language interpreters and compilers provide a wide range of facilities suitable for those wishing to develop commercial and scientific applications. Unfortunately, the facilities offered by a programming language may be somewhat lacking when the software is to be used in conjunction with data acquisition, control hardware, or in an instrumentation application. In such cases, it may be possible to extend the language using proprietary components that are available 'off-the-shelf'. Such components include specialized controls for use with Visual programming languages as well as dynamic link libraries and ready-built procedures that can be used to add functionality to a program. In recent years a number of 'third-party' suppliers have risen to the challenge of producing such extensions and this has added considerably to the pool of resources available to the software developer.

The ActiveX control standard defined by Microsoft describes modular, reusable software components that can be used universally by any environment that supports the standard. For example, without modification, the same controls may be used by Visual BASIC, Visual C++, National Instruments LabVIEW, CEC TestPoint, Borland C++ Builder, Excel, and many more.

As an example, the DATAQ Instruments' ActiveX control library consists of five components, each addressing a different application area. Context-sensitive on-line help is provided for each control. DATAQ Instruments' ActiveX control library supports a wide variety of DATAQ products including the DI-148U, DI-150, DI-151, DI-154RS, DI-158 Series, DI-190, DI-194RS, DI-195B, DI-4xx, DI-5xx, and DI-7xx products.

The universal nature of the ActiveX control standard ensures a consistent and highly simplified software-to-hardware interface that yields programming code that is tremendously reduced in both size and complexity (Figure 10.2). This approach to application development can also be highly cost-effective.

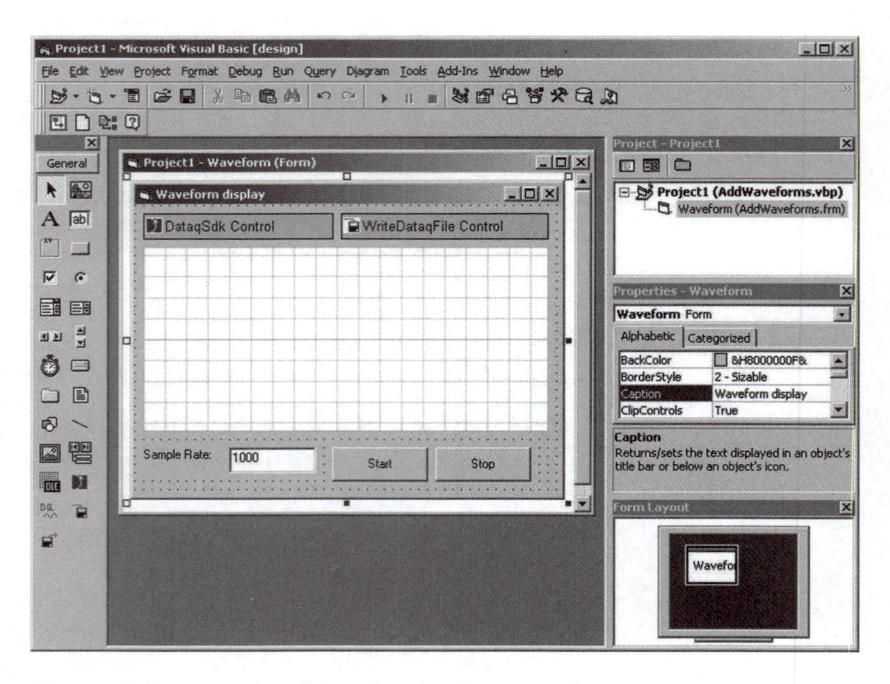

Figure 10.2 *Using Visual Basic to develop an application using an ActiveX control supplied by DATAQ. The control simplifies the task of plotting time-related waveforms of captured data*

Programmable applications

Programming applications are software packages which allow the user to create software that can be used in dedicated (*turnkey*) applications and, while such tools have something in common with a conventional language-oriented development systems, they usually provide an environment in which applications can be developed with minimal programming experience.

LabVIEW

National Instruments LabVIEW is a graphical application development environment designed specifically for engineers who need to create flexible and scalable test, measurement, and control application. LabVIEW is a fully functional graphical programming language which offers a variety of features that simplifies the development of sophisticated applications for control, instrumentation, and data acquisition.

The basic building block of a LabVIEW application is a virtual instrument (VI). This consists of a front panel which contains the user interface and an underlying block diagram which is used to develop the visual code. Furthermore, because LabVIEW's VIs are modular, applications can be developed to any required scale.

Figure 10.3 shows a simple LabVIEW VI which is designed to monitor and control a process based on a heated tank. You might like to compare this with the similar application developed in Visual Basic (see Figure 10.1).

LabVIEW's programming is based on a graphical representation of dataflow. This model frees the developer from the usual sequential architecture of a text-based programming language. An example of a LabVIEW VI block diagram is shown in Figure 10.4. Components are added to this block diagram using drag and drop techniques and then linked together. Application development is thus extremely fast and requires very little programming expertise.

The VI shown in Figure 10.4 is designed to monitor the status of the PC's parallel ports. The corresponding front panel view for the parallel port VI is shown in Figure 10.5. Further examples of LabVIEW VIs are shown in Figures 10.6–10.8.

DASYLab

DASYLab (or 'Data Acquisition Laboratory') from National Instruments provides a graphical environment for developing a range of data acquisition, display and analysis functions (Figure 10.9). A vast range of modules permit customization and little previous programming expertise is required to get the best from this sophisticated yet easy to use package. However, since DASYLab is available in four different versions (Lite, Basic, Full, and Pro), it is essential to select the version that provides the level of functionality required!

The least expensive and powerful version, DASYLab Lite, provides users with the ability to create simple data logging applications and display the results plotted against time. The Lite version offers only a limited number of channels and smaller worksheets. The next level version, DASYLab Basic, can be used to create a smart data loggers which have the ability to reduce the amount of

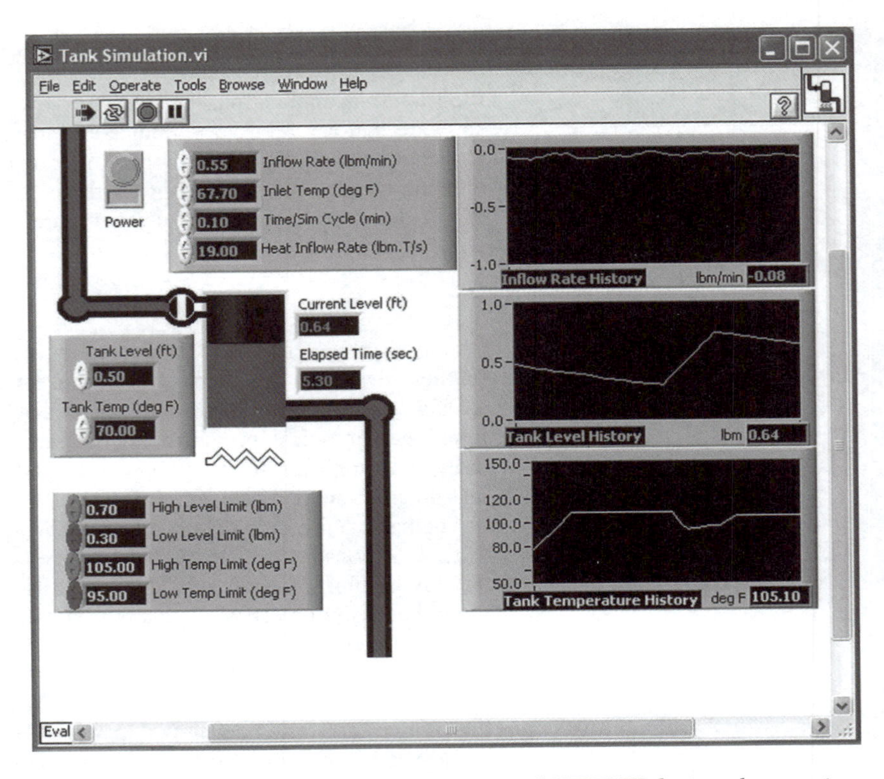

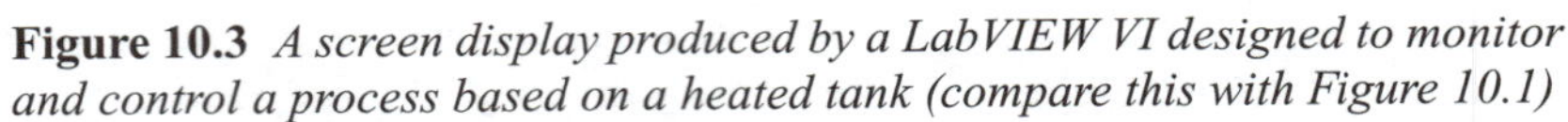

Figure 10.3 *A screen display produced by a LabVIEW VI designed to monitor and control a process based on a heated tank (compare this with Figure 10.1)*

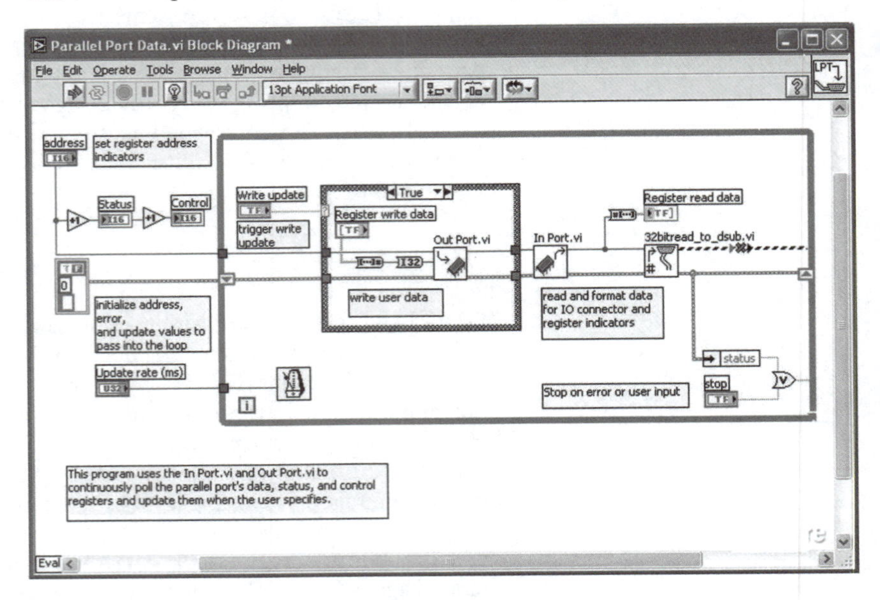

Figure 10.4 *A block diagram of the LabVIEW VI designed to monitor the PC's parallel ports*

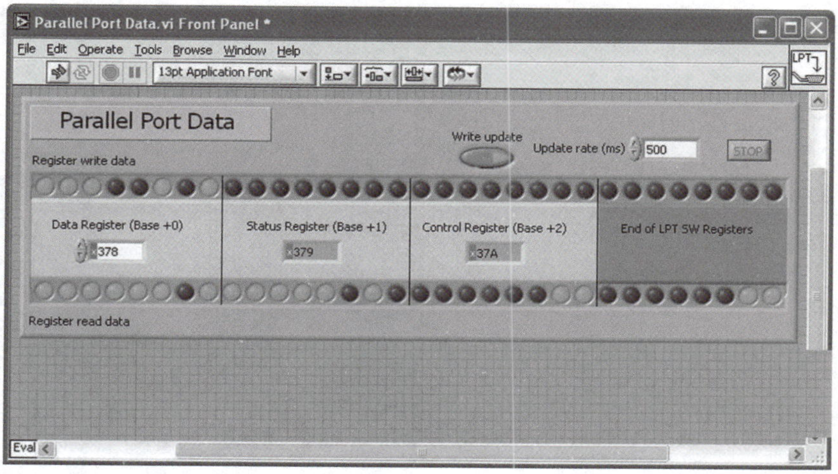

Figure 10.5 *Screen display produced when the parallel port LabVIEW VI is running*

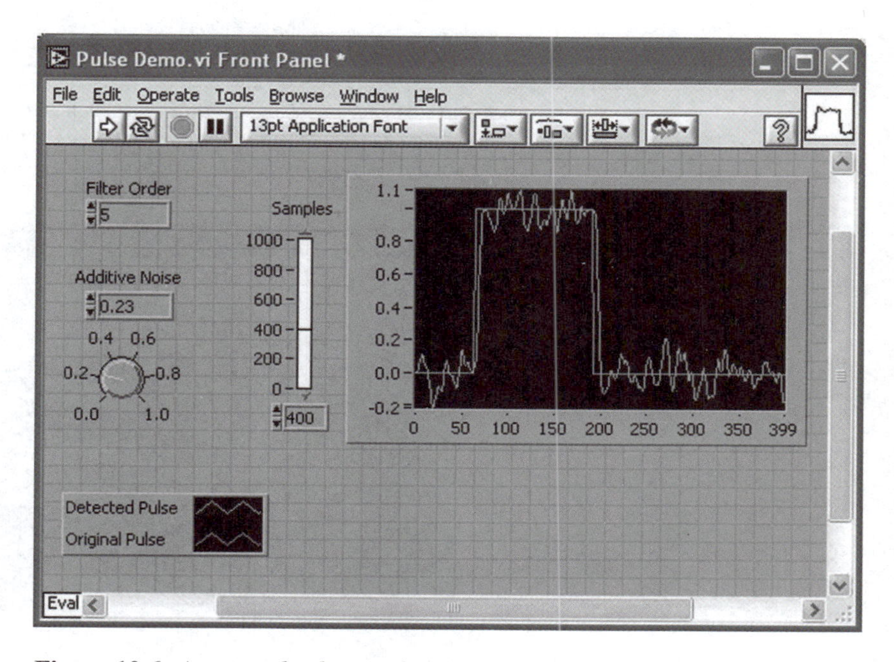

Figure 10.6 *A screen display produced by a LabVIEW VI designed to carry out an analysis of a rectangular pulse in which noise has been added*

data and perform straightforward calculations. DASYLab Basic also has the ability to provide control via analogue and digital outputs (Figure 10.10). This version is intended for users that need to create an application that provides a range of data acquisition features together with the ability to perform some basic

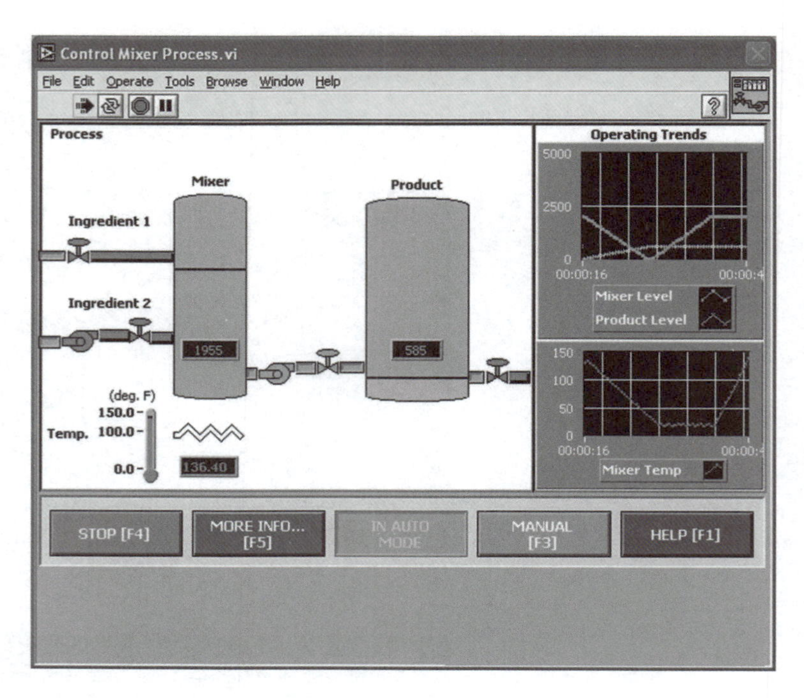

Figure 10.7 *LabVIEW is equally at home in control applications as it is in the field of data acquisition and instrumentation. This screen shows a typical VI used in a process control application*

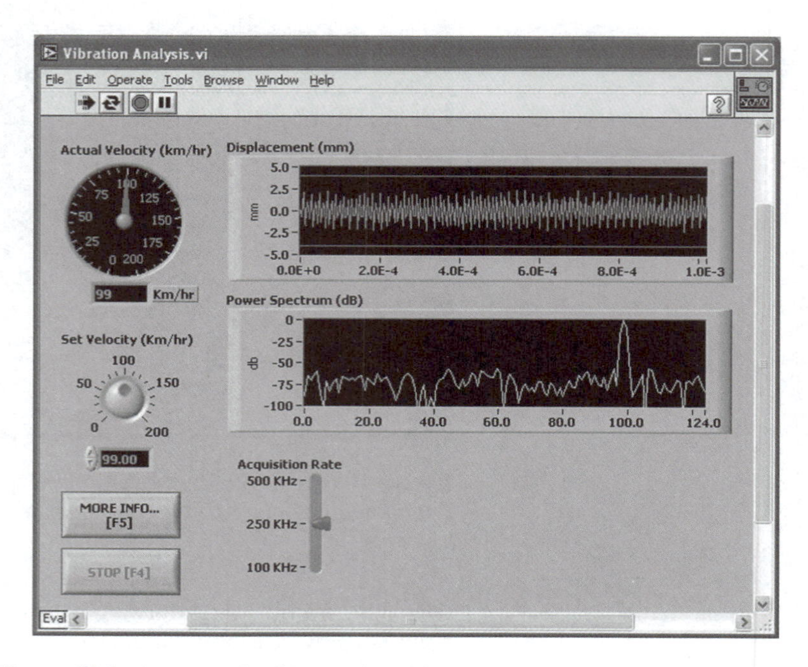

Figure 10.8 *A screen display produced by a LabVIEW VI designed to carry out a vibration study*

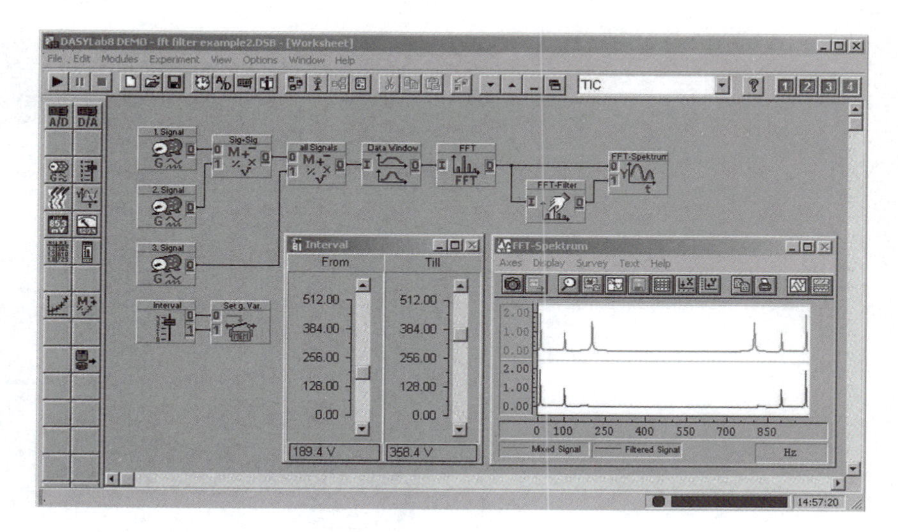

Figure 10.9 *DASYLab combines ease of programming using a simple visual programming interface with sophisticated reporting and display*

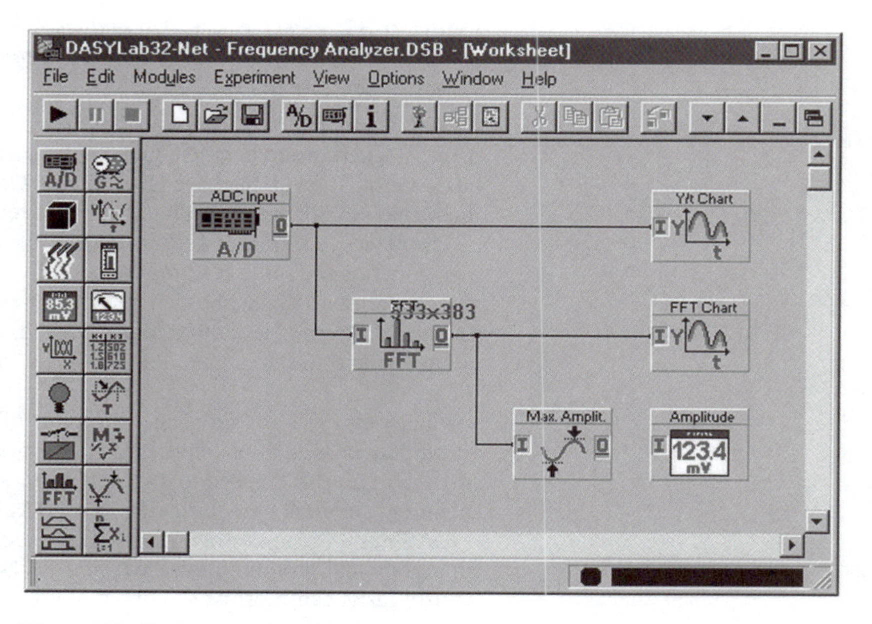

Figure 10.10 *DASYLab programming involves simple 'drag and drop' techniques*

analysis and the ability to save data. The software is well suited to environments where the operator is present and can react to events.

The second highest level is DASYLab Full. This version is capable of performing automated tasks as well as data analysis based on Fast Fourier Transformation (FFT) (Figure 10.11). The automated tasks can involve setting

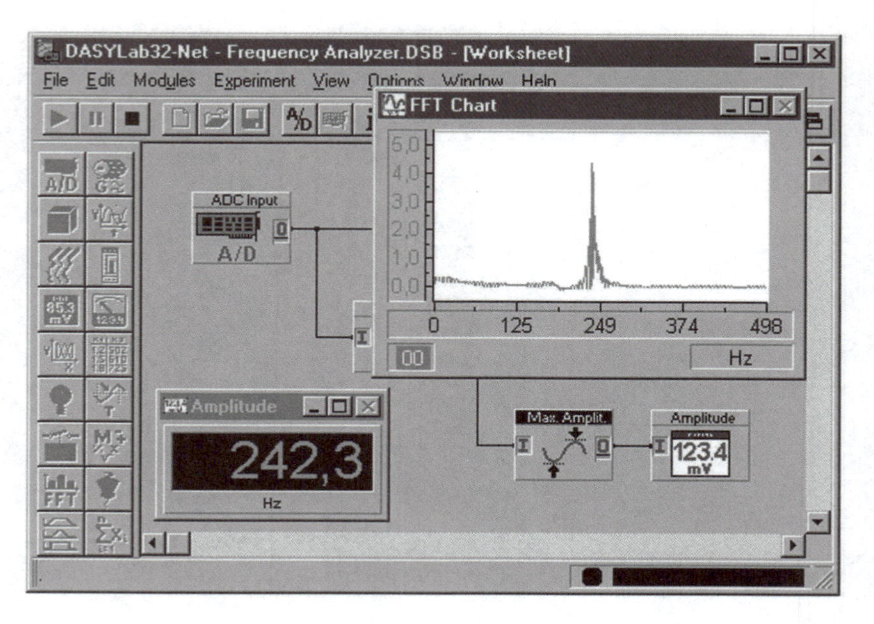

Figure 10.11 *A variety of different display options are available within DASYLab, including the ability to produce Fast Fourier Transformation (FFT)*

trigger-based data acquisition, generating alarms, and other simple automation-based tasks. The Full package is suited to a wide range of applications including those that might have to run unattended or require fully automatic operation.

The most sophisticated and most powerful version of the software, DASYLab Pro, contains the same features as the Full versions with the addition of more signal analysis, a Sequence Generator, high-end or complex frequency analysis, additional filtering, and other tools. DASYLab Pro is ideal for creating stand-alone test and measurement applications with a powerful set of data analysis tools.

Finally DASYLab Net is an extension of DASYLab that uses TCP/IP to communicate with (and control) remote copies of DASYLab Net. This version of DASYLab can be used to remotely start, stop, and load measurement data, including simultaneous starts of several DASYLab Net systems.

DASYLab's programming interface is extremely straightforward and is based on a graphical programming environment. Sophisticated data acquisition and control tasks can be solved without any language-based programming and it is only necessary to insert the appropriate module symbols into the worksheet and connect them by wires. The module symbols represent inputs or outputs, display instruments, or any of the range of operations provided by the program, while the data channels represent the signal flow (Figures 10.12–10.15).

DASYLab provides support for more than 250 data acquisition boards through appropriate drivers. Data acquisition hardware can also be connected to DASYLab through the RS-232 and/or IEEE-488 bus modules. In addition, data can be exchanged between DASYLab and other Windows applications using Dynamic Data Exchange (DDE) techniques.

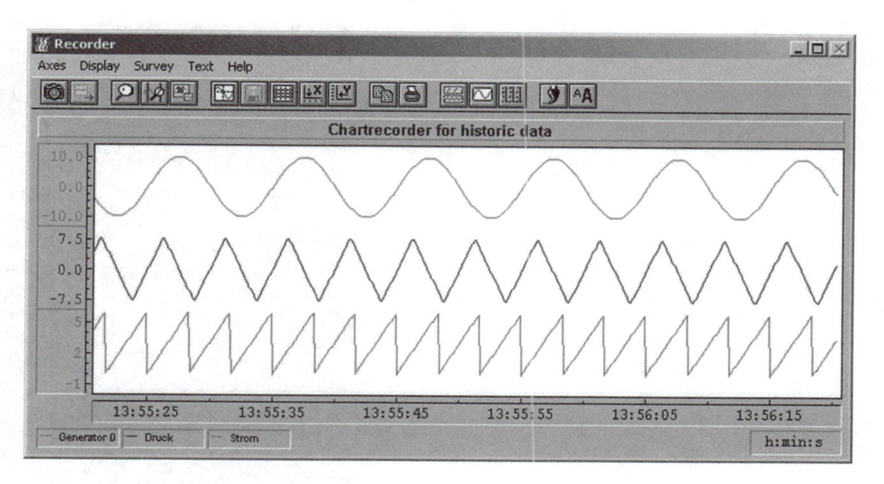

Figure 10.12 *The DASYLab chart recorder display*

List Display

Edit View Help

Number	Generator Volt	Generator bar	Generator A
1492	-4.82	-5.120	4.6
1493	-4.26	-4.480	4.7
1494	-3.68	-3.840	4.9
1495	-3.09	-3.200	5.1
1496	-2.49	-2.560	5.3
1497	-1.87	-1.920	5.5
1498	-1.25	-1.280	5.6
1499	-0.63	-0.640	5.8
1500	-0.00	0.000	0.0
1501	0.63	0.640	0.2
1502	1.25	1.280	0.4
1503	1.87	1.920	0.5
1504	[illegible]	[illegible]	[illegible]

Figure 10.13 *The DASYLab list display*

DADiSP

Originally developed in the early 1980's as part of a research project at MIT to explore the aerodynamics of Formula One racing cars, DADiSP is a visually oriented software package for the display, management, analysis, and presentation

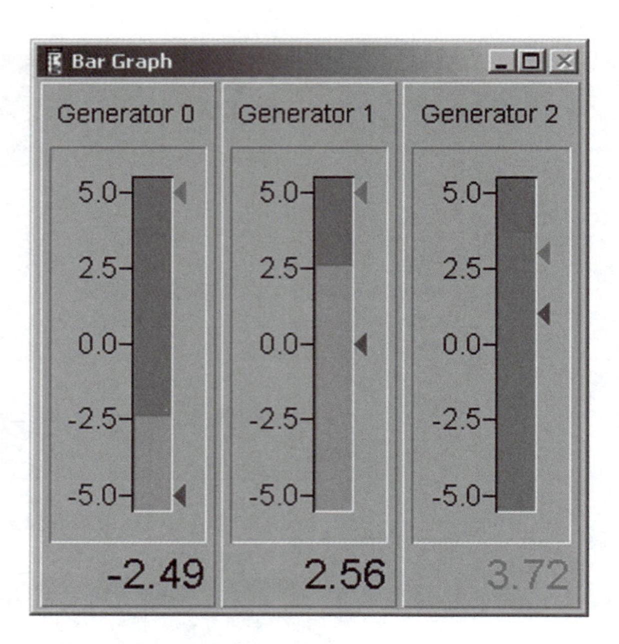

Figure 10.14 *The DASYLab bar graph display*

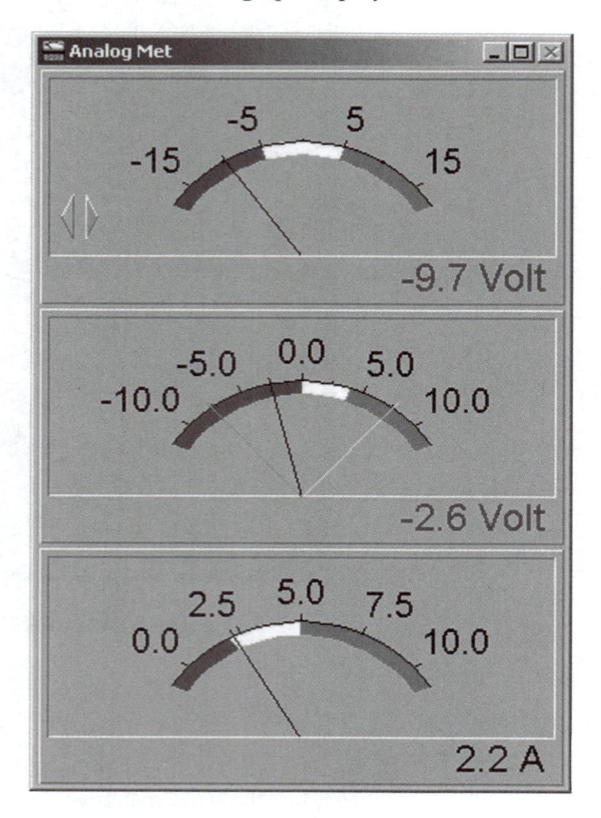

Figure 10.15 *The DASYLab analogue meter display*

of scientific and technical data. DADiSP is designed to handle all of the phases associated with the acquisition and display of data, from initial collection to final analysis.

DADiSP can be used to analyse and display up to 100 windows from the same (or different) data files. The package offers a powerful graphing capability and contains a variety of standard and advanced mathematical and statistical functions ranging from standard arithmetic (addition, subtraction, multiplication, and division) to hyperbolic functions, mean and standard deviations, integration, and differentiation.

DADiSP uses Worksheets to display, manipulate, and analyse data in an integrated graphic environment (Figure 10.16). Each Worksheet can display and process information in an unlimited number of formats. In addition, each Worksheet can be modified, rearranged, and stored as a template for future use with other data.

Commonly used functions are grouped together and made available through pull-down menus. However, display and data manipulation functions can also be entered as text at the command line. DADiSP maintains a full history of commands and lists of commands can be copied and pasted into the command line. Using this technique it is possible to easily produce simple, yet powerful, programs. In fact, DADiSP Worksheets behave as executable programs but eliminate the need for traditional programming. Furthermore, Worksheets will, updated automatically when one or more data items or Window formulae alters so there is no need to re-run an analysis when the data changes.

DADiSP is ideally suited to any application in which captured data is to be analysed and displayed in a graphical format. Typical examples might be X-, Y-, and Z-axis strain developed in a structural member when subjected to repeated cycles of stress or variation in oscillator frequency, amplitude, and noise when

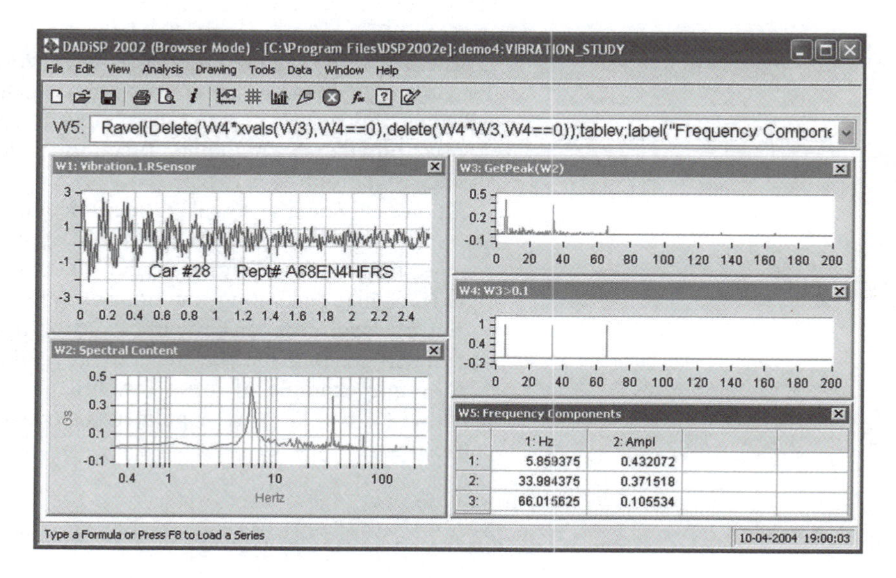

Figure 10.16 *A screen display produced by DADiSP. The screen shows various windows displaying information obtained during a vibration study*

subjected to variations in temperature, pressure, and humidity. In either case, the multiple windowing capability allows the user to form a complete picture of the performance of a system without recourse to a number of discrete graphs and displays.

DADiSP supports a number of advanced mathematical functions including trigonometric and logarithmic functions. These can be applied to any valid signal, scalar, or signal-scalar expression. When applied to a signal, they are applied successively to each point of the signal and the resulting signal is displayed in the current window. When applied to scalars (integers, real numbers, or complex numbers) the resulting value is displayed at the bottom of the screen. The calculus functions are provided for determining derivative and integral functions. Since the signals are discrete, DADiSP provides a means of performing left derivative and right derivative calculations. Four of these functions (`DERIV`, `LDERIV`, `RDERIV`, and `INTEG`) take one signal as an argument and return a new one. `AREA` takes two additional arguments, the starting and ending points within a signal.

Note also that `AREA` returns a scalar whereas the other four calculus functions return signals. The algorithm for calculating the integral is a modification of Simpson's Rule and is more accurate than a simple trapezoidal approximation.

DADiSP's statistical functions provide information about a signal (or two signals in the case of `LINREG2`). `MEAN` and `STDEV` return appropriate values which can be nested in more complicated expressions. `STATS` does not return a signal but displays both the mean and standard deviation at the bottom of the screen. `LINREG` and `LLNREG2` display the regression coefficients and then create a new signal (i.e. the line generated by the linear regression) which is useful for over-plotting. `AMPDIST` generates a new signal which constitutes a bar graph distribution for a signal. The function accepts a real-number argument which is the incremental value (`DELTA X`).

DADiSP contains facilities for frequency domain analysis. Fourier analysis is provided with both the Discrete Transform and the much faster Fast Fourier Transform (FFT). `PARTSUM` creates a new signal which is equivalent to the partial sums of two input signals whilst `SUMS` adds any number of signals together. `MOVAVG` provides a smoothing function while `AVG` takes the point-by-point average of a group of signals.

A powerful range of signal editing functions are also provided. `EXTRACT` creates a new signal by extracting part of an existing signal while `REVERSE` simply changes the polarity of a given signal and `CONCAT` concatenates any number of signals. Signal generating functions are preceded by the letter `G`. An endless variety of waveforms can be synthesized through combination of functions (e.g. `GSINH` will generate the waveform of a hyperbolic arcsine function).

A simple example of using DADiSP is shown in Figure 10.17. The six windows (W1 to W6) are populated with data using the following Worksheet entries:

```
W1: gsin(128, 1/128, 1)
W2: gsin(128, 1/128, 3) /3
W3: gsin(128, 1/128, 5) /5
W4: gsin(128, 1/128, 7) /7
W5: W1+W2+W3+W4
W6: spectrum(W5);sticks
```

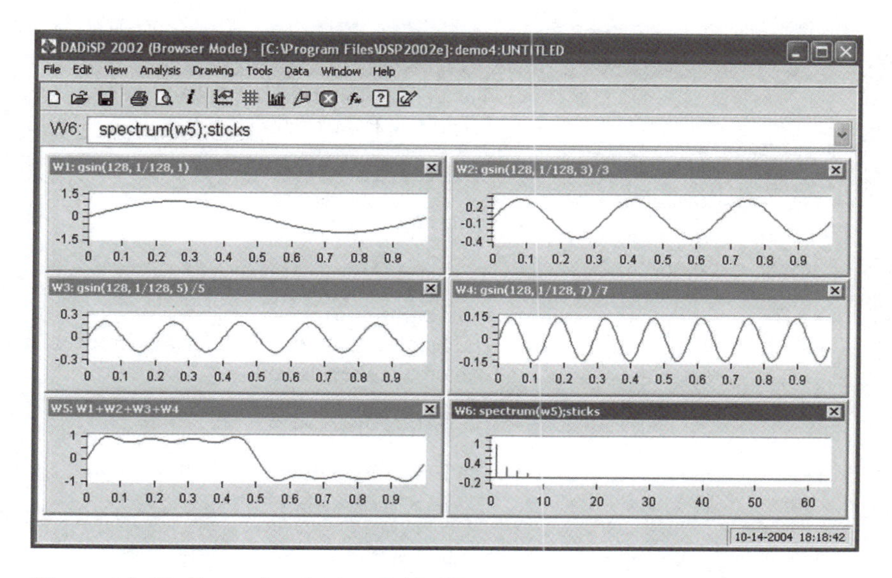

Figure 10.17 *Example of using DADiSP to carry out a simple FFT-based analysis (shown in W6) of a synthesized square wave (shown in W5)*

The resulting frequency spectrum (shown in W6) is generated using a Fourier transform of the data shown in W5. Although this is only a simple example it will hopefully serve to illustrate something of how the software works and its potential for use in applications that require the display and subsequent analysis of data.

Using DADiSP's Series Programming Language

DADiSP's Series Programming Language (SPL) is based on C/C++ and provides users with a means of extending the standard functions provided by DADiSP. SPL files can contain a single function or a collection of functions. As with C/C++, SPL functions are not case sensitive and have the following form:

```
function_name(optional argument list)
{
   local declarations;
   statements;
}
```

As an example, the following SPL function provides the means of returning a temperature in centigrade from an argument given in Fahrenheit:

```
/* Convert Fahrenheit to centigrade */
celsius(f)
{
   local c;
   c = (5.0 / 9.0) * (f - 32.0);
   return(c);
}
```

The celsius function is invoked in the same was as any of the DADiSP built-in functions. For example:

```
celsius(32)
```

will return `0` and

```
celsius(212)
```

will return `100`.

The celsius function can also operate on an entire series. For example, the { } can create a series and

```
celsius({32, 212, 72})
```

returns a three point series with values `{0, 100, 22.22222222}`.

MATLAB

MATLAB stands for matrix laboratory and was originally developed to provide easier programming access to specialized matrix processing routines. MATLAB consists of a command interpreter and a variety of sub-routines that reside in ASCII files known as M files.

MATLAB and its companion products are used in a broad range of applications, including signal and image processing, digital signal processing (DSP), and control design. The MATLAB product family includes tools for:

- Test and measurement
- Data analysis and exploration
- Numeric and symbolic computing
- Plotting and advanced visualization
- Signal and image processing
- Algorithm development
- Deployment of MATLAB applications.

The MATLAB language is designed for interactive or automated computation.

Matrix-optimized functions are used to perform interactive analyses, while the structured language features allow users to develop their own algorithms and applications. The language is applicable to a wide variety of tasks including data acquisition, analysis, algorithm development, system simulation, and application development. The language features include data structures, object-oriented programming, graphical user interface (GUI) development tools, debugging features, and the ability to link with C/C++ routines.

MATLAB offers more than 600 mathematical, statistical, and engineering functions, including:

- Linear algebra and matrix computation
- Fourier and statistical analysis functions
- Differential equation solvers

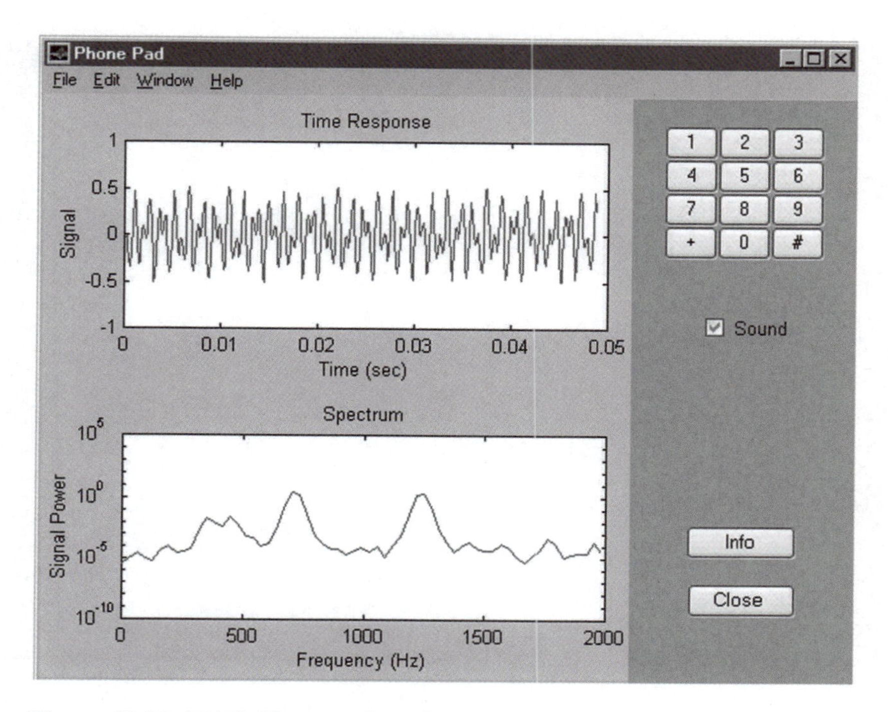

Figure 10.18 *MATLAB spectral analysis of a dual-tone multi-frequency (DTMF) signal*

- Sparse matrix support
- Trigonometric and other fundamental mathematical operations
- Multidimensional data support.

MATLAB is capable of producing 2-D plots, images, and 3-D surfaces and for visualizing volumetric data. Advanced visualization tools include surface and volume rendering, lighting, camera control, and application-specific plot types.

In order to configure MATLAB for use in particular types of application, collections of algorithms, and visual interfaces are provided within a number of MATLAB toolboxes. The most commonly used toolboxes include:

- The Statistics Toolbox includes descriptive statistics, hypothesis testing, probability modelling, and regression functionality.
- The Optimization Toolbox includes minimization tools for linear, quadratic, and non-linear programming, and for solving linear and non-linear least-squares problems.
- The Curve Fitting Toolbox includes routines for pre-processing data, and creating, analysing, and managing models that involve curve fitting.
- The Signal Processing Toolbox includes techniques for time-domain and frequency domain analysis, spectral analysis, and filtering (Figure 10.18).
- The Image Processing Toolbox helps you visualize, process, enhance, and analyse images.

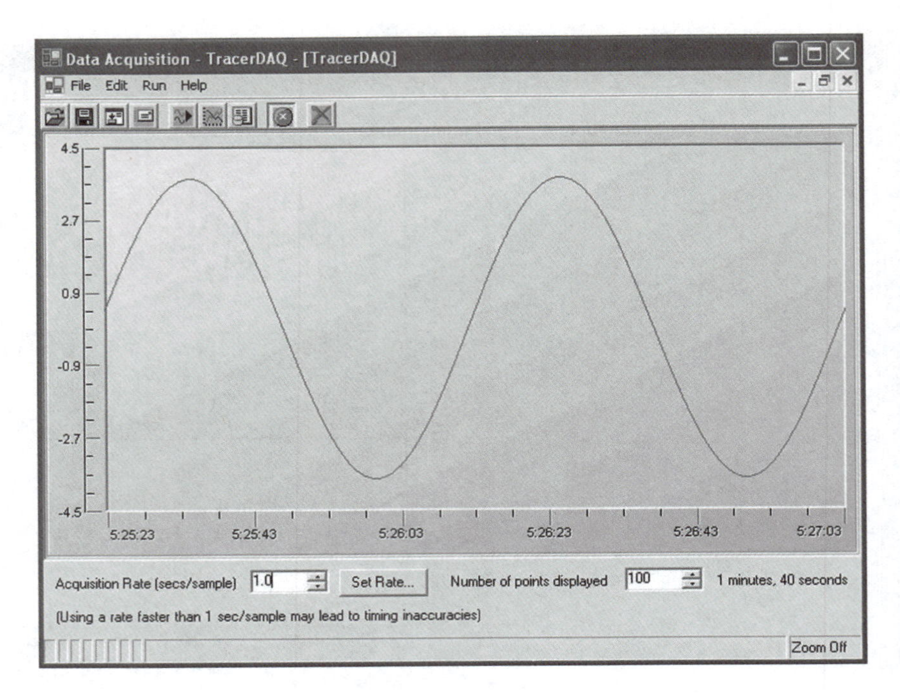

Figure 10.19 *TracerDAQ display showing time-related captured data*

Additional toolboxes, such as Wavelet, Symbolic Math, Fuzzy Logic, and Neural Network, provide complementary, alternative methods for modelling systems and analysing, displaying, and characterizing your data.

MATLAB provides direct access to data from serial ports as well as from MATLAB formatted data files. MATLAB also includes built-in support for popular file formats, including scientific data formats image file formats, and industry-standard formats, such as Microsoft Excel. Additional functions perform ASCII and low-level binary I/O from M-file, C, and Fortran programs.

Dedicated applications

Dedicated applications package are usually designed to solve a particular problem or range of problems. As such, a dedicated applications package may not include programmable features but, instead, are usually configured for use with a particular data source (or range of sources). Such software is often supplied with items of data acquisition hardware such as simple RS-232 and USB-based devices; and usually provides strip chart recording facilities, low-frequency waveform display, and simple data logging and analysis features. TracerDAQ (from Measurement Computing – see Figure 10.19) and WINDAQ (from DATAQ Instruments – see Figure 10.20) are typical examples.

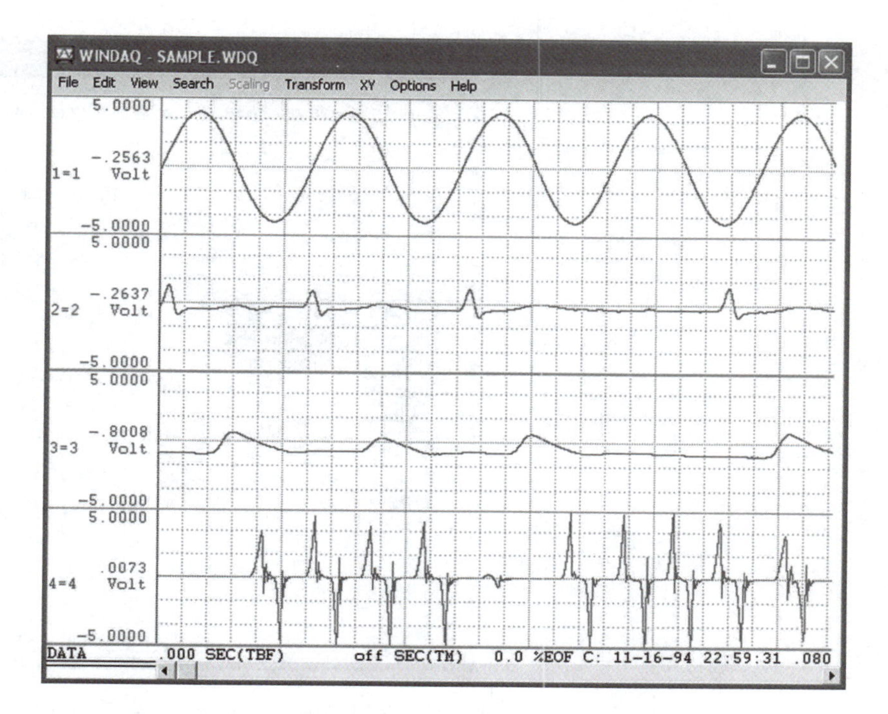

Figure 10.20 *WINDAQ display showing time-related captured data*

Tools and utilities

Tools are aptly named utilities which provide the user with a range of facilities designed to maximize the efficiency of a system. Facilities provided by a tool kit may include any or all of the following:

- Reporting of hardware and software configuration
- System optimization
- Performance measurement
- Diagnostic and troubleshooting facilities
- Data recovery.

Norton SystemWorks

Symantec's widely acclaimed package (originally developed by Peter Norton many years ago) provides a comprehensive suite of programs and utilities designed to cope with a wide range of problems, including hardware and software management. The package offers excellent reporting facilities and is extremely simple to use. In addition, utilities are provided that will monitor aspects of a system's performance. This can be useful when determining how system resources are being used dynamically by applications (Figures 10.21 and 10.22).

Many other specialized tools and utilities are available. These are often dedicated to a very specific area, such as Chris Schroeder's excellent parallel port test utility (see Figure 10.23).

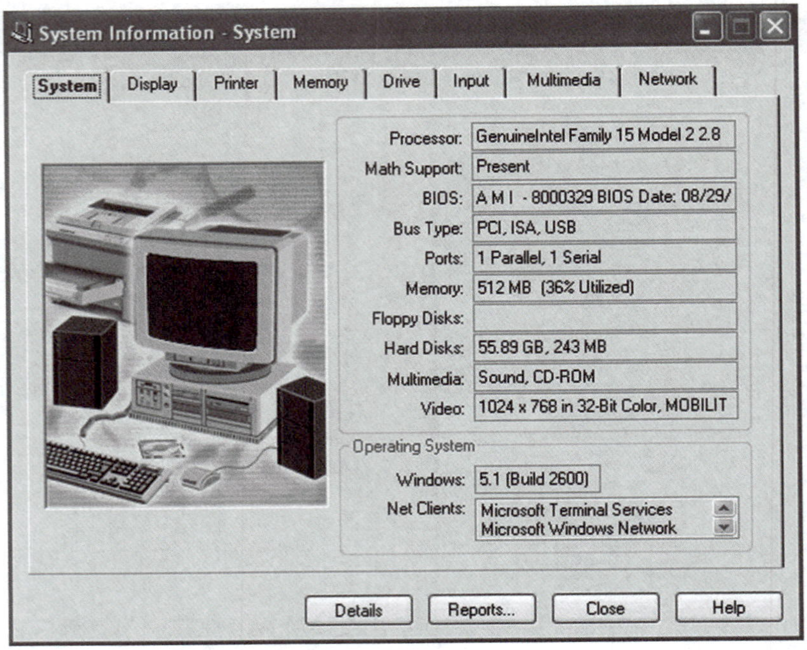

System Information - System

System | Display | Printer | Memory | Drive | Input | Multimedia | Network

Processor:	GenuineIntel Family 15 Model 2 2.8
Math Support:	Present
BIOS:	A M I - 8000329 BIOS Date: 08/29/
Bus Type:	PCI, ISA, USB
Ports:	1 Parallel, 1 Serial
Memory:	512 MB (36% Utilized)
Floppy Disks:	
Hard Disks:	55.89 GB, 243 MB
Multimedia:	Sound, CD-ROM
Video:	1024 x 768 in 32-Bit Color, MOBILIT

Operating System

Windows:	5.1 (Build 2600)
Net Clients:	Microsoft Terminal Services Microsoft Windows Network

Details | Reports... | Close | Help

Figure 10.21 *System Information displayed by Norton Utilities (part of Norton SystemWorks)*

System Information - Memory

System | Display | Printer | Memory | Drive | Input | Multimedia | Network

Windows Memory Usage

Total Physical Memory: 512 MB
Total Windows Memory: 2001 MB
Free: 1762 MB
Used: 238 MB
Selection: 0 KB

Item	Type	Size
WUAUCLT.EXE	32-Bit App	8.63 MB
WINCINEMAMGR.E...	32-Bit App	2.43 MB
EASYSHARE.EXE	32-Bit App	22.1 MB
RUNNER.EXE	32-Bit App	13.3 MB
WLANUTILITY.EXE	32-Bit App	2.24 MB
E_S10IC2.EXE	32-Bit App	896 KB
HPZIPM12.EXE	32-Bit App	1.05 MB
602TEXT.EXE	32-Bit App	27.8 MB
PSP.EXE	32-Bit App	13.2 MB
DLLHOST.EXE	32-Bit App	7.84 MB
MSDTC.EXE	32-Bit App	4.45 MB
SI32.EXE	32-Bit App	12.5 MB
MSMSGS.EXE	32-Bit App	15.1 MB
Windows32	Win32 Sys	61.5 MB

Display Libraries

Details | Reports... | Close | Help

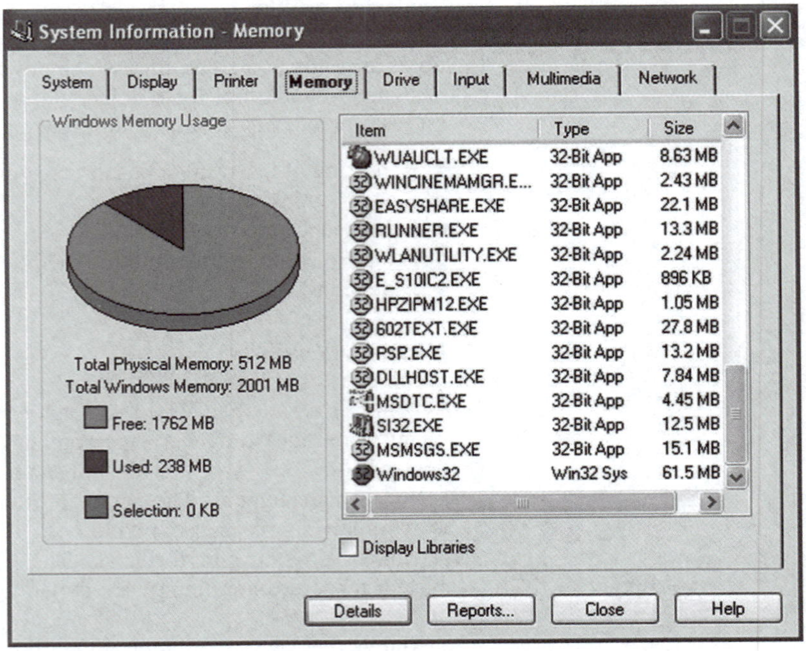

Figure 10.22 *Memory information displayed by Norton Utilities (part of Norton SystemWorks)*

VBPortTest Version 2.0
File Help
Data I/O Register (Base Address)
Data7 Pin 9
Data6 Pin 8
Data5 Pin 7
Data4 Pin 6
Data3 Pin 5
Data2 Pin 4
Data1 Pin 3
Data0 Pin 2
Hex Data
04
Write Data
Strobe Data
Status Input Register (Base Address + 1)
Busy nPin 11
nAck Pin 10
Paper-Out Pin12
Select Pin 13
nError Pin 15
nIRQ Status
Not Used
Not Used
Hex Status
78
Control I/O Register (Base Address + 2)
Not Used
Not Used
Bi-Direct Enable
IRQ Enable
nSelect nPin 17
nInit Pin 16
nAutoLF nPin 14
nStrobe nPin 1
Hex Control
0C
Write Control
Base Address (Hex)
LPT1: 378
LPT2: 278
LPT3: 3BC
Data and Control Registers I/O Mode
Read Only
Byte (Output on Strobe or Write Command Only)
Bit (Output on LED Click only)
Note: Bi-Directional Enable Bit Toggle Active in all Modes

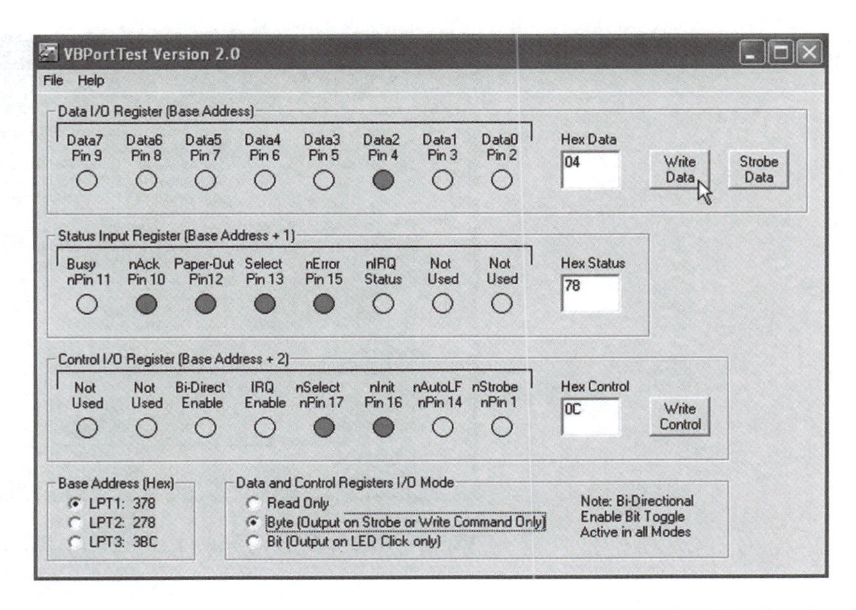

Figure 10.23 *VB Port Test utility developed by Chris Schroeder. This handy program is invaluable for checking the status of the PC's parallel ports*

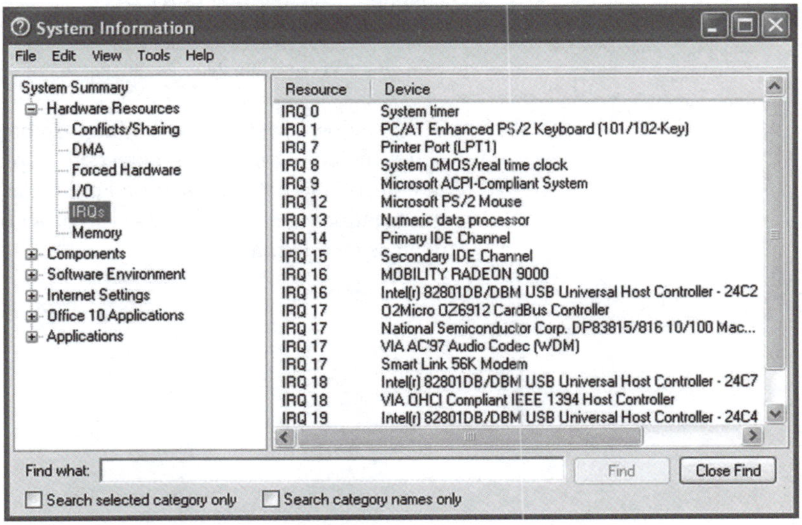

System Information
File Edit View Tools Help
System Summary
Hardware Resources
Conflicts/Sharing
DMA
Forced Hardware
I/O
IRQs
Memory
Components
Software Environment
Internet Settings
Office 10 Applications
Applications
Resource
Device
IRQ 0 System timer
IRQ 1 PC/AT Enhanced PS/2 Keyboard (101/102-Key)
IRQ 7 Printer Port (LPT1)
IRQ 8 System CMOS/real time clock
IRQ 9 Microsoft ACPI-Compliant System
IRQ 12 Microsoft PS/2 Mouse
IRQ 13 Numeric data processor
IRQ 14 Primary IDE Channel
IRQ 15 Secondary IDE Channel
IRQ 16 MOBILITY RADEON 9000
IRQ 16 Intel(r) 82801DB/DBM USB Universal Host Controller - 24C2
IRQ 17 O2Micro OZ6912 CardBus Controller
IRQ 17 National Semiconductor Corp. DP83815/816 10/100 Mac...
IRQ 17 VIA AC'97 Audio Codec (WDM)
IRQ 17 Smart Link 56K Modem
IRQ 18 Intel(r) 82801DB/DBM USB Universal Host Controller - 24C7
IRQ 18 VIA OHCI Compliant IEEE 1394 Host Controller
IRQ 19 Intel(r) 82801DB/DBM USB Universal Host Controller - 24C4
Find what:
Find
Close Find
Search selected category only
Search category names only

Figure 10.24 *IRQ channels information displayed by Windows XP's System Tools utility*

Operating system utilities

Modern operating systems (such as Windows NT or Windows XP) provide a number of built-in utilities that can provide information of value to the engineer. These utilities are normally only for reporting and do not generally allow users

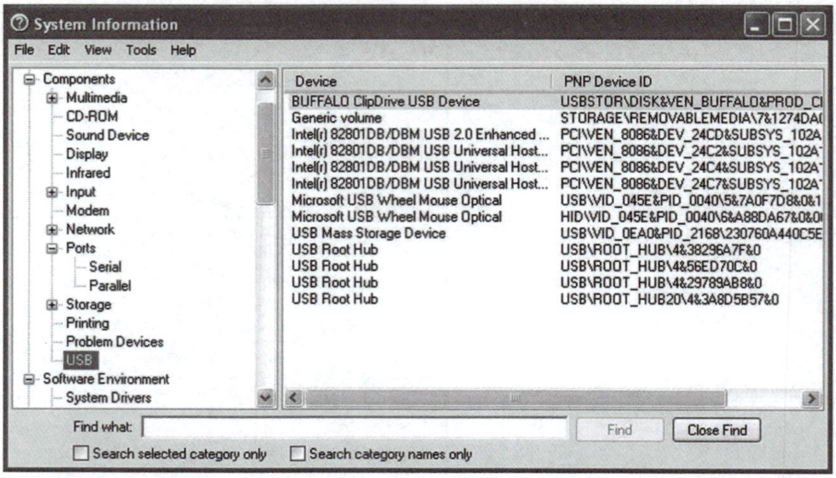

Figure 10.25 *USB devices reported by Windows XP's System Tools utility*

to set parameters or alter the system configuration. A notable exception to this is the System utility (available from the Windows Control Panel) where the Device Manager can be used to change some Resource settings (e.g. the IRQ channel of the parallel port).

Windows System Tools

The Windows System Tools utility can be used to provide detailed information about the configuration of a system. Typical examples of using System Tools in control and data acquisition applications are that of reporting IRQ channel information (see Figure 10.24) and providing information on the installed USB devices (see Figure 10.25).

11 Virtual instruments

PC-based instruments (i.e. *virtual instruments*) are rapidly replacing items of conventional test equipment in many of today's test and measurement applications. This chapter provides readers with an introduction to the principles and practice of these powerful, sophisticated, and highly cost-effective instruments. One obvious use of virtual instruments is that of building automated test systems but general laboratory, bench, and field-service applications are also eminently possible.

Currently available PC instruments include digital storage oscilloscopes (DSO), many of which incorporate additional features, such as spectrum analysis, digital voltmeters and digital frequency meters, counters and timers, waveform generators, arbitrary waveform generators (AWG), and logic analysers. This chapter describes the facilities offered by modern high-specification DSO and also shows how a low-cost sound card can be used to form the basis of a simple oscilloscope for audio and general low-frequency applications.

Selecting a virtual instrument

Several factors need to be considered when selecting a virtual instrument for a particular application. These include:

- *Ease of use*: What level of expertise is required in order to make use of the instrument and also whether the measurements are to be automated or controlled by a human operator?
- *Flexibility*: Can the instrument be easily adapted to differing requirements and can it be easily transferred from one PC to another?
- *Performance*: What performance criteria and specifications must be met?
- *Functionality*: Does the package offer a suitable range of functions and will it interface correctly with an existing hardware/software configuration?
- *Cost*: Has the most cost-effective solution been chosen?

It is important to remember that virtual instruments comprise both hardware and software. The hardware provides a means of interfacing to the PC and of connecting external inputs. Additional controls may also be present (e.g. input selection, trigger selection, attenuation, etc.). The software provides a means of controlling the instrument, collecting data from it, processing the data, and then displaying, analysing, and recording it. When purchasing a virtual instrument it is important to ensure that *both* the hardware *and* the software fully meet your requirements.

Instrument types

The following is a list of the main types of virtual instrument:

- digital storage oscilloscopes;
- digital counters, timers, and frequency meters;

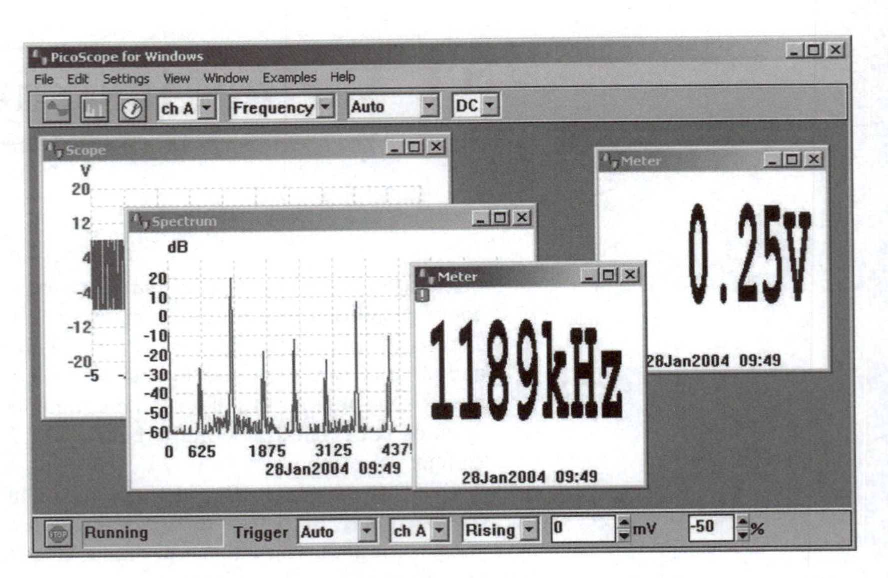

Figure 11.1 *PicoScope software display showing multiple windows providing conventional oscilloscope waveform display, spectrum analyser display, frequency display, and voltmeter display*

- digital voltmeters;
- signal generators (AF, wideband, and RF);
- function generators (sine, square, triangle, and ramp waves);
- arbitrary waveform generators;
- spectrum analysers;
- logic analysers.

Note that several of the above functions may be combined into a single instrument. For example, a DSO can easily provide functions associated with a digital voltmeter, digital frequency meter, and a spectrum analyser (see Figure 11.1).

Instrument connection options

Various connection options for virtual instruments are listed in the table below:

Instrument connection	Comments	Notes
PCI, compact PCI, and PXI adapter cards	High-performance applications based on dedicated PCs. Can be expensive.	See Chapter 2 for further information.
USB ports	Flexible low-cost solution but can have speed limitations.	See Chapter 2 for further information.
PC parallel ports	Usually requires an enhanced parallel port. May require an additional parallel port adapter.	See Chapter 1 for further information.
RS-232 serial ports	Traditional solution allowing connection of multiple instruments but with limited speed.	See Chapter 1 for further information.

(*continued*)

Instrument connection	Comments	Notes
IEEE-488 (via suitable controller card)	Suitable for comple ATE set-ups and test stands (can be limited in speed).	See Chapter 8 for further information.
Firewire and optical connections	Potential for future high-speed instrumentation applications.	
PCMCIA	A future solution for laptop computers in field-service applications?	

Digital storage oscilloscopes

Probably the most well-known computer-based test instrument is the digital storage oscilloscope (DSO). Because of the processing power available from the PC coupled with the mass storage capability, a computer-based DSO is able to provide a variety of additional functions, such as spectrum analysis and digital display of both frequency and voltage. In addition, the ability to save waveforms and measurements for future analysis, or for comparison purposes can be extremely valuable, particularly where evidence of conformance with standards or specifications is required.

Unlike a conventional oscilloscope which is primarily intended for waveform display, a computer-based digital storage oscilloscope (DSO) effectively combines several test instruments in one single package. The functions generally available from a DSO include:

- waveform display;
- precise time and voltage measurement (using adjustable cursors);
- digital display of voltage;
- digital display of frequency and/or periodic time;
- frequency spectrum display and analysis;
- data logging (i.e. storage of waveform data for later analysis);
- ability to save/print waveforms and other information in graphical format (e.g. as .jpg or .bmp files).

The DSO comprises an external hardware unit which is connected to the PC by means of either a conventional 25-pin parallel port connector or by means of a serial USB connector. Some manufacturers also provide dedicated parallel port to USB adapters designed specifically for their own DSO fitted with parallel ports. If in doubt it is wise to contact the supplier for their advice concerning connection to a PC.

The DSO software is usually supplied on CD-ROM (or can be downloaded from the manufacturer's web site). It is important to note that although the DSO hardware cannot usually be used without the appropriate software some manufacturers supply software drivers that will allow you to control the DSO and capture data into your own applications. However, for most of us this is not an option since the supplied software will usually outperform anything that we can write ourselves!

A DSO combines elements of both hardware and software. These must work together to provide all the functionality of a conventional DSO but also those

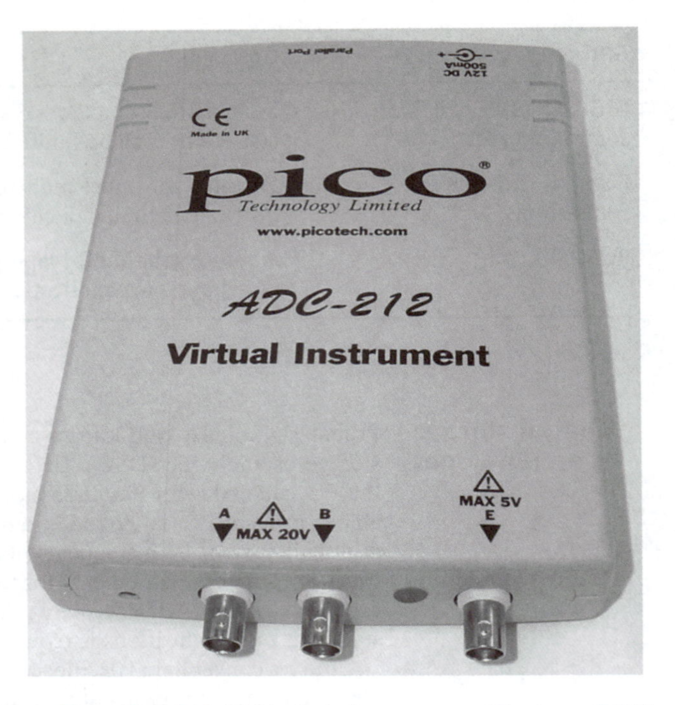

Photo 11.1 *Pico ADC-212 12-bit digital storage oscilloscope (DSO)*

of a spectrum analyser, data logger, digital frequency meter, and voltmeter. In many cases a DSO will be able to replace several items of conventional test equipment. Switching between these instruments is usually quick and easy, and in most cases each instrument is able to have its dedicated window on the PC display.

Multiple views of the same signals and on-screen display voltage and frequency can greatly enhance measurements made with a DSO. In addition, with some DSO waveforms can be annotated with notes and they can subsequently be printed, saved or exported to other applications.

Several types of DSO are currently available. These can be conveniently arranged into three different categories according to their application:

- Low-cost DSO
- High-speed DSO
- High-resolution DSO.

Unfortunately, there is often some confusion between the last two categories. A high-speed DSO is designed for examining waveforms that are rapidly changing. Such an instrument does not necessarily provide high-resolution measurement. Similarly, a high-resolution DSO is useful for displaying waveforms with a high degree of precision, but it may not be suitable for examining fast waveforms. The difference between these two types of DSO should become a little clearer later on but first it is worth explaining some of the more important specifications for this type of equipment.

Sampling rate and bandwidth

The upper signal frequency limit of a DSO is determined primarily by the rate at which it can sample an incoming signal. Typical *sampling rates* for different types of DSO are:

Type of DSO	Typical sampling rate
Low-cost DSO	20–100 KB per second
High-speed DSO	100–1000 MB per second
High-resolution DSO	20–100 MB per second

In order to display waveforms with reasonable accuracy, it is normally suggested that the sampling rate should be at least twice and preferably more than five times the highest signal frequency. Thus, in order to display a 10 MHz signal with any degree of accuracy a sampling rate of 50M samples per second will be required.

The *five times rule* merits a little explanation. When sampling signals in a digital to analogue converter we usually apply the *Nyquist criterion* that the sampling frequency must be at least twice the highest analogue signal frequency. Unfortunately, this no longer applies in the case of a DSO where we need to sample at an even faster rate if we are to accurately display the signal. In practice we would need a minimum of about five points within a single cycle of a sampled waveform in order to reproduce it with approximate fidelity. Hence the sampling rate should be at least five times that of highest signal frequency for a DSO to be able to display a waveform reasonably faithfully.

A special case exists with dual-channel DSOs. Here the sampling rate may be shared between the two channels. Thus an effective sampling rate of 20 MB samples per second might equate to 10 MB samples per second for *each* of the two channels. In such a case the upper frequency limit would not be 4 MHz but only a mere 2 MHz!

The approximate bandwidth required to display different types of signals with reasonable precision is given in the table below:

Signal	Bandwidth required (approx.)
Low frequency and power	DC to 10 kHz
Audio frequency (general)	DC to 20 kHz
Audio frequency (high quality)	DC to 50 kHz
Square and pulse waveforms (up to 5 kHz)	DC to 100 kHz
Fast pulses with small rise times	DC to 1 MHz
Video	DC to 10 MHz
Radio (LF, MF, and HF)	DC to 50 MHz

The general rule is that, for sinusoidal signals the bandwidth should ideally be at least double that of the highest signal frequency whilst the for square wave and pulse signals the bandwidth should be at least 10 times that of the highest signal frequency.

It is worth noting that most manufacturers define the bandwidth of an instrument as the frequency at which a sine wave input signal will be fall to 0.707 of its true amplitude (i.e. the −3 dB point). To put this into context, at the cut-off frequency the displayed trace will be in error by a whopping 29%!

Resolution and accuracy

The relationship between resolution and signal accuracy (not bandwidth) is simply that the more bits used in the conversion process the more discrete voltage levels can be resolved by the DSO. The relationship is as follows:

$$x = 2^n$$

where x is the number of discrete voltage levels and n is the number of bits. Thus, each time we use an additional bit in the conversion process we double the resolution of the DSO, as shown in the table below:

Number of bits (n)	Number of discrete voltage levels (x)
8	256
10	1024
12	4096
16	65 536

A DSO stores its captured waveform samples in a buffer memory. Hence, for a given sampling rate, the size of this memory buffer will determine for how long the DSO can capture a signal before its buffer memory becomes full.

The relationship between sampling rate and buffer memory capacity is important. A DSO with a high sampling rate but small memory will only be able to use its full sampling rate on the top few timebase ranges.

To put this into context, it is worth considering a simple example. Assume that we need to display 10 000 cycles of a 10 MHz square wave. This signal will occur in a time frame of 1 ms. If we are applying the five times rule we would need a bandwidth of at least 50 MHz to display this signal accurately.

To reconstruct the square wave we would need a minimum of about five samples per cycle so a minimum sampling rate would be 5 × 10 MHz = 50 MB samples per second. To capture data at the rate of 50 MB samples per second for a time interval of 1 ms requires a memory that can store 50 000 samples. If each sample uses 16 bits we would require 100 KB of extremely fast memory!

The *measurement resolution* or *measurement accuracy* of a DSO (in terms of the smallest voltage change that can be measured) depends on the actual range that is selected. So, for example, on the 1 V range an 8-bit DSO is able to detect a voltage change of one in two hundred and fifty-sixth of a volt or 1/256 V or about 4 mV. For most measurement applications this will prove to be perfectly adequate as it amounts to an accuracy of about 0.4% of full scale.

Low-cost DSO

Low-cost DSO are primarily designed for low-frequency signals (typically signals up to around 20 kHz) and are usually able to sample their signals at rates of between 10 KB and 100 KB samples per second. Resolution is usually limited to either 8- or 12-bits (corresponding to 256 and 4096 discrete voltage levels, respectively).

Low-cost DSO may be either single- or dual-channel units and, by virtue of their low cost and simplicity, they are ideal for educational purposes as well as basic measurements. Most low-cost DSO provide all of the functionality associated with a conventional 'scope but with many additional software-driven features at a fraction of the price of a comparable conventional test instrument. That said, there are two important limitations of low-cost DSO:

1 Their bandwidth is generally limited so that they are only suitable for examining low-frequency and audio signals.
2 Their resolution is generally limited to 8 bits or 256 discrete steps of voltage.

A typical specification for a low-cost DSO is:

Sampling rate:	20 KB samples per second
Resolution:	8 bits
Number of channels:	1

Low-cost DSO frequently require no external power supply as they draw their power from the parallel port or USB port of a PC. This makes them ideal for use with laptop computers without the need for a mains supply.

High-speed DSO

High-speed DSO are usually dual-channel instruments that are designed to replace conventional general-purpose oscilloscopes (bit with the added advantage that captured data can be stored for subsequent processing and analysis). These instruments have all the features associated with a conventional 'scope including trigger selection, timebase and voltage ranges, and an ability to operate in X-Y mode.

Additional features available with a computer-based instrument that are not usually available from conventional benchtop 'scopes include the ability to capture transient signals (as with a storage 'scope) and save waveforms for future analysis. The ability to analyse a signal in terms of its frequency spectrum is yet another feature that is only possible with a DSO (more of this later).

A typical specification for a high-speed DSO is:

Sampling rate:	50 MB samples per second
Resolution:	8 bits
Number of channels:	2

Autoranging is another very useful feature that is often provided with a DSO. If you regularly use a conventional 'scope for a variety of measurements you will know only too well how many times you need to make adjustments to the vertical sensitivity of the instrument. To have this adjustment performed for you automatically is an absolute boon!

High-resolution DSO

High-resolution DSOs are used for precision applications where it is necessary to faithfully reproduce a waveform, and also to be able to perform an accurate analysis of noise floor and harmonic content. Typical applications include small signal work and high-quality audio.

Unlike the low cost which typically has 8-bit resolution and poor DC accuracy, these units are usually accurate to better than 1% and have either 12- or 16-bit resolution. This makes them ideal for audio, noise, and vibration measurements.

The increased resolution also allows the instrument to be used as a spectrum analyser with very wide dynamic range (up to 100 dB). This feature is ideal for performing noise and distortion measurements on low-level analogue circuits and high-fidelity equipment generally (such as CD and MP3 players).

A typical specification for a high-speed DSO is:

Sampling rate:	33 MB samples per second
Resolution:	12 bits
Number of channels:	2

Choosing a computer-based DSO

Unfortunately, for newcomers to computer-based instruments, choosing a DSO can be an somewhat daunting task. It is extreme to avoid making a costly mistake when choosing an instrument for the first time, and a thought and research at the outset can certainly help!

The first step (and most obvious) step is that of deciding what you want to use the instrument for. This will usually be fairly easy if you are intending to simply replace a conventional stand-alone instrument. It may not be quite so simple if you are starting out from scratch! In either case, it is worth asking the following questions:

- What measurements will you be making on a *regular* basis?
- What addition *measurements* or applications do you wish to perform?
- What signal amplitudes and frequency ranges are you working with?
- Do you need to measure pulse waveforms accurately or do you usually work with sinusoidal signals?
- Are your signals repetitive or are they one-off single-shot signals?
- Do you need to measure small time intervals and precise signal amplitudes?
- Do you need to carry out accurate measurements of noise and harmonic distortion, or are you just interested in displaying waveforms?
- Do you need to analyse signals in the frequency domain (i.e. spectrum analysis) as well as the time domain (conventional waveform display)?
- Will the DSO be used only on the bench or will it be used as a portable item of test equipment with a laptop computer?
- What budget is available for purchasing the instrument?

It is also worth asking yourself a few questions about the PC that you intend to use with the instrument:

- What ports are available?
- Is the parallel port dedicated to a printer or is it available for use by the DSO?
- If the parallel port is unavailable, does the PC have a spare USB port or will you need to purchase a hub? Or, can the PC be fitted with a second parallel port by means of a suitable PCI or EISA adapter card?

In many cases you may find that you are faced with a compromise between resolution and speed. However, a DSO with a 12-bit resolution and sampling rate of 5G samples per second will be more than adequate for most general-purpose applications!

Modern DSOs, with their PC connectivity, can also be fully integrated into automatic test equipment (ATE) systems. In addition, the DSO is often used as the front-end of a highly cost-effective data acquisition system.

Bandwidth alone is not enough to ensure that a DSO can accurately capture a high-frequency signal. The goal of manufacturers is to achieve a flat frequency response. This response is sometimes referred to as a Maximally Flat Envelope Delay (MFED). A frequency response of this type delivers excellent pulse fidelity with minimum overshoot, undershoot, and ringing.

It is important to remember that, if the input signal is not a pure sine wave, it will contain a number of higher-frequency harmonics. For example, a square wave will contain odd harmonics that have levels that become progressively reduced as their frequency increases. Thus, to display a 1 MHz square wave accurately you need to take into account the fact that there will be signal components present at 3, 5, 7, 9, 11 MHz, and so on.

As mentioned earlier, it is wise to purchase a DSO with a bandwidth that is five times higher than the maximum-frequency signal you wish to measure. Note, however, that with some instruments the specified bandwidth is not available on all voltage ranges, so it is worth checking the manufacturer's specification carefully.

Most DSO have two different sampling rates (modes) depending on the signal being measured: real time and equivalent time sampling (ETS) – often called repetitive sampling. However, since this mode works by building up the waveform from successive acquisitions, ETS only works if the signal you are measuring is stable and repetitive.

Basic operation of a DSO

The basic operation of a DSO is extremely straightforward and all of the instrument's controls are accessed through software using a standard Windows interface (see Figure 11.2). It is worth comparing this interface with the controls and adjustments provided on a conventional oscilloscope (see Photo 11.2).

Unlike a conventional oscilloscope, most DSOs will provide you with more than one type of view of the data that you are collecting. Typically these will include:

- an oscilloscope display, with all of the features of a modern storage oscilloscope;

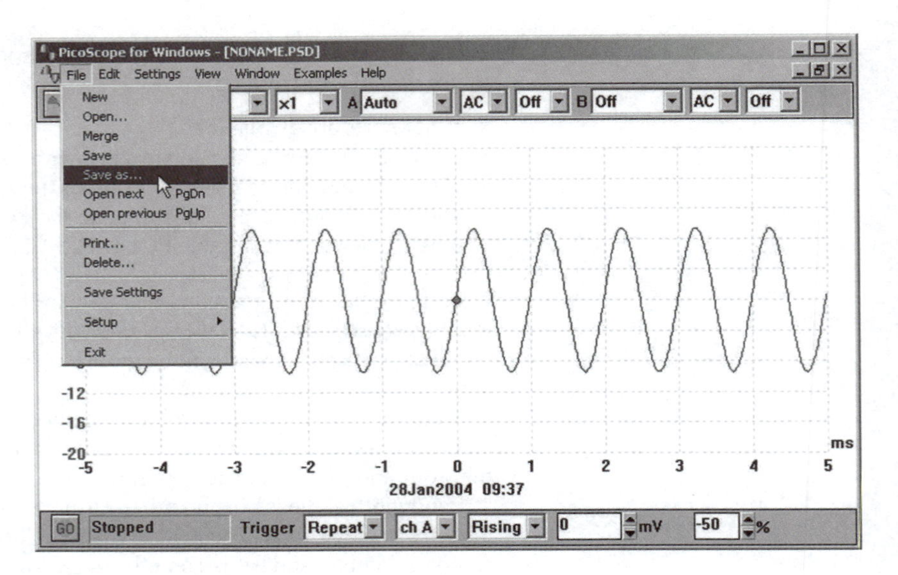

Figure 11.2 *The PicoScope software uses a conventional drop-down menu system. Here the 'Save as' option has been selected from the 'File' menu in order to preserve the waveform data for later viewing or analysis using external software*

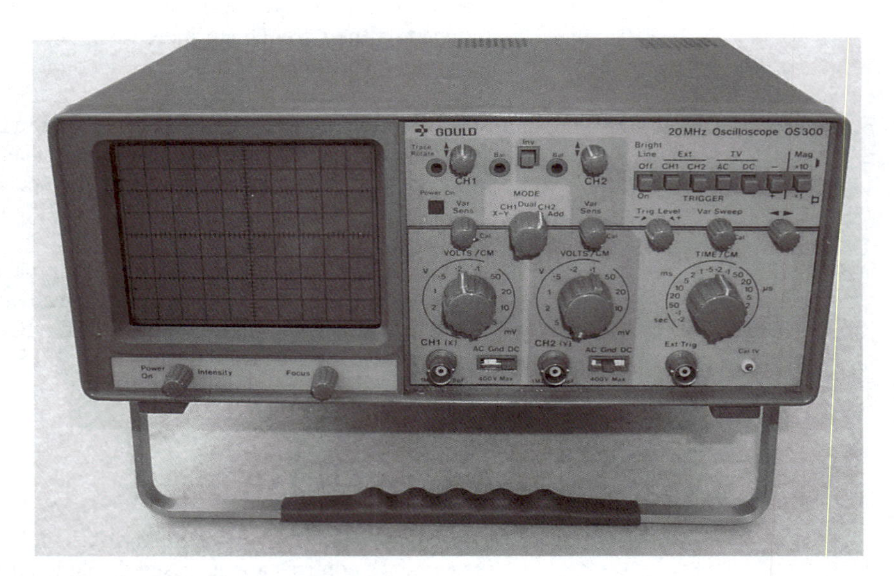

Photo 11.2 *Controls fitted to a conventional oscilloscope (compare this with the virtual interface shown in Figure 11.1)*

- a spectrum analyser, showing the power at each of a range of frequencies;
- a meter, which can show the DC voltage, AC voltage, frequency, or dB;
- an XY oscilloscope display, which shows one channel against another (for Lissajous figures, phase analysis, etc.).

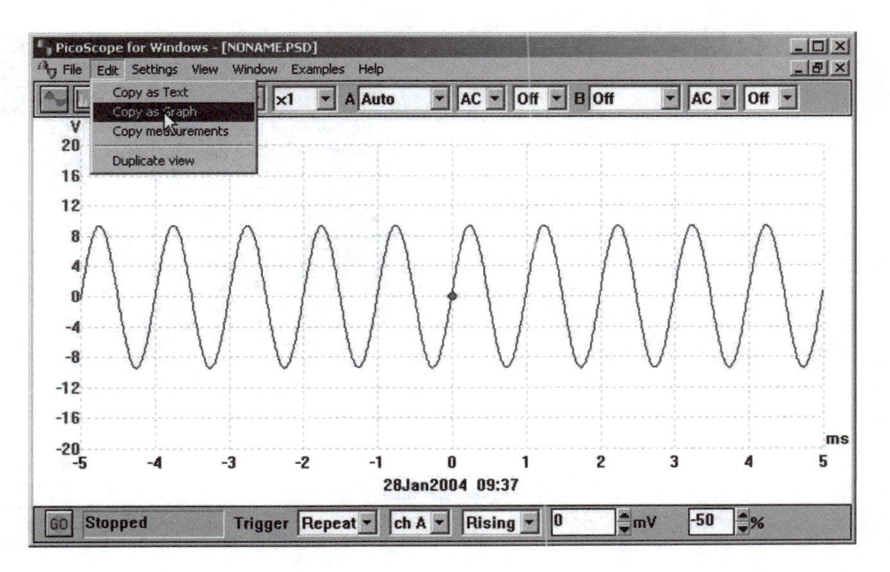

Figure 11.3 *The Windows clipboard can be used to copy waveform and other data from PicoScope to other applications*

In addition to the basic views, it is possible to display a composite view made up of several windows each with different views (see Figure 11.1). Only one of the views on the desktop will actually be 'active' at any time but you can switch between views using all of the usual Windows methods. For example, you can activate a view by clicking the left mouse button over the view.

The settings for data collection or the parameters used for the display of any view can be easily changed, but first the respective window must be activated. To review the settings for any particular view first it needs to be made active and then the Settings Menu can be used to make the required changes. For some views, you may also be able to 'zoom in' on a small area of the display by setting the multiplier for the X- or Y-axis (or both) to a value other than one.

The timebase settings control the time interval as with a conventional oscilloscope display. As with a conventional instrument the control is marked in terms of 'time per division', however, it may also be possible to configure a DSO in terms of 'time per scan' which may make more sense in the case of certain types of measurement.

For dual-channel DSOs you can select which of the two channels (or both) to display on the screen. With two channels, each channel has a separate axis and each trace and its axis can usually displayed in a different colour (unlike a conventional oscilloscope).

Voltage ranges are selected in much the same way as for a conventional oscilloscope (i.e. in terms of 'volts per division') but an autoranging facility may also be included. The autoranging option can be particularly useful if you are switching between different, but consistent, signals. You may also be able to add custom ranges so that the values are displayed in some other units, for example, pressure or acceleration (Figures 11.3 and 11.4).

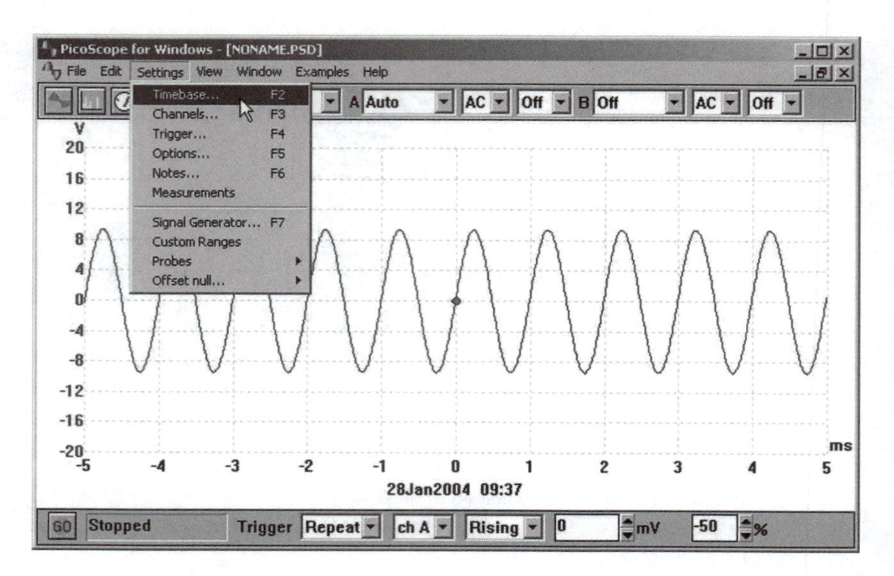

Figure 11.4 *The DSO control settings can be changed by using the drop-down menus as well as the tool buttons and controls provided in the display window*

Waveform display

The most common use of an oscilloscope is that of displaying waveforms. In this respect the operation of a DSO is very similar to that of a conventional oscilloscope. You simply need to select the required voltage scale and timebase settings, and this can be done by pointing and clicking on the control buttons and drop-down selection boxes. Thus, for basic waveform measurements there is usually no need to use the menu system.

Figure 11.5 shows a 10 V peak–peak sine wave at 1 kHz displayed with the voltage scale on its 'Auto' (i.e. autoranging) setting and the timebase set to 1 ms per division. Note that this results in a full-screen timebase scan of 10 ms. In other words, we are looking at a 10 ms sample of the sine wave.

The Trigger control has also been set to 'Auto' and the trigger point set to 'Channel A', 'Rising' (i.e. positive going trigger), and 0 mV (i.e. the zero voltage axis crossing point). Note that the trigger point is shown on the display as a small grey circle at the origin of the axis (i.e. at $t = 0$ ms and $V = 0$ V).

If only a single cycle of the sine wave is to be displayed the timebase needs to be set to 100 μs per division (as shown in Figure 11.6). Once again, the trigger point appears at $t = 0$ ms and $V = 0$ V. Note that it is very important to be able to control the trigger point, particularly where a waveform is non-repetitive in nature.

Parameter measurement

For accurate parameter measurement cursors may be added to the display in the form of horizontal or vertical rulers that can be moved using the mouse or cursor keys. Figure 11.7 shows how a cursor can be added in order to determine

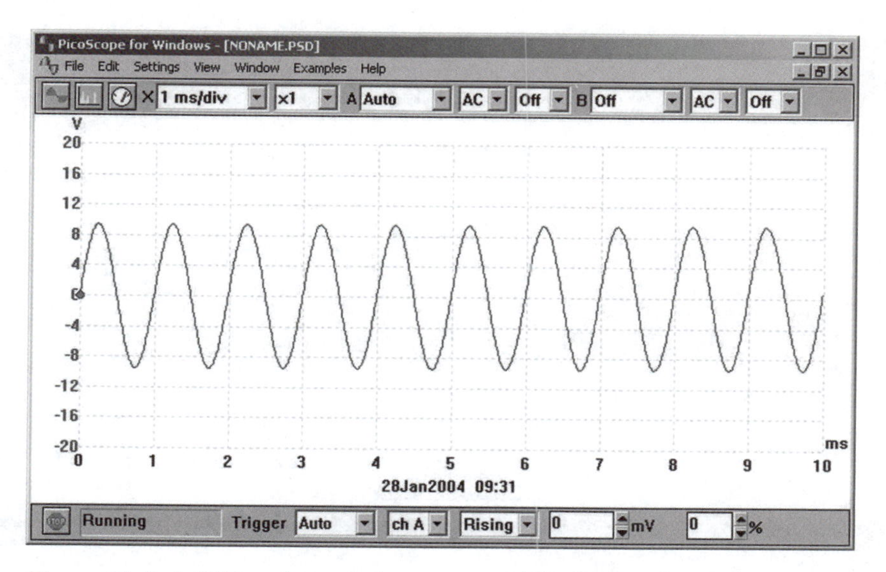

Figure 11.5 *A 10 V peak–peak sine wave at 1 kHz displayed with the timebase set to 1 mV/div. and the vertical scale set to autoranging*

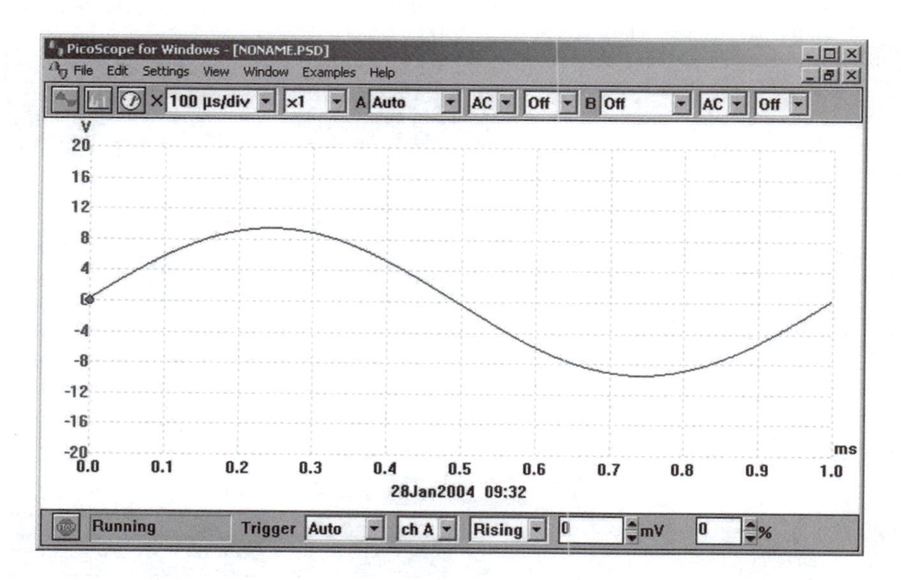

Figure 11.6 *The DSO adjusted as for Figure 11.6 but with the timebase set to 100 μs/div.*

the exact peak value of a nominal 10 V peak sine wave. The peak value has here been measured at 9.625 V. A further cursor has been added in order to determine the time that the waveform takes to reach the first peak. This is found to be 0.2467 ms. It should be clear from this example just how useful the cursors can be if you need to make precise measurements of voltage and

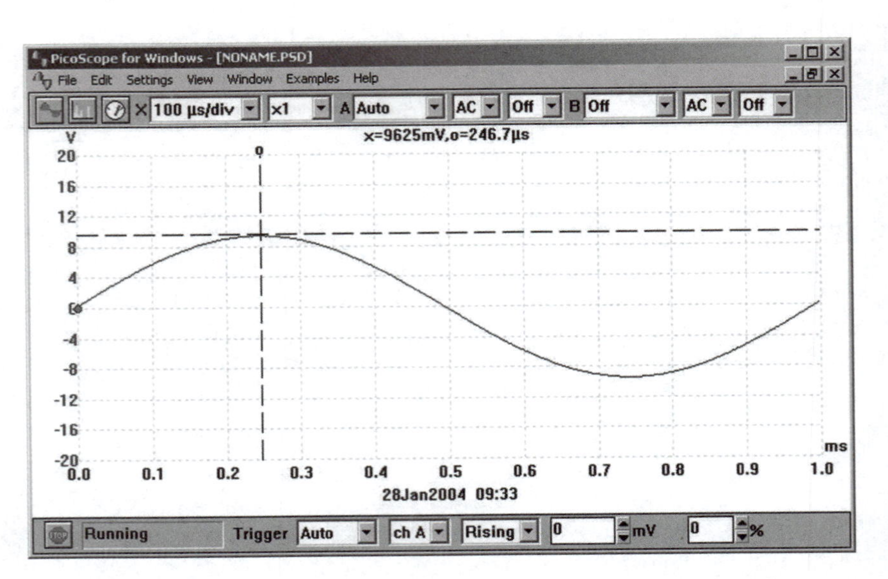

Figure 11.7 *Adjustable cursors make it possible to carry out extremely accurate measurements. Here the peak value of the (nominal 10 V peak) waveform is measured at precisely 9625 mV (9.625 V). The time to reach the peak value (from 0 V) is measured as 246.7 µs (0.2467 ms)*

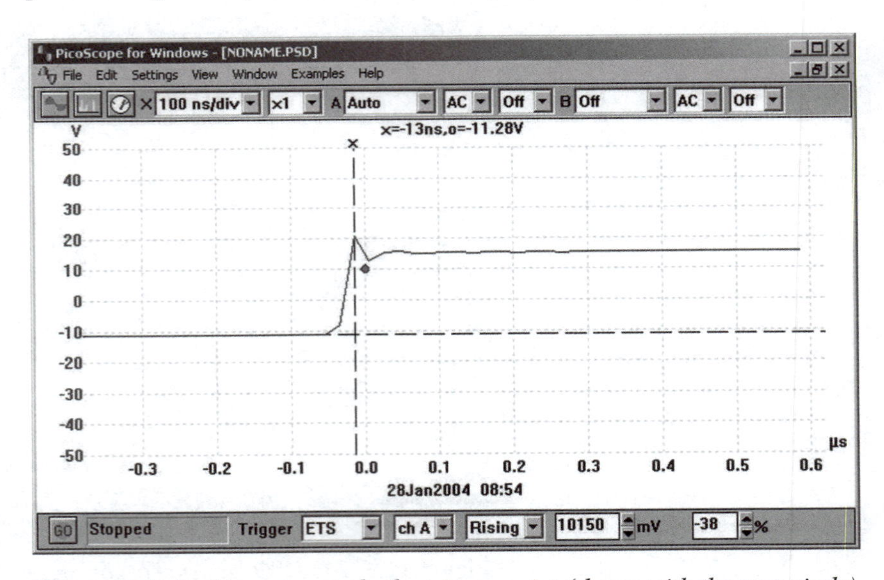

Figure 11.8 *In this screen grab, the trigger point (shown with the grey circle) has been adjusted to just over 10 V (10 150 mV) and the time scale shifted in order to display the leading edge of a fast pulse*

time. Gone are the times when you had to estimate the position of a point on a waveform using a dimly lit graticule!

A further example is shown in Figure 11.8. In this example the trigger point (the grey circle) has been adjusted to just over 10.15 V and the time scale shifted

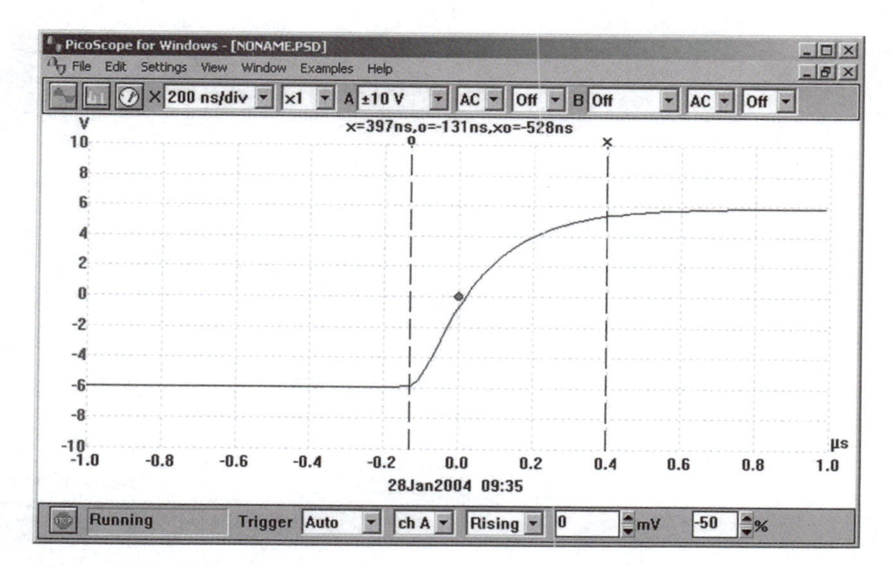

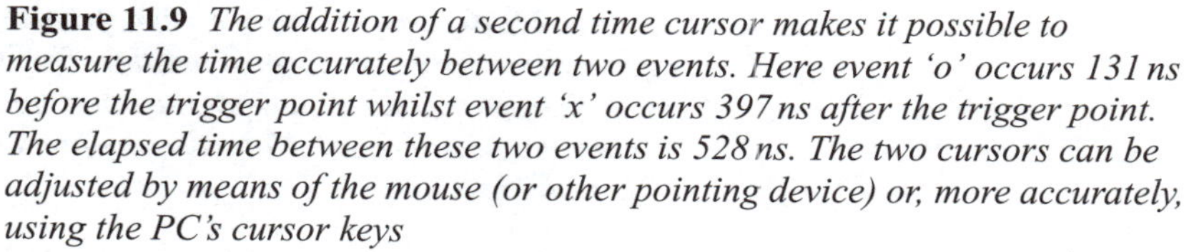

Figure 11.9 *The addition of a second time cursor makes it possible to measure the time accurately between two events. Here event 'o' occurs 131 ns before the trigger point whilst event 'x' occurs 397 ns after the trigger point. The elapsed time between these two events is 528 ns. The two cursors can be adjusted by means of the mouse (or other pointing device) or, more accurately, using the PC's cursor keys*

in order to display the leading edge of a fast pulse. By manipulating the cursors it is possible to accurately determine the rise time of the pulse as well as the overshoot, undershoot, and damping factor.

Further cursors can be added in order to determine the difference between two discrete events or points on a continuous waveform. In the example shown in Figure 11.9, the addition of a second time cursor has allowed us to make an accurate measurement of the elapsed time between the start of a pulse and the 95% point. The start of the pulse has occurred 131 ns before the trigger point whilst the 95% point has been reached 397 ns after the trigger point. The time difference between these two events has been calculated and displayed as 520 ns.

Spectrum analysis

The technique of Fast Fourier Transformation (FFT) calculated using software algorithms using data captured by a DSO has made it possible to produce frequency spectrum displays. Such displays can be to investigate the harmonic content of waveforms as well as the relationship between several signals within a composite waveform.

Figure 11.10 shows the frequency spectrum of a 1 kHz square wave derived from a low-cost waveform generator. The DSO has been set to capture data at a rate of 256 samples per second over the frequency range DC to 12.2 kHz. As expected, the odd harmonics (3, 5, 7 kHz, and so on) are present at amplitudes that decay progressively with harmonic order. The display indicates that the

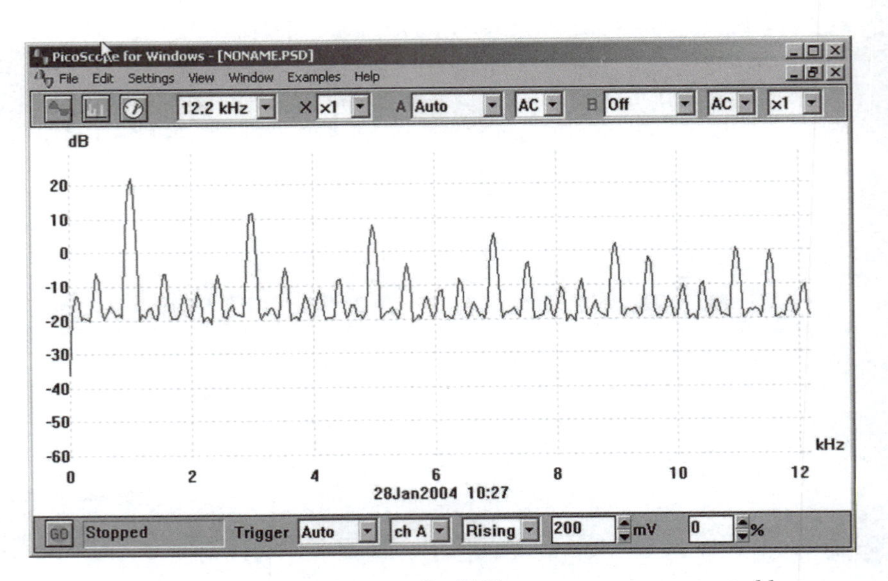

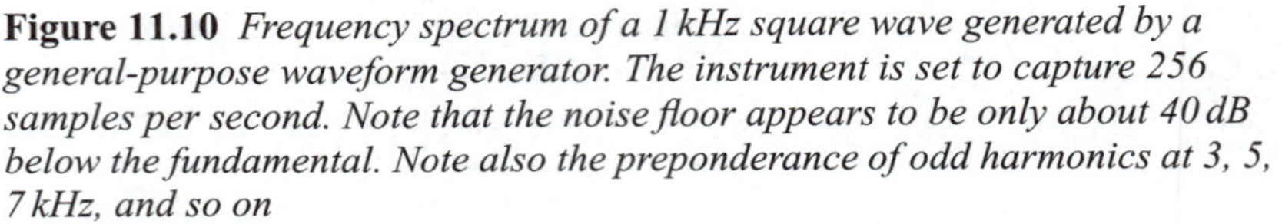

Figure 11.10 *Frequency spectrum of a 1 kHz square wave generated by a general-purpose waveform generator. The instrument is set to capture 256 samples per second. Note that the noise floor appears to be only about 40 dB below the fundamental. Note also the preponderance of odd harmonics at 3, 5, 7 kHz, and so on*

noise floor is at a level of −20 dB (or 40 dB down on the level of the fundamental which is +20 dB).

Figure 11.11 shows the effect of increasing the sampling rate to 512 samples per second. The increased sampling rate makes it possible to make a more accurate assessment of the levels of the individual harmonic components but clearly there is more present in this spectrum than just the expected harmonics!

Figure 11.12 shows the effect of a further increase in sampling rate. At 1024 samples per second, it is possible to see the noise floor a little more accurately and also to detect the presence of multiple signal components between each of the harmonics of the 1 kHz fundamental.

At 2048 and 4096 samples per second we obtain a rather different view of what's going on (see Figures 11.13 and 11.14). In fact, the noise floor is nearer −60 dB and what we are looking at is a series of harmonics each with a set of side-frequency components that result from modulation by unwanted signal components with a spacing of about 250 Hz. This display very effectively shows the advantages of using a high sampling rate and having sufficient buffer memory available to actually store the captured data!

Finally, and in contrast with the previous examples, Figure 11.15 shows a much purer signal. This signal is a 1 kHz sine wave derived from a low-distortion AF signal generator. Here the DSO has been set to capture samples at a rate of 4096 per second within a frequency range of DC to 12.2 kHz. The display clearly shows the second harmonic (at a level of −50 or 70 dB relative to the fundamental), plus further harmonics at 3, 5, and 7 kHz (all of which are greater than 75 dB down on the fundamental).

When cursors are added to a frequency spectrum display, it is possible to make extremely accurate measurements. Figure 11.16 shows the frequency

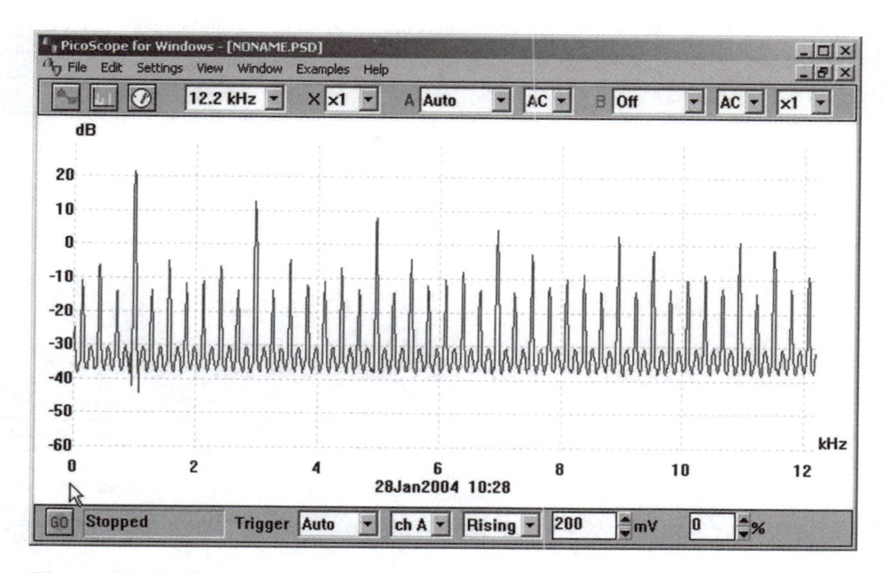

Figure 11.11 *Frequency spectrum of a 1 kHz square wave generated by a waveform generator. The instrument is set to capture 512 samples per second. Note that it is now possible to make a more accurate assessment of the various frequency components present. Note also that the noise floor can be more accurately measured*

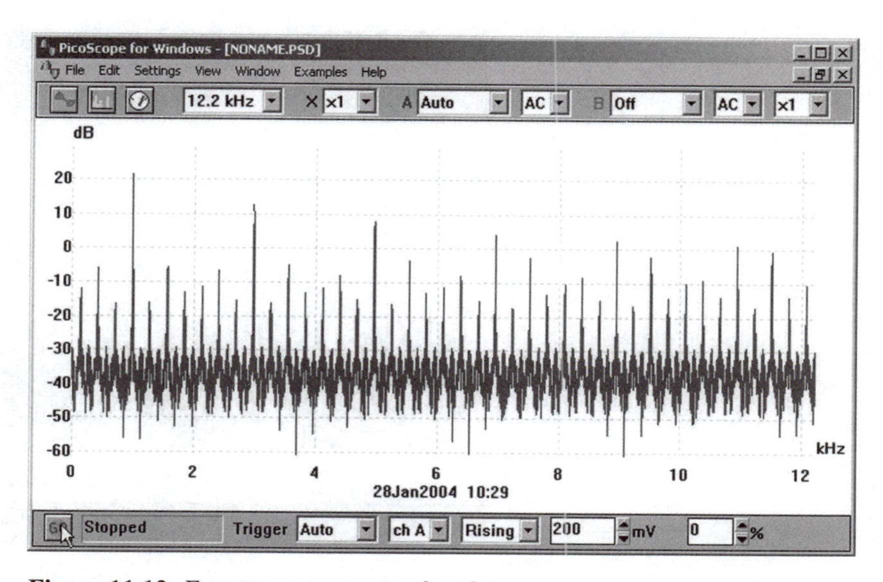

Figure 11.12 *Frequency spectrum of a 1 kHz square wave generated by a waveform generator. The instrument is set to capture 1024 samples per second. Note how the frequency resolution of the instrument has increased considerably compared with the previous frequency spectra*

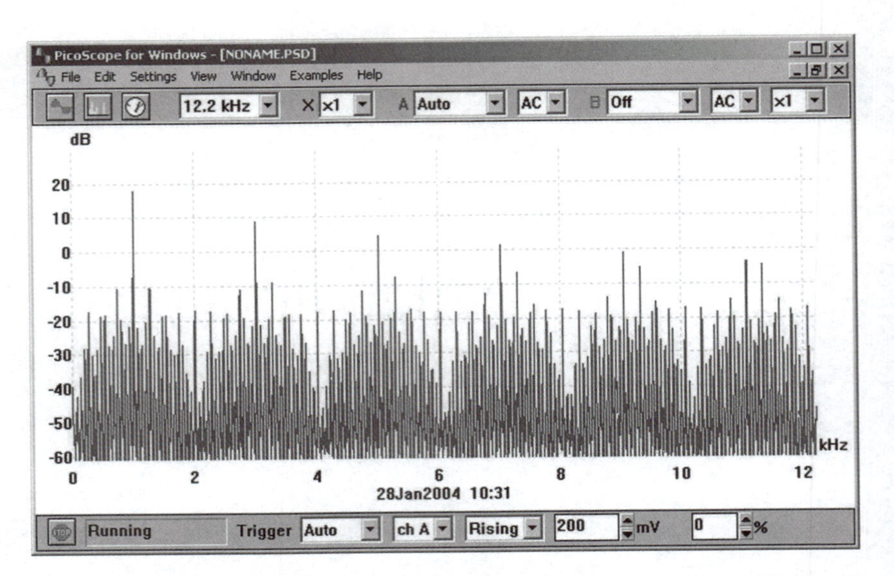

Figure 11.13 *Frequency spectrum of a 1 kHz square wave generated by a waveform generator. The instrument is set to capture 2048 samples per second. The increased frequency resolution of the instrument is now showing us how the noise spectrum is actually modulated onto the fundamental and harmonic components. This was simply not apparent in the spectra shown in Figures 11.11 and 11.12*

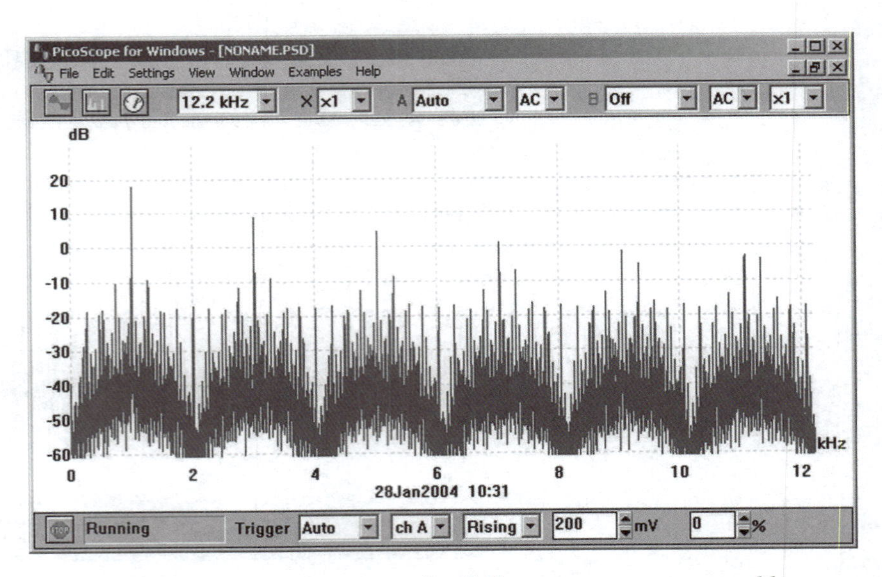

Figure 11.14 *Frequency spectrum of a 1 kHz square wave generated by a waveform generator. The instrument is set to capture 4096 samples per second. It is now easy to see how the modulated spectrum varies*

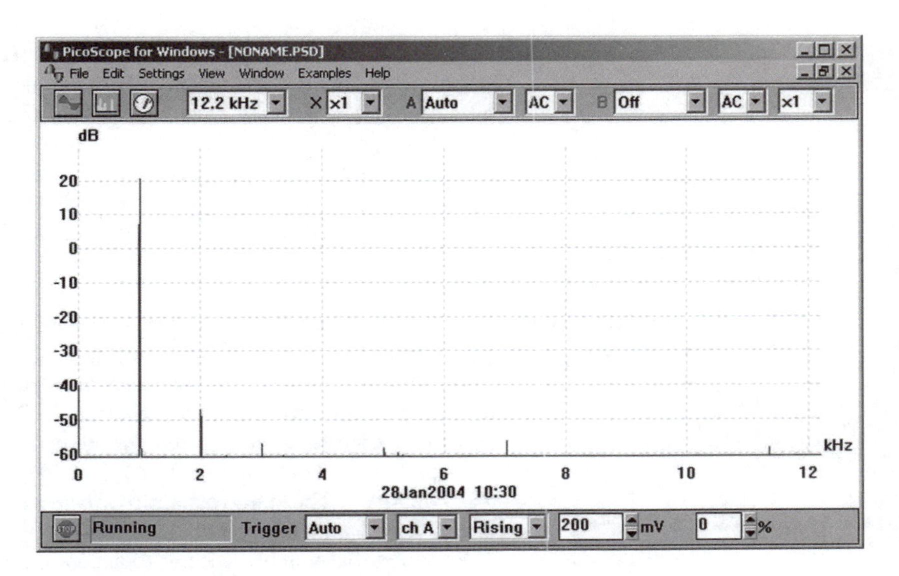

Figure 11.15 *Frequency spectrum of a 1 kHz sine wave generated by a low-distortion signal generator. The instrument is set to capture 4096 samples per second. Notice how the second harmonic is approximately 70 dB down on the fundamental. It is also possible to see the third, fifth, and the seventh harmonic*

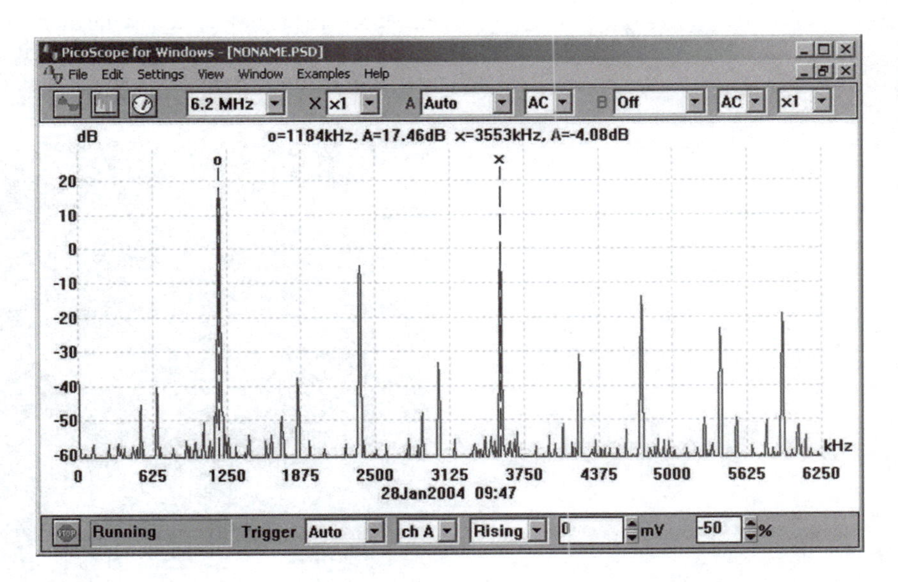

Figure 11.16 *Spectrum analysis of a radio signal at 1184 kHz (1.184 MHz). The cursors have been adjusted in order to display the amplitude and frequency of the fundamental and its third harmonic. The third harmonic at 3553 kHz (3.553 MHz) is approximately 21 dB down on the fundamental*

spectrum of a radio-frequency signal at 1184 kHz (1.184 MHz). Numerous other signal components are present including the second, third, fourth, and fifth harmonics. The cursors have been set to the fundamental (1.184 MHz) and third harmonics (3.553 MHz). The level and frequency of each harmonic has been displayed. Note that the fundamental has an amplitude of 17.46 dB whilst the third harmonic is at −4.08 dB. Needless to say, this type of measurement would be impossible with a conventional oscilloscope!

Sound card oscilloscopes

Any PC with a standard sound card (or equivalent interface integrated with the motherboard) can be used to record and playback analogue signals. In order to do this, the sound card interface incorporates 16- or 32-bit analogue-to-digital and digital-to-analogue converters (ADC and DAC, respectively). Sound cards can usually be configured for stereo or mono operation with sampling rates of 11 025 Hz ('voice quality'), 22 050 Hz ('tape quality'), and 44 100 Hz ('CD quality'). The stereo capability makes it possible to have two independent channels (Y1 and Y2) as with a conventional oscilloscope.

Provided that you are willing to accept the bandwidth limitation (i.e. 20 Hz to 20 kHz) typical of audio signals, the ability to convert an analogue signal into digital stored data makes, it possible to use a sound card as the hardware component of a digital storage oscilloscope (DSO).

With the aid of appropriate software (available at minimal cost) the stored data produced by a sound card can be made to display on a screen in much the same way as it appears on the screen of a conventional oscilloscope. Furthermore, the use of Fast Fourier Transformation (FFT) makes it possible to produce a frequency domain display as well as the more conventional time domain display.

Whilst it is possible to use a sound card without any external interface (most sound cards have ample sensitivity so additional gain from a pre-amplifier stage

Photo 11.3 *A typical low-cost sound card supplied in PCI format*

is unlikely to be required) there are several good reasons for using an input interface:

- The 50 kΩ input impedance of a sound card (which is appropriate for use with audio equipment) is too low for general-measurement applications.
- The 50 kΩ input impedance is inappropriate for use with standard 'scope probes which are designed to work with standard 1 MΩ oscilloscope inputs.
- The sensitivity of most sound cards varies according to software gain settings and the size/resolution of PC displays also tends to vary widely hence some method of calibration is essential if accurate measurements are to be made.

These limitations can be overcome by means of a simple interface of the type shown in Figure 11.17. This circuit incorporates two identical channels for the Y1 and Y2 inputs. The input signal is fed to a switched potential divider attenuator (R10 to R15 and R20 to R265) which has a constant input impedance of 1 MΩ. Junction gate field-effect transistors (TR10 and TR20) are connected as unity gain source followers with the output voltage developed across the two variable gain controls (VR10 and VR20). Capacitors, C10 and C20, are used to provide DC isolation at the input whilst C11 and C21 provide DC isolation at the output.

The sound card input interface incorporates its own AC mains supply which also provides a 1 V peak–peak calibration signal at 50 Hz. Diodes, D1 and D2, provide full-wave bi-phase rectification with an output of approximately 8.5 V appearing across the reservoir capacitor, C1. Secondary current from T1 is also fed to the anti-parallel diode clamp, D3 and D4. A potential divider, R2 and R3, provides the 1 V peak–peak calibration signal which is applied to the input attenuator when switch, S2, is set to the 'calibrate' position. Light emitting diodes, D5 and D6, indicate whether S2 the instrument is set to the 'calibrate' or the 'operate' position.

As mentioned earlier, the upper signal frequency limit of a DSO is determined primarily by the rate at which it can sample an incoming signal. Using a maximum sound card sampling rate of 44.1 kHz a sinusoidal signal at 20 kHz can be displayed with acceptable accuracy (the Nyquist criterion).

However, in order to display non-sinusoidal signals faithfully, we need to sample at an even faster rate if we are to accurately display the signal. In practice we would need a minimum of about five points within a single cycle of a sampled waveform in order to reproduce it with approximate fidelity. Hence the sampling rate should be at least five times that of highest signal frequency for a DSO to be able to display a waveform with reasonable accuracy.

With a fixed 44.1 kHz sampling rate this would suggest that audio-frequency signals of up to 4 kHz will be faithfully displayed using a sound card; however, it is also necessary to take into account the fact that a sound card interface is AC coupled and therefore is unable to respond to DC levels!

Windows Oscilloscope 2.51

Windows Oscilloscope 2.51 was written by Konstantin Zeldovich who is a research associate in the Physics Department at Moscow State University.

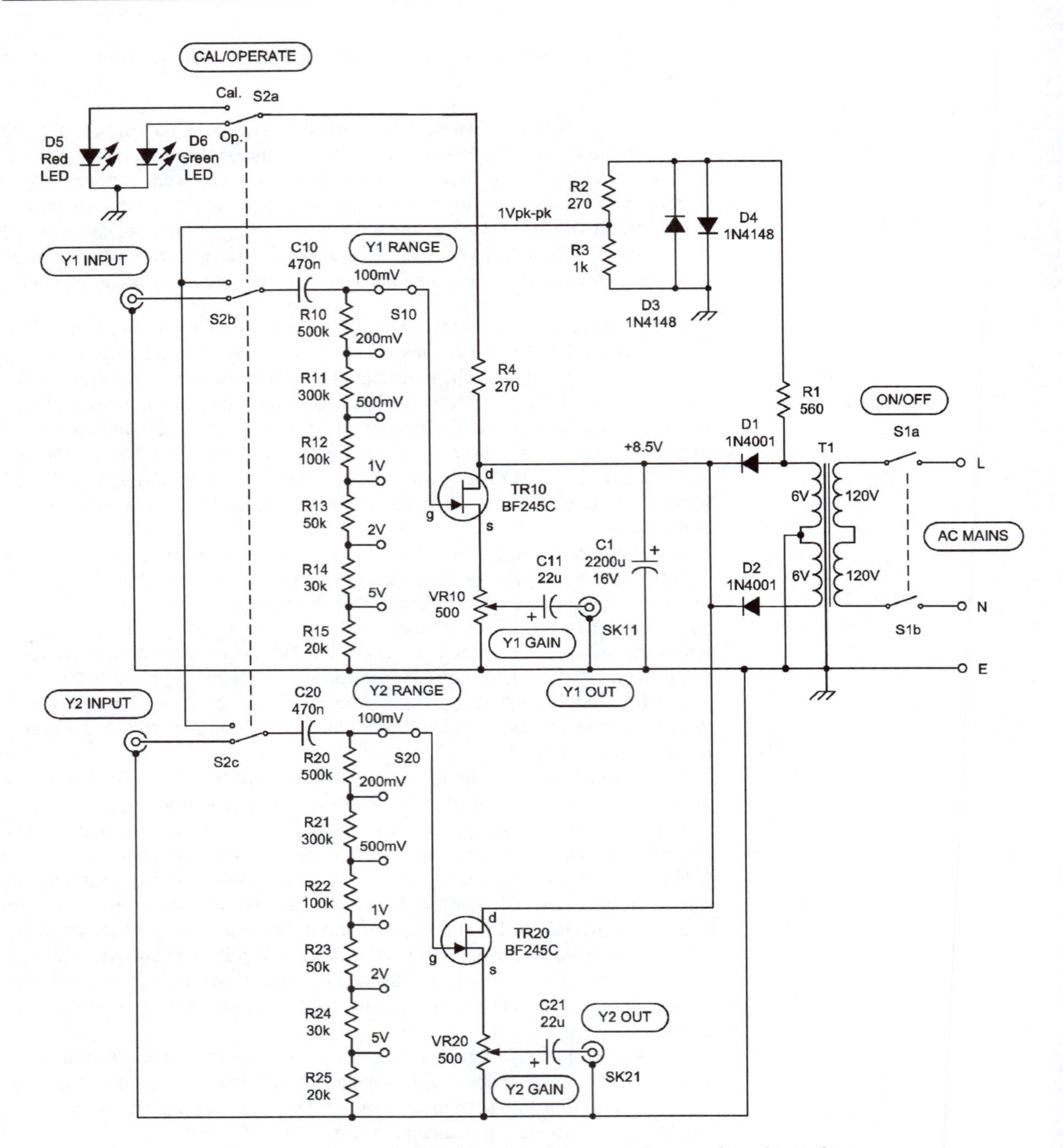

Figure 11.17 *Complete circuit of the sound card interface*

This competent program works extremely well and is provided as freeware. The main features of Windows Oscilloscope 2.51 are as follows:

- single trace, dual trace, and XY modes;
- spectrum analyser (real-time, built-in);

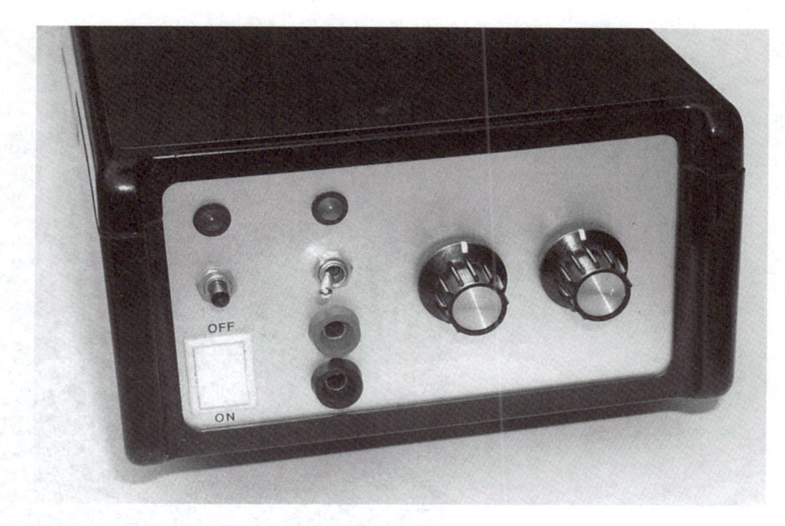

Photo 11.4 *The prototype sound card oscilloscope interface*

- 20 kHz bandwidth;
- trigger adjustment;
- point-and-click measurement function;
- storage mode;
- data export to the Windows clipboard or to a disk file.

A typical Windows Oscilloscope 2.51 display is shown in Figure 11.18. The software operates in dual-channel mode with different colour traces used to differentiate the two channels. Separate sliding position and gain controls are provided. Both the trigger level and the trigger delay are adjustable using slider control. The timebase is also set by means of a slider control.

Clicking the meter button in the toolbar of Windows Oscilloscope 2.51 toggles Meter Mode on and off. In Meter Mode, you can measure time and level of a waveform simply by clicking left and right mouse buttons on the Oscilloscope display. A left button click sets cursor 1 and right click sets cursor 2. When both cursors are set, the difference between the cursors' positions is shown in the right three parts of the status bar.

For convenience, the reciprocal value of time/frequency difference is shown in the 1/dt (1/dF) status window. This allows you to quickly measure the frequency of a waveform by clicking left and right buttons on the successive maximums and looking at the 1/dt value displayed.

Software Oscilloscope

The second of our duo of sound card oscilloscopes originates in Japan. Unlike the previous two packages, Software Oscilloscope operates with a single channel, and it provides *simultaneous* time domain and frequency domain displays (see Figure 11.19).

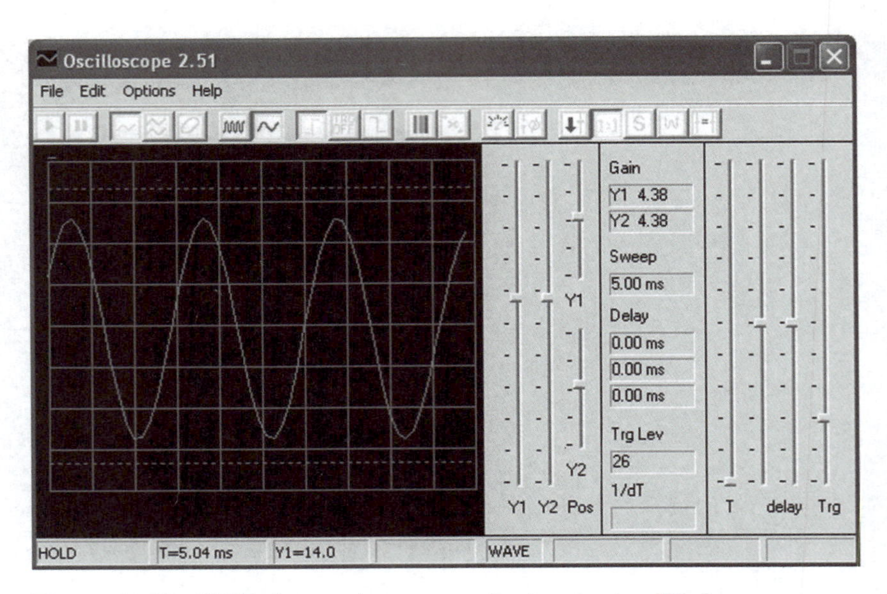

Figure 11.18 *600 Hz (approx.) sine wave displayed using Windows Oscilloscope 2.51 and the sound card interface*

Software Oscilloscope uses selectable sampling rates of 11, 22 and 44.1 kHz (note that the highest rate is only available to registered users). The sound card needs to operate in full-duplex mode, allowing concurrent recording and playback. Software Oscilloscope runs with most versions of Windows (but not Windows NT or Windows 3.x) but it does require DirectX support (see below).

Software Oscilloscope requires the DirectX Runtime module. Various versions of DirectX are available including:

Windows version	DirectX compatibility
Windows 98	DirectX 9.0
Windows Me	DirectX 9.0
Windows 2000	DirectX 9.0
Windows 95	DirectX 8.0
Windows XP	Not required
Windows NT	Not supported

Waveform display

Waveform display with a sound card oscilloscope is extremely straightforward. Sinusoidal signals in the frequency range 20 Hz to 20 kHz can be displayed with a high degree of accuracy. Figure 11.18 shows a typical sine wave displayed using Windows Oscilloscope 2.51. Note that the Y1 and Y2 gains are set to 4.38 and that the time for a complete sweep is set to 5 ms (the signal frequency is

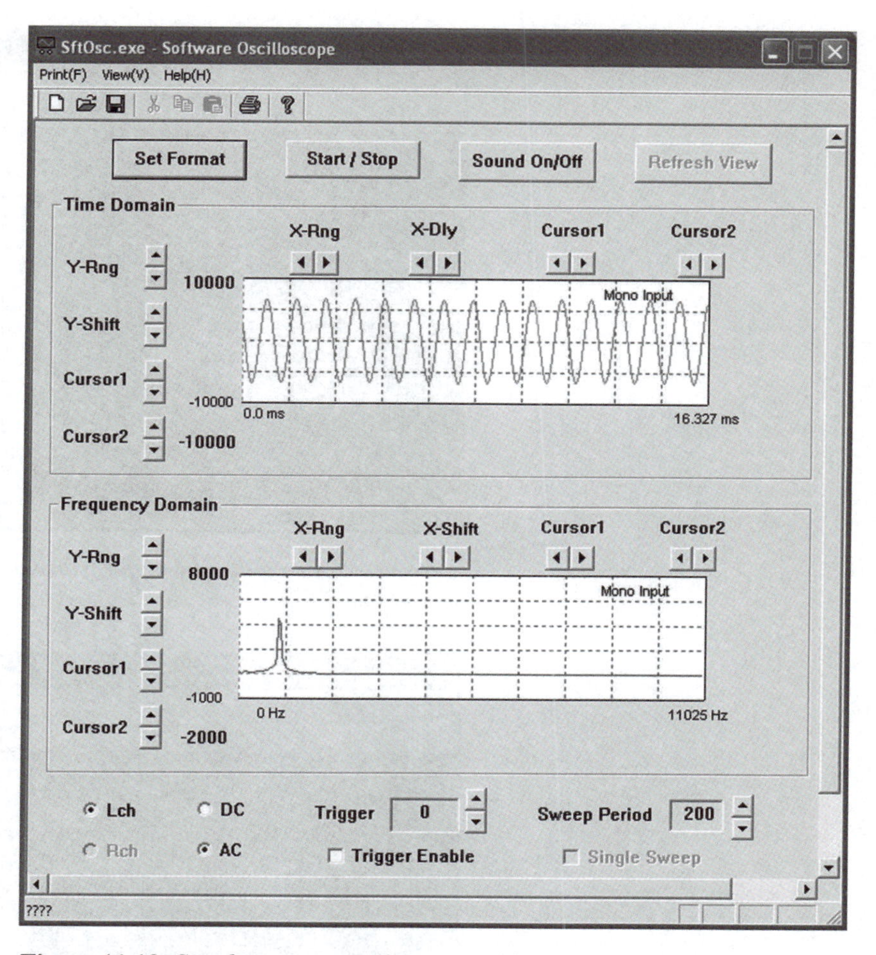

Figure 11.19 *Simultaneous time domain and frequency domain displays provided by Software Oscilloscope*

approximately 600 Hz). Figure 11.20 shows a modulated waveform displayed using Windows Oscilloscope 2.51. The waveform consists of a sine wave at 440 Hz modulated at 100 Hz to a depth of about 25%.

A video cross-hatch waveform (vertical scan) displayed using Windows Oscilloscope 2.51 is shown in Figure 11.21. Note that the sweep time has been set to 36 ms and, as a consequence, 1.8 (i.e. 36/20) cycles of the 50 Hz vertical scan waveform are displayed. The signal amplitude is approximately 1.3 V peak–peak. A vertical scan video colour bar waveform is shown in Figure 11.22. Note that the absolute DC level cannot be correctly shown due to the AC coupling used in the sound card input interface.

Figure 11.23 shows an expanded view of the vertical sync pulse shown in Figure 11.22. To obtain this waveform the timebase has been set to produce a horizontal sweep time of 5.4 ms. Notice that it is just possible to distinguish individual line scans in Figure 11.23.

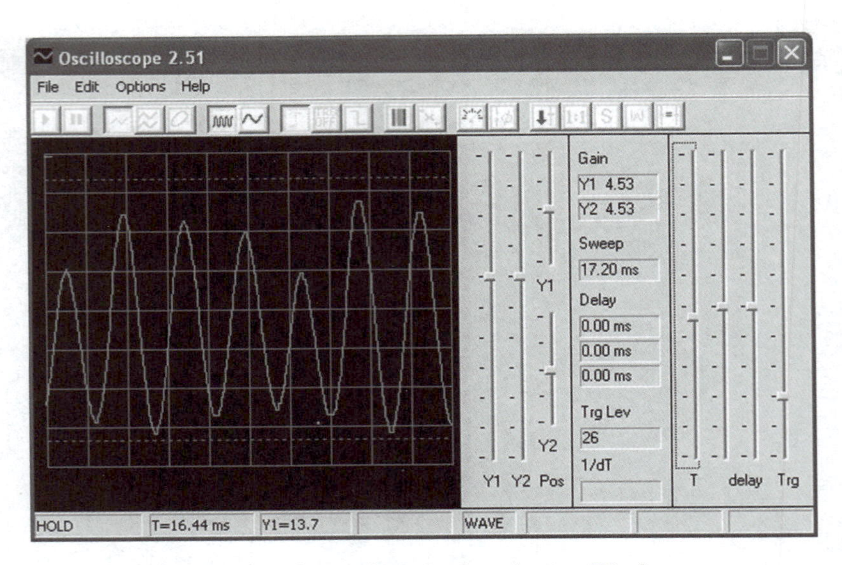

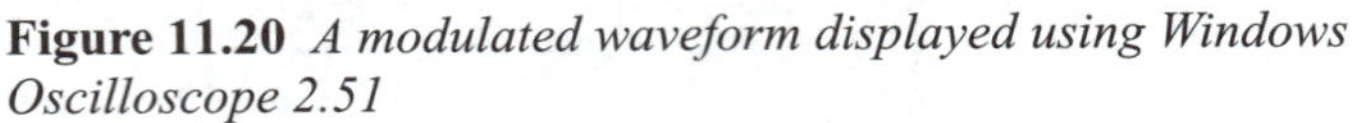

Figure 11.20 *A modulated waveform displayed using Windows Oscilloscope 2.51*

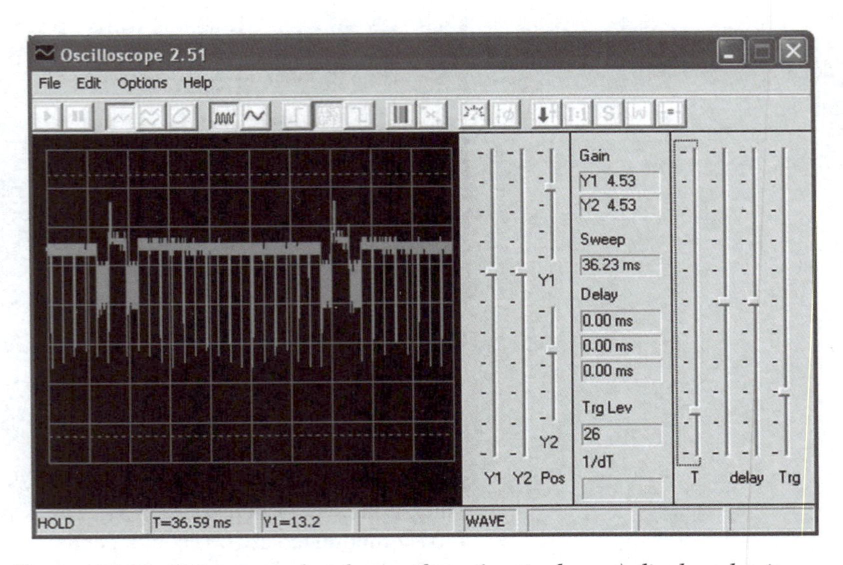

Figure 11.21 *Video cross-hatch waveform (vertical scan) displayed using Windows Oscilloscope 2.51*

Parameter measurement

With sound card oscilloscopes, accurate parameter measurement can be made possible in various ways, for example:

- Continuously adjustable cursors (Software Oscilloscope).
- Dedicated measurement modes with cross-hairs that can be placed using a mouse (Windows Oscilloscope 2.51).

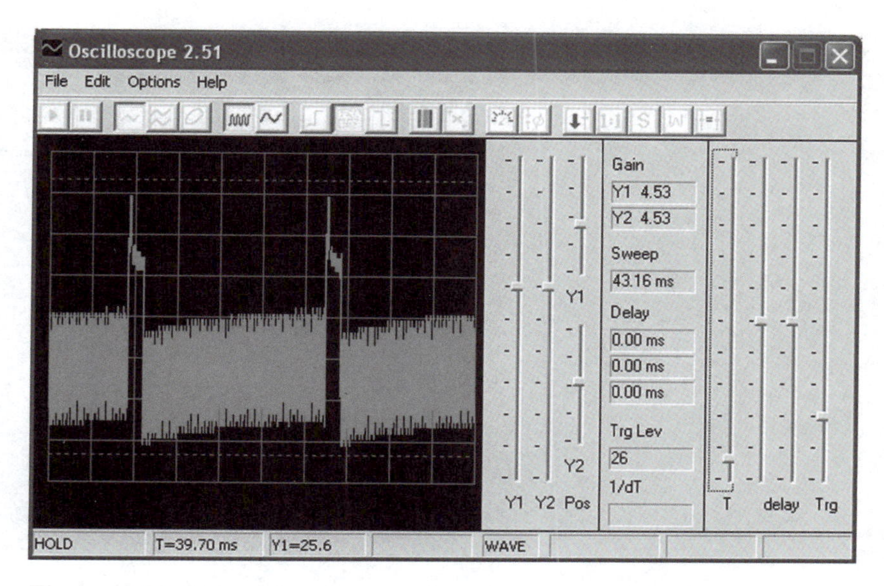

Figure 11.22 *Video colour bar waveform (vertical scan) displayed using Windows Oscilloscope 2.51*

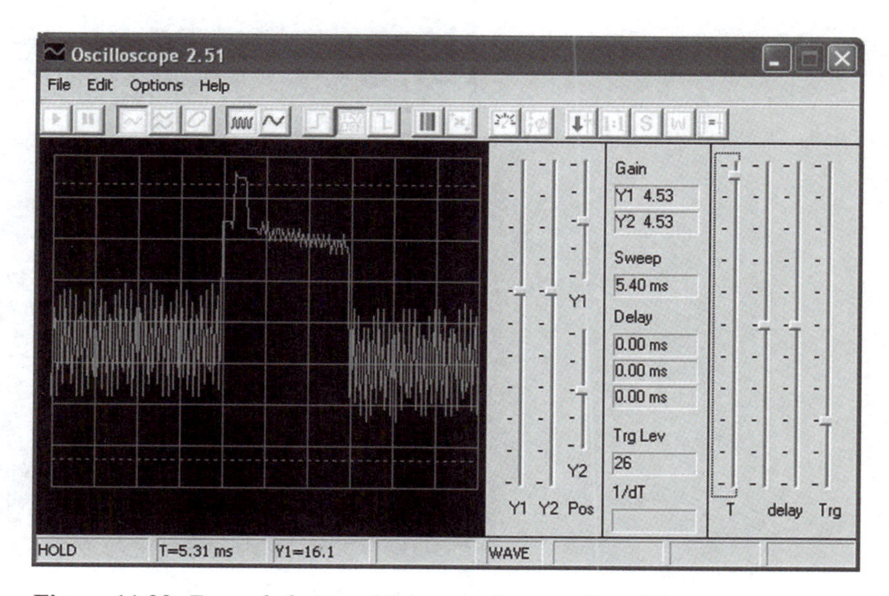

Figure 11.23 *Expanded view of the vertical sync pulse of the waveform shown in Figure 11.22*

Figure 11.24 shows the parameters of a square wave being measured using Software Oscilloscope. The horizontal (X) cursors are placed at equivalent points on successive cycles and the time difference between then is calculated and displayed (approximately 20 ms). A similar process using the vertical (Y) cursors can be used to determine the peak and peak–peak voltage. Cursors can

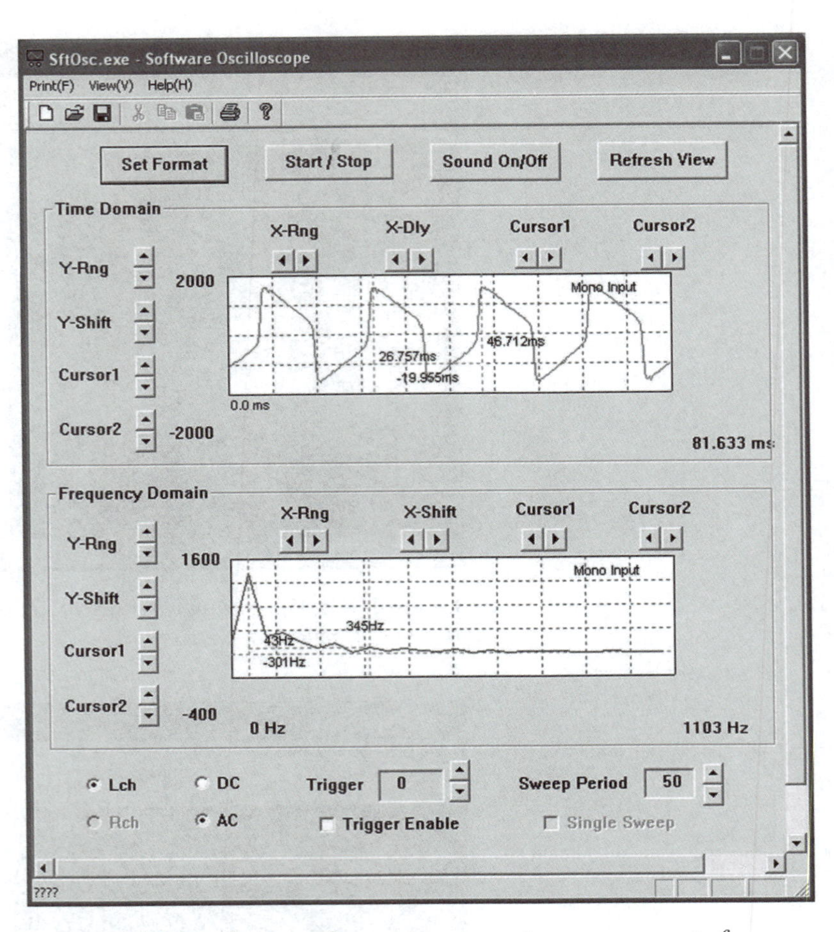

Figure 11.24 *Use of cursors for precise parameter measurement of a repetitive waveform using Software Oscilloscope*

also be used in conjunction with Software Oscilloscope's frequency domain display (as shown in Figure 11.24).

A further example is shown in Figure 11.25. Here the parameters of a pulse are being accurately measured. The pulse width is found to be 1.596 ms (i.e. 2.249 − 0.653 ms) and its peak–peak value is 9400 mV (i.e. 7000–(−2400 mV)).

Spectrum analysis

The technique of Fast Fourier Transformation (FFT) makes it possible to produce frequency spectrum displays. Such displays can be to investigate the harmonic content of waveforms as well as the relationship between several signals within a composite waveform.

Figure 11.26 shows the frequency spectrum for the sine wave shown previously in Figure 11.18. It should be noted that the sine wave is extremely pure with only one single-frequency component evident.

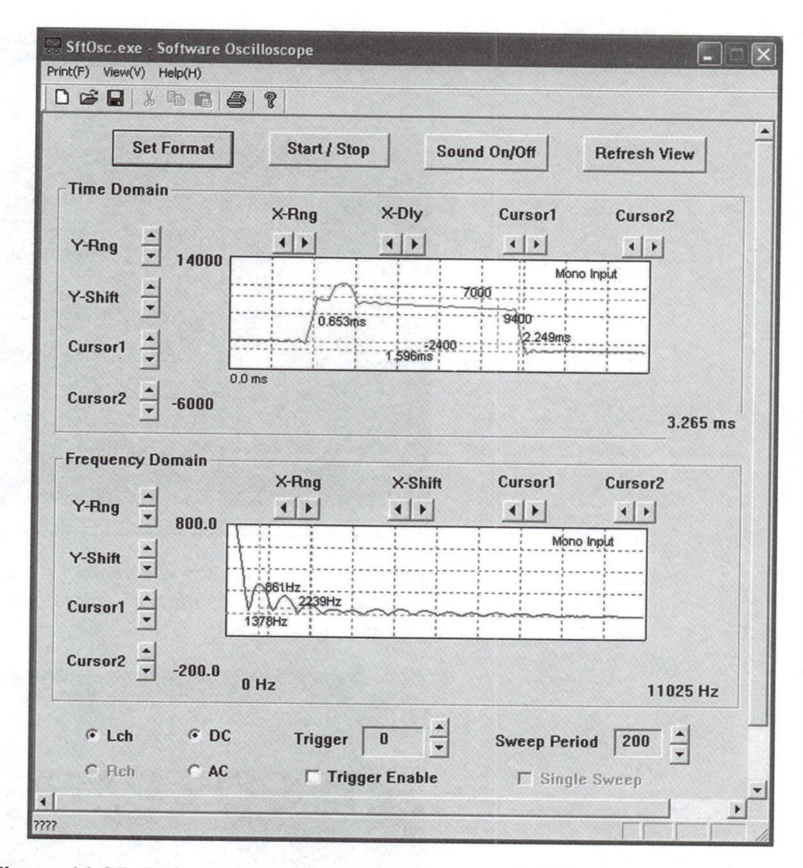

Figure 11.25 *Pulse parameter measurement using Software Oscilloscope*

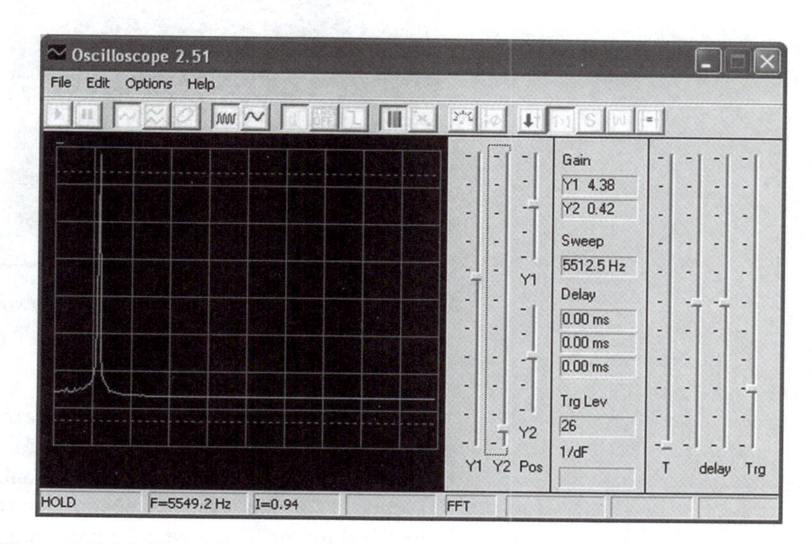

Figure 11.26 *Frequency spectrum for the sine wave shown in Figure 11.18*

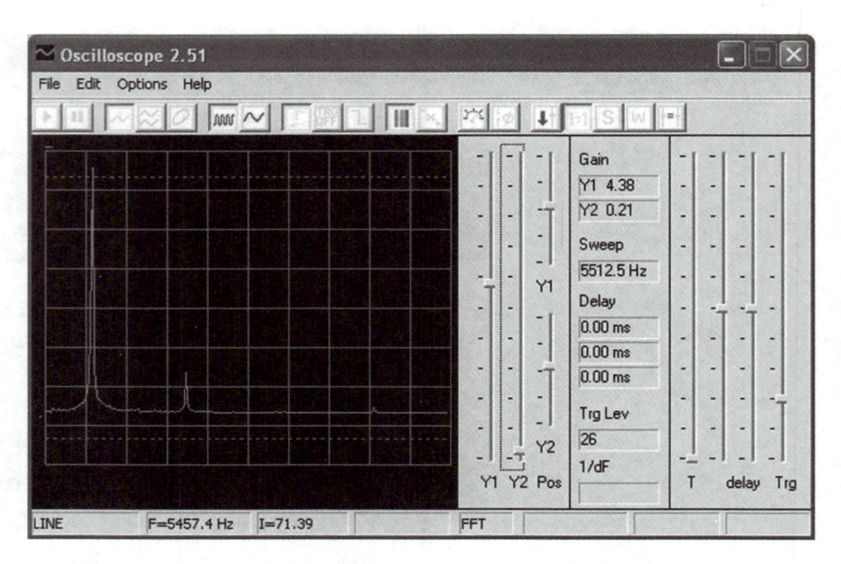

Figure 11.27 *Frequency spectrum for a sine wave with mild distortion present (third and seventh harmonic components are visible)*

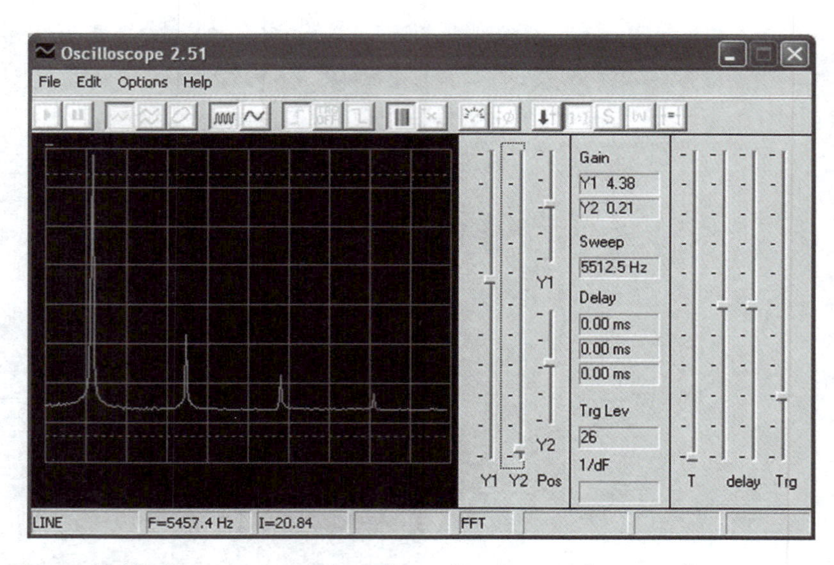

Figure 11.28 *Frequency spectrum for a sine wave with severe distortion present (third, fifth, and seventh harmonic components can be clearly seen)*

Figure 11.27 shows the same waveform as that shown in Figure 11.26 but with mild distortion present. The waveform is no longer a pure sinusoid and the third and seventh harmonic components have become visible.

Figure 11.28 shows the frequency spectrum of a severely distorted sine wave in which third, fifth, and seventh harmonics are present (the ninth harmonic is also present but cannot be seen due to the sweep setting).

12 Applications

The PC is a potential prime mover in a huge variety of process control and instrumentation applications ranging from simple stand-alone machine controllers to fully integrated production control systems. This chapter aims to provide readers with an introduction to the procedure for selecting and specifying hardware and software for a PC-based instrumentation or process control system. In addition, several representative applications of PC-based systems are discussed.

Expansion cards

The range of PC expansion boards currently available from a large number of manufacturers includes:

- Analogue I/O cards with up to 16 analogue inputs and up to four buffered analogue outputs.
- Digital I/O cards with direct TTL-compatible inputs and outputs.
- Digital I/O cards with opto-isolated inputs and outputs.
- Digital I/O cards with buffered I/O lines.
- Digital output cards fitted with relays or solid-state devices for AC or DC power control.
- EPROM programmers.
- IEEE-488/GPIB interface cards.
- Network adapter cards.
- Modem cards.
- Prototyping cards (these may include the necessary PC ISA/PCI bus interface logic and provide the user with an area for soldering components fitted into a 0.1″ matrix of plated through holes).
- Serial communications cards for RS-232, RS-422, RS423, or RS-485 serial ports.
- Stepper motor controllers.
- Multi-function I/O cards (offering mixed analogue and digital I/O facilities).
- Thermocouple interface cards.
- High-speed data acquisition cards.
- Bus expansion cards (which interface with external card frames or motherboards).
- PC instrument cards (e.g. function generators, counters/digital frequency meters, spectrum analysers, etc.).

In addition, the system builder is able to select from a large range of signal conditioning cards which provide the necessary interfacing circuitry for a wide range of popular sensors and output devices. It is thus eminently possible to construct a PC-based process control system simply by selecting ‘off-the-shelf’ modules. Only when dealing with very specialized applications it is necessary to manufacture ones own dedicated I/O cards and/or external signal conditioning boards. Appendix J lists a number of major suppliers of PC expansion cards and associated signal conditioning equipment.

Approaches

The system designer can select from a range of options depending upon the complexity and individual requirements of a particular application. The following general approaches are available:

- Stand-alone PC systems based on internally fitted expansion cards, rack modules, or separately enclosed units).
- PC systems based on standard PC expansion cards (and I/O processing cards, where appropriate) fitted into external card frame modules.
- Industrial PC systems (using a ruggedized PC functioning as a dedicated process controller or data-gathering device) fitted with internal or external expansion cards, and housed in a rack or freestanding enclosure.
- RS-232-based systems with the PC as controller (peripheral hardware connected via an asynchronous serial link).
- IEEE-488-based systems with the PC as controller.
- Backplane bus-based systems with a PC bus master/controller and a card frame bus.
- Networked/distributed PC systems (e.g. based on Ethernet or BITBUS) with enclosures and expansion cards to meet local requirements.

PC instruments

In addition to the vast range of expansion cards currently available, several manufacturers have developed a range of dedicated PC instruments (see Chapter 11) that emulate conventional items of test equipment (such as oscilloscopes, counters, function generators, and digital frequency meters).

A PC instrument offers many advantages over its conventional counterpart. It is flexible and adaptable and, in many cases, measurements may be automated under programmed control. Furthermore, considerable savings can be achieved from the elimination of redundant hardware (such as displays, operator controls, power supplies, etc.).

PC-based instruments can also offer very significant cost savings when compared with simple IEEE-488 bus-based instrumentation systems. A typical PC-based system for the acquisition of analogue voltages can, for example, be realized for less than 50% of the cost of a similarly specified system based on IEEE-488 hardware and software.

PC-based instruments are available in three general formats (Figure 12.1):

1. Using internally fitted expansion cards (plugged into a free slot in the PC).
2. Using an external rack with plug-in PC expansion cards.
3. Using separately enclosed modules (which may, if desired, be stacked) based on RS-232, USB, or IEEE-488 bus systems.

All three of these approaches have their own particular virtues and the system builder should include all three in his/her portfolio of potential engineering solutions.

Internally fitted cards generally offer the lowest cost approach to building a PC instrument. The disadvantage of this technique is that it necessitates internal

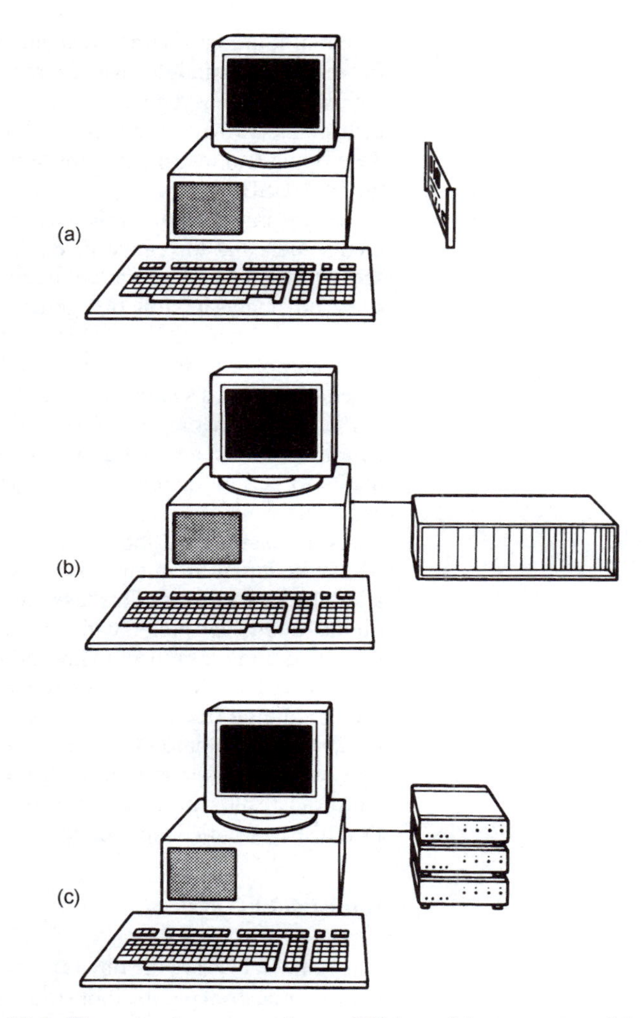

Figure 12.1 *Three basic approaches to PC-based instruments: (a) internally fitted PC expansion cards; (b) PC expansion cards in an external card frame; and (c) separately enclosed PC instruments linked via a parallel port, RS-232 serial port, USB, or via the IEEE-488 bus*

fitting and, since there may be a limited number of slots available, the expansion capability may be somewhat limited.

An external rack system allows the PC bus to be extended so that standard PC expansion cards may be fitted into an external card frame. This system is, however, relatively expensive and generally only appropriate where large-scale expansion is required. An alternative to that of extending the PC bus beyond the confines of the system enclosure is that of making use of a proprietary I/O bus. Such systems generally provide for between 1 and 32 I/O boards mounted in standard rack enclosures.

Separately enclosed modules (which may be interfaced to a PC by means or the RS-232, USB, or IEEE-488 bus) provide the third of this trio of potential solutions. With the exception of a front panel display controls, separate PC instrument modules usually resemble the conventional stand-alone instruments which they replace. However, in a relatively recent development, USB instruments provide a low-cost solution and several manufacturers are actively developing test instruments for budget conscious application that make use of the PC's USB port.

Several manufacturers have risen to the challenge of producing PC instruments such that the range of equipment currently available includes oscilloscopes, digital multimeters, universal counters, timers and frequency meters, spectrum analysers, function generators, pulse generators, voltage/current generators, and logic analysers. Individual instruments can be combined to provide more complex instrumentation facilities. A data logger, for example, can be assembled from a scanner and multimeter, and controlled flexibly from the PC.

PC instruments are ideal for making repetitive measurements during which data must be accumulated over a period time. The PC allows such measurements to be automated with the data acquired being sent to a file for future analysis.

As an example of the use of a PC instrument, consider an application in which the output frequency of an oscillator has to be monitored accurately over a long period of time. This task can be accomplished by means of a dedicated digital frequency meter with readings taken at appropriate intervals, logged on paper, and a graph showing the long-term variation of frequency can then be drawn. The alternative approach using a 'PC instrument' simply involves fitting a digital frequency meter expansion card (such as the Guide Technology GT200) to a standard PC compatible and using simple software (in conjunction with the driver(s) supplied with the card) to automate the measurement, and store the results in a data file for import into an analysis package (such as DADiSP). A typical application is discussed later in this chapter.

Industrial PC systems

Ruggedized PCs are the obvious choice for use in the harsh environment found in most industrial plants. Industrial PCs usually offer the same range of facilities associated with conventional PCs and compatibles, and invariably support the industry standard bus architecture. Hence an industrial PC will generally accept the same range of expansion cards as mentioned under the previous heading. Alternatively, where additional expansion beyond the limit imposed by the available free slots, industrial PCs may be fitted with bus extenders which are normally based on an external rack assembly.

In difficult environments it is possible to implement a completely *diskless system* using solid-state read/write memory devices (e.g. Flash, SD, or XD cards) inserted into IDE slots in order to provide a bootable operating system together with one or more application programs.

Backplane bus-based systems

A backplane bus system offers a reasonable compromise between a standard PC-based system at one extreme and a specialized industrial PC system at the other. Backplane bus systems are inherently flexible and reliable, and can simply be fitted with a PC processor card in order to make use of standard PC software packages.

Networked/distributed PC systems

Networked distributed PC systems are appropriate in large-scale applications where several processes are carried out concurrently. Each individual PC will be responsible for part of the process and data will be shared between the PCs by means of the network. As an example, consider the case of a packaging plant which manufactures and fills cardboard boxes on a continuous basis. One PC may be dedicated to the cutting and folding operation whilst another may be responsible for controlling glueing and stapling. A third PC would be responsible for filling and sealing the boxes. Data from all three PCs would then be collected by a fourth PC which oversees the entire process. Such a system provides an alternative to conventional solutions based on distributed programmable logic controllers (PLC).

Larger-scale systems are possible using bus systems based on Process Field Bus (Profibus), Actuator Sensor Interface (AS-Interface), Interbus, Modbus-1, BITBUS, and Ethernet. These systems provide remote I/O with a maximum number of nodes ranging from around 32–512 depending upon the standard concerned (Ethernet is theoretically unlimited). These arrangements are used for use in large-scale manufacturing and process control systems, and are thus somewhat beyond the scope of this book.

Intel's BITBUS (IEEE-1118) provides a simple and elegant solution to applications that require the services of a *multi-drop network*. BITBUS is a serial data bus based on the RS-485 physical and electrical interface standard (RS-485 is a multi-drop version of RS-422), and the datalink protocol employed is a subset of SDLC/HDLC.

BITBUS complements Manufacturing Automation Protocol (MAP) which has gained widespread recognition as the industrial standard for the upper level of factory data communications. At the machine and process level, however, where time critical data from sensors, actuators, and alarms has to be transmitted, the response time of MAP, though guaranteed, is inadequate. BITBUS, on the other hand, is well suited to the transfer of short 'Field Data' messages.

BITBUS is configured as a single-master, multi-slave network, and operates in one of two modes: synchronous and self-clocked. Synchronous operation permits speeds of up to 2.4 megabits/s but requires twin twisted-pair cables and is restricted to transmission over distances less than 300 m. Furthermore, since repeaters cannot be used in this mode, a maximum of 31 nodes is possible. Self-clocked mode, on the other hand, requires only single pair cable, can operate at either 62.5 or 375 kilobits/s and, with repeaters, can cater for up to 250 nodes at distances not exceeding 13 km. Access times of 3–4 ms per command are easily achieved.

Interfacing with BITBUS is usually made possible with the use of an Intel 8044/80154 compatible micro-controller which implements the BITBUS protocol using an on-chip SDLC controller and ROM-based firmware. An interface of this type may be incorporated within a processor card or may be provided as part of an auxiliary communications interface.

Specifying hardware and software

When specifying hardware and software to be used in a given PC bus application, it is essential to adopt a 'top-down' approach. An important first stage in this process is that of defining the overall aims of the system before attempting to formalize a detailed specification. The aims should be agreed with the end-user and should be reviewed within the constraints of available budget and time. Specifications should then be formalized in sufficient detail for the performance of the system to be measured against them and should include such items as input and output parameters, response time, accuracy, and resolution.

Having set out a detailed specification, it will be possible to identify the main hardware elements of the system as well as the types of sensor and output device required (see Chapter 9). The following checklist, arranged under six major headings, should assist in this process:

1 Performance specification
- What are the parameters of the system?
- What accuracy and resolution are required?
- What aspects of the process are time critical?
- What environment will the equipment be used in?
- What special contingencies should be planned for?
- What degree of fault tolerance is required?

2 I/O devices
- What sensors will be required?
- What output devices will be required?
- What I/O and signal conditioning boards will be required?
- Will it be necessary to provide high-current or high-voltage drivers?
- Should any of the inputs or outputs be optically isolated?

3 Displays and operator inputs
- What expertise can be assumed on the part of the operator?
- What alarms and status displays should be provided?
- What inputs are required from the operator?
- What provision for resetting the system should be incorporated?

4 Program/data storage
- What storage medium and format are to be employed?
- How much storage space will be required for the operating system and/or control program?
- How much storage space will be required for data?
- How often will the control program need updating?
- Will stored data be regularly updated during program execution?
- What degree of data security and integrity must be achieved?

5 Communications
- What existing communications standards are employed by the end-user?
- Will a standard serial data link based on RS-232 be sufficient or will a faster, low-impedance serial data communications standard be needed?
- What data rates will be required?
- What distances are involved?
- Will it be necessary to interface with automatic test equipment?
- Will a networking capability be required?

6 Expansion
- What additional facilities are envisaged by the end-user?
- What additional facilities could be easily incorporated?
- Will expansion necessitate additional hardware, additional software, or both?
- What provision should be made for accommodating additional hardware?

Hardware design

Start by identifying the principal elements of the system including PC, card frame, power supply, etc. as appropriate. Then itemize the input devices (such as keypads, switches, and sensors) and output devices (such as motors, actuators, and displays). This process may be aided by developing a diagram of the system showing the complete hardware configuration and the links which exist between the elements. This diagram will subsequently be refined and modified but initially will serve as a definition of the hardware components of the system.

Having identified the inputs required, a suitable sensor or input device should be selected for each input (see Chapter 9). It should then be possible to specify any specialized input signal conditioning required with reference to the manufacturer's specification for the sensor concerned. Input signal conditioning should then be added to the system diagram mentioned earlier.

Next, a suitable driver or output interface should be selected for each output device present (see Chapter 9). Any additional output signal conditioning required should also be specified and incorporated in the system diagram.

Software design

Software design should mirror the 'top-down' approach adopted in relation to the system as a whole. At an early stage, it will be necessary to give some consideration to the overall structure of the program, and identify each of the major functional elements of the software and their relationship within the system as a whole. It is important to consider the constraints of the system imposed by time critical processes and hardware limitations (such as the size of available memory). Furthermore, routines to cope with input and output may require special techniques (e.g. specialized assembly language routines).

The software should be designed so that it is easy to maintain, modify, and extend. Furthermore, the programmer should use or adapt modules ported from other programs. These modules will already have been proven and their use should be instrumental in minimizing development time.

When developing software, it is advisable to employ only 'simple logic' (i.e. that which has been tried and understood). The temptation to produce untried and over-complicated code should be avoided. Simple methods will usually produce code which is easy to maintain and debug, even if the code produced requires more memory space or executes more slowly (see Chapter 4). If the process is time critical or memory space is at a premium then code can later be refined and optimized. It is also important to consider all eventualities which may arise, not just those typical of normal operation. The following are particularly important:

- Will the system initialize itself in a safe state? Will there be momentary unwanted outputs during start-up?
- What will happen if the user defaults an input or if an input sensor becomes disconnected?
- What will happen if the power fails? Will the system shut down safely?
- What input validation checks are required? What steps should be taken if an 'out-of-range' input is detected?

Applications

The remainder of this chapter provides details of eight representative PC-based applications. These applications are not particularly novel but they do address problems that are typical of those which face the instrumentation and control engineer. The applications have been chosen to illustrate contrasting aspects of design and, while it would be impossible to describe any of these applications in their entirety, they should provide a feel for various aspects within the process of designing and implementing a PC-based system.

Monitoring oscillator stability

The client is a manufacturer of synthesized HF radio transceivers and wishes to develop a prototype voltage-controlled oscillator (VCO) which operates in the range 40–60 MHz for use within the frequency-generating circuitry. Several circuits have been constructed and the client wishes to ascertain the short- and long-term frequency stability of each unit.

Specification

The manufacturer requires that the output frequency is measured at appropriate intervals (e.g. every 100 ms for the short-term stability measurement and every 10 s for the long-term stability measurement). The results of each set of measurements are to be stored in an ASCII file for later graphical analysis. The software is, however, required to determine a number of simple performance indicators for each prototype unit including:

- Maximum frequency during the measurement period.
- Minimum frequency during the measurement period.
- Mean frequency over the measurement period.
- Total frequency drift during the measurement period.

The manufacturer also requires that the entire set of measurements and statistical calculations should be repeated at ambient temperatures of 0°C, 10°C, 20°C, 30°C, and 40°C.

This task would require considerable manual effort if it were to be carried out using a conventional digital frequency meter. It is, however, an ideal candidate for automated measurement using a PC and appropriate expansion card.

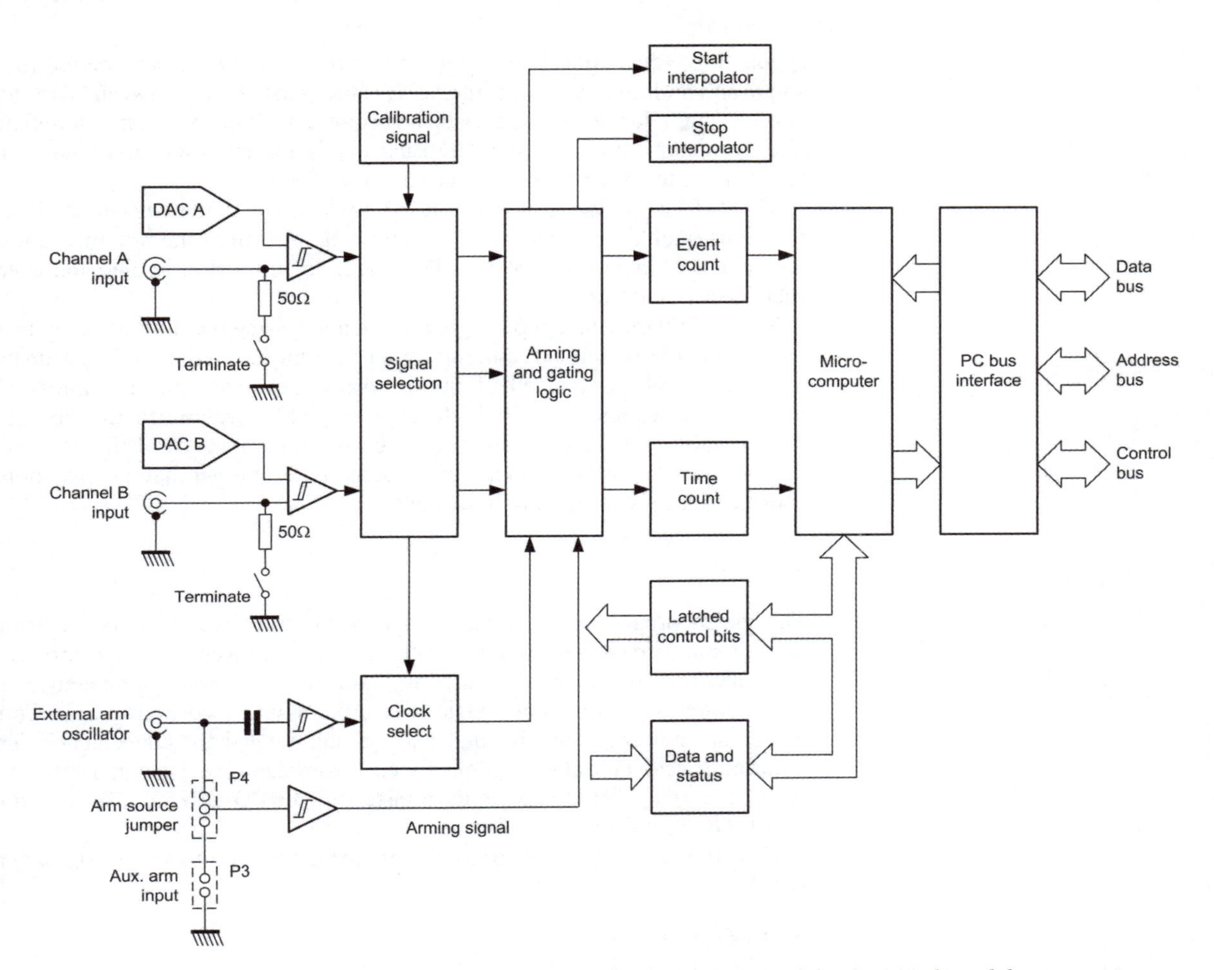

Figure 12.2 *Simplified block schematic of the GT200 digital frequency meter*

Hardware

The Guide Technology GT200 Universal Counter was chosen to provide the frequency measuring facility in conjunction with a Samsung AT-compatible microcomputer which already resides in the client's RF laboratory. The GT200 takes the form of a full-size PC-compatible expansion card which is supplied together with a device driver (GT200.SYS) and virtual front panel software (VIRT.EXE) on a floppy disk. The simplified block schematic of the GT200 is shown in Figure 12.2.

The GT200 is supplied together with a device driver (GT200.STS) and virtual front panel software (VIRT.EXE) on floppy disk. The disk also contains software which assists with setting the base address switch and includes a program which allows users to test the GT200's programming commands.

The GT200 offers a variety of measuring facilities including frequency measurement (from DC to 100 MHz, with automatic pre-scaling above about 1 MHz), fast frequency measurement (a special mode for high-speed data acquisition which allows up to 2300 measurements per second), period (both single and multiple), and time interval (i.e. the elapsed time between 'start' and 'stop' events). In addition, a direct data acquisition mode places measurements into a memory array without the usual overheads required to communicate results back to an application program via DOS).

The GT200 software is capable of performing a number of statistical functions (including mean, standard deviation, maximum and minimum measurements within a sample block). These are ideal for determining parameters, such as drift and 'jitter'.

The GT200 measures input signal frequencies using the most accurate technique available, reciprocal counting coupled to time interpolation. There are two primary benefits of this method: improved accuracy and reduced measurement time. Fast measurements with high accuracy yield more information concerning the stability of a signal. The GT200 is able to compute the drift rate, mean, and peak–peak jitter of a signal in the same time interval that a conventional counter is simply measuring frequency.

Software

The control program sends commands to the GT200 driver as character strings through standard DOS file write operations. Several conventions must be obeyed when incorporating commands into programs (e.g. individual commands must be separated by semicolon, carriage return, or line feed delimiters). Commands are not case sensitive and may be abbreviated for convenience. The minimum acceptable abbreviations for each command are listed in the manual. As an example, FREQ may be used instead of FREQUENCY, FU instead of FUNCTION, and so on.

GT200 commands are incorporated in normal program statements, such as BASIC:

```
PRINT#, "fu freqa; gate 0.01"
```

or in C:

```
fprintf(COUNTER, "fu freqa; gate 0.01");
```

A simple program, like the QBASIC program shown below, can be easily developed to meet the client's requirements for the long-term stability measurement (involving 100 readings taken at 10-s intervals).

```
REM Oscillator test program
REM Declare sub-programs
DECLARE SUB max ()
DECLARE SUB min ()
DECLARE SUB mean ()
REM Dimension array for collected data
DIM freq(100)

REM Get oscillator reference
CLS
INPUT "Enter oscillator reference: "; ref$
LET ref$ = LEFT$ (ref$, 6)
```

```
INPUT "Enter ambient temperature: "; temp$
osc$ = ref$ + temp$

REM Initialise digital frequency meter
OPEN "GT200$" FOR OUTPUT AS #1
OPEN "GT200$" FOR INPUT AS #2
PRINT #1, "init; timeout 4; function frequency A; gate 0.2"

REM Start collecting readings
PRINT "Hit (RETURN) to start measurement..."
WHILE r$ = ""
   r$ = INKEY$
WEND
FOR time% = 0 TO 99
PRINT #1, "reset"
' INPUT #2, freq(time%)
PRINT "Time = "; 10 * time%; " sec. Frequency = ";
 freq(time%); " Hz"
PRINT #1, "wait 10"
NEXT time%
CLOSE #1
CLOSE #2

REM Calculate and print statistics
PRINT
PRINT "Performance data for oscillator ref: "; ref$
PRINT
PRINT "Performance measured at: "; temp$; " deg.C"
max
PRINT "Maximum frequency; "; maxfreq; " Hz"
min
PRINT "Minimum frequency: "; minfreq; " Hz"
mean
PRINT "Mean frequency: "; meanfreq; " Hz"
PRINT "Frequency drift: "; maxfreq - minfreq; " Hz"
PRINT

REM Save data in an ASCII file
LET file$ = osc$ + ".DAT"
OPEN file$ FOR OUTPUT AS #3
FOR time% = 0 TO 99
PRINT #3, freq(time%)
NEXT time%
CLOSE #3


END


SUB max
SHARED freq()
SHARED maxfreq
maxfreq = 0
FOR i% = 0 TO 99
IF freq(i%) > maxfreq THEN maxfreq = freq(i%)
NEXT i%
END SUB
```

```
SUB mean
SHARED freq()
SHARED meanfreq
total = 0
FOR i% = 0 TO 99
total = total * freq(i%)
NEXT i%
meanfreq = total / 100
END SUB

SUB min
SHARED freq()
SHARED minfreq
minfreq = 1E+09
FOR i% = 0 TO 99
IF freq(i%) < minfreq THEN minfreq = freq(i%)
NEXT i%
END SUB
```

Three subprograms, `max()`, `min()`, and `mean()` are declared at the beginning of the program. The array, `freq()` (which will contain the returned data from the GT200 card) is then dimensioned for a total of 100 values.

The user is then prompted to enter the oscillator reference (which is truncated to include only the first six characters) and the ambient temperature used for the measurement.

The GT200 digital frequency meter is then associated with channel 1 for output and channel 2 for input by means of the OPEN statements. The instrument is initialized to measure frequency using input A with a timeout and gate times of 4 and 0.2 s, respectively.

The program then waits for the user to indicate that he/she is ready to begin a measurement by hitting the RETURN key. Once the key has been hit, the program takes 100 readings of frequency, placing each returned reading into the `freq()` array. The time between readings is set at l0 s by means of the `wait` command. Times and corresponding frequency readings are displayed on the screen on each pass through the main `FOR ... NEXT` loop so that the user is kept informed of the current state of measurement.

When the main loop has been completed, the two communications channels are closed. Thereafter, the performance data for the oscillator in question is printed with calls to the three subprograms which determine the maximum, minimum, and mean frequency values. The total frequency drift is calculated simply by subtracting the minimum frequency from the maximum frequency.

The three subprograms, `max()`, `min()`, and `mean()`, are quite straightforward and need no comment. A typical résumé of oscillator performance (printed by the program) data is shown in Figure 12.3.

Finally, the data is stored in an ASCII file. Note that the filename is constructed from the concatenation of the first six (or less) characters of the oscillator reference and the ambient temperature which was entered by the user, together with the file extension, .DAT. The file is opened for output (via channel 3) and all 100 values stored in the array are written to it. The channel is then closed.

```
C:\QBASIC\QB.EXE

Performance data for oscillator ref: HXO

Performance measured at: 22 deg.C
Maximum frequency;   5.016176E+07  Hz
Minimum frequency:   5.015923E+07  Hz
Mean frequency:      5.016055E+07  Hz
Frequency drift:     2532   Hz
```

Figure 12.3 *Sample printed oscillator performance data*

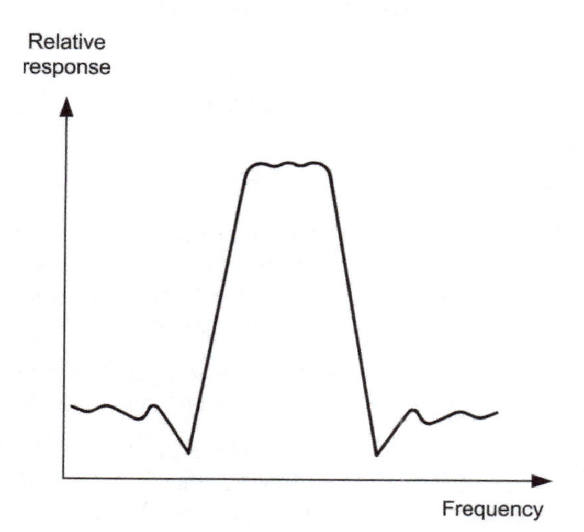

Figure 12.4 *Typical crystal filter response characteristic*

Testing crystal filters

The client is a manufacturer of RF passive components. Part of the company's product range includes 10.7 MHz crystal filters of various types which are manufactured to close tolerance in a batch process. Each filter is checked (on a test jig) to determine whether it meets the design specification which includes bandwidth (measured at −6 and −40 dB) and pass-band ripple. It is also considered desirable to display the response of the filter graphically in order that the ultimate stop-band attenuation can be gauged. Figure 12.4 shows a typical filter response characteristic.

Specification

The company wishes to automate the process of filter measurement and, at the same time, generate statistical information which can be used to check the manufacturing process.

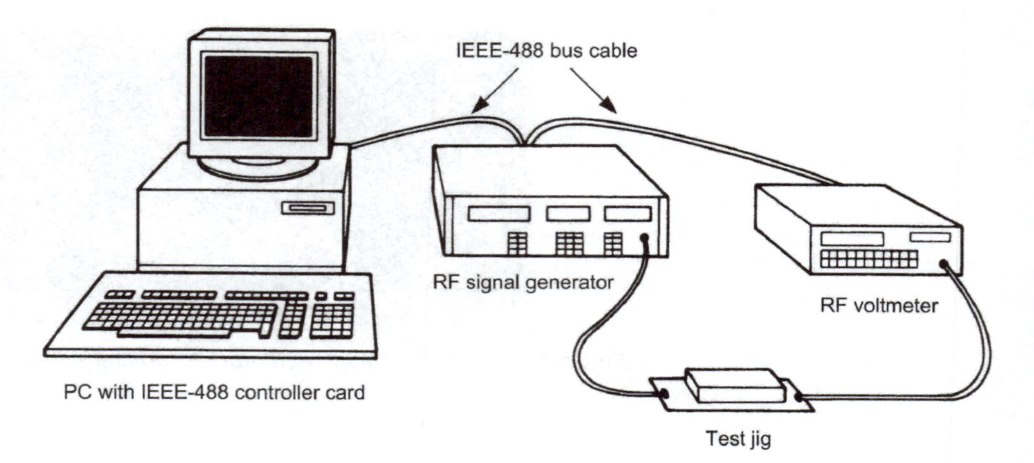

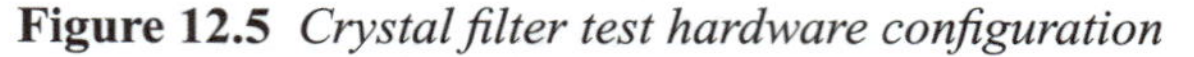
Figure 12.5 *Crystal filter test hardware configuration*

Hardware

This application is ideally suited to an IEEE-488-based system (based on test instruments fitted with the requisite IEEE-488 interface which are already available in the company's test department). Apart from the PC controller (which will require an IEEE-488/GPIB interface card) the two instruments required are:

- an RF voltmeter (Marconi 2610 with GPIB interface);
- an RF signal generator (Marconi 2018A with GPIB module).

The RF signal generator will be configured as a 'listener' whilst the RF voltmeter will be a 'talker'. A test jig will have to be constructed to accommodate the filter under test. Furthermore, since the filter source and load impedances are critical, appropriate matching components must be incorporated into the test jig. The simplified block schematic of the hardware is shown in Figure 12.5.

Software

The control software is again easily written in QuickBASIC (or equivalent) and the required program can be based on the following algorithm (expressed in a form of structured English):

```
INITIALISE SYSTEM
DISPLAY WELCOME SCREEN
DO
  GET SYSTEM PARAMETERS
  CONFIGURE IEEE-488 SYSTEM
  ENTER FILTER REFERENCE
  DO
    READ-VOLTAGE LEVEL
    INCREMENT GENERATOR FREQUENCY
  LOOP UNTIL FINAL FREQUENCY
  CALCULATE FILTER SPEC
  DISPLAY FILTER SPEC
  STORE FILTER SPEC
```

```
  PRINT FILTER SPEC
  PRINT FILTER LABEL
  LOOP UNTIL LAST FILTER
END
```

Most of the statements within the algorithm are coded as procedures. As an example, the procedure which prompts the user for values which will be used to set the system parameters (`GET SYSTEM PARAMETERS`) is itself described by the algorithm:

```
PROCEDURE GET SYSTEM PARAMETERS
  GET INITIAL FREQUENCY
  GET FINAL FREQUENCY
  GET FREQUENCY INCREMENT
  GET RF LEVEL
END PROCEDURE
```

Having decomposed each procedure, it is possible to translate each structured English statement into equivalent BASIC program statements. As an example, `GET INITIAL FREQUENCY` could be coded (in minimal form) as follows:

```
INPUT "Start frequency (kHz) "; start
```

In practice, a range check is desirable on this input since the normal range of start frequencies will lie within the range 400–450 kHz. The final code for `GET INITIAL FREQUENCY` was therefore:

```
DO
 INPUT "Start frequency (kHz) "; start
LOOP WHILE start < 400 OR start > 450
```

A speech enunciator

The client is a manufacturer of 'user-friendly' data entry devices and requires a low-cost system capable of recording and playing back analogue speech signals. This system will then be incorporated into an existing terminal based on a PC-compatible motherboard and fitted with a solid-state disk. The prototype speech enunciator card is shown in Photo 12.1.

Specification

The client requires that speech of up to 30-s duration and nominal bandwidth 6 kHz be available within the system. The speech signal (input from a microphone) is to be converted to digital information and stored in one or more data files within a reserved partition on the hard disk. The speech data is then to be made available for replay (as required) by the terminal control program.

Hardware

This system requires a fast A/D and D/A interface together with additional analogue signal filtering in order to reduce the effects of aliasing. No card of

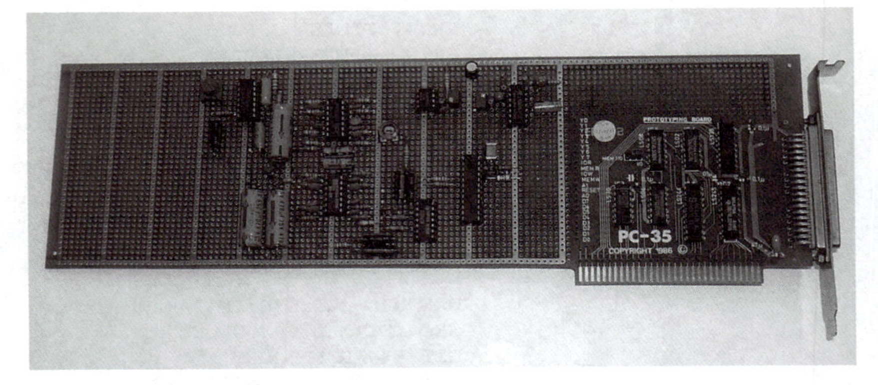

Photo 12.1 *Prototype speech enunciator card*

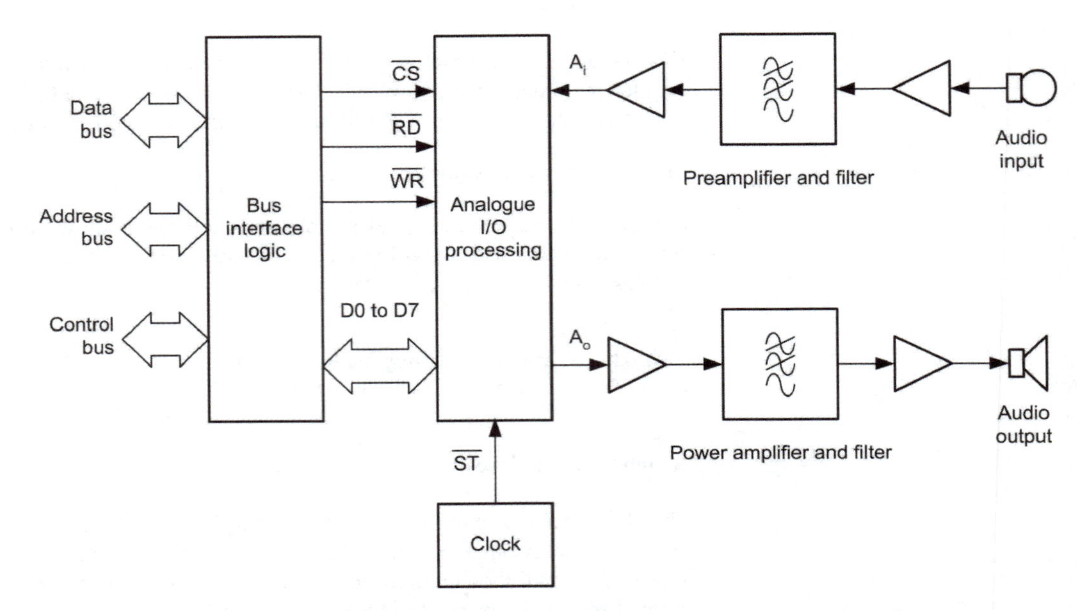

Figure 12.6 *Simplified block schematic of the prototype speech enunciator*

this type is available 'off the shelf' and thus a board must be prototyped from scratch. The prototype is built using a full-size ISA prototyping card which incorporates the necessary bus interface logic (see Chapter 2). Figure 12.6 shows a simplified block schematic of the hardware arrangement.

The need for A/D and D/A conversion can be realized by using a complete analogue I/O system in the form of the Analogue Devices AD7569. This unit offers 8-bit resolution (adequate for this simple speech application) coupled with a 2 μs ADC track/hold time, and on-chip band-gap 1.25 V voltage reference. The device is fabricated in linear-compatible CMOS (LC^2MOS) and is supplied in a 24-pin 'skinny' DIP package. The internal architecture of the AD7569 is shown in Figure 12.7 while the simplified circuit of the prototype interface card is shown in Figure 12.8.

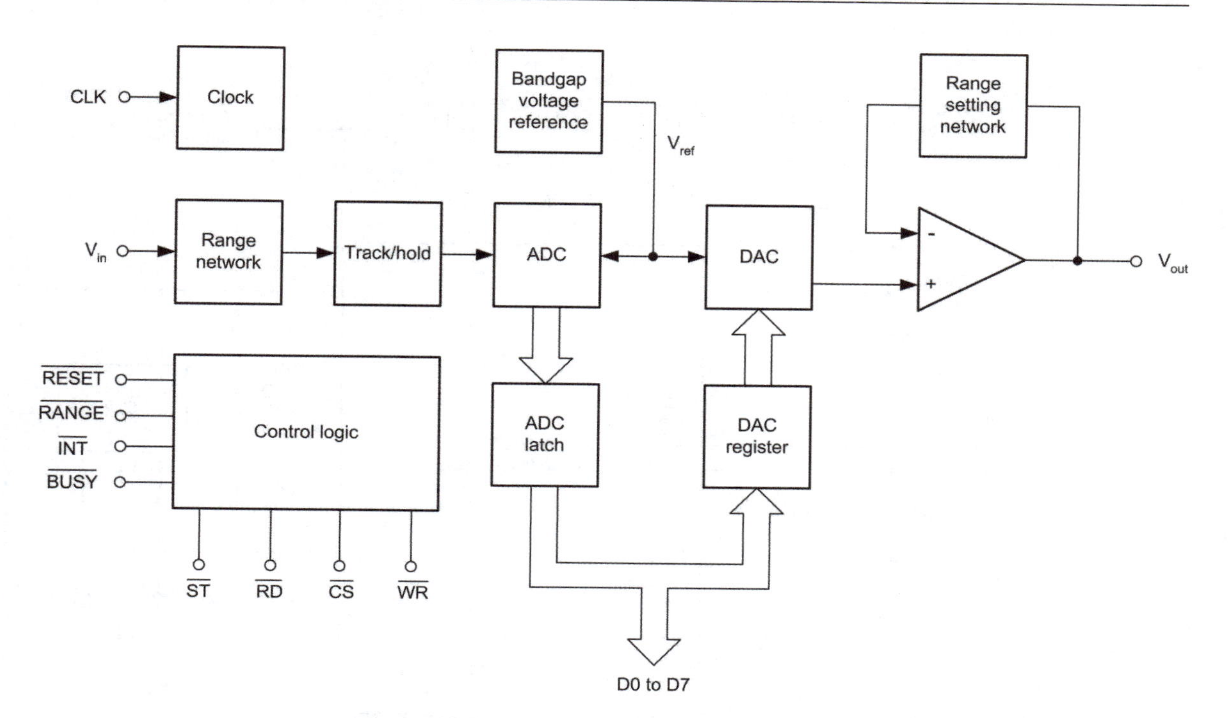

Figure 12.7 *Internal arrangement of the Analogue Devices AD7569 8-bit analogue I/O system*

Software

The software for the speech enunciator can usefully take advantage of the mixed language interface which is provided within the Microsoft suite of programming languages. Time critical routines (such as those which drive the ADC and DAC) can be written in assembly language while those which deal with disk filing and screen displays can be quickly and easily developed in QuickBASIC.

The assembly language module shown below is responsible for the recoding and playback process. These routines are liberally commented and should thus be reasonably self-explanatory (Chapter 5 provides more details of assembly language programming).

```
.MODEL          MEDIUM
.STACK          100H
.CODE
                ; This routine records data from the ADC in a
                ; 128k byte buffer - starting at 70000H
                ; Registers used: AX,BX,CX,CX,DX,DI,DS
                ; Parameters passed: 16-bit delay in stack frame
                ; Parameter returned: none

                PUBLIC Rec

Rec PROC
                PUSH   BP            ; save old base pointer
                MOV    BP,SP         ; set stack frame pointer
```

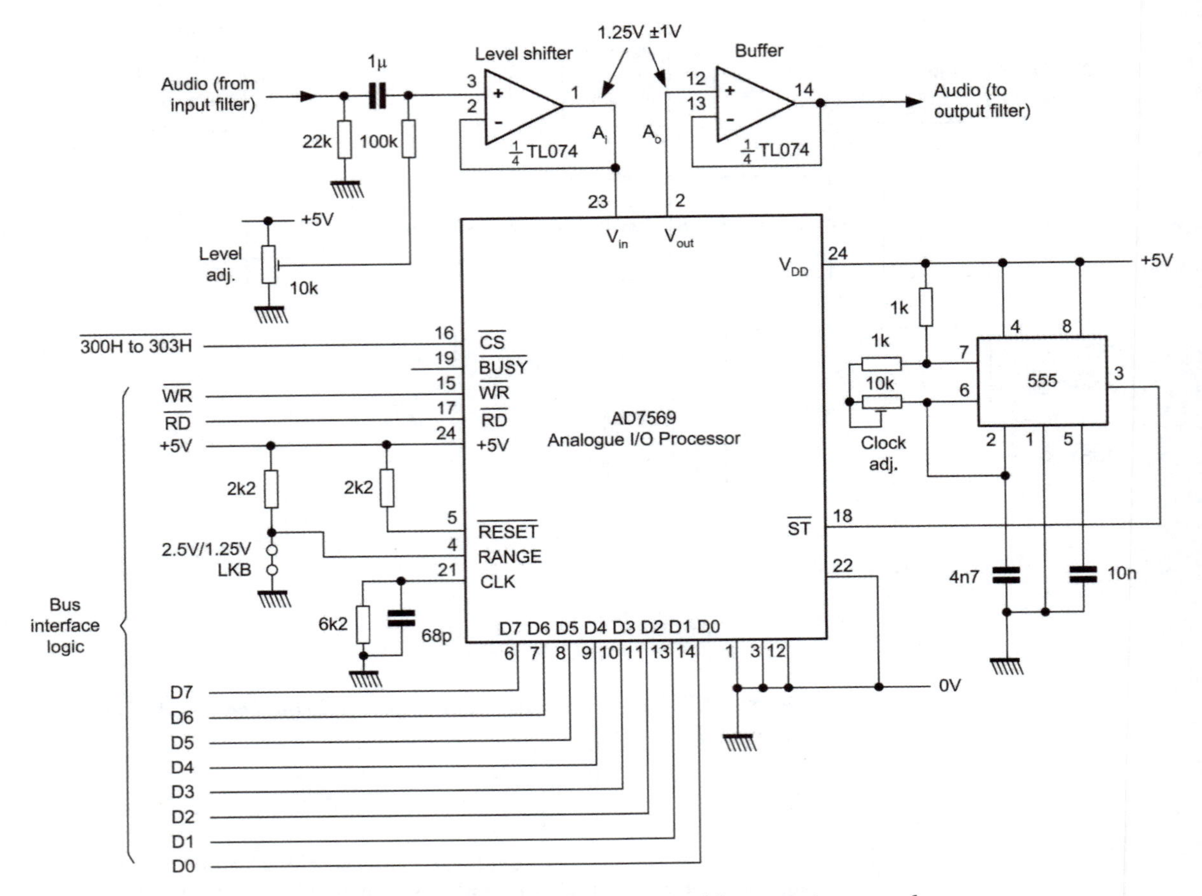

Figure 12.8 *Simplified circuit of the enunciator card*

```
                MOV     BX,[BP+6]       ; get argument passed
                MOV     AX,[BX]         ; and preserve in BX
                MOV     BX,AX

                PUSH    SI
                PUSH    DI
                PUSH    SS
                PUSH    DS

                MOV     DX,0300H        ; port used for analogue input
                MOV     AX,7000H        ; block 0 is at 70000H
                MOV     DS,AX
                MOV     DI,0            ; first location
                MOV     CX,0FFFFH       ; buffer size 64k

Rloop1:         IN      AL,DX           ; get a byte
                MOV     [DI],AL         ; and save it to the buffer
                INC     DI              ; point to next location
                CALL    Sdelay          ; sampling delay
```

```
                LOOP  Rloop1      ; go back for more
                MOV   AX,8000H    ; block 1 is at 80000H
                MOV   DS,AX
                MOV   DI,0        ; first location
                MOV   CX,0FFFFH   ; buffer size 64k

Rloop2          IN    AL,DX       ; get a byte
                MOV   [DI],AL     ; and save it to the buffer
                INC   DI          ; point to next location
                CALL  Sdelay      ; sampling delay
                LOOP  Rloop2      ; go back for more

                POP   DS
                POP   SS
                POP   DI
                POP   SI

                POP   BP          ; restore base pointer
                RET   2           ; bye!
Rec ENDP

                ; This routine records data from the ADC in a
                ; 128k byte buffer - starting at 70000H
                ; Registers used: AX,BX,CX,CX,DX,DI,DS
                ; Parameters passed: 16-bit delay in stack
                  frame
                ; Parameter returned: none

                PUBLIC Playb

Playb PROC
                PUSH  BP          ; save old base pointer
                MOV   BP,SP       ; set stack frame pointer
                MOV   BX,[BP+6]   ; get argument passed
                MOV   AX,[BX]     ; and preserve in BX
                MOV   BX,AX

                PUSH  SI
                PUSH  DI
                PUSH  SS
                PUSH  DS

                MOV   DX,0300H    ; port used for analogue input
                MOV   AX,7000H    ; block 0 is at 70000H
                MOV   DS,AX
                MOV   DI,0        ; first location
                MOV   CX,0FFFFH   ; buffer size 64k

Ploop1:         MOV   AL,[DI]     ; get a byte
                OUT   DX,AL       ; and output it
                INC   DI          ; point to next location
                CALL  Sdelay      ; sampling delay
                LOOP  Ploop1      ; go back for more
                MOV   AX,8000H    ; block 1 is at 80000H
                MOV   DS,AX
                MOV   DI,0        ; first location
                MOV   CX,0FFFFH   ; buffer size 64k
```

```
Ploop2      MOV   AL,[DI]     ; get a byte
            OUT   DX,AL       ; and save it to the buffer
            INC   DI          ; point to next location
            CALL  Sdelay      ; sampling delay
            LOOP  Ploop2      ; go back for more

            POP   DS
            POP   SS
            POP   DI
            POP   SI

            POP   BP          ; restore base pointer
            RET   2           ; bye!
Playb ENDP

            ; Delay routine to determine sampling rate
            ; called by Rec and Playb
            ; Registers used: BX,CX
            ; Parameters passed: none

Sdelay:     PUSH  CX          ; save current byte count
            MOV   CX,BX       ; sets time delay
Sloop:      LOOP  Sloop       ;
            POP   CX          ; restore byte count
            RET               ; back to the main loop
            END
```

Strain measurement and display

The client is a manufacturer of aircraft undercarriage components and wishes to carry out a series of strain measurements on structures when a stress is suddenly applied. In addition, the company wishes to display the response to an impulse force in real-time using a conventional oscilloscope-type display on the screen of a PC.

Specification

The measurement interval is to range from approximately 200 ms to 3 s, and the strain gauges and associated signal conditioning circuitry are expected to produce signals in the range ±250 mV. Eight sets of strain gauges are fitted to the structural member under test.

Software

The quasi-real-time oscilloscope display can easily be developed in C or Quick-BASIC. An unrefined (but nevertheless functional) routine is shown below. The routine displays the analogue signal returned from the strain gauge fitted to channel 0 (I/O address 300 hex.).

```
' Transient strain display
' PowerBASIC 3.5
' Runs in full screen mode
```

```
declare sub sweepdelay (count%)

' Set up the screen and graphics viewport
screen 8
view (0, 20)-(639, 199)

' Set initial timebase rate
dly% = 50

' Get initial voltage level
v% = inp(&H300)
if v% > 127 then v% = v% - 256
q% = 350 + v% - 255

' Main loop
do
   cls

   ' Plot the axes

   line (0, 0)-(0, 179), 5
   line (0, 179)-(640, 179), 5

   ' and the grid
   for i% = 0 to 179 step 12
   line (0, i%)-(640, i%), 5
   next i%

   for i% = 0 to 640 step 25
   line (i%, 179)-(i%, 0), 5
   next i%

   ' Update the status display
   sweeptime$ = str$(int(sweeptime!/25000))
   locate 2, 1
   print "X = ";sweeptime$; " ms/div "
   locate 2, 20
   print "Y = 50 mV/div"
   locate 1, 1
   print "Press <SPACE> to abort, <X> to freeze, ";
   print "<+> or <-> to change timebase setting"

   ' Get initial voltage level and plot the starting point
   v% = inp(&H300)
   if v% > 127 then v% = v% - 256
   q% = 85 + v%
   pset (0, q%), 10

   ' Scan across the screen from left to right
   mtimer ' Reset the timer
   for x% = 0 to 639
      v% = inp(&H300)
      if v% > 127 then v% = v% - 256
      q% = 85 + v%
      line -(x%, q%), 10
      call sweepdelay(dly%)
```

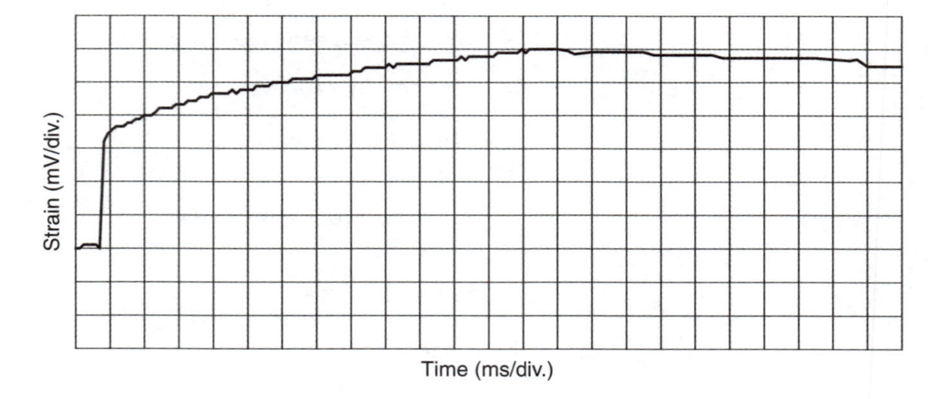

Figure 12.9 *Typical (strain plotted against time) display produced by the oscilloscope program*

```
    next x%
    sweeptime! = mtimer ' Get the sweep time

    ' Check to see if the user wishes to alter the scan rate
    r$ = inkey$
    if r$ = "+" or r$ = "=" then dly% = dly% - 10
    if r$ = "-" or r$ = "_" then dly% = dly% + 10
    if dly% < 10 then dly% = 10
    if dly% > 100 then dly% = 100

    ' Check to see whether the user wishes to freeze the
    screen while r$ = "X" or r$ = "x"
       ' Erase previous status line
       locate 1, 1
       print string$(80, 32)
       ' Tell the user how to resume
       locate 1, 1
       print "Display frozen - press <C> to continue"
       do
         r$ = inkey$
       loop until r$ = "C" or r$ = "c"
       wend

loop until r$ = " " ' Does the user want to quit?
end

sub sweepdelay (count%)
cal% = 9875 ' Calibrate sweep delay
for z% = 0 to count%
for k% = 0 to cal%: next k%
next z%
end sub
```

Figure 12.9 shows a typical display produced by the software.

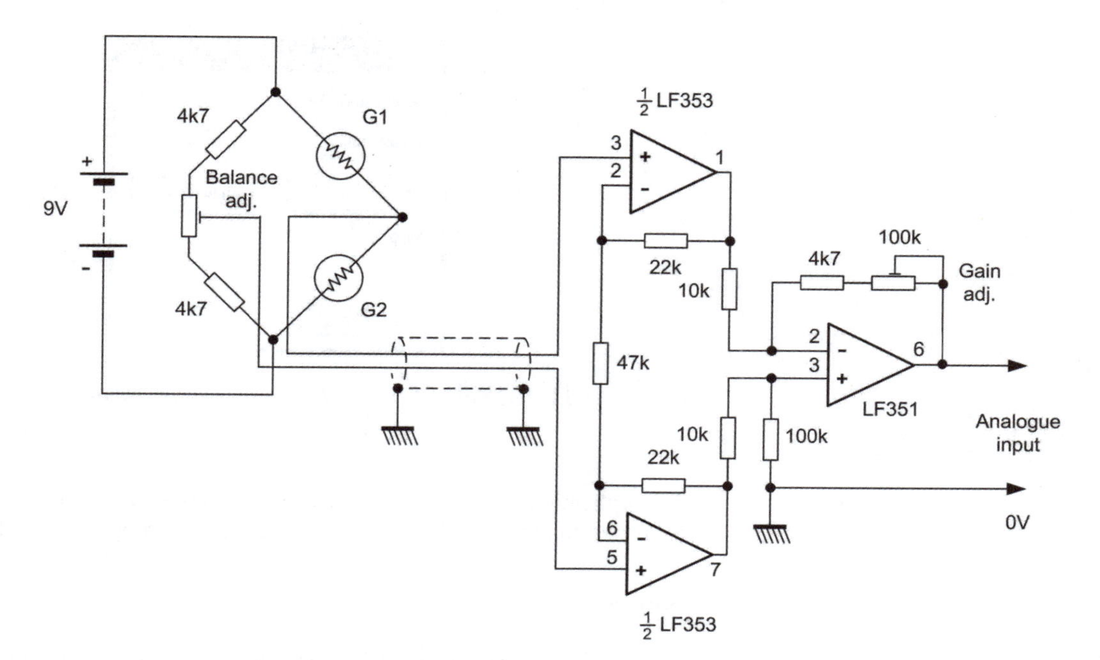

Figure 12.10 *Strain gauge input signal conditioning circuitry*

Hardware

The system can be based on a PC fitted with almost any 8-channel analogue input card (e.g. Arcom's PCAD12/16 which is a 16-channel 12-bit ISA card). The signal conditioning circuitry (replicated eight times) is based on a conventional temperature compensated half-bridge with operational amplifiers to provide voltage gain (variable from approximately 500 to 1500). To minimize noise, the input cable from each strain gauge bridge is balanced and shielded. Figure 12.10 shows the signal conditioning circuitry associated with each strain gauge bridge.

Backup battery load test

The client is a manufacturer of low-power FM radio relays. Each relay is fitted with a standby battery comprising four 2 V sealed lead–acid cells, each rated at 2 V 80 Ah.

Specification

The battery load test is to capture backup battery voltage data at regular intervals ranging from 10 ms to 100 s for periods of between 1 min (accelerated load test) and 10 days (prolonged load test). Voltage readings are to be within the range 0–10 V DC and they are to be accurate to within ±50 mV (±0.05 V). Data is to captured in a form that is compatible with a standard spreadsheet (e.g. MS Excel) for subsequent display and analysis. A dedicated PC is unavailable for

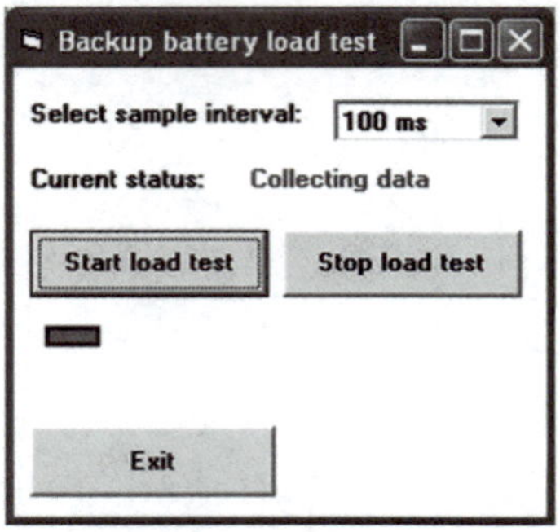

Figure 12.11 *Backup battery load test application written in Visual Basic*

this application so the hardware interface is required to be external and available for fitting to any one of several bench PCs in the production test laboratory.

Hardware

This application makes use of the Measurement Computing PMD-1208LS USB Personal Measurement Device. This device (see Chapter 2) has four differential or eight single-ended analogue input channels and is easily moved from one PC to another. Screw terminals permit connection of test leads and no further adjustment or configuration is necessary other than ensuring that the software is loaded and appropriate Measurement Computing library is installed on the host computer.

Software

The software was written using MS Visual Basic (see below) and the application is shown in Figure 12.11. A combo-box provides a means of selecting the sampling rate (from 10 ms to 100 s) with buttons provided to start and stop the load test. A virtual LED and text field provides status indication. When the stop button is operated data is sent to a data file in a format that can subsequently be imported into MS Excel (see Figure 12.12).

```
'============================================================
' Name:                    loadtest
' Purpose:                 collects backup battery voltage data
' Library calls:           cbAIn%() and cbErrHandling%()
' Hardware:                PMD-1208LS USB HID
'============================================================

Const BoardNum% = 1              ' Board number
Dim Index As Integer
Dim Record As Integer
' Dimension data array
Dim data_array(10000)
Dim Gain As Integer
```

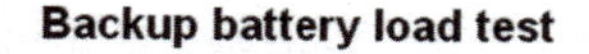

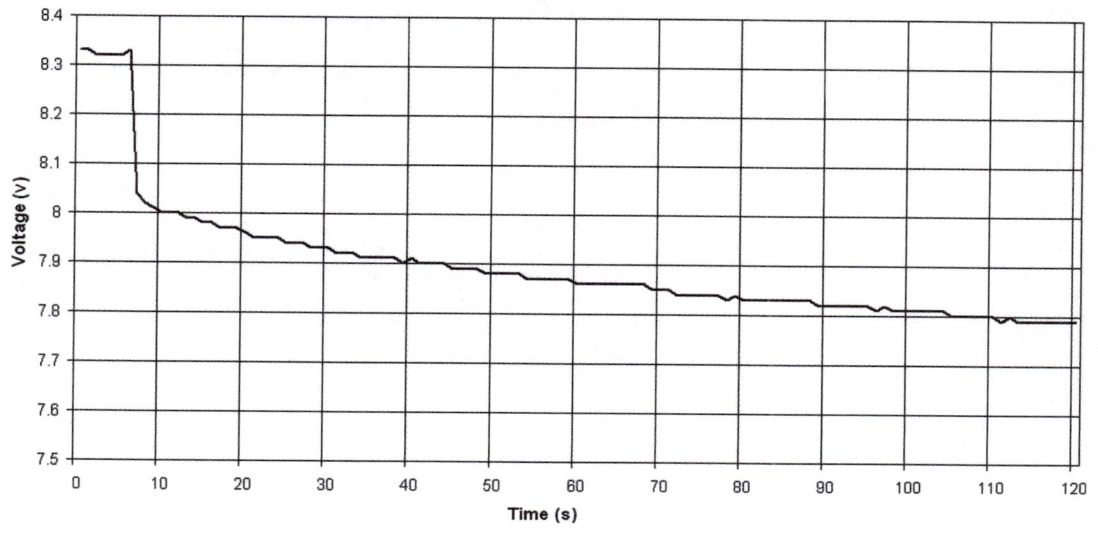

Figure 12.12 *Backup battery load test results (graph produced from data exported to Excel). Note that the load is applied at $t = 7\,s$*

```
Private Sub cmdExit_Click()
End
End Sub

Private Sub cmdStart_Click()
tmrConvert.Enabled = True
' Turn LED indicator on
sample_led.FillColor = "&H000000FF"
Index% = 0
lblStatus.Caption = "Collecting data"
End Sub

Private Sub cmdStop_Click()
tmrConvert.Enabled = False
' Turn LED indicator off
sample_led.FillColor = "&H00E0E0E0"
' Initial status message
lblStatus.Caption = Format$(Index, "0") + " samples collected"
' Prepare to write data file
CommonDialog1.DialogTitle = "File Save"
CommonDialog1.InitDir = App.Path
CommonDialog1.DefaultExt = "dat"
CommonDialog1.FILTER = "Data (*.dat)"
CommonDialog1.FileName = "sample.dat"
CommonDialog1.ShowSave
If CommonDialog1.FileName <> " " Then
    Open CommonDialog1.FileName For Output As #1
    For Record% = 1 To Index%
    Write #1, Record%, data_array(Record%)
    Next Record%
    Close #1
    lblStatus.Caption = "Data file written"
```

```
' Turn LED indicator green
sample_led.FillColor = "&H0000FF00"
Else
    lblStatus.Caption = "Data file not written"
' Turn LED indicator grey
sample_led.FillColor = "&H00E0E0E0"
End If
End Sub

Private Sub Form_Load()
' Declare revision level of Universal Library
ULStat% = cbDeclareRevision(CURRENTREVNUM)
' Initiate error handling
ULStat% = cbErrHandling(PRINTALL, DONTSTOP)
If ULStat% <> 0 Then Stop
' Set channel number and gain
Chan% = 0
' Set default range on start-up
Gain = BIP20VOLTS
' Set default maximum number of samples
max_samples = 100000
Index% = 0
' Disable timer
tmrConvert.Enabled = False
cmbInterval.Text = "10 ms"
' Initial status message
lblStatus.Caption = "Waiting for Start button"
' Sample LED set to off
sample_led.FillColor = "&H00E0E0E0"
' Default filename
file_name = "sample.dat"
End Sub

Private Sub tmrConvert_Timer()
Index% = Index% + 1
If cmbInterval.Text = "10 ms" Then tmrConvert.Interval = 10
If cmbInterval.Text = "100 ms" Then tmrConvert.Interval = 100
If cmbInterval.Text = "1 s" Then tmrConvert.Interval = 1000
If cmbInterval.Text = "10 s" Then tmrConvert.Interval = 10000
If cmbInterval.Text = "100 s" Then tmrConvert.Interval = 100000
' Collect the data
ULStat% = cbAIn(BoardNum%, Chan%, Gain, DataValue%)
If ULStat% = 30 Then MsgBox "Gain setting not valid", 0,
    "Unsupported Gain"
If ULStat% <> 0 Then Stop
ULStat% = cbToEngUnits(BoardNum%, Gain, DataValue%, EngUnits!)
If ULStat% <> 0 Then Stop
data_array(Index%) = Format$(EngUnits!, "0.00")
End Sub
```

Load sequencer

The client uses a manufacturing process based on eight devices that operate from a nominal 8 A 115 V AC supply. Unfortunately, the momentary surge current taken by each device (each of which involves a degaussing component)

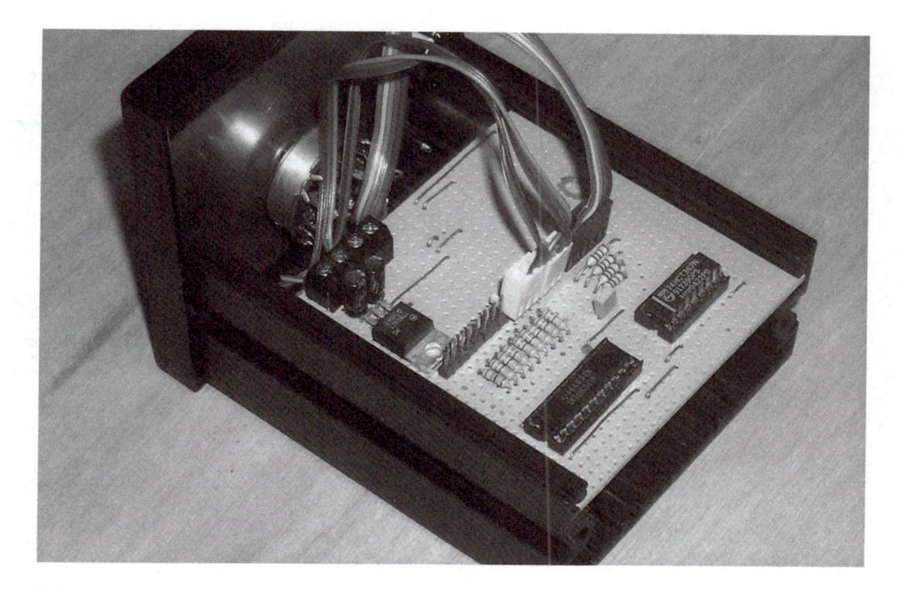

Photo 12.2 *Prototype parallel port interface module*

is greatly in excess of the rated current. Furthermore, when all devices operate simultaneously (or within a few milliseconds of one another) the surge current will invariably trip out the mains supply. This, in turn, causes disruption to the manufacturing process because each device has to be individually turned off before the mains trip can be manually reset. The client requires a simple and reliable means of automatically sequencing the application of power to the loads.

Specification

The time delay in applying the AC mains supply to each device is to be configurable to within 1 s up to a maximum of 30 s. The operator is to be provided with a simple graphical interface that shows the status of each load and allows the delay to be set using a simple slider control.

Hardware

Since there are eight loads and they are only required to be switched on and off, this application requires a simple 8-bit parallel port interface module. However, it is expected that the production system may be expanded at some point in the future and it could be advantageous to provide a solution that can be easily expanded on a modular basis (see Photo 12.2).

The circuit of one 8-bit parallel interface module is shown in Figure 12.13. The module is connected to the PC by means of a standard parallel port (see Chapter 2). In order to cater for future expansion, the module can be assigned to one of four controlled groups by means of a group channel select switch. Each channel group (A , B, C, and D) will then have eight controlled channels (channels 1–8) and each of these channels will correspond to a particular device.

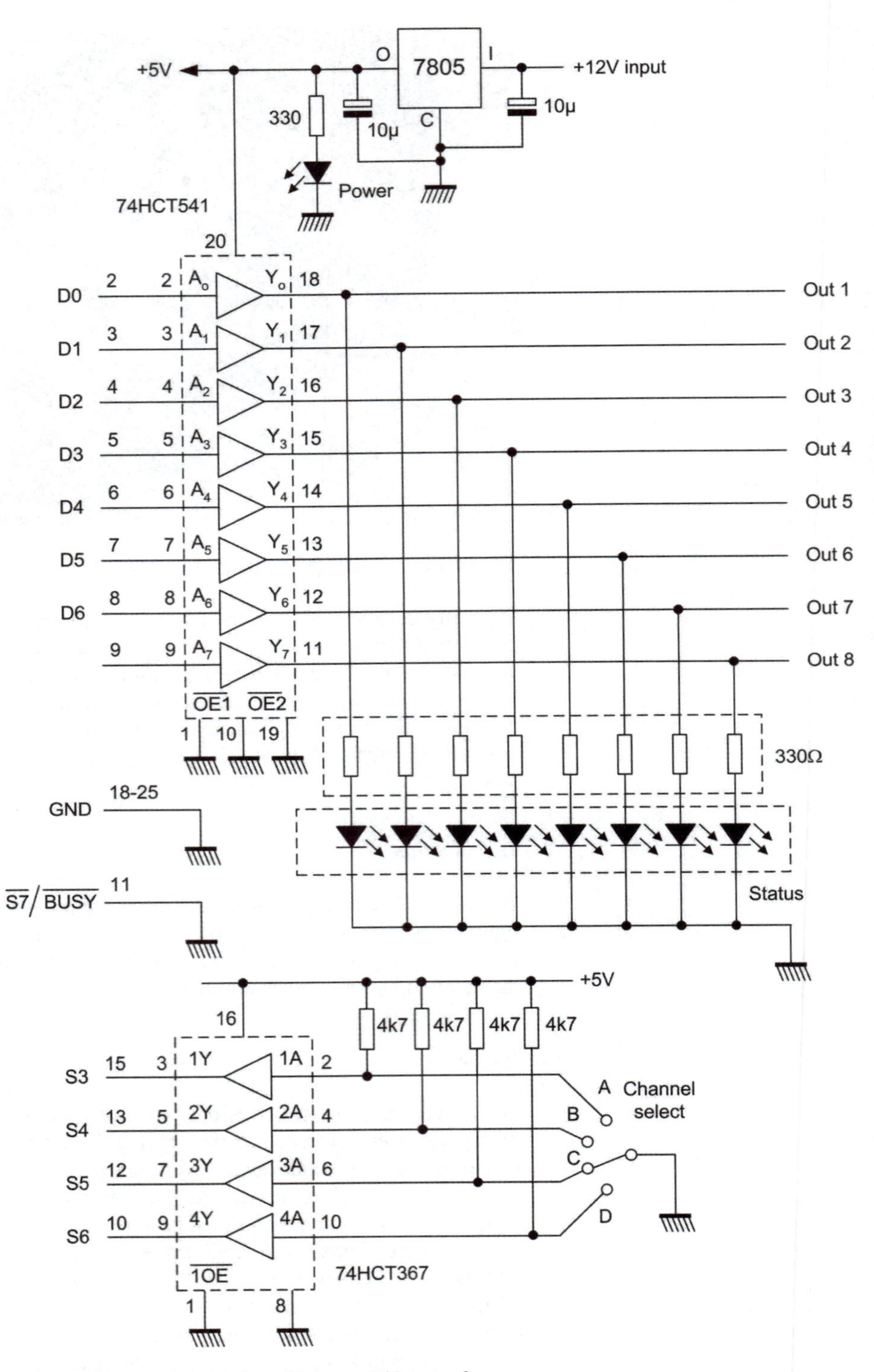

Figure 12.13 *Parallel port I/O interface*

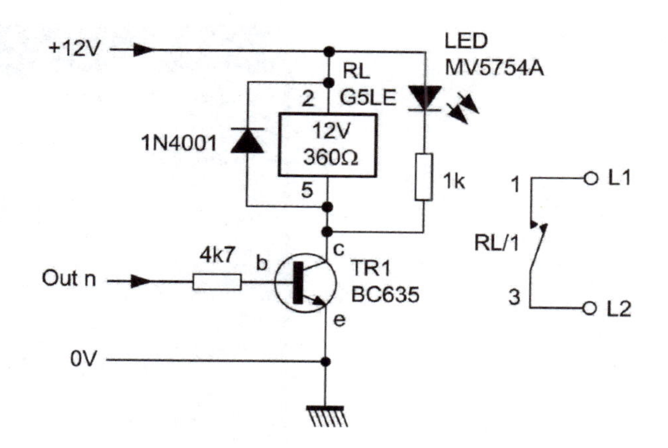

Figure 12.14 *Relay driver (a high output from a port line enables the load)*

Hence the system has potential for controlling up to 32 devices using four identical interface modules.

Each channel output from the interface module is connected to a relay driver (see Figure 12.14). This circuit is capable of switching a load of up to 10 A at 115 V AC. A status LED is included for test purposes.

Software

The application uses a simple Visual Basic routine (see below). The user interface is designed so that the operator can set the delay on any channel to any time between 0 and 30 s using a simple slider control. Each channel is fitted with a virtual LED indicator so that the operator knows which loads have become active. In addition, a further status field shows the elapsed time (see Figure 12.15). This application makes extensive use of the Visual Basic Timer (Chapter 6 contains more information on Visual Basic programming).

```
'==============================================================
' Name:           seqcon2
' Purpose:        controls switching sequence on channels
                      1 to 8
' Library calls:  requires inpout32.bas for I/O
' Hardware:       parallel port with relay modules
'==============================================================

Dim Port1 As Integer
Dim Port2 As Integer
Dim Port3 As Integer
Dim OutData As Integer
Dim ETime As Integer

Private Sub Start_Click()
Timer1.Enabled = True
Timer2.Enabled = True
Timer3.Enabled = True
Timer4.Enabled = True
```

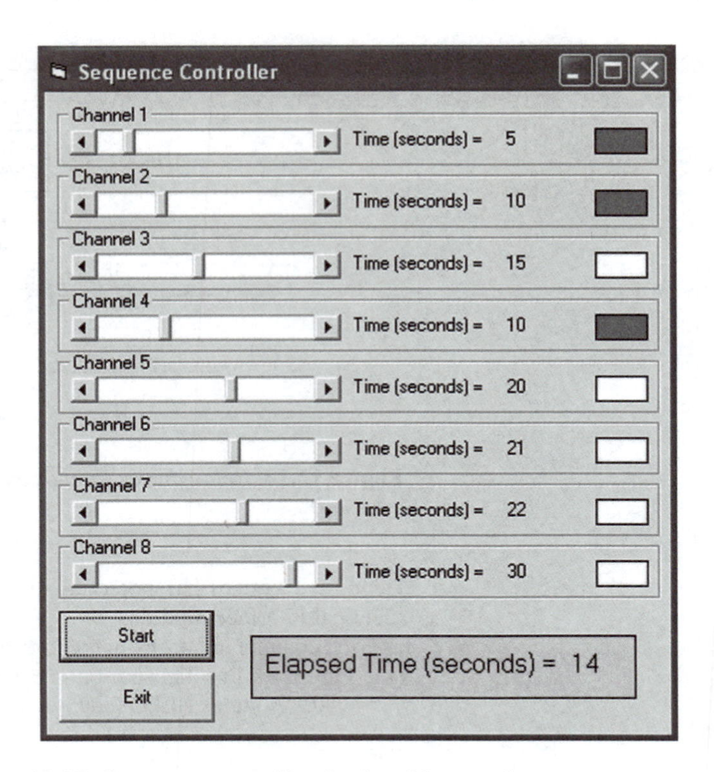

Figure 12.15 *Sequence controller display 14 s into the power-up sequence*

```
Timer5.Enabled = True
Timer6.Enabled = True
Timer7.Enabled = True
Timer8.Enabled = True
MasterClock.Enabled = True
End Sub

Private Sub Exit_Click()
End
End Sub

Private Sub Form_Load()
Port1 = 888
Port2 = 889
Port3 = 890
OutData = 0
Out Port1, OutData
SetTime1.Caption = 10
SetTime2.Caption = 10
SetTime3.Caption = 10
SetTime4.Caption = 10
SetTime5.Caption = 10
SetTime6.Caption = 10
SetTime7.Caption = 10
SetTime8.Caption = 10
```

```
Timer1.Interval = 10000
Timer2.Interval = 10000
Timer3.Interval = 10000
Timer4.Interval = 10000
Timer5.Interval = 10000
Timer6.Interval = 10000
Timer7.Interval = 10000
Timer8.Interval = 10000
ETime = 0
End Sub

Private Sub HScroll1_Change()
Timer1.Interval = HScroll1.Value
SetTime1.Caption = Int(Timer1.Interval / 1000)
End Sub

Private Sub HScroll2_Change()
Timer2.Interval = HScroll2.Value
SetTime2.Caption = Int(Timer2.Interval / 1000)
End Sub

Private Sub HScroll3_Change()
Timer3.Interval = HScroll3.Value
SetTime3.Caption = Int(Timer3.Interval / 1000)
End Sub

Private Sub HScroll4_Change()
Timer4.Interval = HScroll4.Value
SetTime4.Caption = Int(Timer4.Interval / 1000)
End Sub

Private Sub HScroll5_Change()
Timer5.Interval = HScroll5.Value
SetTime5.Caption = Int(Timer5.Interval / 1000)
End Sub

Private Sub HScroll6_Change()
Timer6.Interval = HScroll6.Value
SetTime6.Caption = Int(Timer6.Interval / 1000)
End Sub

Private Sub HScroll7_Change()
Timer7.Interval = HScroll7.Value
SetTime7.Caption = Int(Timer7.Interval / 1000)
End Sub

Private Sub HScroll8_Change()
Timer8.Interval = HScroll8.Value
SetTime8.Caption = Int(Timer8.Interval / 1000)
End Sub

Private Sub Timer1_Timer()
Shape1.FillColor = "&H000000FF"
OutData = Inp(Port1)
Out Port1, (OutData Or 1)
End Sub
```

```
Private Sub Timer2_Timer()
OutData = Inp(Port1)
Out Port1, (OutData Or 2)
Shape2.FillColor = "&H000000FF"
End Sub

Private Sub Timer3_Timer()
OutData = Inp(Port1)
Out Port1, (OutData Or 4)
Shape3.FillColor = "&H000000FF"
End Sub

Private Sub Timer4_Timer()
OutData = Inp(Port1)
Out Port1, (OutData Or 8)
Shape4.FillColor = "&H000000FF"
End Sub

Private Sub Timer5_Timer()
OutData = Inp(Port1)
Out Port1, (OutData Or 16)
Shape5.FillColor = "&H000000FF"
End Sub

Private Sub Timer6_Timer()
OutData = Inp(Port1)
Out Port1, (OutData Or 32)
Shape6.FillColor = "&H000000FF"
End Sub

Private Sub Timer7_Timer()
OutData = Inp(Port1)
Out Port1, (OutData Or 64)
Shape7.FillColor = "&H000000FF"
End Sub

Private Sub Timer8_Timer()
OutData = Inp(Port1)
Out Port1, (OutData Or 128)
Shape8.FillColor = "&H000000FF"
End Sub

Private Sub MasterClock_Timer()
ETime = ETime + 1
Clock.Caption = ETime
End Sub
```

Environmental monitoring

The client is a company that specializes in heating and ventilation of commercial buildings. The company wishes to have a means of regularly capturing temperature data from different points in a building and of later analysing this

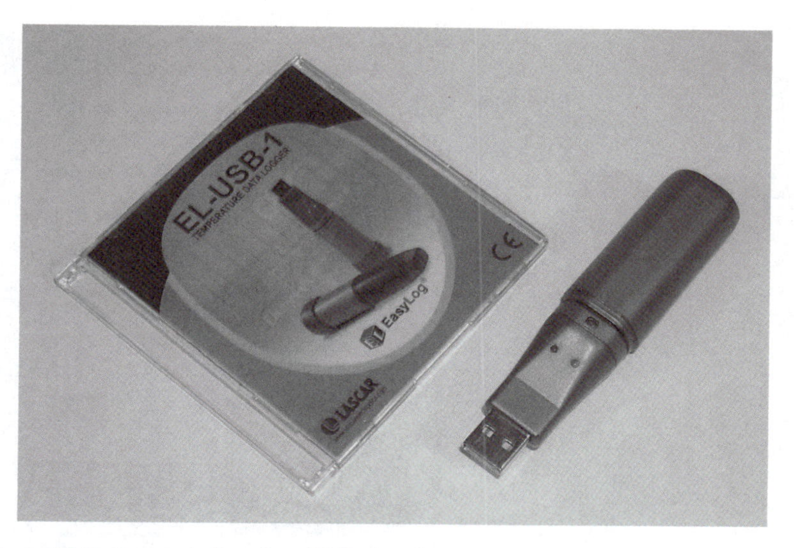

Photo 12.3 *Lascar's EasyLog USB data logger*

data to assess the effectiveness of various heating and ventilation solutions. The sensors must be left in position for long periods (usually between 2 and 6 weeks) and data must be sampled throughout this period at intervals of 5 min, or less.

Specification

The temperatures collected are to be in the range −20°C to +50°C with a resolution of ±0.5°C and an accuracy of ±1°C. Because sensing devices are to be purchased and deployed in quantity (between 10 and 40 sensors per project) sensors must be low cost and require minimal effort in deployment.

Hardware

The Lascar USB data logger was chosen for this application (see Photo 12.3). This is a low-cost device which is interfaced to a PC through a standard USB port. Once set, the device can be removed from the PC and left *in situ* to collect data.

The data logger measures and stores up to 16 382 temperature readings over the range −25°C to +80°C (−13°F to +176°F). The data logger uses a long-life 3.6 V lithium battery and will operate for approximately 12 months before battery replacement is required. Logging rates can be set to 10 s, and 1, 5, 30 min, and 1, 6, 12 h. The data logger also offers high and low alarms (not used in this application).

Software

Lascar's own data logging software was found to be perfectly adequate for this application and no further bespoke software was required. Figures 12.16–12.19

EasyLog USB
Options Help
Please select the procedure you wish to perform:
Set up and start the USB Data Logger.
Stop the USB Data Logger and download data.
View previously saved data.
EasyLog®

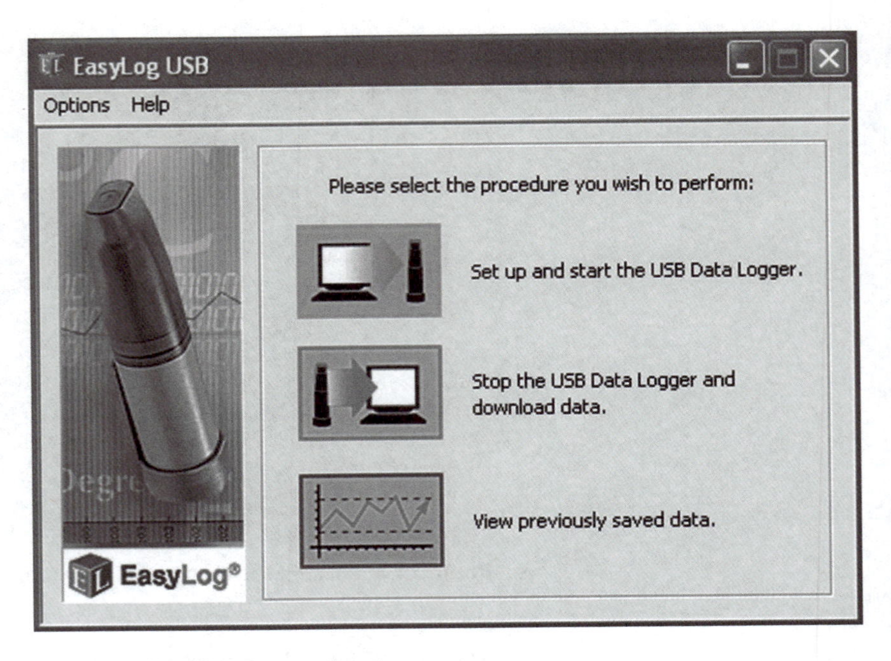

Figure 12.16 *EasyLog application*

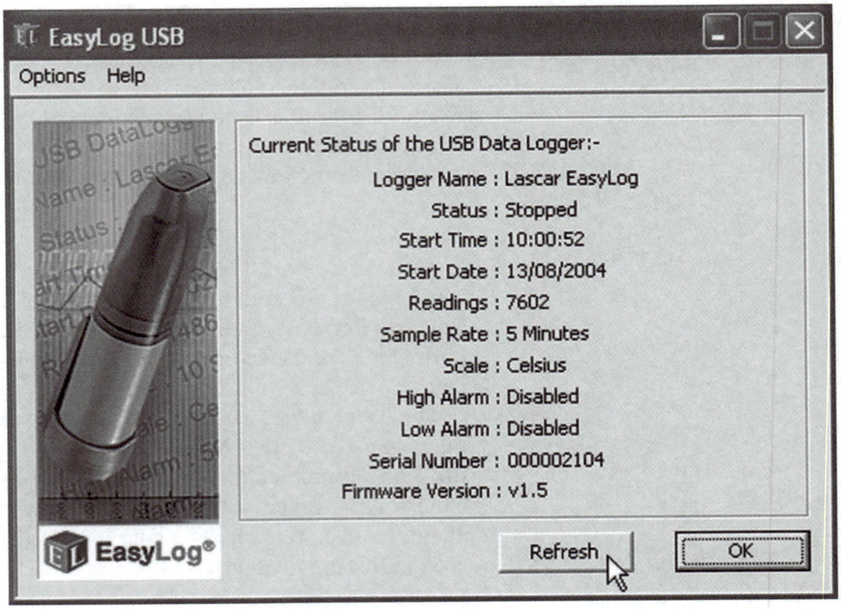

EasyLog USB
Options Help
Current Status of the USB Data Logger:-
Logger Name : Lascar EasyLog
Status : Stopped
Start Time : 10:00:52
Start Date : 13/08/2004
Readings : 7602
Sample Rate : 5 Minutes
Scale : Celsius
High Alarm : Disabled
Low Alarm : Disabled
Serial Number : 000002104
Firmware Version : v1.5
Refresh
OK
EasyLog®

Figure 12.17 *EasyLog status display*

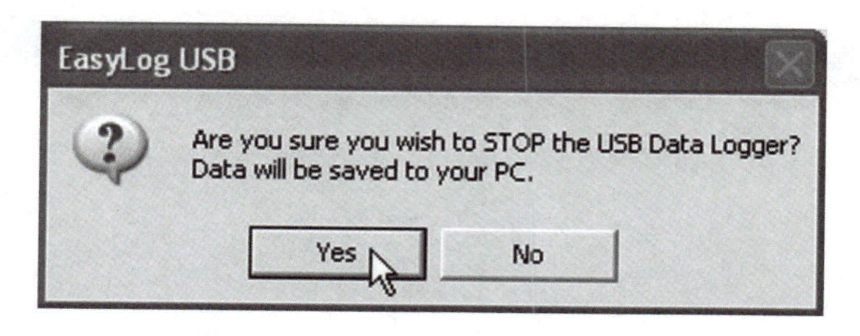

Figure 12.18 *EasyLog message box requiring confirmation of stopping data logging*

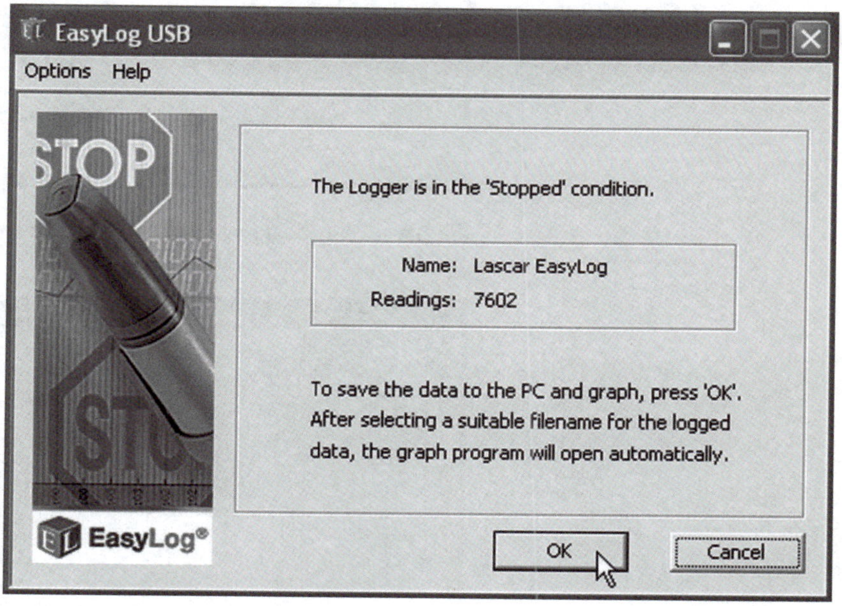

Figure 12.19 *EasyLog confirmation display*

shows how the data logger is configured, started and stopped, whilst Figures 12.20 and 12.21 show typical examples of captured data.

Icing flow tunnel

A college department is engaged in research into the effectiveness of various methods of aircraft deicing based on the application of anti-icing fluids. The department has a wind tunnel capable of speeds of up to 80 m/s (Mach 0.28) supplied by a fan driven by a 10 HP variable speed DC motor. The test section can be adjusted through a pitch angle of ±20° and instrumentation can be attached to parts and components mounted in this section. The moving air stream is cooled by means of a refrigerated cooling unit such that airflow temperatures of between −18°C and ambient can be produced.

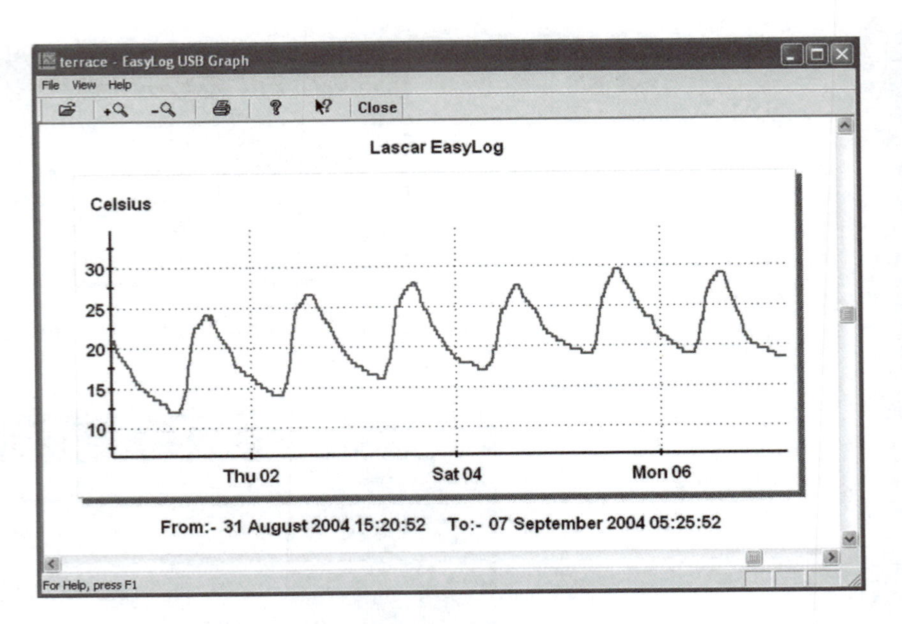

Figure 12.20 *EasyLog display of captured data*

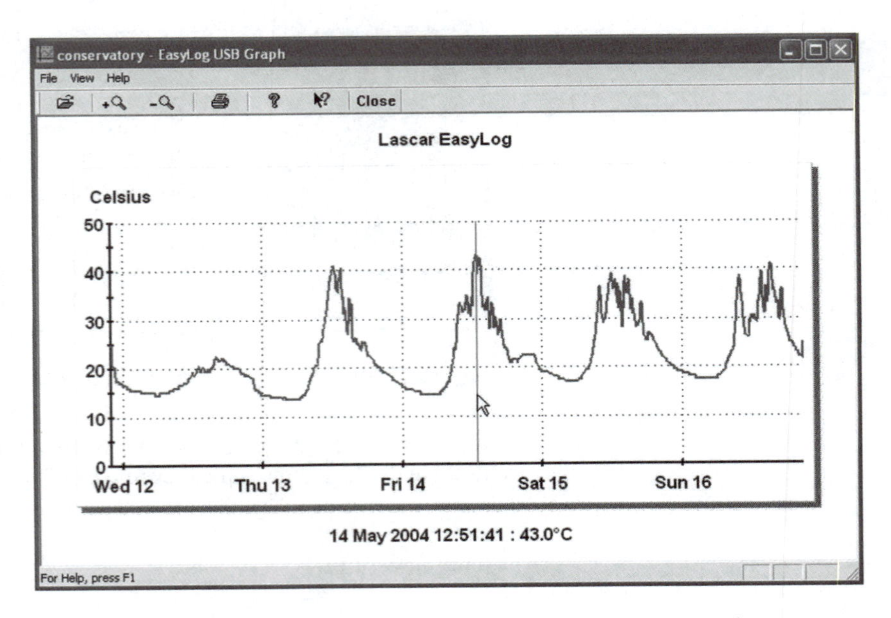

Figure 12.21 *EasyLog's cursors can be moved to obtain precise data*

Specification

The system must provide control for the variable speed fan motor (to an accuracy of ± 1 m/s), air temperature (to an accuracy of $\pm 1°C$), and pitch angle (to an accuracy of $\pm 2°$). The system is to provide a graphical display of the controlled variables with digital readout of the controlled variables. In addition, an on/off

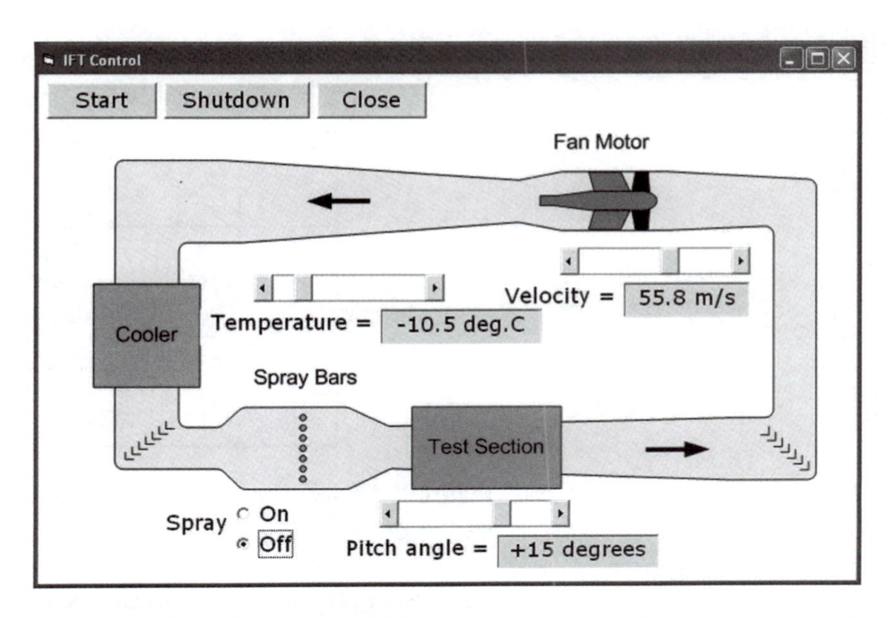

Figure 12.22 *Ice flow tunnel (IFT) control application (the form was created using MS Visio and Visual Basic)*

spray control is to be provided. The system is to be controlled from a low-cost dedicated PC controller. The test component is to be fitted with temperature sensors so that differential readings are available for display. The overall range of differential measurement is to be from 0°C to 20°C with a resolution of better than 0.5% of reading and an accuracy of better than ±0.5°C.

Software

Two Visual Basic 6 applications are used concurrently in this application. Visual Basic 6 was chosen for the software development because of the ease of creating visual controls and because the language was already being used extensively within the department. One of the Visual Basic applications provides control for the ice flow tunnel (IFT control) whilst the other (TDC Control) is responsible for collecting data from the AD590 differential sensing arrangement and then storing this for later analysis. The graphical displays were first produced using MS Visio and then imported into the Visual Basic forms. The Visual Basic controls were then superimposed.

The IFT control application provides slider controls for setting the air temperature, velocity, and pitch angle of the component on test (see Figure 12.22).

Hardware

The PC is fitted with an ISA I/O card which has eight analogue inputs and two analogue outputs. A further ISA I/O card provides 48 digital I/O lines arranged in six groups of eight.

The variable speed drive (VFD) for the 10 HP fan motor requires an input of 10 V DC for frequency adjustment (over the range 0.1–400 Hz) and an airflow

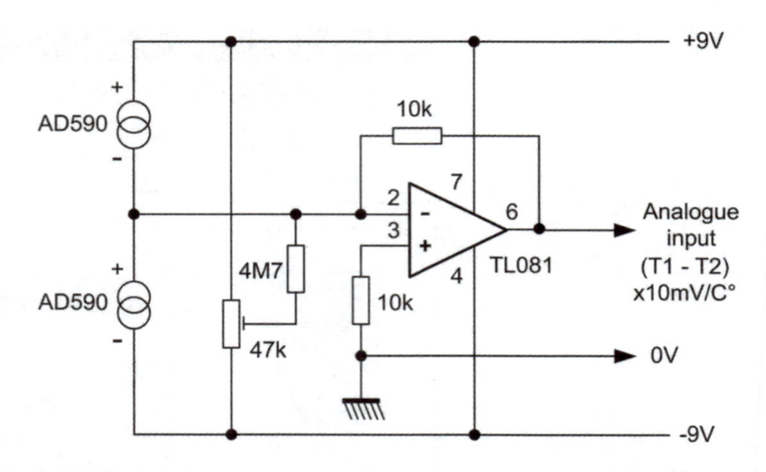

Figure 12.23 *Arrangement of AD590 temperature sensors for differential measurements*

sensor is used to determine the actual air velocity produced. The sensor produces an output of 0–10 V. The speed control thus requires an analogue input port and an analogue output port. Both ports are to a full-scale range of 0–10 V.

The refrigeration unit is controlled with on/off digital control using a dedicated I/O card and a comparator controlled from a DC 0 to 5 V control signal. A temperature sensor is suspended in the airflow output from the refrigeration unit and a signal (10 mV/°C) is fed to a further analogue input port.

The pitch angle control uses a digital output port with a stepper motor (see Chapter 9) and the spray bar control uses a single bit on a further digital output port to provide simple on/off control.

Temperature sensing within the test section is based on a differential sensing arrangement with pairs of AD590 temperature sensors (see Chapter 9). The AD590 is well suited to this application as it offers excellent linearity (better than ±3°C over the entire range) and the ability to operate well in remote sensing applications with simple twisted-pair connections. Lead wire compensation filters and circuits to ensure linearity are unnecessary with this type of sensor.

The output voltage from the differential sensing arrangement (see Figure 12.23) is 10 mV for every 1°C difference in temperature. Hence an output of 100 mV will result from a temperature difference of 10°C. Additional signal gain is applied within the analogue input card.

13 Reliability and fault-finding

The principal goal of the designer of an instrumentation or process control system is that of optimizing system performance within the constraints imposed by time and a given budget. At the same time, he or she will not wish to compromise the overall quality or reliability of the system. This final chapter deals with quality and reliability in the context of PC-based instrumentation and process control systems, and also sets out to examine some basic fault-finding and troubleshooting techniques which can be instrumental in reducing system down-time.

Quality procedures

In a general engineering context, quality is often defined as the degree to which a product or its components conform to the standards specified by the designer. Such standards generally relate to identifiable characteristics relating to materials, dimensions, tolerances, performance, and reliability. In a production engineering environment, the degree of effectiveness in meeting these standards can be assessed by conventional acceptance tests, sampling, and statistical analysis. In the case of a one-off process control system, quality control procedures will generally involve the following tests:

- Functional tests under normal (or simulated normal) operating conditions.
- Functional tests under extreme (or simulated extreme) operating conditions.
- Overload tests to determine the behaviour of the system under abnormal or totally unexpected operating conditions.
- Environmental testing to determine the performance of the system under various extreme conditions of humidity, temperature, vibration, etc.

The instrumentation and process control specialist must inevitably undertake some or all of the functions performed by the quality engineer in a production environment. Not only will he be involved with specifying, designing, building and installing a system but he must also ensure that the overall quality of the system is assured and that the system meets the standard and criteria laid down in the initial specification. The quality assurance function requires an ongoing involvement with the project from design to subsequent installation and use.

Reliability and fault-tolerance

Reliability of a process control system is often expressed in terms of its percentage 'up-time'. Thus, a system which is operational for a total of 950 h in a period of 1000 h is said to have a 95% up-time. An alternative method of expressing reliability involves quoting a mean time before failure (MTBF). The MTBF is equivalent to the estimated number of hours that a system is expected to operate before it encounters a failure requiring a period of 'down-time'.

Various techniques can be used to make PC-based instrumentation and process control systems inherently fault-tolerant. Such techniques can be classified under the general categories of 'hardware' and 'software'. We shall discuss these techniques separately.

Hardware techniques

Hardware methods generally involve the use of a 'watchdog methods'. These are based upon hardware devices for monitoring the performance of the system. Typical techniques include:

- Configuring external hardware such that it generates a status byte which is periodically read (typically every 2 to 10s) by the control program in order to ascertain the state of the system. If the status byte is not read within a pre-determined period, the PC controller assumes that a fault condition has been encountered and then takes appropriate action (such as generating an error message, sounding an alarm, or invoking redundant backup hardware). Watchdog techniques can be useful in overcoming a system 'hang' which may occur when the PC fails to access a malfunctioning item of peripheral hardware.
- Monitoring a power rail and generating appropriate signals when the voltage present fails to meet the defined tolerance limits for the rail concerned. Typical actions involve closing down the system in an orderly fashion or invoking the changeover to a backup supply.
- Fitting an uninterruptable power supply to the PC and important items of peripheral hardware.
- Using a backup control system and, where necessary, duplicating critical I/O circuitry attached to independent signal-conditioning boards.

Software techniques

Software techniques generally involve incorporating software routines, procedures, or functions which will:

1. Perform full system diagnostic tests during initialization.
2. Perform periodic diagnostic tests during program execution (e.g. periodically reading a status byte).
3. Ensure that out of range indications are recognized and erroneous data is ignored.
4. Generate error and warning messages to alert the user to the presence of a malfunction.
5. Log faults as they occur together, where possible, with sufficient information (including date and time) so that the user can determine the point at which the fault occurred and the circumstances prevailing at the time.

The resident system software invariably incorporates simple diagnostic routines of the type mentioned in (1). These routines check the major hardware components within the PC (including ROM and RAM) and are described in further detail later. Where necessary (particularly when a system is in constant operation) it may be desirable to make further checks of the system available as a menu option. The necessary routines are quite straightforward.

As an example, a ROM checksum can be produced simply by reading each byte in turn, adding the values returned (ignoring any overflow) and comparing the result with the known checksum for the ROM. Any difference will indicate a ROM error and appropriate action can be taken. In the case of the RAM, a somewhat different technique is employed. Here the process involves writing and reading each byte of RAM in turn, checking, in each case, that the desired change has been effected. Where a particular bit refuses to be changed, the diagnostic procedure is temporarily halted and an appropriate error message is generated (this may also provide sufficient information for it to be possible to locate the individual device which has failed).

It is, of course, desirable that RAM diagnostics can also be carried out on a non-destructive basis. In such cases, the byte read from RAM is replaced immediately after each byte has been tested. It is thus possible to perform a major RAM diagnostic routine without destroying data stored in read/write memory.

The Power On Self Test (POST)

The Power On Self Test (POST) checks the hardware system during initialization and performs the following checks:

- System motherboard
- Memory
- Keyboard
- Drives

If the Power On Self Test fails, the normal operating system boot sequence is halted and an error message is displayed. The error message varies according to the BIOS type and reference should be made to the BIOS manufacturer's data in order to determine the appropriate course of action.

Once the system is booted (either into DOS or Windows) it is a relatively easy matter to determine the hardware configuration using simple diagnostic software (see Figure 13.1) or using in-built utilities (see page 323). However, in order to make changes to the low-level system configuration it is necessary to make use of facilities that are available from within the BIOS (as described in the next section).

System BIOS

The Basic Input and Output System (BIOS) is a program stored in a read-only memory (ROM) chip on the motherboard. When a computer is first powered-up the BIOS program performs a number of functions including performing the Power On Self Test (POST) and loading the operating system. The BIOS assists with the management of PC hardware via a set of BIOS run-time service routines.

In order to configure a PC's settings, a BIOS setup program is provided in order to optimize and configure the system. Various settings and options are provided, including:

- adding additional floppy or hard drives
- changing a systems boot sequence (e.g., allowing a system to check for a boot CD before booting from a hard drive)

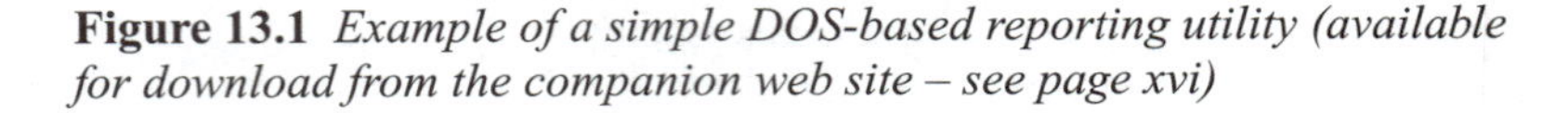

Figure 13.1 *Example of a simple DOS-based reporting utility (available for download from the companion web site – see page xvi)*

- changing the system's date and time
- enabling special features to enhance memory and read/write performance
- setting a BIOS password.

Part of the system's hardware configuration is saved in a small area of Complementary Metal Oxide Semiconductor (CMOS) memory. This memory comprises 64 bytes of battery-backed read/write memory that contains, amongst other things, settings for the PC's system clock, information on memory speed, whether the CPU cache is enabled or disabled, and how fast the PCI bus communicates with adaptor cards. The data contained in CMOS memory will become lost if the CMOS battery fails but the settings can be reinstated by re-entering data using the BIOS Setup program. Sometimes it may be necessary

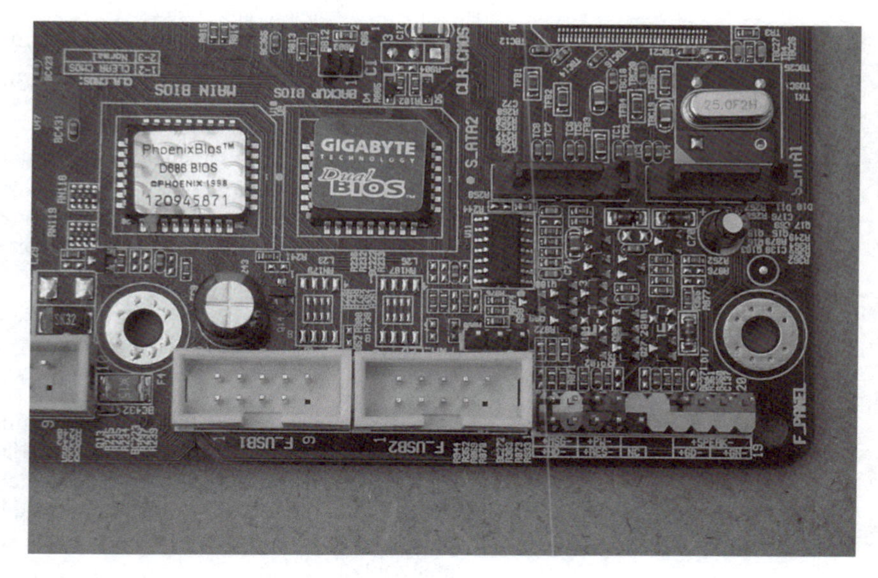

Photo 13.1 *Dual BIOS ROM (main and backup) fitted to a modern PC*

Photo 13.2 *CMOS batteries may need replacing if CMOS errors are reported by the BIOS during system initialization*

to clear the data stored in CMOS memory. This may be required when making a flash upgrade to the BIOS data or when the BIOS password is forgotten (or simply not passed on from one owner to another).

When the BIOS POST fails it will generate a beep code that motherboard will send to the system speaker. Beep codes can be invaluable when a system refuses to boot. Indeed, this will be the *only* information that you have to work on unless you have access to a hardware diagnostic aid!

Two of the most common BIOS POST error beep codes are as follows:

- one long and three short beeps – video fault
- continuous beeps – memory fault.

If the system produces no display and no sound (i.e. no beep) the fault is likely to be CPU or power supply related.

The BIOS date is shown during the first screen when power is applied to the motherboard and is normally displayed as part of the BIOS copyright message, for example:

```
Award Modular BIOS v4.51PG, An Energy Star Ally
Copyright (C) 1984-99, Award Software Inc.
03/08/1999 For SIS530 PCI/AGP 3D VGA Chipset
```

This message indicates that we are dealing with an Award BIOS dated '03/08/1999' designed for the SIS PCI/AFP 3D VGA chipset. The BIOS version is '4.51PG'. Further messages will follow relating to BIOS extensions that may be present. Note that BIOS dates are invariably displayed in [month/day/year] format. Thus the BIOS date in the previous example is 8th March 1999 *not* 3rd August 1999.

If you are unable to read the BIOS information from the screen at power-up or you would prefer to access the BIOS version and date from within Windows, you can use various diagnostic tools and utilities to collect this information.

It is advisable to keep a backup copy of the CMOS data so that it can be restored in the event of failure of the backup battery or loss/corruption of the information held in the CMOS memory. This can often save time (and guess-work!) if/when you find it necessary to restore the CMOS data the hard way!

BIOS upgrading

In recent years it has become possible to determine BIOS information (and also to upgrade a BIOS where appropriate) using the Internet or by means of a remote dial-up. Several manufacturers now provide this facility as part of their after-sales service and it should be used wherever possible.

Where a manufacturer does not provide a BIOS upgrade service it may still be possible to upgrade the BIOS using one of several Internet BIOS resellers. These companies can also remotely interrogate a PC to determine the current BIOS version and whether an upgrade would be appropriate, or not.

Upgrading a modern flash BIOS is much easier than it used to be when BIOS chips had to be replaced manually. That said, it is important to think very carefully before you decide to upgrade a BIOS. In particular you need to have good reason to upgrade (e.g., because hardware conflicts have occurred or some new hardware has become available since the original BIOS was supplied). If a system is working satisfactorily *without* the benefit of an upgraded BIOS you should leave it that way!

The procedure for performing a flash BIOS upgrade is described below. However, since the procedure can be somewhat complex (and the consequence of a failed upgrade is serious) it is important to be sure that you are confident that you know what is going on before you start! In particular you should always ensure that you have a startup disk, a copy of your CMOS and BIOS settings, and that you save a copy of your original flash BIOS data as you perform the upgrade. Important data should, of course, be regularly backed up in the normal course of events!

The typical steps required to perform a flash BIOS upgrade are as follows:

1. Identify your motherboard model number and BIOS version. Ensure that you have all the information (if necessary use several of the methods described earlier). It is also worth noting down the BIOS setup configuration if you don't already keep a record of it. You can do this by booting the system and pressing the F2 or Delete keys (as appropriate to the system) and then viewing the BIOS setup screens. Exit these screens without saving any of the changes.
2. Connect to the Internet and locate the BIOS manufacturer's web site. Investigate the BIOS upgrade, technical support, or software download sections of the manufacturer's web site. In some cases software will be available for downloading to the PC that will perform an automatic update. If this is not the case you may need to perform the upgrade manually by downloading files and then executing them as directed in the steps that follow. If an automatic update is available you should always follow the manufacturer's instructions to download the required software and start the upgrade (refer to sections 13–19 below).
3. Locate the motherboard and download the latest version of the Phoenix BIOS or Award FLASH.EXE program from the manufacturer's site. You will normally find this in the same section as the BIOS upgrade BIN file. Some manufacturers provide you with a search facility that you may (or may not!) find useful.
4. Locate the most up to date version of the BIOS available for the particular motherboard type. Follow the instructions given to download this from the site.
5. Click on Start and select Control Panel. Next click on Add/Remove Programs and select the Startup Disk tab. Insert a blank formatted 1.44 MB floppy disk into the floppy drive and create a startup disk which you will later use to boot the system.
6. Copy the downloaded Phoenix BIOS or Award flash upgrade program to the newly created boot disk.
7. Copy the downloaded BIOS to the floppy disk and unzip or extract the file. This will create a BIN file with the new BIOS flash data to be written to the PC's flash memory. If a 'disk full' message appears you will need to free up some space on the boot disk by removing some files. The following files, although useful for an emergency startup disk, are not required for the flash upgrade: FDISK, FORMAT, and MSCDEX. However, if you have to do this it is strongly recommended that you have at least one 'full' startup disk in case things go wrong!
8. Check to ensure that all files are in place and have been unzipped or expanded.
9. Check that the system is set to boot from the floppy disk drive. If you suspect that this is not the case you will need to temporarily remove the floppy disk from the drive and reboot the system in order to enter and change the BIOS settings. Use F2 or the Delete key (as appropriate) to interrupt the boot sequence and enter the BIOS setup screen. Once you have done this change

the BIOS settings to select a boot sequence that starts with A: and then C: (rather than the normal C: and then A:).

10 Once the BIOS boot sequence has been changed, insert the BIOS upgrade boot disk into the floppy drive and reboot the system.
11 During the boot sequence press the F5 key in order to display the DOS prompt (i.e. A:>). Then type DIR in order to display the contents of the newly created BIOS upgrade boot disk. Locate the name of the executable flash BIOS upgrade program (it will have an EXE file extension) and the name of the BIOS data file (it will have a BIN file extension).
12 Enter the name of the upgrade program (e.g. FLASH.EXE or AWD-FLASH.EXE). This will prompt you to enter the name of the BIOS upgrade. When you have entered this name the upgrade process will commence (ensure that you enter this correctly as the filename can be case sensitive).
13 Next you will be asked whether you wish to save a copy of the old BIOS. You should answer 'Yes' and follow the instructions given.
14 At this point, you will be asked whether you wish to 'flash' the BIOS. You should answer 'Yes' to complete the upgrade. If successful, you will see a message informing you that the upgrade has been completed without error. If unsuccessful, an error message will be displayed. Do not attempt to continue with an upgrade if such a message appears!
15 Finally, remove the flash upgrade boot disk and restart the computer. If necessary, reset the boot sequence so that the system boots first from C: and then from A:.
16 As the system boots note the new BIOS copyright message. If the new BIOS has been flashed correctly this should display the new BIOS version.
17 Next enter the BIOS setup screen once again by hitting the Delete or F2 key. Then select the option to set the BIOS to its default setting.
18 Restart the system and re-enter the BIOS setup screen one more time. Now enter BIOS settings that you previously noted down or select new settings as required by the system's current hardware configuration.
19 Restart the computer and let it complete the full boot sequence. Check that the system operates as you would expect. If necessary, BIOS settings can be changed to improve the system's performance. In exceptional cases you may find it necessary to revert to the saved BIOS data (using the same procedure as before).

If you encounter problems while updating the new BIOS *do not turn off or remove power from the system* and this may prevent your system from subsequently booting up. Instead, you should repeat the process but if the problem persists, it will be necessary to revert to the original saved BIOS data. You may also find that you have to clear or reset the CMOS data when you perform a flash BIOS upgrade (or if you don't have the password required to enter the BIOS setup program). This task is usually performed by changing the position of a jumper located close to the CMOS battery. It is also worth noting that some motherboards have two different sized versions of BIOS data files. The flash upgrade software will usually report a mismatch in file size by displaying a message such as 'File size does not match'.

It should go without saying that, when saving an original BIOS data file, it is important to use a *different filename* from that of the upgrade BIN data file! I suggest that you name the original data OLD.BIN or OLDBIOS.BIN. It is also worth noting that some browsers may rename the BIOS BIN file with an EXE file extension during the download. If this happens you will need to rename the file with the correct extension (BIN) *before* using the FLASH upgrade program. It can also be important to ensure that memory managers (such as HIMEM.SYS and EMM386.EXE) are not resident when the flash upgrade is running.

Troubleshooting Windows problems

Anyone who has been involved with PC's at anything more than the basic user level will almost certainly have come across the unhelpful (and occasionally totally incomprehensible) error messages that Windows, in all its incarnations, is capable of generating! Windows problems can be arranged into the following main categories:

- Invalid page faults
- General protection faults
- Fatal exceptions
- Protection errors
- Kernel errors
- Dynamic link library (DLL) faults.

At this stage it's worth noting that modern CPUs are designed to detect situations in which an executable program attempts to do something that is nonsensical or 'invalid' in terms of the hardware and software configuration of the system. The most common problems are stack faults, invalid instructions, divide errors (divide by zero), and general protection faults. These can often be caused by malfunctioning or badly constructed code in a program.

Invalid page faults

Invalid page faults can occur for any of the following reasons:

- An unexpected event has occurred in Windows. An invalid page fault error message often indicates that a program improperly attempted to use random access memory (RAM). For example, this error message can occur if a program or a Windows component reads or writes to a memory location that is not allocated to it. When this happens the program can potentially overwrite and corrupt other program code in that area of memory.
- A program has requested data that is not currently in virtual memory, and Windows attempts to retrieve the data from a storage device and load it into RAM. An invalid page fault error message can occur if Windows is unable to locate the data. This is often the case when the virtual memory area has become corrupted for some reason.
- The virtual memory system has become unstable because of insufficient physical memory (RAM).
- The virtual memory system has become unstable because of a insufficient free disk space.
- The virtual memory area has been corrupted by a program.
- A program is attempting to access data that is being modified by another program that is running.

If you are using Windows 95 or Windows 98, you may receive the following error message:

This program has performed an illegal operation and will be shut down. If the problem persists, contact the program vendor.

If you subsequently click on Details, you may receive an error message of the form:

[Program] caused an invalid page fault in module at [location].

This type of error is 'unrecoverable' and hence, after you click OK, the program somewhat unhelpfully shuts down!

Note that if you are using Windows ME (Millennium Edition), you will receive an error message of the form:

[Program] has caused an error in [address]. [Program] will now close.

If you continue to experience this type of error message you should restart the computer. To view the details of the problem you should press ALT+D, or open the Faultlog.txt file in the Windows folder.

To resolve this problem it is important to identify when, and in what situation, the error message *first* occurred. Also, determine if you recently made changes to the computer, for example, if you installed software or changed the hardware configuration. In either case, you should use a clean boot troubleshooting procedure (see later) to help you identify the cause of the error message.

General protection faults

All protection violations that do not cause another exception result in a *general protection fault* (GPF). These can be caused by:

- Exceeding the segment limit when using the CS, DS, ES, FS, or GS segments. This is a very common problem in programs and it is usually caused when a program miscalculates how much memory is required in an allocation.
- Transferring execution to a segment that is not executable (e.g., jumping to a location that contains garbage).
- Writing to a read-only area or to a Code segment.
- Loading a bad value into a Segment Register.
- Using a null pointer. A value of zero (i.e. 0) is defined as a null pointer. When operating in Protected Mode, it is always invalid to use a Segment Register that contains zero.

A *general protection fault* often indicates that there is a problem with the software that you are using or that you need to update a device driver installed on the PC. The Dr. Watson utility (see page 410) can often help you to identify the cause of the error message by taking a snapshot of the system at the point at which the fault occurs.

Because general protection faults can be caused by software or hardware, the first step is to restart the PC computer in Safe mode in order to narrow down the source of the error. Restarting in Safe mode will allow you to check whether the problem is attributable to hardware of results from a fault in a driver or an application program.

Restarting in *Safe mode* (see page 407) allows you to test your computer in a state in which only essential components of Windows are loaded. If you restart your computer in Safe mode and the error message does not occur, the origin is more likely to be a driver or program. If you restart in Safe mode and then test your computer and the error message does occur, the issue is more likely to be hardware or damaged Windows core files.

Safe mode starts Windows with a basic VGA video driver. To determine if the issue you are experiencing is related to your video driver you will need to change to the appropriate VGA driver for testing purposes. Note, however, that if you have removed the Protected Mode drivers in order to isolate conflicts you will have already reverted back to the basic VGA video driver.

When you start Windows in Safe mode the registry is only partially read. Damage to the registry may not therefore be evident when running in Safe mode and you may need to replace the existing registry data file (System.dat) with a recent backup in order to see if this resolves the problem in which case the cause is likely to be a damaged registry data file. The following procedure is required in order to troubleshoot a damaged registry:

1 Boot to a DOS command prompt.
2 Remove the file attributes from the backup of the registry by typing the following DOS command:

```
c:\windows\command\attrib -h -s -r c:\system.1st
```

3 Remove the file attributes from the current registry by typing the following DOS command:

```
c:\windows\command\attrib -h -s -r c:\windows\system.dat
```

4 Rename the registry by typing the following command:

```
ren c:\windows\system.dat *.dax
```

5 Copy the backup file to the current registry by typing the following command:

```
copy c:\system.1st c:\windows\system.dat
```

6 Restart the computer.

Note that the System.1st file is a backup of the registry that was created during the final stage of the *original* Windows Setup. Therefore, the 'Running Windows for the first time' banner is displayed and Windows will finalize its settings as if it is being installed for the first time.

If replacing the System.dat file with the System.1st file resolves the issue, the problem may be related to a damaged Windows registry. Any programs and device drivers that were subsequently installed may require reinstallation to update the new registry. For this reason it is essential to keep all of your original installation disks in a safe place!

If you determine that the problem is not caused by a faulty registry data file you will need to restore the original registry data file. The procedure is as follows:

1 Restart the computer to a command prompt.
2 Type the following commands, pressing ENTER after each command:

```
c:\windows\command\attrib -s -h -r c:\windows\system.dat
copy c:\windows\system.dax c:\windows\system.dat
```

3 Overwrite the existing System.dat file if you are prompted to do so.
4 Restart the computer.
5 If the problem is still unresolved the next stage is that of re-installing the Windows core files. You will need the original installation CD-ROM and you should install Windows in a 'clean' folder. If the new installation resolves the problem this usually indicates that either one or more of your Windows core files has been damaged, or that there is an error in the configuration of your original installation. You can choose to use the new installation of Windows, but you will have to reinstall any application programs so that they are correctly recognized by Windows.
6 If the problem is not resolved with a 'clean' installation, the condition is probably attributable to faulty hardware. In such a case you may need to contact the motherboard manufacturer as well as the manufacturer of any adapter cards that are fitted to the system. If you have access to a similar system that is fault-free, you should, of course, be able to carry out substitution tests.

Fatal exceptions

Fatal exceptions occur in the following situations:

- If access to an illegal instruction has been encountered
- If invalid data or code has been accessed
- If the privilege level of an operation is invalid.

When any of these situations occur, the processor returns an exception to the operating system, which in turn is handled as a fatal exception error message. In many situations, the exception is non-recoverable and you must either shut down or restart the computer, depending on the severity of the error.

Fatal exceptions are likely to be encountered when:

- you attempt to shut down the computer
- you start Windows
- you start an application or other program from within Windows.

In either of these cases, an error message like that shown below will appear:

A fatal exception [code] has occurred at [location].

In order to distinguish the type of fatal exception that has occurred these errors are given codes that are returned by a program. The value of the code represents the enhanced Instruction Pointer to the Code Segment; the 32-bit address is the actual address where the exception occurred.

It is important to appreciate that, whilst Windows does not actually cause these errors, it has the exception-handling routine for that particular processor exception and this, in turn, is what actually displays the error message.

For those with some experience of low-level architecture, the various fatal exception error codes (in hexadecimal) are listed below:

1 *00: Divide fault*
The processor returns this exception when it encounters a divide fault. A divide fault occurs if division by zero is attempted or if the result of the operation does not fit in the destination operand.
2 *02: NMI Interrupt*
Interrupt 2 is reserved for the hardware non-maskable-interrupt condition. No exceptions trap via interrupt 2.
3 *04: Overflow trap*
The overflow trap occurs after an INTO instruction has executed and the 0F bit is set to 1.
4 *05: Bounds check fault*
The BOUND instruction compares the array index with an upper and lower bound. If the index is out of range, then the processor traps to interrupt 05.
5 *06: Invalid Opcode fault*
This error is returned if any one of the following conditions exists:
   - The processor tries to decode a bit pattern that does not correspond to any legal computer instruction.
   - The processor attempts to execute an instruction that contains invalid operands.
   - The processor attempts to execute a protected-mode instruction while running in virtual 8086 mode.
   - The processor tries to execute a LOCK prefix with an instruction that cannot be locked.
6 *07: Coprocessor not available fault*
This error occurs if the computer does not have a math coprocessor and the EM bit of register CR0 is set indicating that Numeric Data Processor emulation is being used. Each time a floating point operation is executed, an interrupt 07 occurs.
   This error also occurs when a math coprocessor is used and a task switch is executed. Interrupt 07 tells the processor that the current state of the coprocessor needs to be saved so that it can be used by another task.
7 *08: Double fault*
Processing an exception sometimes triggers a second exception. In the event that this occurs, the processor will issue a interrupt 08 for a double fault.
8 *09: Coprocessor Segment overrun*
This error occurs when a floating point instruction causes a memory access that runs beyond the end of the segment. If the starting address of the floating point operand is outside the segment, then a General Protection Fault occurs (interrupt 0D).
9 *10 (0Ah): Invalid Task State Segment fault*
Because the Task State Segment contains a number of descriptors, any number of conditions can cause exception 0A. Typically, the processor can gather

enough information from the Task State Segment to issue another fault pointing to the actual problem.

10 *11 (0Bh): Not Present fault*
The Not Present interrupt allows the operating system to implement virtual memory through the segmentation mechanism. When a segment is marked as 'not present', the segment is swapped out to disk. The interrupt 0B fault is triggered when an application needs access to the segment.

11 *12 (0Ch): Stack fault*
Stack fault occurs with error code 0 if an instruction refers to memory beyond the limit of the stack segment. If the operating system supports expand-down segments, increasing the size of the stack should alleviate this problem. Loading the Stack Segment with invalid descriptors will result in a general protection fault.

12 *13 (0Dh): General protection fault*
Any condition that is not covered by any of the other processor exceptions will result in a general protection fault. The exception indicates that this program has been corrupted in memory, usually resulting in immediate termination of the program.

13 *14 (0Eh): Page fault*
The Page fault interrupt allows the operating system to implement virtual memory on a demand-paged basis. An interrupt 14 usually is issued when an access to a page directory entry or page table with the present bit set to 0 (not present) occurs. The operating system makes the page present (usually retrieves the page from virtual memory) and re-issues the faulting instruction, which then can access the segment. A page fault also occurs when a paging protection rule is violated (when the retrieve fails, or data retrieved is invalid, or the code that issued the fault broke the protection rule for the processor). In these cases the operating system takes over for the appropriate action.

14 *16 (10h): Coprocessor Error fault*
This interrupt occurs when an unmasked floating-point exception has signalled a previous instruction. (Because the 80386 does not have access to the floating point unit, it checks the ERROR pin to test for this condition.) This is also triggered by a WAIT instruction if the Emulate math coprocessor bit at CR0 is set.

15 *17 (11h): Alignment Check fault*
This interrupt is only used on the 80486 CPUs. An interrupt 17 is issued when code executing at ring privilege 3 attempts to access a word operand that is not on an even-address boundary, a double-word operand that is not divisible by four, or a long real or temp real whose address is not divisible by eight. Alignment checking is disabled when the CPU is first powered up and is only enabled in protected mode.

Because there are many conditions that can cause a fatal exception error, the first step in resolving the issue is to narrow the focus by using the clean boot procedure described earlier. It is also worth noting that many problems occur because of conflicting drivers, terminate-and-stay-resident programs (TSRs), and other settings that are loaded when the computer first starts.

Protection errors

Windows protection error message occurs when a computer attempts to load or unload a virtual device driver (VxD). This error message is a way to let you know that there is a problem with the device driver. In many cases, the VxD that did not load or unload is mentioned in the error message. In other cases, you may not be able to determine the VxD that caused the behaviour; however, you should be able to find the cause of the error message if you use clean boot troubleshooting.

Windows Protection error messages can occur in any of the following situations:

- If a real-mode driver and a protected-mode driver are in conflict.
- If the registry is damaged.
- If either or both the Win.com file or the Command.com file are infected with a virus, or if either of the files has become corrupted or damaged.
- If a protected-mode driver is loaded from the System.ini file and the driver is already initialized.
- If there is a physical input/output (I/O) address conflict or a random access memory (RAM) address conflict.
- If there are incorrect Complementary Metal Oxide Semiconductor (CMOS) settings for a built-in peripheral device (such as cache settings, CPU timing, hard disks, and so on).
- If the Plug and Play feature of the Basic Input/Output System (BIOS) on the computer is not working correctly.
- If the computer contains a malfunctioning cache or malfunctioning memory.
- If the motherboard on the computer is not working properly.

When you start Windows, you may receive one of the following error messages:

```
While initializing device [device name] Windows Protection Error
```

or, the even more succinct (and somewhat less helpful) message:

```
Windows Protection Error
```

The following procedure is recommended when investigating Windows Protection errors:

1. First enter Safe mode, as follows:
   - For Windows 95, restart your computer, press F8 when you see the 'Starting Windows 95' message, and then choose Safe Mode.
   - For Windows 98 (and Windows 98 Second Edition), restart the computer, press and hold down the CTRL key until you see the Windows 98 Startup menu, and then choose Safe Mode.
   - For Windows Millennium Edition (ME), press and hold down the CTRL key while you restart the computer, and then choose Safe Mode on the Windows ME Startup menu.
2. If you do not receive the error message when you start the computer in Safe mode (or when you shut down the computer from Safe mode) you should

follow the described earlier in order to check that the computer is correctly configured and that the system hardware and associated drivers is operating correctly.

3 If you receive the error message when you attempt to start the computer in Safe mode, you follow the steps listed below to restore the registry:
   (a) Boot to a command prompt.
   (b) Remove the file attributes from the backup of the registry by typing the following DOS command:

   ```
   c:\windows\command\attrib -h -s -r c:\system.1st
   ```

   (c) Remove the file attributes from the current registry by typing the following DOS command:

   ```
   c:\windows\command\attrib -h -s -r c:\windows\system.dat
   ```

   (d) Rename the registry by typing the following command:

   ```
   ren c:\windows\system.dat *.dax
   ```

   (e) Copy the backup file to the current registry by typing the following command:

   ```
   copy c:\system.1st c:\windows\system.dat
   ```

4 Restart the computer and verify that the computer's current CMOS settings are correct.
5 Install a 'clean' copy of Windows in an empty folder. If the new installation resolves the problem this usually indicates that either one or more of your Windows core files has been damaged, or that there is an error in the configuration of your original installation. You can choose to use the new installation of Windows, but you will have to reinstall any application programs so that they are correctly recognized by Windows.
6 If the problem is not resolved with a 'clean' installation, the condition is probably attributable to faulty hardware. In such a case you may need to contact the motherboard manufacturer as well as the manufacturer of any adapter cards that are fitted to the system. If you have access to a similar system that is fault-free, you should, of course, be able to carry out substitution tests.

The virtual device driver (VxD) that is generating the error message can be any VxD, either a default VxD that is installed, or a third-party .386 driver that is loaded from the System.ini file. If you do not know which driver is causing the error message, create a Bootlog.txt file, and then check to see which driver is the last driver that is initialized. This is typically the driver that is causing the problem.

Kernel errors

The Kernel32.dll file is a 32-bit dynamic link library (DLL) file that handles memory management, input/output operations, and interrupts. When you start Windows, Kernel32.dll is loaded into a protected memory space. An invalid page fault (IPF) error message will occur when a program tries to access the protected memory space allocated to Kernel32.dll. Occasionally, the error message is caused by one particular program whilst on other occasions it may be generated by several programs.

If the problem results from running one program, the program should be replaced. If the problem occurs when you access multiple files and programs, the damage is likely caused by damaged hardware. You may want to clean boot the computer to help you identify the particular third-party memory-resident software. Note that programs that are not memory-resident can also cause IPF error messages.

The following faults can cause Kernel32.dll error messages:

- Damaged swap file
- File allocation damage
- Damaged password list
- Damaged or incorrect version of the Kernel32.dll file
- Damaged registry
- Hardware, hot CPU, over-clocking, faulty broken power supply, RF noise, or a defective hard disk controller
- BIOS settings for Wait states, RAM timing, or other BIOS settings
- Third-party software that is damaged or incorrectly installed .dll files that are saved to the desktop
- A non-existent or damaged Temp folder
- A corrupted Control Panel (.cpl) file
- Incorrect or damaged hardware driver
- Incorrectly installed printer drivers (or HP Jetadmin drivers)
- Damaged Java Machine
- Damaged .log files
- Damaged entries in the History folder
- Incompatible or damaged dynamic link library files
- Viruses
- Damaged or incorrect Msinfo32.exe file
- Low disk space.

If you are using Windows 95 or Windows 98, you may receive the following error message:

> This program has performed an illegal operation and will be shut down. If the problem persists, contact the program vendor.

When you click Details, you may receive the following error message:

> [Program] caused an invalid page fault in module at [location]

After you click OK, the program shuts down.

If you are using Windows Millennium Edition (ME), you may receive the following error message:

> [Program] has caused an error in [location].
> [Program] will now close.

To view the details, press ALT+D, or open the Faultlog.txt file in the Windows folder. If you continue experiencing problems, you should try restarting your computer.

Dynamic link library faults

A *dynamic link library* (DLL) file is an executable file that allows programs to share code and other resources necessary to perform particular tasks. Microsoft Windows provides DLL files that contain functions and resources that allow Windows-based programs to operate in the Windows environment.

It is important to be aware that, whilst dynamic link libraries usually have a .DLL extension, they may also have an .EXE or other extension. For example, Shell.dll provides the Object Linking and Embedding (OLE) drag and drop routines that Windows and other programs use whilst Kernel.exe, User.exe, and Gdi.exe are all examples of DLLs with .EXE extensions and they all provide code, data, or routines to programs running under the Windows operating system. In Windows, an installable driver is also a DLL.

DLLs are usually placed in the Windows directory, Windows\System directory or in the directory in which an application resides. If a program is started and one of its DLL files is missing or damaged, you may receive an error message like:

```
Cannot find [filename.dll]
```

If a program is started with an outdated DLL file or mismatched DLL files, the error message

```
Call to undefined dynalink
```

may be displayed. In these situations, the DLL file must be obtained and placed in the proper directory in order for the program to run correctly.

The following procedure can be used to determine the version number, company name or other information about a dynamic link library file:

1. Click Start, point to Find, and then click Files or Folders.
2. In the Name box, type the name of the file you want to find, for example, 'shell32.dll' (but without the quotation marks).
3. Click Local Hard Drives (or the drive letter you want to search) in the Look In box, and then click Find Now.
4. Right-click the file in the list of found files, click Properties, and then click the Version tab.

Using Dr. Watson

The diagnostic tool, Dr. Watson, is supplied as part of the Windows operating system yet rarely is it ever referred to and most Windows users don't know that it exists! If a program fault occurs, Dr. Watson will generate a snapshot of the current software environment which can provide invaluable information of what was happening at the point at which the fault occurred.

To start Dr. Watson, you can either:

1. Click Start, click Run.
2. Enter `drwatson` in the box and then click on OK.

or

1. Click Start, select Programs and Accessories, and then click on System Tools.
2. Click System Information, and then click Dr. Watson on the Tools.

When Dr. Watson is running in the background you will see an additional icon displayed on your taskbar.

You can click the Details button in the error message to view the information that is gathered by Dr. Watson. However, in most cases you will want to have a record of what was happening at the point at which the fault occurred. If this is the case, you can generate a log file by double-clicking the Dr. Watson icon on the taskbar. In either case, Dr. Watson gathers information about the operating system and then a Dr. Watson dialog box is displayed.

The log files produced by Dr. Watson have a .wlg extension and they are stored in the \Windows\Drwatson folder. The log file provides a great deal of useful information including the name of the program that has created the fault, the program that the fault occurred in (not necessarily the same), and the memory address where the fault occurred. It is important to note that Dr. Watson cannot create a snapshot if the program does not respond (i.e. if it hangs).

Dr. Watson collects detailed information about the state of the operating system at the time of a program fault. Dr. Watson then intercepts the software faults, identifies the software that has produced the fault, and then provides a detailed description of the cause. When this feature is enabled, Dr. Watson automatically logs this information.

When Dr. Watson is loaded, click any tab to move out of the text box. The Dr. Watson window closes if you press ENTER. To view the advanced tabs in Dr. Watson, follow these steps:

1 Double-click the Dr. Watson icon.
2 On the View menu, click Advanced View.

The following tabs will then be displayed (see Figure 13.2) providing detailed information about the system:

System — Includes information that you would see on the General tab of System Properties.

Tasks — Includes information about the tasks that were running when the snapshot was taken. This tab also includes information about the program, the version, the manufacturer, the description, the path, the type, and the program that this program is related to (when this information is available) (See Figure 13.3).

Startup — Includes information about the programs that are configured to load during Startup. This tab includes the program name, and information about where the program was loaded from, and the command line that is used to load the program (See Figure 13.4).

Hooks — Provides information about modules that have intercepted (i.e. 'hooked') various aspects of the system. This tab can be used to show the hook type, the application, and the path (See Figure 13.5).

Kernel Drivers — Includes information about where the Kernel-mode drivers are installed, including the name of the driver, the version, the manufacturer, the description, the likely path, information about where the driver is

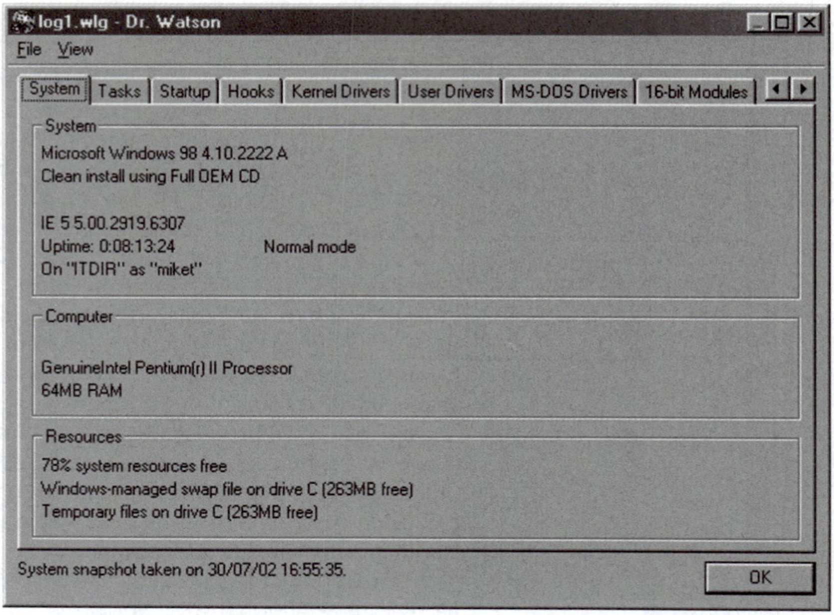

Figure 13.2 *The Dr. Watson dialogue box. Dr. Watson has captured system information in a file named log1.wlg (see top left of window). The System tab displays version data relating to Windows and its installation (in this case, a clean installation using a full OEM CD), the version of Internet Explorer, the current log-in information (user name), the hardware platform (Pentium II processor with 64 MB RAM), and the available resources (78% free, 263 MB free space on the C: drive, etc.)*

loaded from, the type of driver, and the program that the driver related to (when information is available) (See Figure 13.6).

User Drivers — Includes information about the User-mode drivers that are installed, including the name of the driver, the version, the manufacturer, the description, the likely path, the type of driver, and the program that the driver is related to (when information is available) (See Figure 13.7).

MS-DOS Drivers — Includes information about the MS-DOS drivers that are installed (See Figure 13.8).

16-bit Modules — Includes information about the 16-bit modules that were in memory when the snapshot was taken, including the name of the module, the version, the manufacturer, the description, the likely path, the type of driver, and the program that the driver is related to (when information is available) (See Figure 13.9).

Details — Lists the events that occurred before and during the fault, in progressive order. Note that this tab is only displayed when Dr. Watson has captured a fault.

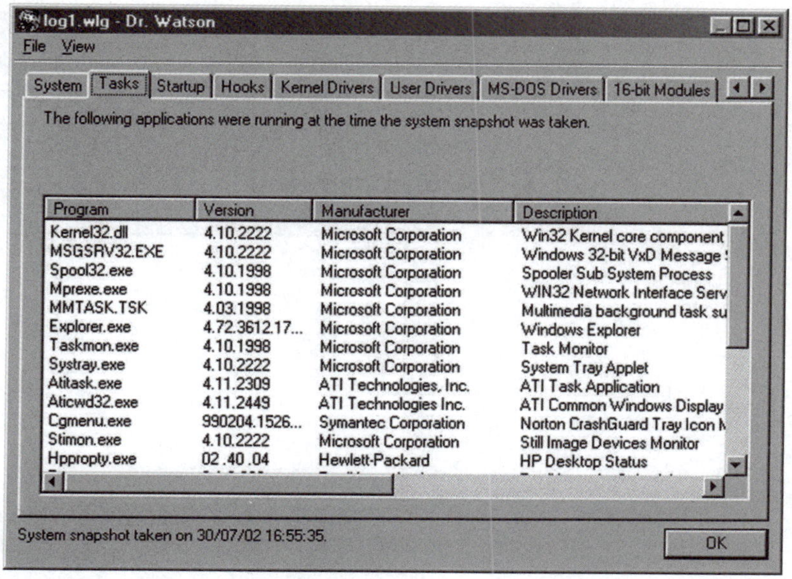

log1.wlg - Dr. Watson

File View

System | Tasks | Startup | Hooks | Kernel Drivers | User Drivers | MS-DOS Drivers | 16-bit Modules

The following applications were running at the time the system snapshot was taken.

Program	Version	Manufacturer	Description
Kernel32.dll	4.10.2222	Microsoft Corporation	Win32 Kernel core component
MSGSRV32.EXE	4.10.2222	Microsoft Corporation	Windows 32-bit VxD Message
Spool32.exe	4.10.1998	Microsoft Corporation	Spooler Sub System Process
Mprexe.exe	4.10.1998	Microsoft Corporation	WIN32 Network Interface Serv
MMTASK.TSK	4.03.1998	Microsoft Corporation	Multimedia background task su
Explorer.exe	4.72.3612.17...	Microsoft Corporation	Windows Explorer
Taskmon.exe	4.10.1998	Microsoft Corporation	Task Monitor
Systray.exe	4.10.2222	Microsoft Corporation	System Tray Applet
Atitask.exe	4.11.2309	ATI Technologies, Inc.	ATI Task Application
Aticwd32.exe	4.11.2449	ATI Technologies Inc.	ATI Common Windows Display
Cgmenu.exe	990204.1526...	Symantec Corporation	Norton CrashGuard Tray Icon
Stimon.exe	4.10.2222	Microsoft Corporation	Still Image Devices Monitor
Hppropty.exe	02.40.04	Hewlett-Packard	HP Desktop Status

System snapshot taken on 30/07/02 16:55:35.

OK

Figure 13.3 *The Tasks tab displays a list of the programs that were running at the point at which the snapshot was taken. This important information shows the filename of the executable as well as its version number, its manufacturer, and a brief description that tells you what it does*

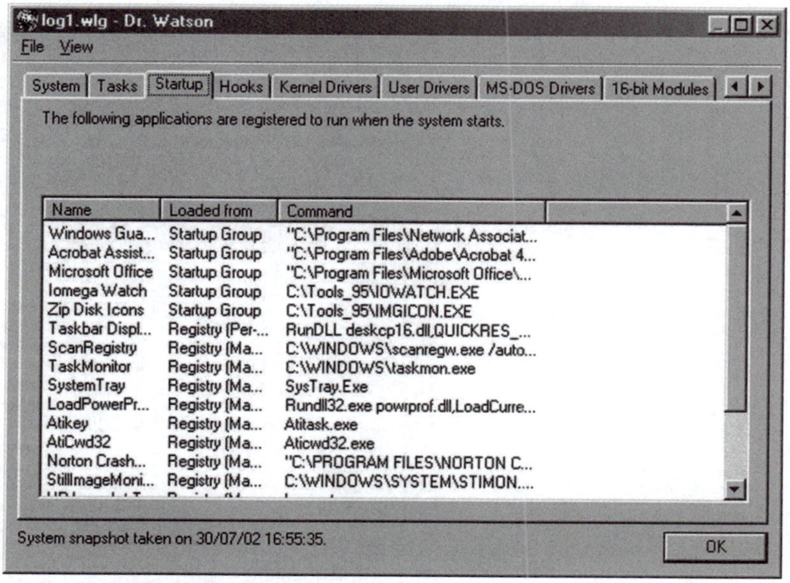

log1.wlg - Dr. Watson

File View

System | Tasks | Startup | Hooks | Kernel Drivers | User Drivers | MS-DOS Drivers | 16-bit Modules

The following applications are registered to run when the system starts.

Name	Loaded from	Command
Windows Gua...	Startup Group	"C:\Program Files\Network Associat...
Acrobat Assist...	Startup Group	"C:\Program Files\Adobe\Acrobat 4...
Microsoft Office	Startup Group	"C:\Program Files\Microsoft Office\...
Iomega Watch	Startup Group	C:\Tools_95\IOWATCH.EXE
Zip Disk Icons	Startup Group	C:\Tools_95\IMGICON.EXE
Taskbar Displ...	Registry (Per-...	RunDLL deskcp16.dll,QUICKRES_...
ScanRegistry	Registry (Ma...	C:\WINDOWS\scanregw.exe /auto...
TaskMonitor	Registry (Ma...	C:\WINDOWS\taskmon.exe
SystemTray	Registry (Ma...	SysTray.Exe
LoadPowerPr...	Registry (Ma...	Rundll32.exe powrprof.dll,LoadCurre...
Atikey	Registry (Ma...	Atitask.exe
AtiCwd32	Registry (Ma...	Aticwd32.exe
Norton Crash...	Registry (Ma...	"C:\PROGRAM FILES\NORTON C...
StillImageMoni...	Registry (Ma...	C:\WINDOWS\SYSTEM\STIMON....

System snapshot taken on 30/07/02 16:55:35.

OK

Figure 13.4 *The Startup tab displays a list of the applications that are registered to run when the system starts. This information indicates whether the program is run from and entry in the Startup group of whether it is from the registry*

log1.wlg - Dr. Watson
File View
System Tasks Startup Hooks Kernel Drivers User Drivers MS-DOS Drivers 16-bit Modules
The following modules have intercepted ("hooked") various aspects of the system.
Hook type Hooked by Application DLL path
Shell Imghook.dll IMGICON.EXE C:\TOOLS_95\Imghook.dll
System snapshot taken on 30/07/02 16:55:35.
OK

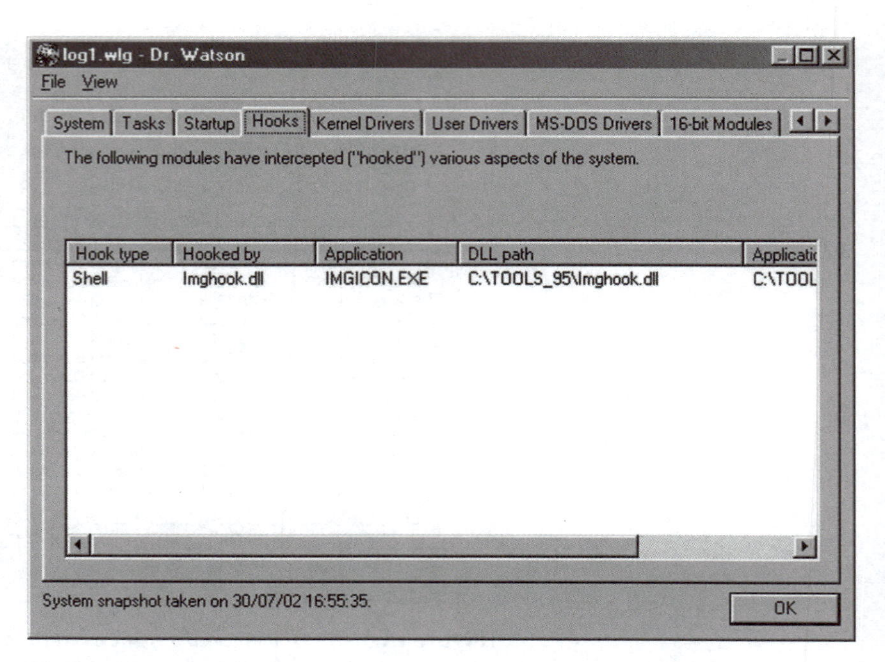

Figure 13.5 *The Hooks tab provides information about modules that have intercepted (i.e. 'hooked') various aspects of the system. In this screen, Dr. Watson is reporting a single hooked application, IMGICON.EXE*

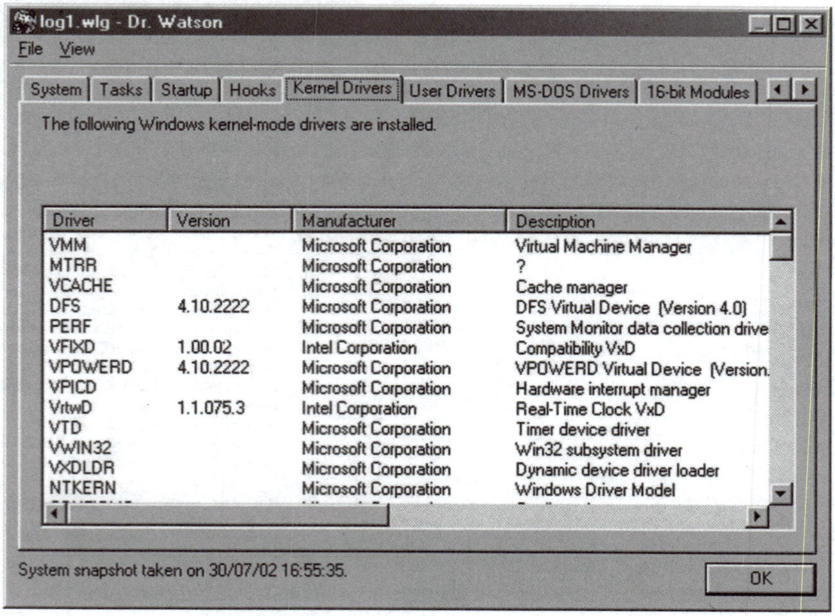

log1.wlg - Dr. Watson
File View
System Tasks Startup Hooks Kernel Drivers User Drivers MS-DOS Drivers 16-bit Modules
The following Windows kernel-mode drivers are installed.
Driver Version Manufacturer Description
VMM Microsoft Corporation Virtual Machine Manager
MTRR Microsoft Corporation ?
VCACHE Microsoft Corporation Cache manager
DFS 4.10.2222 Microsoft Corporation DFS Virtual Device (Version 4.0)
PERF Microsoft Corporation System Monitor data collection drive
VFIXD 1.00.02 Intel Corporation Compatibility VxD
VPOWERD 4.10.2222 Microsoft Corporation VPOWERD Virtual Device (Version
VPICD Microsoft Corporation Hardware interrupt manager
VrtwD 1.1.075.3 Intel Corporation Real-Time Clock VxD
VTD Microsoft Corporation Timer device driver
VWIN32 Microsoft Corporation Win32 subsystem driver
VXDLDR Microsoft Corporation Dynamic device driver loader
NTKERN Microsoft Corporation Windows Driver Model
System snapshot taken on 30/07/02 16:55:35.
OK

Figure 13.6 *The Kernel Drivers tab displays a list of kernel-mode drivers, including their manufacturer and version number*

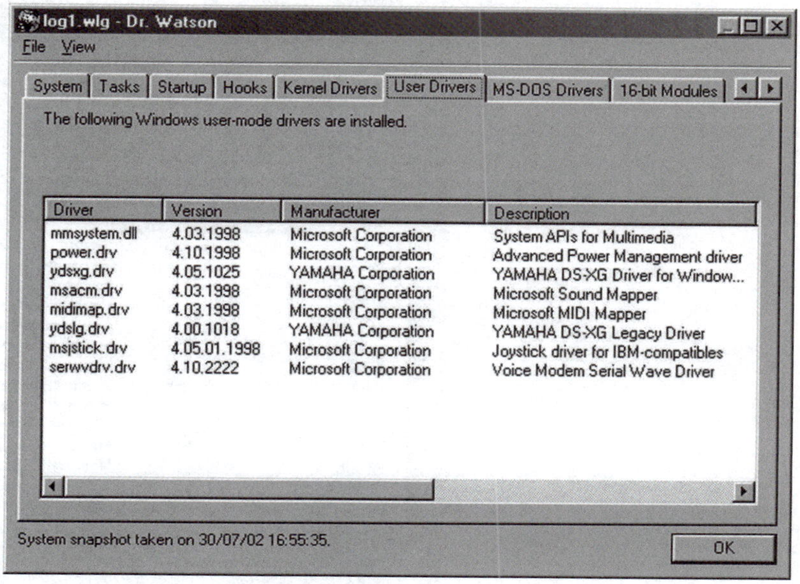

log1.wlg - Dr. Watson

File View

System | Tasks | Startup | Hooks | Kernel Drivers | User Drivers | MS-DOS Drivers | 16-bit Modules

The following Windows user-mode drivers are installed.

Driver	Version	Manufacturer	Description
mmsystem.dll	4.03.1998	Microsoft Corporation	System APIs for Multimedia
power.drv	4.10.1998	Microsoft Corporation	Advanced Power Management driver
ydsxg.drv	4.05.1025	YAMAHA Corporation	YAMAHA DS-XG Driver for Window...
msacm.drv	4.03.1998	Microsoft Corporation	Microsoft Sound Mapper
midimap.drv	4.03.1998	Microsoft Corporation	Microsoft MIDI Mapper
ydslg.drv	4.00.1018	YAMAHA Corporation	YAMAHA DS-XG Legacy Driver
msjstick.drv	4.05.01.1998	Microsoft Corporation	Joystick driver for IBM-compatibles
serwvdrv.drv	4.10.2222	Microsoft Corporation	Voice Modem Serial Wave Driver

System snapshot taken on 30/07/02 16:55:35.

OK

Figure 13.7 *The User drivers tab provides information on current user-drivers. In this screen, Dr. Watson is reporting on the various multimedia driver components. Once again, note the clarity and level of reporting provided by this excellent free tool*

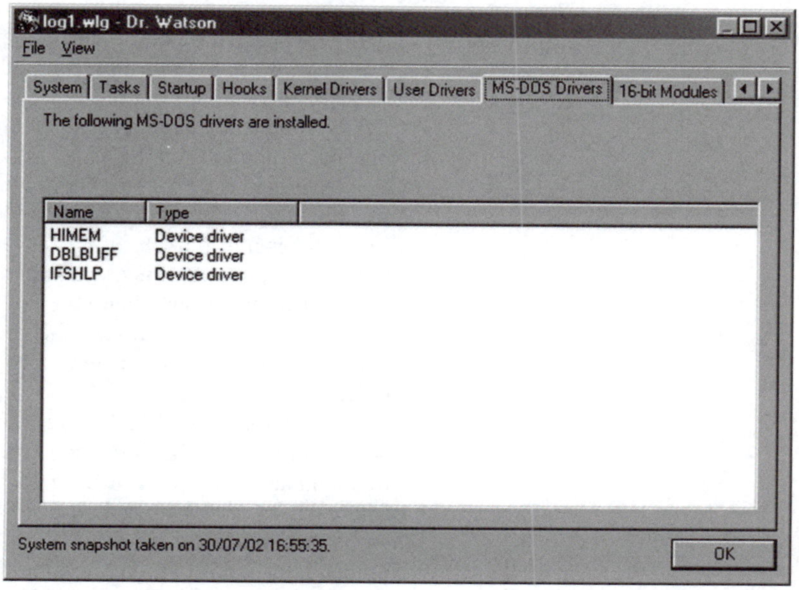

log1.wlg - Dr. Watson

File View

System | Tasks | Startup | Hooks | Kernel Drivers | User Drivers | MS-DOS Drivers | 16-bit Modules

The following MS-DOS drivers are installed.

Name	Type
HIMEM	Device driver
DBLBUFF	Device driver
IFSHLP	Device driver

System snapshot taken on 30/07/02 16:55:35.

OK

Figure 13.8 *The MS-DOS drivers tab reports on any MS-DOS drivers that happen to be present. These drivers are only used by DOS applications and not directly by Windows*

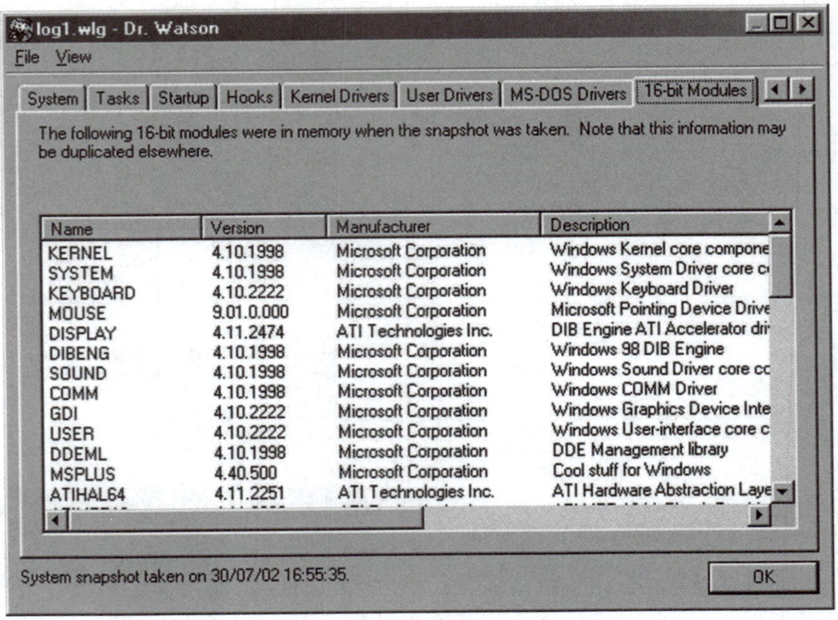

Figure 13.9 *The 16-bit Modules tab provides information on Windows core components and modules such as the display driver*

If you experience a program fault, and you want to use Dr. Watson, follow these steps:

1 Try to reproduce the fault to verify that it is not a random failure.
2 Click Start, point to Programs, point to Accessories, and then click System Tools.
3 Click System Information, and then on the Tools menu, click Dr. Watson.
4 Reproduce the fault.
5 Click Details in the Program Fault window.
6 View the Diagnosis window to determine the source of the fault.
7 If the issue is intermittent or not easy to reproduce, put Dr. Watson in your Startup folder so that it is always running and will be ready to capture the fault information as and when the fault recurs.
8 When the fault next occurs examine the information captured in the log file. To save the information generated by Dr. Watson, click Save on the File menu.

   You may also wish to add a few comments of your own stating under what circumstances the fault occurred. When you have done this, select the File menu and click Save or Save As to save the file. Note that if you only click OK in the Dr. Watson dialog box, the information that you enter in the text box is not saved.
9 You can later view a Dr. Watson log file by using the Dr. Watson program or by using Microsoft System Information (MSInfo). To view Dr. Watson log files by using MSInfo, follow these steps:
   (a) Click Start, point to Programs, point to Accessories, point to System Tools, and then click System Information.

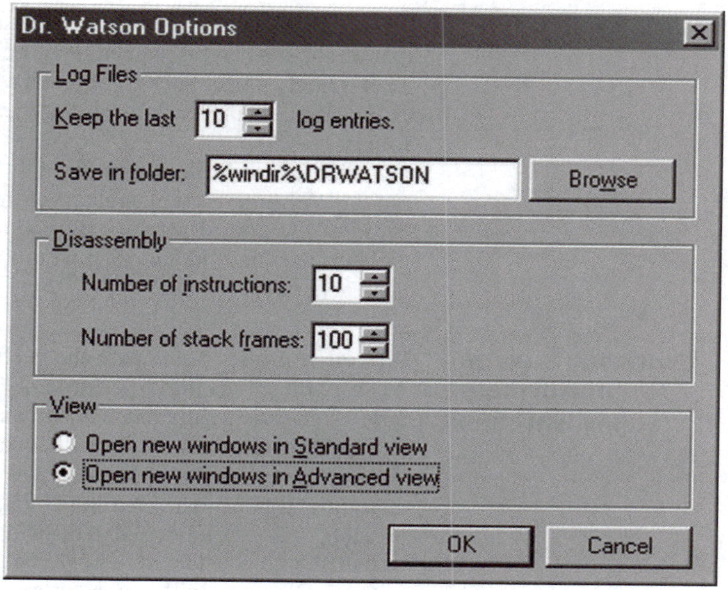

Figure 13.10 *Dr. Watson provides a limited range of configuration options (see text)*

   (b) On the File menu, click Open.
   (c) Open the folder where the Dr. Watson log is saved.
   (d) In the Files of type list, click Dr. Watson Log File (*.wlg).
   (e) Click the file, and then click Open.

10 To print Dr. Watson log files, click on Print from the File menu. To print only specific information, you can use Microsoft System Information to view the log file, and then copy the specific information to an ASCII text editor, such as Microsoft Notepad. (Note that, depending on the software that happens to be running, a typical Dr. Watson log file can amount to more than 15 pages of A4 text!)

Dr. Watson can be configured using the limited number of options available (see Figure 13.10). The procedure for customizing Dr. Watson to your own requirements is as follows:

1 Select the View menu and click Options.
2 Click on Log Files to configure the number of log files that are able to be stored on the computer and the folder that the log files will be saved in.
3 Click on Disassembly to configures the number of CPU instructions and stack frames that are to be reported in the log file.
4 Click on View to configure the view that Dr. Watson is displayed in (either Standard View or Advanced View).

Dr. Watson can be configured so that it loads automatically when Windows starts. To do this, create a shortcut to Drwatson.exe in the Startup folder. This configuration is useful when an issue is not easily reproducible. When Dr. Watson traps the program fault and creates the log, you can contact technical support for further assistance.

The Dr. Watson dialog box includes a text box that you can use to enter information about what was happening when the fault occurred. This information can be extremely useful later – particularly when the same machine next produces errors. By default, Dr. Watson log files are saved to the \Windows\Drwatson folder.

Finally, it is worth noting that Dr. Watson is best used with reproducible faults. With intermittent faults, you may often not be able to determine the cause of the fault in which case you should follow the procedures described earlier depending on the exact nature of the Windows error message that has been generated.

Benchmarking and performance measurement

It is often useful to compare the performance of one PC with another or to measure the comparative performance of a PC over a period of time, particularly when changes are made to software, hardware, and system configuration. Several software packages offer benchmarking checks but one of the best is a suite of programs and utilities known as Fresh Diagnose. Fresh Diagnose can analyze and benchmark the individual parts of a computer system making it possible to detect individual items of hardware that are not configured correctly or that should considered to be prime candidates for upgrading.

Fresh Diagnose will scan a system and produce a comprehensive report on the hardware and software, including information on:

- motherboard type and configuration
- CPU type and clock settings
- video system
- PCI/AGP bus peripheral devices (e.g. keyboard, mouse, and printer)
- network connections.

Fresh Diagnose will also perform a series of tests in order to measure the performance of a system. These tests include:

- CPU performance
- hard disk performance
- CD/DVD ROM performance.

In addition to absolute measurement of performance, Fresh Diagnose can provide a comparison of the current system with others. This information can be invaluable in confirming (or otherwise!) that a PC is performing according to expectation. Fresh Diagnose will operate successfully with systems that use Windows 95, Windows 95 OSR2, Windows 98, Windows 98 Second Edition, Windows ME, Windows NT 4.0, Windows 2000, and Windows XP (Figure 13.11).

System information

Fresh Diagnose incorporates a large number of individual program modules that can be used to provide information on both the system hardware and its software. The modules provide comprehensive information on:

- the operating system version and configuration (Figure 13.12)
- advanced power monitor (APM)
- CMOS settings

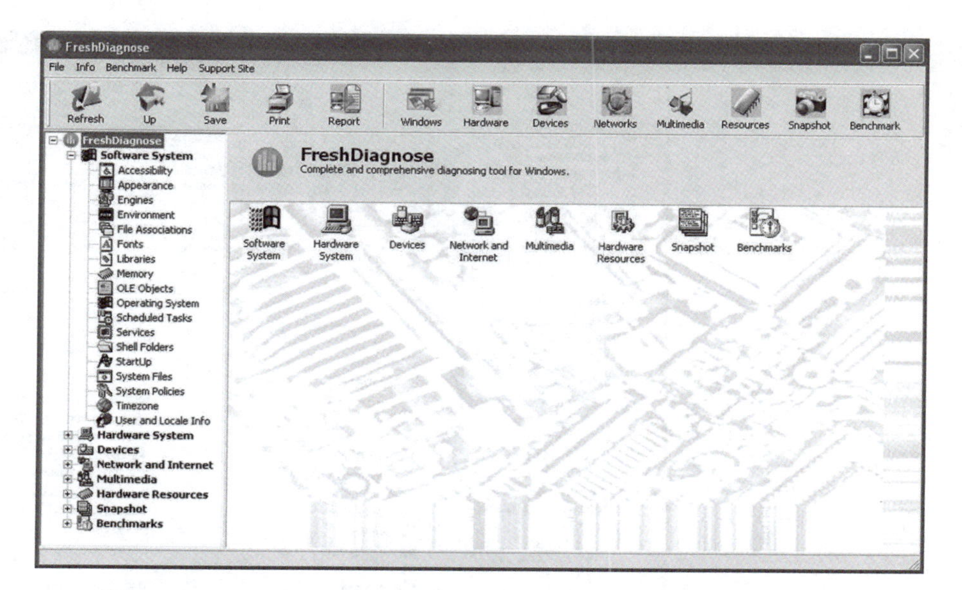

Figure 13.11 *The Fresh Diagnose opening screen showing the eight main options that may be selected by pointing and clicking on the icons. Other options, such as Print and Report, may be selected directly from the tool bar. The window on the left provides a means of selecting individual modules*

FreshDiagnose
File Info Benchmark Help Support Site
Refresh Up Save Print Report Windows Hardware Devices Networks Multimedia Resources

FreshDiagnose
Software System
Accessibility
Appearance
Engines
Environment
File Associations
Fonts
Libraries
Memory
OLE Objects
Operating System
Scheduled Tasks
Services
Shell Folders
StartUp
System Files
System Policies
Timezone
User and Locale Inf
Hardware System
BIOS
Busses
Cache Memory
CMOS
Memory

Memory
This module contains information about operating system memory and the memory manager.

Item	Value
Memory General	
Memory Load	33 %
Total Physical Memory	536,199,168 Bytes (511.36 MB)
Available Physical Memory	355,700,736 Bytes (339.22 MB)
Total Page File	1,311,137,792 Bytes (1.22 GB)
Available Page File	1,029,111,808 Bytes (981.44 MB)
Total Virtual Memory	2,147,352,576 Bytes (2.00 GB)
Available Virtual Memory	2,084,306,944 Bytes (1.94 GB)
Extended Virtual Memory	None
Swap File Size	282,025,984 Bytes (268.96 MB)
Swap File Usage	78 %
Page Size	4,096 Bytes (4 KB)
Allocation Granularity	65,536 Bytes (64 KB)
Min. Application Address	$10000
Max. Application Address	$7FFEFFFF

Figure 13.12 *The Software System Memory report provides comprehensive information on operating system memory usage and on the memory manager. In this example, the total physical memory reported is 511 MB of which 339 MB is currently available to applications*

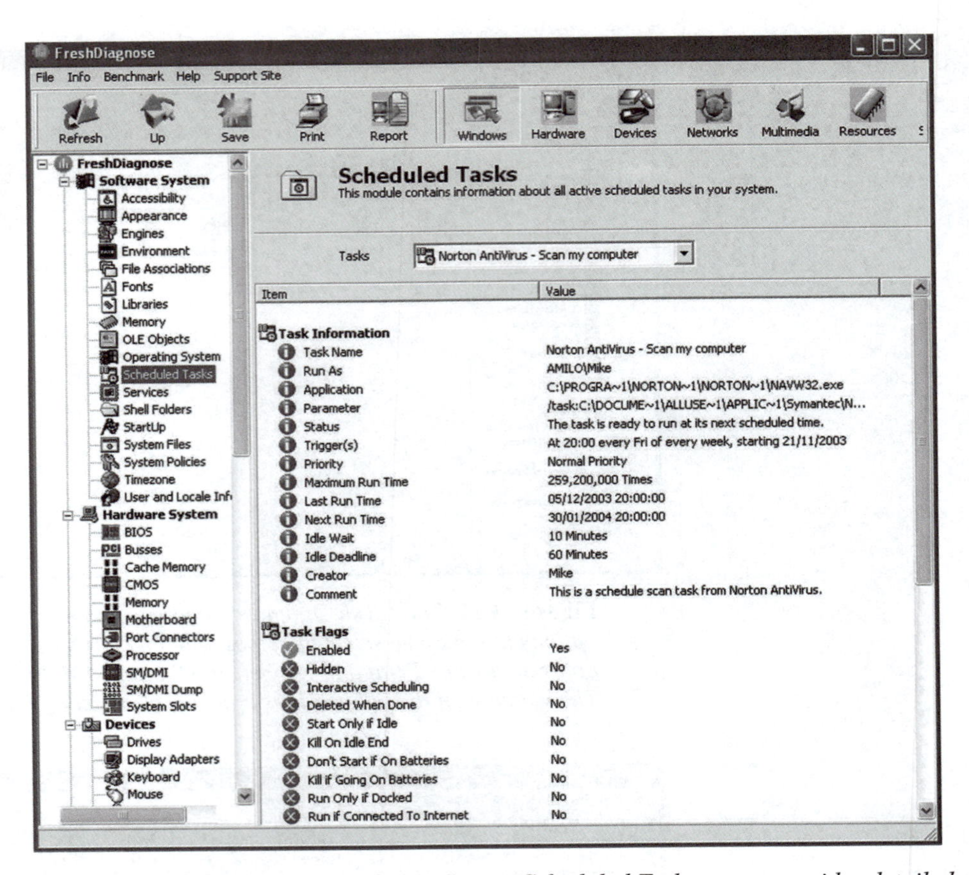

Figure 13.13 *The Software System Scheduled Tasks report provides detailed information on each scheduled task. In this example, the Norton Antivirus task is being reported on. The task is scheduled to run every Friday at 20:00 and it will be next run on 30/01/2004*

- processes, services and media control settings
- user and location (including time zone) settings (Figure 13.13)
- fonts and system files processor and memory resources drives and display adapters
- keyboard, mouse, joystick, and MIDI settings
- ports and port settings network and Internet settings
- games software extensions (DirectX, DirectDraw, and DirectSound)
- interrupt requests (IRQ) and DMA channels (Figure 13.14).

Benchmarking

Fresh Diagnose incorporates six benchmarking modules. These are as follows:

- Processor Benchmark
- Multimedia Benchmark
- Memory Benchmark
- Hard Disk Benchmark
- CD Drive Benchmark
- Network Benchmark.

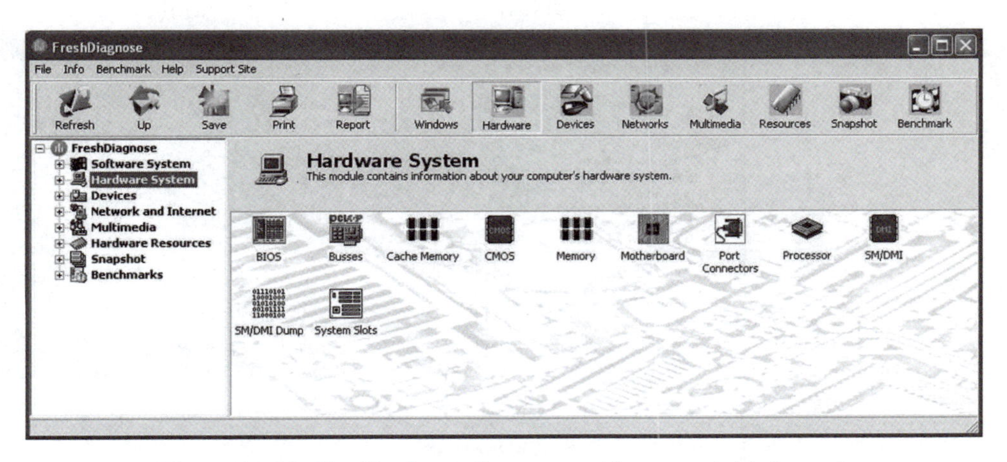

Figure 13.14 *The Hardware System provides essential information on interrupts, direct memory access (DMA) channels, port settings, and the available memory resources*

Processor Benchmark

The Processor Benchmark performs continuous and complex calculations based on the industry standard *Whetstone* and *Dhrystone* algorithms. These provide a measure of the performance of a CPU when carrying out floating point and integer arithmetic operations, respectively.

The Whetstone benchmarking algorithm was created by Harold Curnow in 1972 and optimized for floating point arithmetic. The Dhrystone benchmarking algorithm is the standard for measuring integer performance. This was developed by Reinhold P. Weicker and is similar to the Whetstone algorithm but without floating point arithmetic. As well as producing a speed rating in terms of MHz, Fresh Diagnose produces benchmarks expressed in terms of Millions of Whetstone Instructions Per Second (MWIPS) and Millions of Dhrystone Instructions Per Second (MDIPS).

CPU Multimedia Benchmark

This benchmark performs a set of Intel SSE, SSE2, x87, and AMD 3DNow! instructions including binary and logical operations. When carrying out CPU benchmarking it is important to be aware that the results of measurements will often be different for identical processors operating with different operating systems. This is due to minor differences in the way that individual operating systems support a processor's instruction set. Generally (but not in every case) the later operating systems will yield faster benchmark results (Figure 13.15).

Memory Benchmark

This benchmark performs a set of memory operations (at least 100 KB) including array assignment and splitting. The measurement produces memory speeds

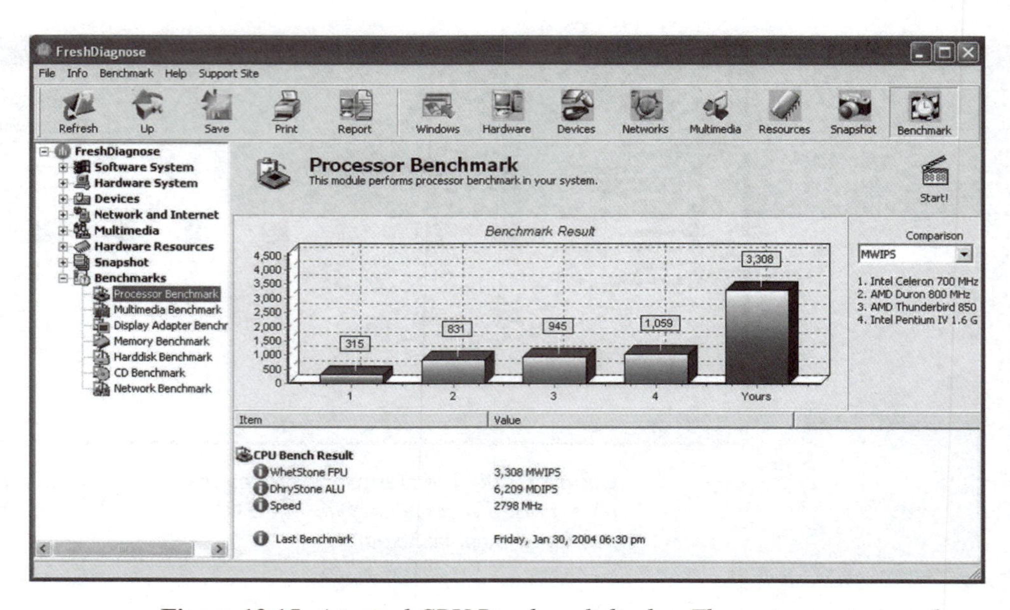

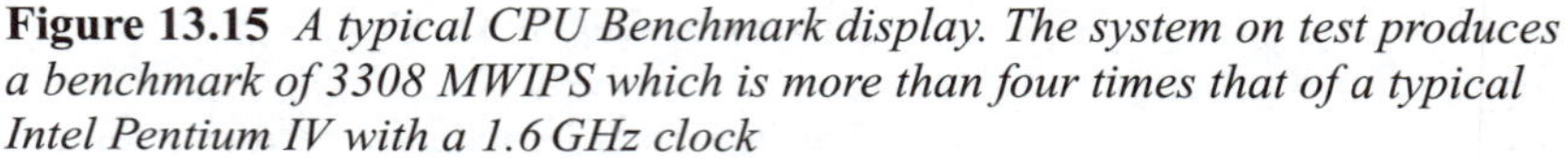

Figure 13.15 *A typical CPU Benchmark display. The system on test produces a benchmark of 3308 MWIPS which is more than four times that of a typical Intel Pentium IV with a 1.6 GHz clock*

for integer array handling (Integer Assignment) and integer splitting (Integer Split) (Figure 13.17).

Hard Disk Benchmark

The Hard Disk Benchmark performs both read and write tests on the hard disk drive. The module creates a temporary file called 'sysinfo.bch' in the root directory and then uses this to perform subsequent read and write tests. The results appear in MB/s (Figure 13.16).

CD Benchmark

The CD Benchmark performs a single read test to the CD drive. In order to perform this test Fresh Diagnose requires the insertion of a CD audio, CD data, VCD, or DVD to use as the basis of the measurement. The media used should be a reliable CD which does not auto-run. Once again, the result is in MB/s.

Network Benchmark

The Network Benchmark performs a ping instruction and both read and write tests to a specified connection. If the selected connection is a read-only one, the measurement will only perform a ping test. The results of read and write tests are in MB/s.

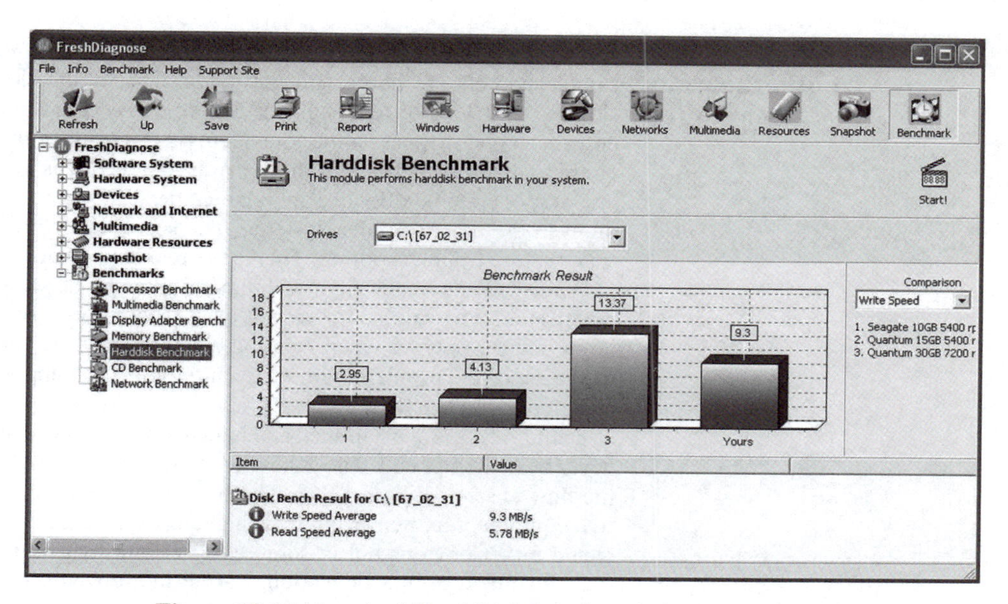

Figure 13.16 *A typical Hard Disk Benchmark display. The system on test produces a write speed of 9.3 MB/s. This is around 40% slower than that of a 30 GB Quantum drive which rotates at 7200 rpm*

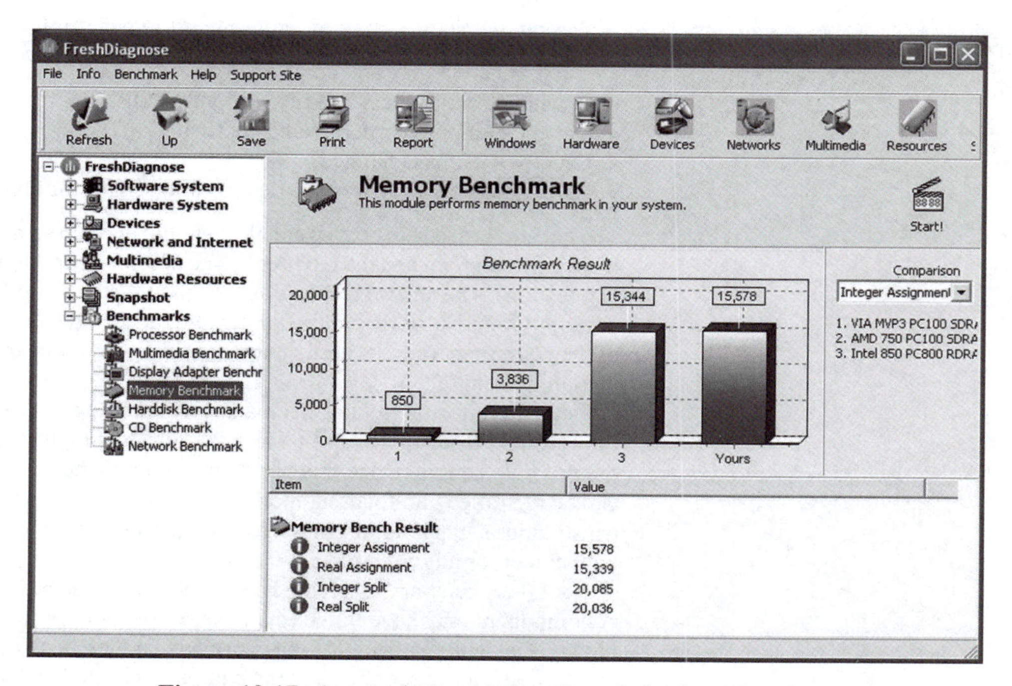

Figure 13.17 *A typical Memory Benchmark display. This shows that the performance of the system on test is virtually identical to that of an Intel 850 PC800 RDRAM*

Fault-finding and troubleshooting techniques

A popular misconception concerning electronic fault finding is that good troubleshooters are borne and not made. The implication of this is that the skills of a service or test engineer cannot be acquired unless the person concerned happens to possess the equivalent (in electronic terms) of 'green fingers'. Nothing could be further from the truth – indeed it is quite possible for anyone of moderate intelligence and manual dexterity to successfully locate faults on even the most complex systems. The secret lies with adopting the correct *approach* to troubleshooting. This is the real key to successful fault finding.

With experience, the right technique will come as second nature. Indeed, a practised service engineer may not even be conscious of the technique which he or she is applying when tackling a fault. They may appear to get right to the cause of the problem without even thinking. By applying a little logic and reasoning, you can do the same.

Fault finding is a disciplined and logical process in which 'trial fixing' should never be contemplated. The generalized process of fault finding is illustrated in the flow chart of Figure 13.18. The first stage is that of identifying the defective equipment and ensuring that the equipment really is defective! This may sound rather obvious but in some cases a fault may simply be attributable to maladjustment or misconnection. Furthermore, where several items of equipment are connected together, it may not be easy to pinpoint the single item of faulty equipment. For example, take the case of a process control system in which the user simply states that there is 'no output'. The fault could be almost anywhere in the system; computer, display, printer, or any one of several connecting cables.

The second stage is that of gathering all relevant information. This process involves asking questions such as:

- In what circumstances did the equipment fail?
- Has the equipment operated correctly before?
- Exactly what has changed?
- Has there been a progressive deterioration in performance?

The questions used are crucial and they should explore all avenues and eventualities (particularly when the repairer has no previous experience of the equipment in question). The answers to the questions will help to build a conceptual model of the symptoms – before and after the fault occurred. Coupled with knowledge of the equipment (e.g. its performance specification) this model can often point to a unique cause.

Once the information has been analysed, the next stage involves separating the 'effects' from the 'cause'. Here the aim is simply that of listing each of the *possible* causes. Once this has been accomplished, the *most probable* case can be identified and focused upon. Corrective action should be applied (to this cause alone). Such action may require component removal and replacement, adjustment, or alignment, etc.

Next it is necessary to decide whether the fault has been correctly identified. A component may have failed (open circuit or short circuit) or a fuse may have blown. This will confirm that the cause has, in fact, been correctly identified. If so, the fault can be rectified and the equipment brought back into service. If not, any new information that has been generated can be evaluated before reverting to the selection of the next most probable cause. In practice, the loop may have to be executed several times until the fault is correctly identified and rectified.

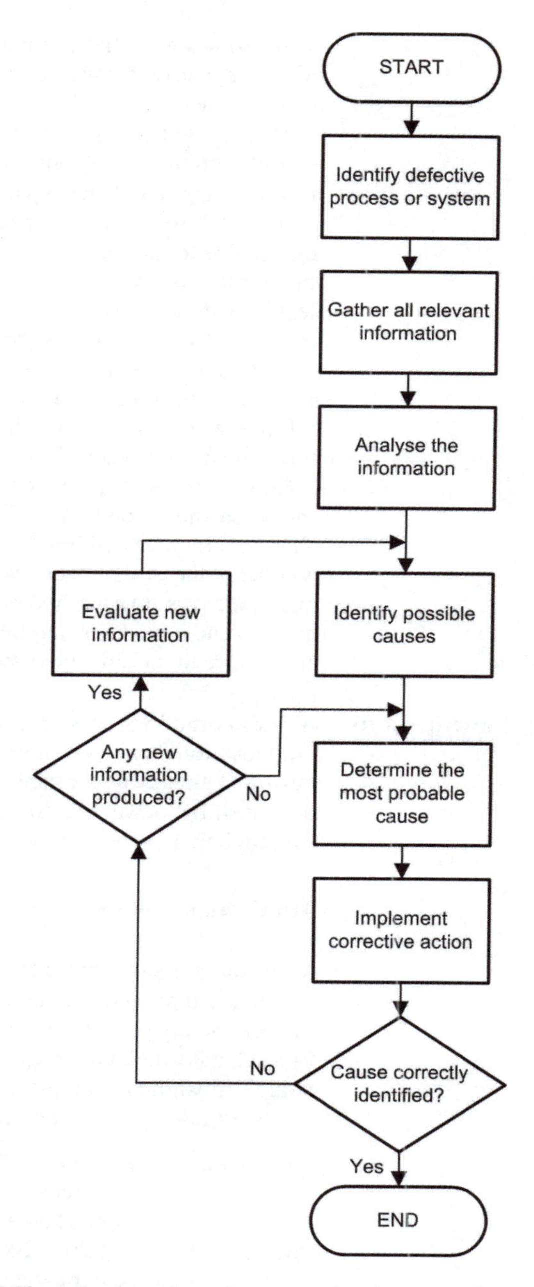

Figure 13.18 *Flow chart to show the generalized approach to troubleshooting (a series of more detailed charts appears later in this chapter)*

Instrumentation and process control specialists will rarely wish to deal with fault-finding down to component level. In order to avoid a prohibitive investment in test equipment and technical expertise, it is generally considered more cost-effective to have such repairs carried out by specialists. Despite this, it is sometimes essential to minimize the time taken to correct the failure of a PC-based instrumentation or process control system. An ability to make on-site repairs, at least to board level, is thus highly desirable.

At first sight, the prospect of fault-finding a PC-based instrumentation or control system can be somewhat daunting. This is especially true when those having to carry out the repairs may be relatively unfamiliar with electronic circuitry. However, in the author's experience, the vast majority of faults are attributable to failure of external devices (such as sensors, cables, and connectors) rather than with the board and cards themselves. Furthermore, even when dealing with boards within the system enclosure, most faults can be detected without recourse to sophisticated test gear.

When component rather than board level servicing has to be undertaken, it is useful to obtain a circuit diagram and service information on the equipment before starting work. This information will be invaluable when identifying components and establishing their function within the system as a whole.

Certain 'stock faults' (such as chip failure) may be prevalent on some boards and these should be known to manufacturers and their service agents. A telephone enquiry, describing the symptoms and clearly stating the type and version number of the card or board, will often save much time and effort. Furthermore, manufacturers are usually very receptive to information which leads to improvement of their products and may also be prepared to offer retrofit components and/or circuit modifications to overcome commonly identified problems.

Test equipment

A few items of basic test gear will be required by anyone attempting to perform fault location on bus systems. None of the basic items is particularly costly and most will already be available in an electronics laboratory or workshop. For the benefit of the newcomer to electronics we will briefly describe each item and explain how it is used in the context of PC-based system fault-finding.

Multi-range meters

Multi-range meters provide either analogue or digital indications of voltage, current, and resistance. Such instruments are usually battery-powered and are thus eminently portable. Connection to the circuit under test is made via a pair of test leads fitted with probes or clips. Controls and adjustments are extremely straightforward and a typical meter layout is shown in Figure 13.19.

The following specification is typical of a modern digital multi-range meter:

DC voltage	200 mV, 2 V, 20 V, 200 V, and 1.5 kV full-scale Accuracy ±0.5% Input resistance 10 MΩ
AC voltage	2 V, 20 V, 200 V, and 1 kV full-scale Accuracy ±2% Input resistance 10 MΩ
DC current	200 μA, 2 mA, 20 mA, 200 mA, and 2 A full-scale Accuracy ±1%
AC current	200 μA, 2 mA, 20 mA, 200 mA, and 2 A full-scale Accuracy ±2%
Resistance	200 Ω, 2 kΩ, 20 kΩ, 200 kΩ, 2 MΩ full-scale Accuracy ±2%

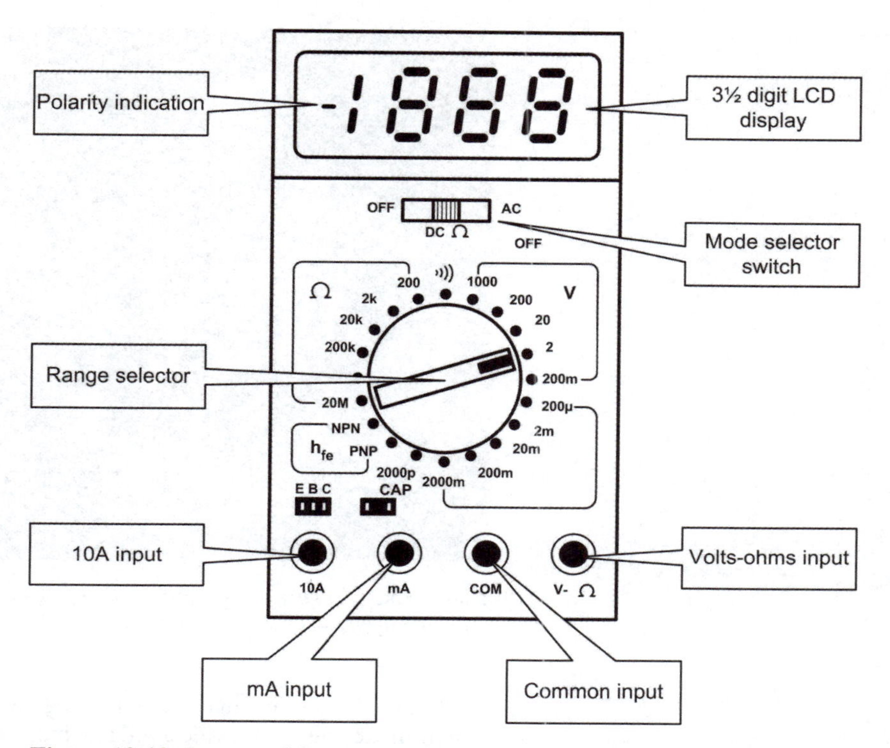

Figure 13.19 *Layout of the controls and adjustments of a typical digital multi-meter*

A typical application for a multi-range meter is that of checking the various supply voltages present within the PC. For an operational system the supply voltages should be within the range given below:

Nominal value (V)	Acceptable value (V)	
	Minimum	Maximum
+5	+4.75	+5.25
−5	−4.75	−5.25
+12	+11.4	+12.6
−12	−11.4	−12.6

Multi-range meters may also be used for checking the voltages present on the supply rails within individual expansion cards. Particular points of interest will be those associated with the supplies to individual chips. In such cases, PC bus extension card frames can be employed in order to gain access to a 'live' expansion card. Alternatively, the expansion card in question can be fitted to the left-most slot within a PC in order to provide access to the printed wiring of the card.

Multi-range meters may even be used to display logic states on signal lines which remain static for long periods. This is often the case when dealing with

Photo 13.3 *A drive power connector makes a convenient test point for measuring the +12 V and +5 V power rails*

I/O lines however, in situations where logic levels are continuously changing, a multi-meter cannot provide a reliable indication of the state of a line.

Where logic levels do remain static for several seconds, the multi-range meter may be used on the DC voltage ranges to sense the presence of logic 0 or 1 states according to the following table which gives the conventional voltage levels associated with TTL logic:

Logic level	Voltage present (V)
1	>2.0
0	<0.8
indeterminate	0.8 to 2.0

It should be noted that an 'indeterminate' logic level may result from a tri-state condition in which bus drivers are simultaneously in a high impedance state. Modern high-impedance instruments will usually produce a misleading fluctuating indication in such circumstances and this can sometimes be confused with an actively pulsing bus line.

Logic probes

The simplest and most convenient method of tracing logic states involves the use of a logic probe rather than a multi-range meter. This invaluable tool comprises a hand-held probe fitted with LEDs to indicate the logical state of its probe tip.

Unlike multi-range meters, logic probes can generally distinguish between lines which are actively pulsing and those which are in a permanently tri-state

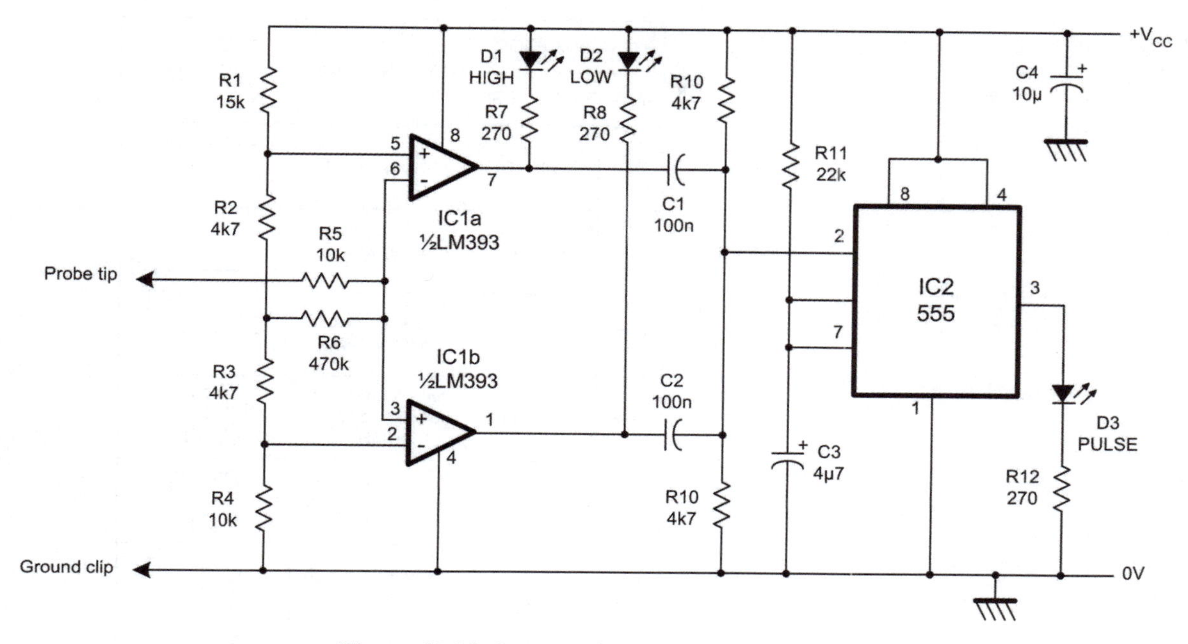

Figure 13.20 *Logic probe circuit*

condition. In the case of a line which is being pulsed, the logic 0 and 1 indicators will both be illuminated (though not necessarily with the same brightness) whereas, in the case of a tri-state line neither indicator should be illuminated.

Logic probes generally also provide a means of displaying pulses having a very short duration which may otherwise go undetected. A pulse stretching circuit is usually incorporated within the probe circuitry so that an input pulse of very short duration is elongated sufficiently to produce a visible indication on a separate pulse LED.

Logic probes invariably derive their power supply from the circuit under test and are connected by means of a short length of twin flex fitted with insulated crocodile clips. While almost any convenient connecting point may be used, the leads of an electrolytic +5 V rail decoupling capacitor fitted to an expansion card make ideal connecting points which can be easily identified.

A typical logic probe circuit is shown in Figure 13.20. This circuit uses a dual comparator to sense the logic 0 and 1 levels and a timer which acts as a monostable pulse stretcher to indicate the presence of a pulse input rather than a continuous logic 0 or 1 condition. Typical logic probe indications and waveforms are shown in Figure 13.21.

Figure 13.22 shows how a logic probe can be used to check a typical combinational logic arrangement. The probe is moved from node to node, and the logic level is displayed and compared with the expected level.

Logic pulsers

It is sometimes necessary to simulate the logic levels generated by a peripheral device or sensor. A permanent logic level can easily be generated by pulling

LED INDICATOR			STATE INDICATED	WAVEFORM
LOW	PULSE	HIGH		
OFF	OFF	ON	Steady logic 1	1 0
ON	OFF	OFF	Steady logic 0	1 0
OFF	OFF	OFF	Open circuit or undefined level	1 0
OFF	BLINK	OFF	Pulse train of near 50% duty cycle at >1MHz	1 0
ON	BLINK	ON	Pulse train of near 50% duty cycle at <1MHz	1 0
OFF	BLINK	ON	Pulse train of high mark:space ratio	1 0
ON	BLINK	OFF	Pulse train of low mark:space ratio	1 0

Figure 13.21 *Logic probe indications and test waveforms*

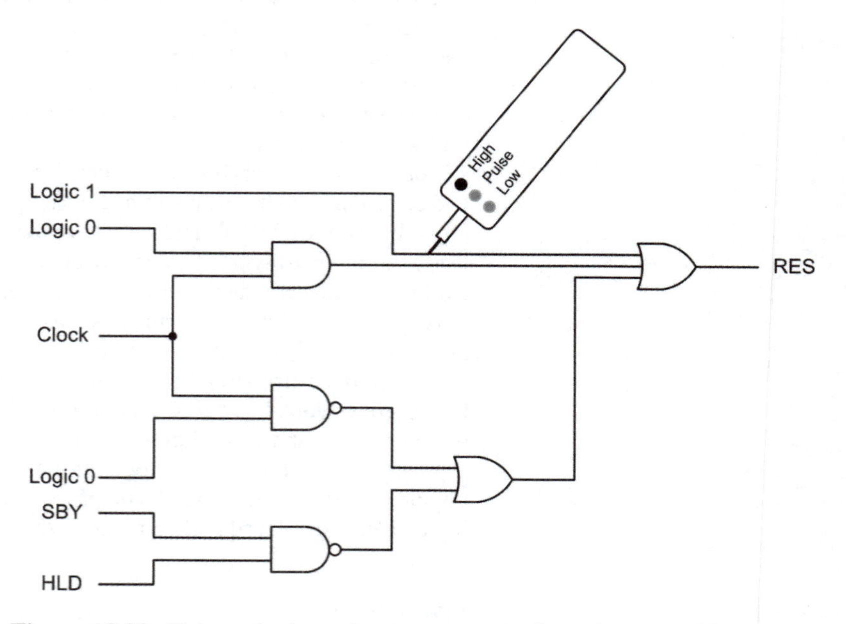

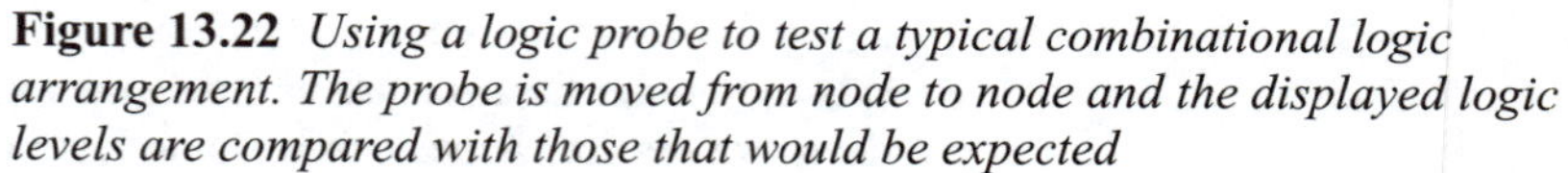

Figure 13.22 *Using a logic probe to test a typical combinational logic arrangement. The probe is moved from node to node and the displayed logic levels are compared with those that would be expected*

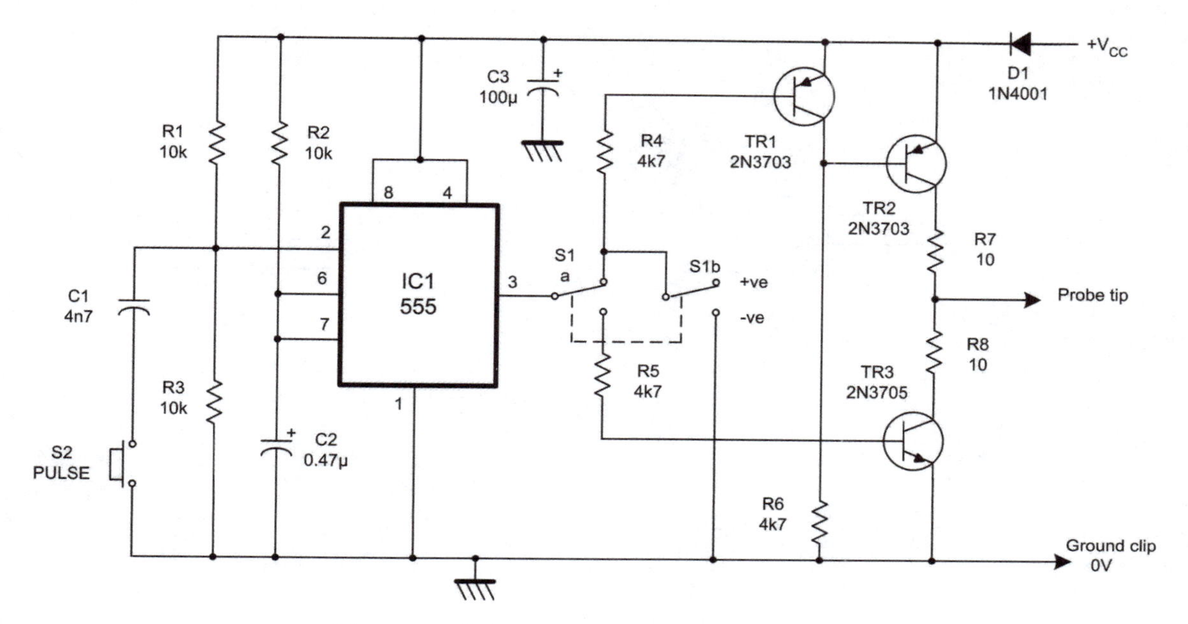

Figure 13.23 *Logic pulser circuit*

a line up to +5 V via a 1 kΩ resistor or by temporarily tying a line to 0 V. However, on other occasions, it may be necessary to simulate a pulse rather than a permanent logic state and this can be achieved by means of a logic pulser.

A logic pulser provides a means of momentarily forcing a logic level transition into a circuit regardless of its current state and thus overcomes the need to disconnect or de-soldering any of the devices. The polarity of the pulse (produced at the touch of a button) is adjusted so that the node under investigation is momentarily forced into the opposite logical state. During the period before the button is depressed and for the period after the pulse has been completed, the probe tip adopts a tri-state (high impedance) condition. Hence the probe does not permanently affect the logical state of the point in question.

Logic pulsers derive their power supply from the circuit under test in the same manner as logic probes. Here again, the leads of an electrolytic decoupling capacitor or the +5 V and GND terminals fitted to an expansion card make suitable connecting points.

A typical logic pulser circuit is shown in Figure 13.23. The circuit comprises a 555 monostable pulse generator triggered from a push-button. The output of the pulse generator is fed to a complementary transistor arrangement in order to make it fully TTL-compatible. As with the logic pulser, this circuit derives its power from the circuit under test (usually +5 V).

Figure 13.24 shows an example of the combined use of a logic pulser and a logic pulser for testing a simple J-K bistable. The logic probe is used to check the initial state of the Q and /Q outputs of the bistable (see Figure 13.24(a) and (b)). Note that the Q and /Q outputs should be complementary. Next, the logic pulser is applied to the clock (CK) input of the bistable (Figure 13.24(c)) and the Q output is checked using the logic probe. The application of a pulse (using the trigger button) should cause the Q output of the bistable to change state (see Figure 13.24(d)).

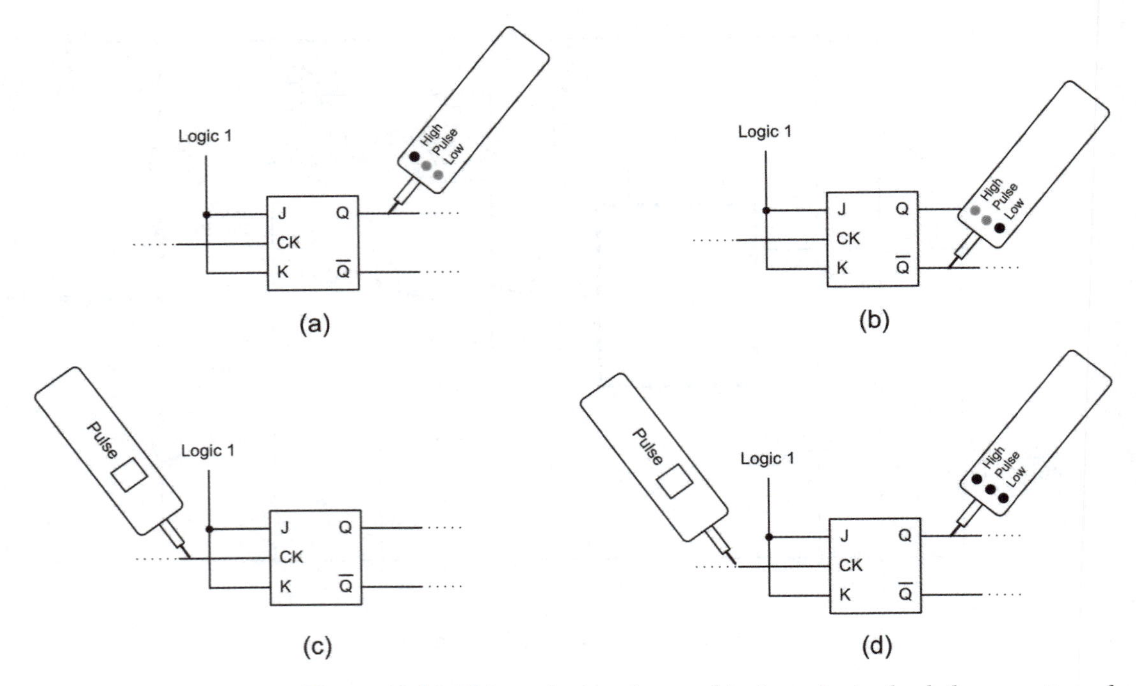

Figure 13.24 *Using a logic pulser and logic probe to check the operation of a J-K bistable. The pulser is used to force a change of logic level at the clock (CK) input of the bistable*

Oscilloscopes

The use of an oscilloscope in the examination of time-related signals (waveforms) is well known. Such instruments provide an alternative means of tracing logic states present in a PC-based system and may also be used for detecting noise and unwanted AC signals which may be present on power-supply rails. It must, however, be stressed that, since low-cost oscilloscopes generally do not possess any means of storing incoming signals, severe triggering problems arise when signals are non-repetitive. This is an important point since many of the digital signals present on a bus are both asynchronous and non-repetitive.

Apart from displaying the shape of waveforms present in a bus system, oscilloscopes can also be used to make reasonably accurate measurements of voltage and time. In such cases, measurements are made with reference to a graticule fitted to the CRT and scale factors are applied using the time and voltage range switches. However, before attempting to take measurement from the graticule it is essential to check that any variable front panel controls are set to the calibrate (CAL) position. Failure to observe this simple precaution may result in readings which are at best misleading or at worst grossly inaccurate.

Since modern oscilloscopes employ DC coupling throughout the vertical amplifier stages, a shift along the vertical axis will occur whenever a direct voltage is present at the input. When investigating waveforms in a circuit one often encounters AC signals superimposed on DC levels; the latter may be removed by inserting a capacitor in series with the input using the 'AC-GND-DC' switch. In the 'AC' position the capacitor is inserted at the input, whereas in the 'DC' position the capacitor is shorted. If 'GND' is selected the vertical input is taken to common (0 V) and the input terminal is left floating. In order to measure the DC level of an input signal, the 'AC-GND-DC' switch must first be placed in the 'GND' position. The 'vertical position' is then adjusted so that the trace is coincident with the central horizontal axis. The 'AC-GND-DC' switch is then placed in the 'DC' position and the shift along the vertical axis measured in order to ascertain the DC level.

Most dual-beam oscilloscopes incorporate a 'chopped-alternate' switch to select the mode of beam splitting. In the 'chopped' position, the trace displays a small portion of one vertical channel waveform followed by an equally small portion of the other. The traces are thus sampled at a fast rate so that the resulting display appears to consist of two apparently continuous traces. In the 'alternate' position, a complete horizontal sweep is devoted to each channel on an alternate basis.

Chopped mode operation is appropriate to signals of relatively low frequency (i.e. those well below the chopping rate) where it is important that the display accurately shows the true phase relationship between the two displayed signals. Alternate mode operation, on the other hand, is suitable for high-frequency signals where the chopping signal would otherwise corrupt the display. In such cases it is important to note that the relative phase of the two signals will not be accurately displayed.

Most modern oscilloscopes allow the user to select one of several signals for use as the timebase trigger. These 'trigger source' options generally include an internal signal derived from the vertical deflection system, a 50 Hz signal derived from the AC mains supply, and a signal which may be applied to an 'external trigger input'. As an example, the 50 Hz trigger source should be selected when checking for mains-borne noise and interference whereas the external trigger input may usefully be derived from a processor clock signal when investigating the synchronous signals present within the PC expansion bus.

Figure 13.25 shows the typical control layout of a modern dual-beam bench oscilloscope.

Fault location procedure

To simplify the process of fault location on a PC and associated expansion bus, it is useful to consider the system as a number of interlinked sub-systems. Each sub-system can be further divided into its constituent elements. Fortunately, the use of a standard expansion bus makes fault finding very straightforward since it is eminently possible to isolate a fault to a particular part of the system just by removing a suspect board and substituting one which is known to be functional.

The following eight-point checklist may prove useful; the questions should be answered *before* attempting to make any measurements or remove any suspect boards.

1. Has the system operated in similar circumstances without failure? Is the fault inherent in the system?
2. If an inherent fault is suspected, why was it not detected by normal quality procedures?
3. If the fault is not considered inherent and is attributed to component failure, in what circumstances did the equipment fail?

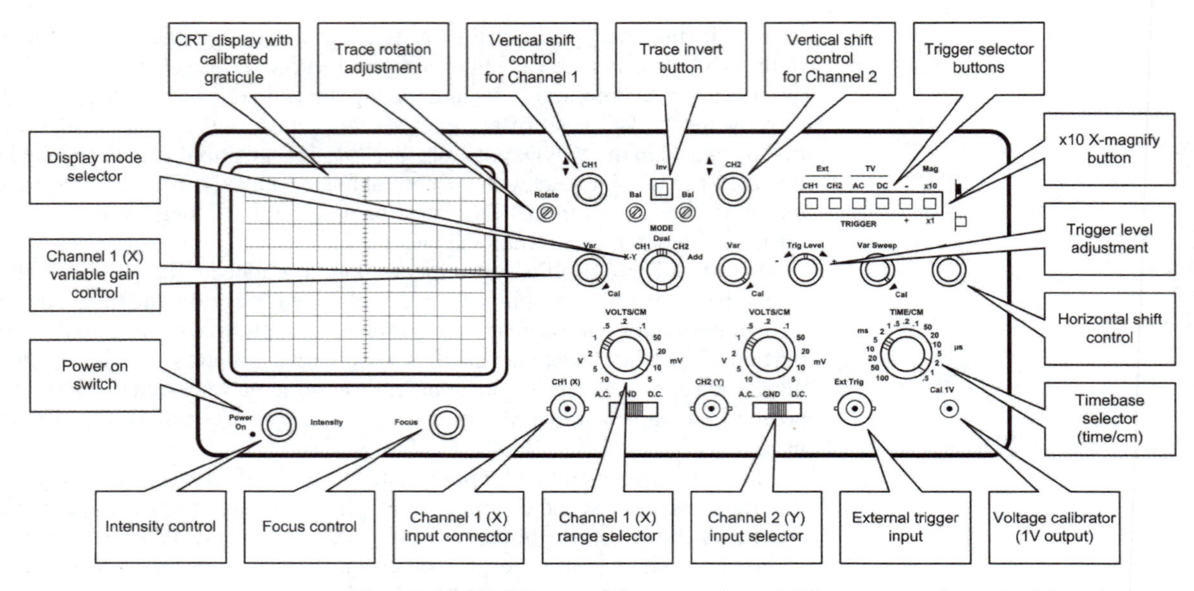

Figure 13.25 *Layout of the controls and adjustments of a typical bench oscilloscope*

4 Is the fault intermittent or is it present at all times?
5 If the fault is intermittent, in what circumstances does it arise? Is it possible to predict when the fault will occur?
6 To facilitate testing and diagnosis, can conditions be reproduced so that the fault manifests itself permanently?
7 What parts of the equipment are known to be functioning correctly? Is it possible to isolate the fault be isolated to a particular part?
8 Is the fault a known 'stock fault'? Has the fault been documented elsewhere?

Having answered the foregoing questions, and assuming that one is confronted with a system which is totally unresponsive, the first step is that of checking the power-supply rails using a multi-range meter. Where any one of the supply rails is low (or missing altogether) the power supply should be disconnected from the backplane and the measurement should be repeated in order to establish whether the absence of power is due to failure of the power supply or whether the fault can be attributed to excessive loading. This, in turn, can either be due to a short-circuit component failure within an expansion card or a similar fault within the system motherboard.

The system power supply employs switched mode techniques and it should be borne in mind that such units generally require that a nominal load be present on at least one of their output rails before satisfactory regulation can be achieved. Failure to observe this precaution can lead the unsuspecting test engineer to conclude that a unit is not regulating correctly when it has been disconnected from a system. In any event, it is advisable to consult the manufacturer's data before making measurements on individual supplies.

Having ascertained that the system is receiving its correct power-supply voltages, the next stage is that of activating the system reset switch and noting whether any changes are produced. After each of the initial diagnostic

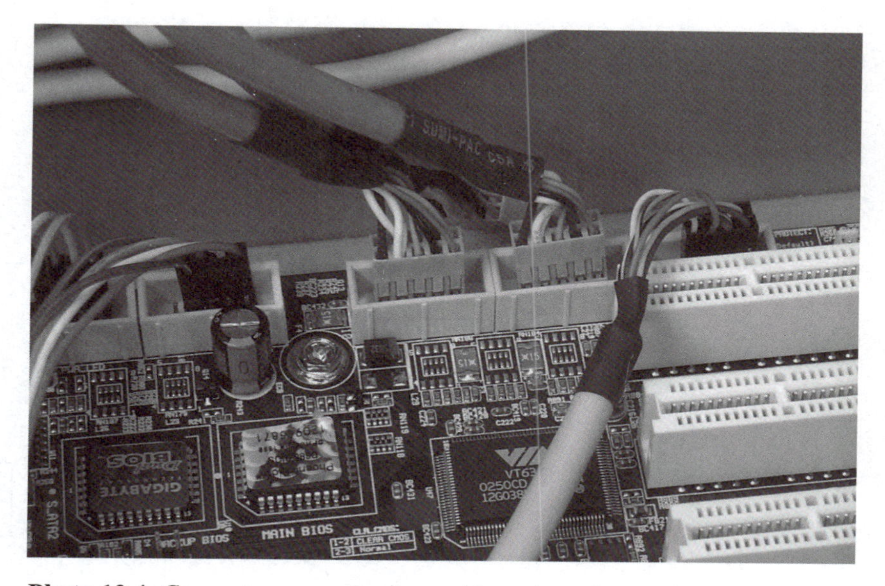

Photo 13.4 *Connectors can often be problematic and are a frequent cause of intermittent hardware faults. In this case, the rightmost serial ATA (SATA) connector is not correctly seated. Simply pushing the connector fully home was sufficient to clear the fault!*

procedures are completed an appropriate message is printed on the screen. Furthermore, once the initial procedures have been completed, any disk drive fitted to the system will normally become active as the system is 'booted'. If neither of these indications is produced, the system motherboard must be suspected as the fault will probably be attributable to failure of the CPU or one of the major VLSI support devices present.

At this stage it may be worth replacing the system motherboard with a known functional unit. If this is not possible, checks should be performed on each of the VLSI devices starting with the CPU. Where a fault is intermittent (e.g. the system runs for a time before stopping) it is worth checking connectors and also investigating the cleanliness of the supply. It is also worth checking for devices that may have become overheated after a period of operation. Check also, that CPU voltages have been correctly set. Attempts to overclock a CPU will often result in overheating – particularly if the CPU core voltage has been raised.

Connectors are often prone to failure and, if the principal chips are socketed these, too, can sometimes cause problems. Intermittent faults can sometimes be corrected simply by pressing each of the larger chips into its socket. In some cases it may be necessary to carefully remove the chips before replacing them; the action of removal and replacement can sometimes be instrumental in wiping the contacts clean.

Where a fault is permanently present and one or more of the supply rail voltages is lower than normal, chip failure may be suspected. In such an event, the system should be left running for some time and the centre of each chip should be touched in turn in order to ascertain its working temperature. If a chip that is not fitted with a heat removal device is running distinctly hot (i.e. very warm or too hot to comfortably touch) it should be considered a prime suspect.

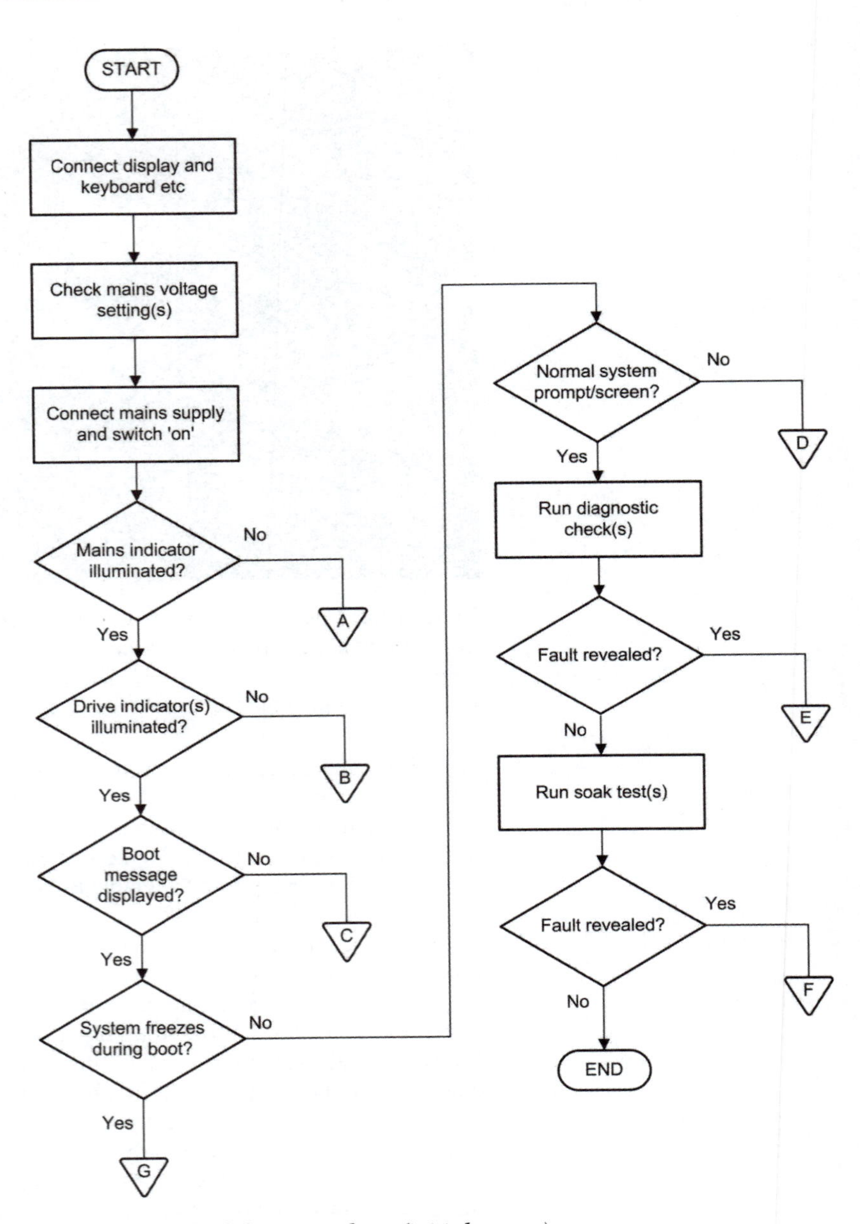

Figure 13.26 *Fault location chart (initial stages)*

Where possible the temperature should be compared with that generated by a similar chip fitted in the same board or that present in another functional module. Where the larger chips have been fitted in sockets, each should be carefully removed and replaced in turn (disconnecting the power, of course, during the process) before replacing it with a known functional device.

Figures 13.26–13.32 show a series of fault location charts that can be used to pinpoint faults in a generic PC.

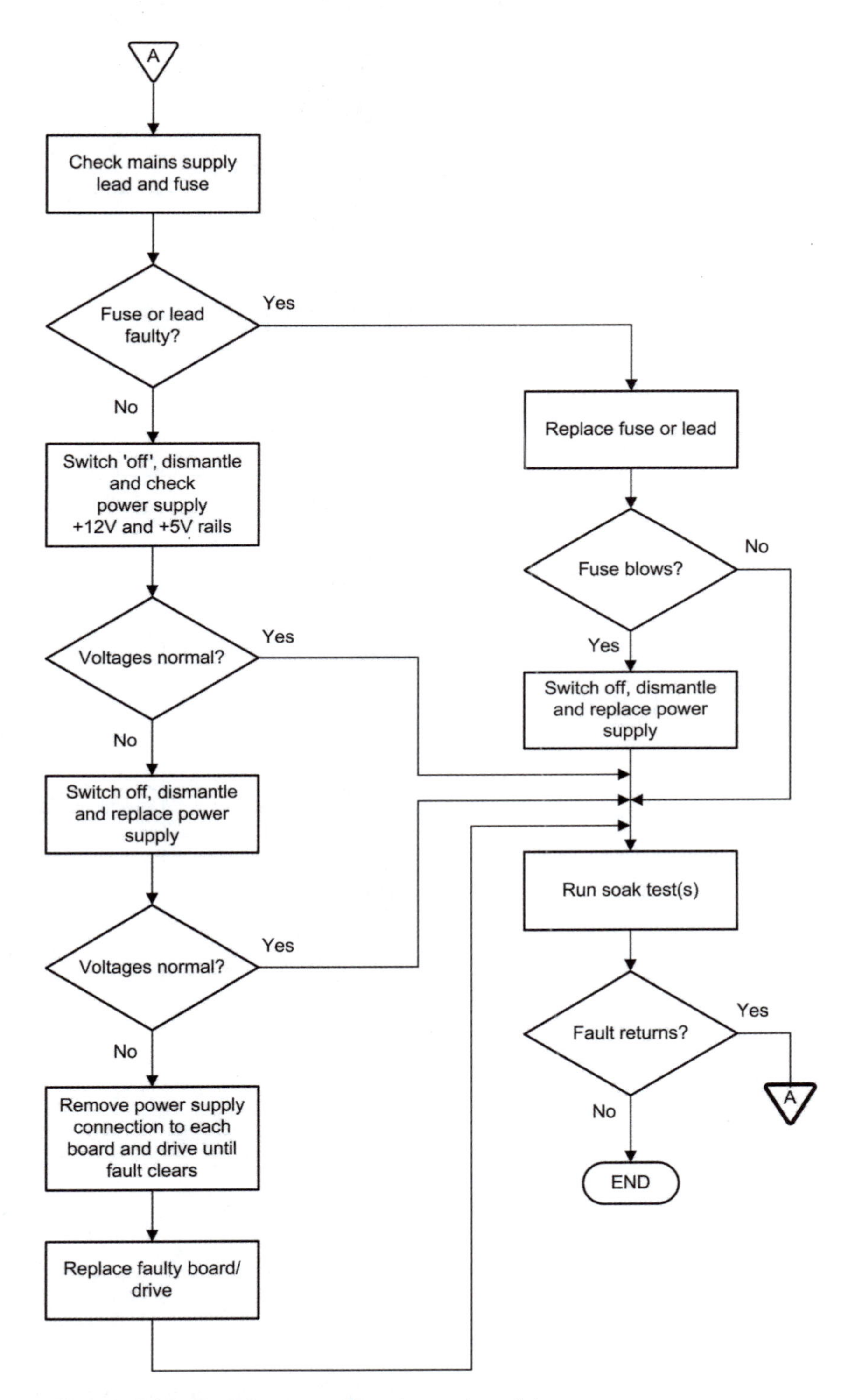

Figure 13.27 *Fault location chart (power supply)*

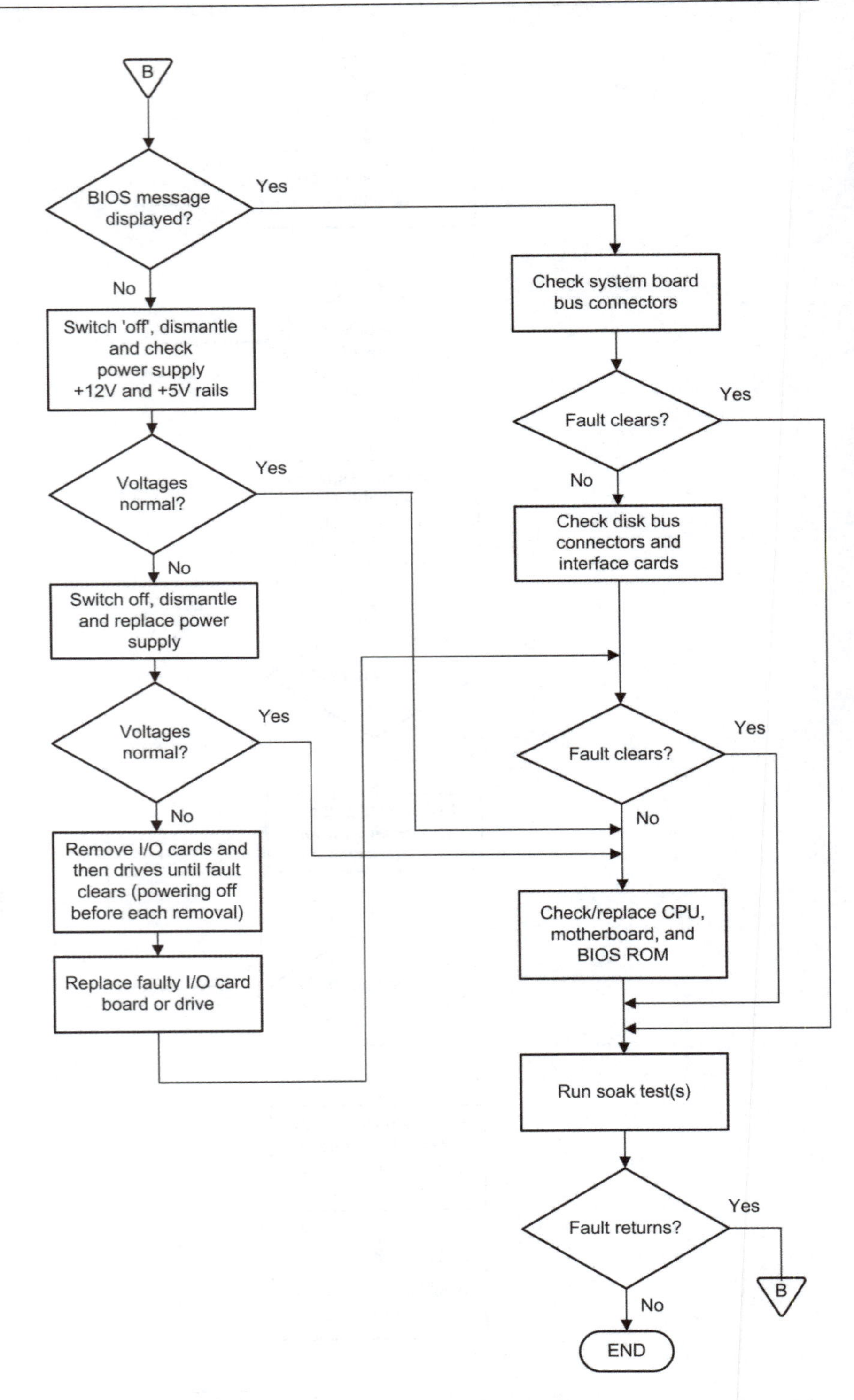

Figure 13.28 *Fault location chart (motherboard, CPU, and I/O cards)*

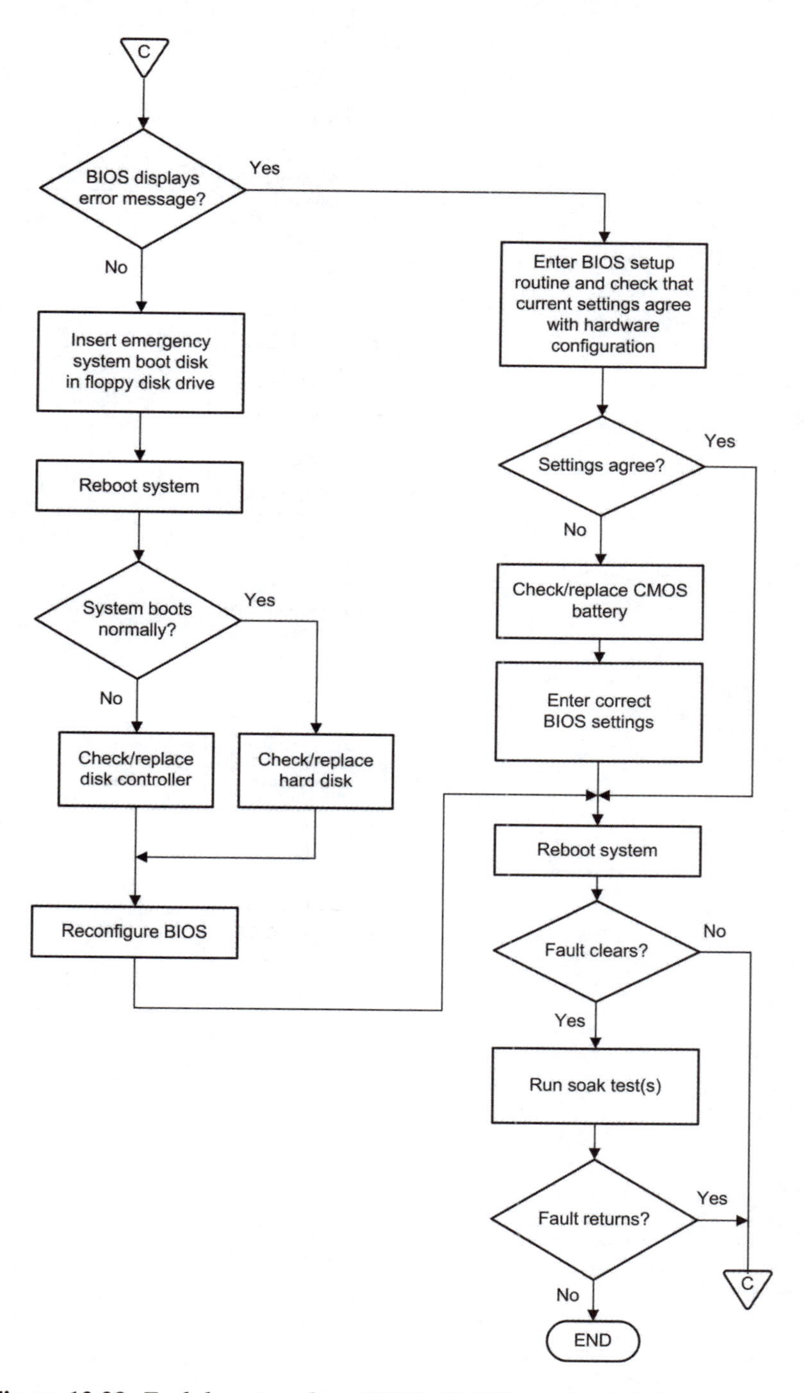

Figure 13.29 *Fault location chart (BIOS, CMOS, and hard disk)*

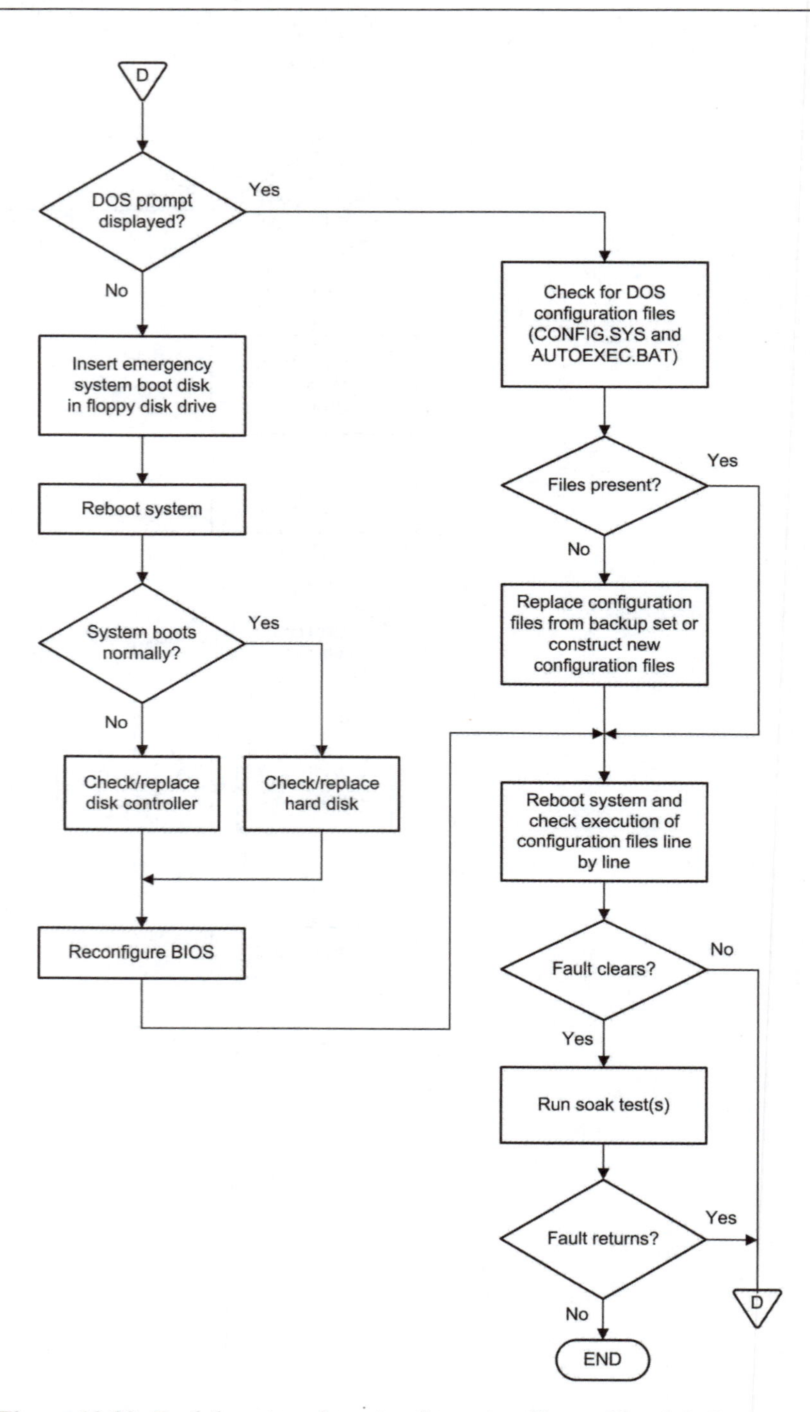

Figure 13.30 *Fault location chart (configuration files and hard disk)*

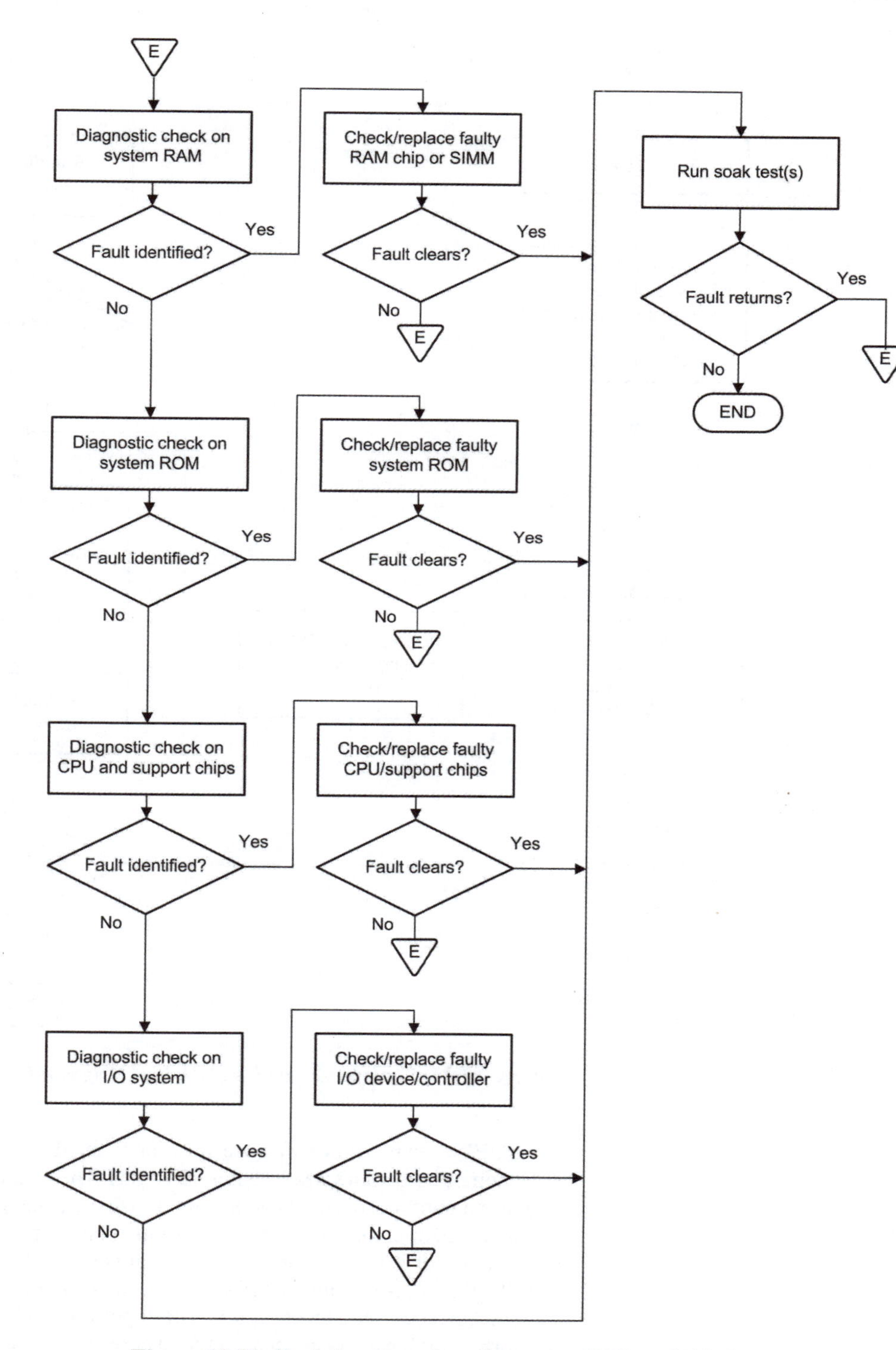

Figure 13.31 *Fault location chart (memory, CPU, and I/O diagnostics)*

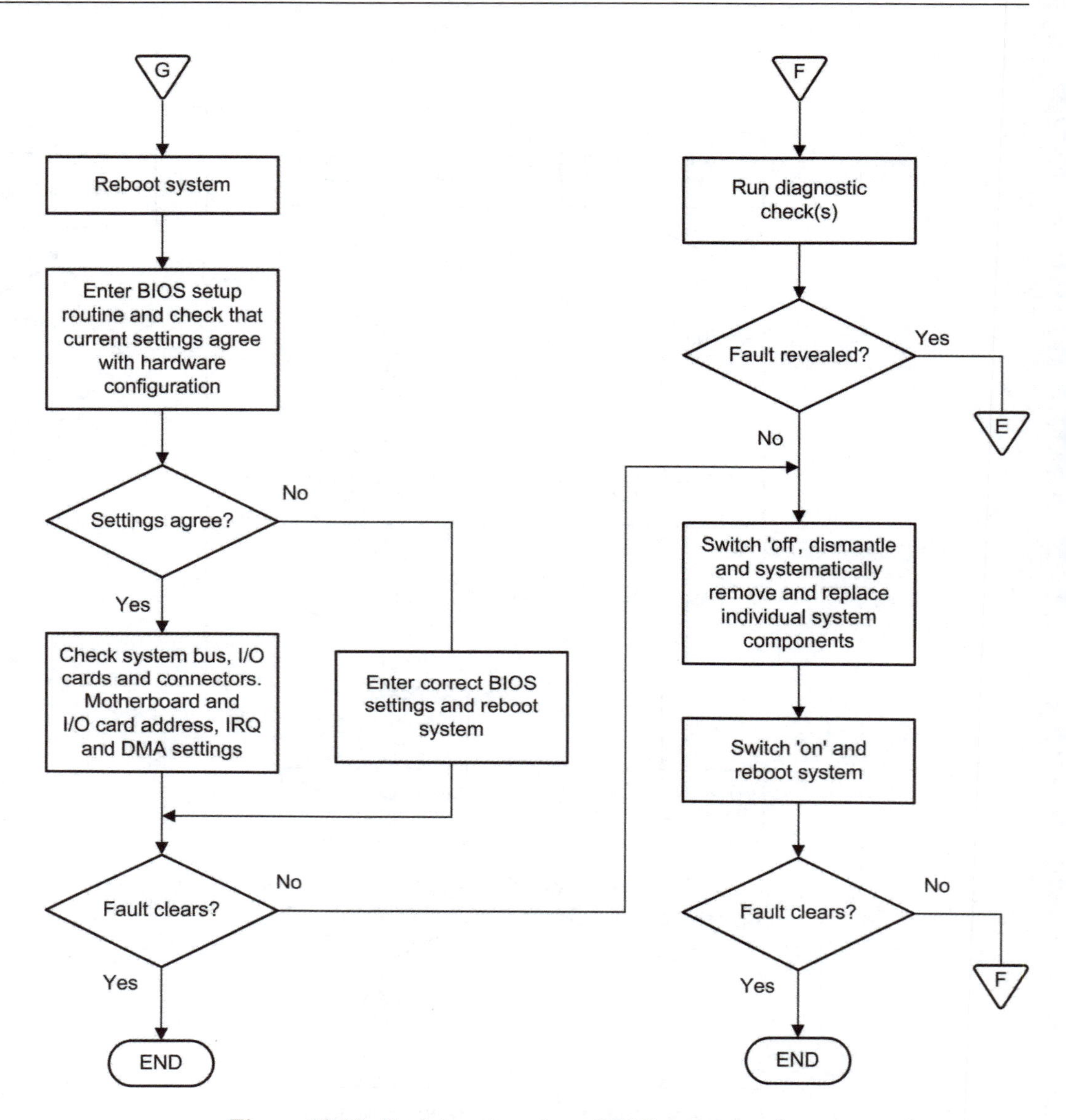

Figure 13.32 *Fault location chart (BIOS settings and system configuration)*

ISA/EISA expansion cards are invariably fitted with links which provide selection of base addresses (see Figure 13.33). These links *must* be configured so that no conflicts occur. This precaution is particularly important when new or replacement cards are fitted to a system. This is, perhaps, a rather obvious precaution but it is nevertheless one which is easily forgotten! PCI cards are usually plug-and-play and therefore there is less risk of links, and DIP switches being incorrectly set. However, it is still important to check that board settings are correct for the address range and hardware configuration in which the board is used. This is even more important where several boards are fitted since default (factory) settings may be identical.

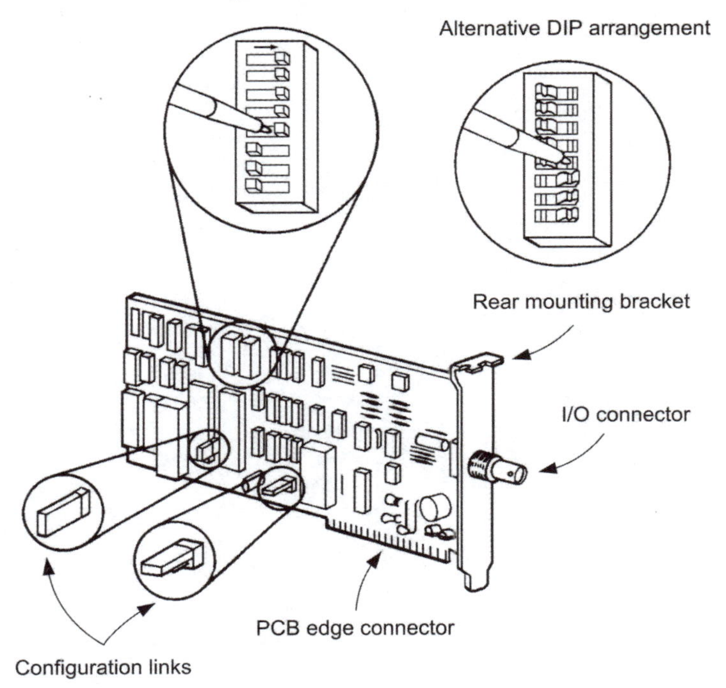

Figure 13.33 *Typical bus adapter card showing links and settings*

Finally, in a perfect world there would be no uncertainty nor any ambiguity about the logic levels present in a digital system. Unfortunately, this is seldom the case since *spurious signals* and *noise* are invariably present to some degree. The ability to reject noise is thus an important requirement of PC-based control systems. This is particularly true where a system is to be used in a particularly noisy environment (such as a shipyard or steelworks). In such a situation, special precautions may be necessary in order to avoid corruption of signals and data, and one or more of the following techniques may be applied:

- Using a 'clean' AC supply for the PC controller and peripheral devices (where appropriate). If such a supply is not available, a supply filter or AC power conditioner should be fitted.
- Screening all signal cables (particularly those used to connect remote transducers) and returning the outer braid screen to earth (note that noise rejection is sometimes enhanced if the screen is only earthed at one point).
- Ensuring that the PC system enclosure is adequately earthed and that none of the outer panels or metal chassis parts of external card frames or enclosures are allowed to 'float'.
- Decoupling supply rails at the point at which they enter each external signal conditioning board (where appropriate).
- In extreme cases, making use of optical fibres (and appropriate interface hardware) rather than twisted pairs or co-axial cables for the asynchronous transmission of digital signals.

Appendix A Glossary of terms

Accelerated Graphics Port (AGP) A high speed interface for video cards. AGP typically runs at 1× (66 MHz), 2× (133 MHz), or 4× (266 MHz).

Access time The time taken to retrieve data from a memory/storage device, i.e. the elapsed time between the receipt of a read signal at the device and the placement of valid data on the bus. Typical access times for semiconductor memory devices are in the region 100–200 ns whilst average access times for magnetic disks typically range from 10–50 ms.

Accumulator A register within the processor (or CPU) in which the result of an operation is placed.

Acknowledge (ACK) A signal used in serial data communications which indicates that data has been received without error.

Active high A term used to describe a signal which is asserted in the high (logic 1) state.

Active low A term used to describe a signal which is asserted in the low (logic 0) state.

Address A reference to the location of data in memory or within I/O space. The processor (or other controlling device) places addresses (in binary coded form) on the address bus (see also **Address bus**).

Address bus The set of lines used to convey address information. The original PC bus had twenty address lines (A0 to A19) and these were capable of addressing a linear address space with a little more than a million address locations. One byte of data may be stored at each address.

Address decoder A hardware device (often a single integrated circuit) which provides chip select or chip enable signals from address patterns which appear on an address bus.

Address selection The process of selecting a specific address (or range of addresses). In order to prevent conflicts, expansion cards must usually be configured (by means of DIP switches or links) to unique addresses within the I/O address map.

Amplifier A circuit or device which increases the power of an electrical signal.

Analogue The representation of information in the form of a continuously variable quantity (e.g. voltage).

AND Logical function which is asserted (true) when all inputs are simultaneously asserted.

ANSI character set The American National Standard Institute's (ANSI) character set which is based on an 8-bit binary code and which provides 256 individual characters. See also **ASCII**.

Archive A device or medium used for storage of data which need not be instantly accessible (e.g. a tape cartridge).

ASCII A code which is almost universally employed for exchanging data between microcomputers. Standard ASCII is based on a 7-bit binary code and caters for alphanumeric characters (both upper and lower case), punctuation, and special control characters. Extended ASCII employs an 8th bit to provide an additional 128 characters (often used to represent graphic symbols).

Assembly language A low-level programming language which is based on mnemonic instructions. Assembly language is often unique to a particular microprocessor or microprocessor family.

Asserted A term used to describe a signal when it is in its logically true state (i.e. logic 1 in the case of an active high signal or logic 0 in the case of an active low signal).

Asynchronous transmission A data transmission method in which the time between transmitted characters is arbitrary. Transmission is controlled by start and stop bits (no additional synchronizing or timing information is required).

ATAPI The ATAPI (or Advanced Technology Attachment Packet Interface) standard provides a simple means of connecting a CD-ROM drive to an EIDE adapter. Without such an interface, a CD-ROM drive will require either a dedicated interface card or an interface provided on a sound card.

AUTOEXEC.BAT A file which contains a set of DOS commands and/or program names which is executed automatically whenever the system is initialized and provides the means of configuring a system.

Backup A file or disk copy made in order to avoid the accidental loss, damage, or erasure of programs and/or data.

Basic Input Output System (BIOS) The BIOS is the part of the operating system which handles communications between the microcomputer and peripheral devices (such as keyboard, serial port, etc.). The BIOS is supplied as firmware and is contained in a read-only memory (ROM).

Batch file A file containing a series of DOS commands which are executed when the file name is entered after the DOS prompt. Batch files are given a BAT file extension. A special type of batch file (AUTOEXEC.BAT) is executed (when present) whenever a system is initialized. See also **AUTOEXEC.BAT**.

Baud rate The speed at which serial data is transferred between devices.

Binary file A file which contains binary data (i.e. a direct memory image). This type of file is used for machine readable code, program overlays, and graphics screens.

Bit A contraction of 'binary digit'; a single digit in a binary number.

Boot The name given to the process of loading and initializing an operating system (part of the operating system is held on disk and must be loaded from disk into RAM on power-up).

Boot record A single-sector record present on a disk which conveys information about the disk and instructs the computer to load the requisite operating system files into RAM (thus booting the machine).

Buffer In a hardware context, a buffer is a device which provides a degree of electrical isolation at an interface. The input to a buffer usually exhibits a much higher impedance than its output (see also 'Driver'). In a software context, a buffer is a reserved area of memory which provides temporary data storage and thus may be used to compensate for a difference in the rate of data flow or time of occurrence of events.

Bus An electrical highway for signals which have some common function. Most microprocessor systems have three distinct buses; an address bus, data bus, and control bus. A local bus can be used for high-speed data transfer between certain devices (e.g. processor, graphics processor, and video memory).

Byte A group of 8 bits which are operated on as a unit.

Cache A high-speed random-access memory which is used to store copies of the data from the most recent main memory or hard disk accesses. Subsequent accesses fetch data from this area rather than from the slower main memory or hard disk.

Central processing unit (CPU) See **Processor**.

Channel A path along which signals or data can be sent.

Character set The complete range of characters (letters, numbers, and punctuation) which are provided within a system. See also ANSI and ASCII.

Checksum Additional binary digits appended to a block of data. The value of the appended digits is derived from the sum of the data present within the block. This technique provides the means of error checking (validation).

Chip The term commonly used to describe an integrated circuit.

Chipset The chipset is the name given to the two or more integrated circuits which control the interface between the processor, RAM, I/O devices, bus expansion, and adapter cards. Different chipsets provide support for different processors and motherboard configurations.

CISC The term CISC refers to a 'Complex Instruction Set Computer' – the standard Intel family of CPUs all conform to this model rather than the alternative 'Reduced Instruction Set Computer' (RISC). There is much debate about the pro's and con's of these two design methodologies but, in fact, neither of these two contrasting approaches has actually demonstrated clear superiority over the other. See also **RISC**.

Clock A source of timing signals used for synchronizing data transfers within a microprocessor or microcomputer system.

Cluster A unit of space allocated on the surface of a disk. The number of sectors which make up a cluster varies according to the DOS version and disk type (see also **Sector**).

Command An instruction (entered from the keyboard or contained within a batch file) which will be recognized and executed by a system (see also **Batch file**).

Common A return path for a signal (often ground).

CONFIG.SYS A file which contains DOS configuration commands which are used to configure the system at start-up. The CONFIG.SYS file specifies device drivers which are loaded during initialization and which extend the functionality of a system by allowing it to communicate with additional items of hardware. See also **Device Driver**.

Controller A sub-system within a microcomputer which controls the flow of data between the system and an I/O or storage device (e.g. a CRT controller, hard disk controller, etc.). A controller will generally be based on one, or more, programmable VLSI devices.

Coprocessor A second processor which shares the same instruction stream as the main processor. The coprocessor handles specific tasks (e.g. mathematics) which would otherwise be performed less efficiently (or not at all) by the main processor. Note that maths coprocessors are no longer needed as all modern processors have internal registers required for mathematics processing.

Cylinder The group of tracks which can be read from a hard disk at any instant of time (i.e. without steeping the head in or out). In the case of a floppy disk (where there are only two surfaces), each cylinder comprises two tracks. In the case of a typical IDE hard disk, there may be two platters (i.e. four surfaces) and thus four tracks will be present within each cylinder.

Daisy chain A method of connection in which signals move in a chained fashion from one device to another. This form of connection is commonly used with disk drives.

Data A general term used to describe numbers, letters, and symbols present with a computer system. All such information is ultimately represented by patterns of binary digits.

Data bus A highway (in the form of multiple electrical conductors) which conveys data between the different elements within a microprocessor system.

Data file A file which contains data (rather than a program) and which are used by applications such as spreadsheet and database applications. Note that data may or may not be stored in directly readable ASCII form.

Device A hardware component such as a memory card, sound card, modem, or graphics adapter.

Device driver A term used to describe memory resident software (specified in the CONFIG.SYS system file) which provides a means of interfacing specialized hardware (e.g. a graphics adapter). See **CONFIG.SYS**.

Digital storage oscilloscope (DSO) A Digital storage oscilloscope (or DSO) combines elements of both hardware and software. These must work together to provide all the functionality of a conventional DSO but also those of a spectrum analyzer, data logger, digital frequency meter, and voltmeter. In many cases a modern DSO will be able to replace several items of conventional test equipment. Switching between these instruments is usually quick and easy, and in most cases each instrument is able to have its dedicated window on the PC display.

Direct memory access A method of fast data transfer in which data moves between a peripheral device (e.g. a hard disk) and main memory without direct control of the processor.

Directory A catalogue of disk files (containing such information as filename, size, attributes, and date/time of creation). The directory is stored on the disk and updated whenever a file is amended, created, or deleted. A directory entry usually comprises 32 bytes for each file.

DIP switch A miniature PCB mounted switch that allows configuration options (such as IRQ or DMA settings) to be selected.

Disk operating system (DOS) A group of programs which provide a low-level interface with the system hardware (particularly disk I/O). Routines contained within system resident portions of the operating system may be used by the programmer. Other programs provided as part of the system include those used for formatting disks, copying files, etc.

Double word A data value which comprises a group of 32 bits (or two words). See also **Word**.

DRAM DRAM (or Dynamic Random Access Memory) refers to the semiconductor read/write memory of a PC. DRAM requires periodic 'refreshing' and therefore tends not to offer the highest speeds required of specialized memories (such as cache memory). DRAM is, however, relatively inexpensive.

Driver In a software context, a driver is a software routine which provides a means of interfacing a specialized hardware device (see also **Device driver**). In a hardware context, a driver is an electrical circuit which provides an electrical interface between an output port and an output transducer. A driver invariably provides power gain (i.e. current gain and/or voltage gain), see also **Amplifier**.

EIDE EIDE (or 'Enhanced Integrated Drive Electronics') is the most widely used interface for connecting hard disk drives to a PC. Most motherboards now incorporate an on-board EIDE controller rather than having to make use of an adapter card. This allows one or two hard disk drives to be connected directly to the motherboard.

Expanded memory (EMS memory) Memory which is additional to the conventional 'base' memory available within the system. This memory is 'paged' into the base memory space whenever it is accessed. The EMS specification uses four contiguous 16 K pages of physical memory (64 K total) to access up to 32 M of expanded memory space. See also **Expanded memory manager**.

Expanded memory manager An expanded memory manager (such as EMM386.EXE included with MS-DOS 5.0 and later) provides a means of establishing and controlling the use of expanded memory (i.e. memory above the DOS 1 MB limit). Unlike DOS and Windows 3.1, Windows 9x incorporates its own memory management and thus EMM386 (or its equivalent) is not required. See also **Expanded memory**.

Extended memory (XMS memory) Memory beyond the 1 MB range ordinarily recognized by MS-DOS. The XMS memory specification resulted from collaboration between Lotus, Intel, and Microsoft (sometimes known as LIM specification).

File Information (which may comprise ASCII encoded text, binary coded data, and executable programs) stored on a floppy disk, hard disk, or other disk-oriented storage device (such as a solid-state USB drive). Files may be redirected from one logical device to another using appropriate DOS commands.

File allocation table (FAT) The file allocation table (or 'FAT') provides a means of keeping track of the physical location of files stored on a floppy disk or hard disk. Part of the function of DOS is to keep the FAT up to date whenever a file operation is carried out. DOS does not necessarily store files in physically contiguous clusters on a disk and it is the FAT that maintains the addresses of clusters occupied by a particular file. These clusters may, in fact, be scattered all over the surface of the disk (in which case we describe the file as having been 'fragmented').

File attributes Information which indicates the status of a file (e.g. hidden, read-only, system, etc.).

Filter In a software context, a filter is a software routine which removes or modifies certain data items (or data items within a defined range). In a hardware context, a filter is an electrical circuit which modifies the frequency distribution of a signal. Filters are often categorized as low-pass, high-pass, band-pass, or band-stop depending upon the shape of their frequency response characteristic.

Firmware A program (software) stored in read-only memory (ROM). Firmware provides non-volatile storage of programs.

Fixed disk A disk which cannot be removed from its housing. Note that, whilst the terms 'hard' and 'fixed' are often used interchangeably, some forms of hard disk are exchangeable.

Font A set of characters (letters, numbers, and punctuation) with a particular style and size.

Format The process in which a magnetic disk is initialized so that it can accept data. The process involves writing a magnetic pattern of tracks and sectors to a blank (uninitialized) disk. A disk containing data can be reformatted, in which case all data stored on the disk will be lost. An MS-DOS utility program (FORMAT.COM) is supplied in order to carry out the formatting of floppy and hard disks. Note that, in the case of hard drives, there are differences between the logical organization of tracks and sectors and the underlying physical arrangement of the data stored on the drive.

Graphics adapter An option card which provides a specific graphics capability (e.g. CGA, EGA, HGA, and VGA). Graphics signal generation is not normally part of the functionality provided within a system motherboard.

Handshake An interlocked sequence of signals between peripheral devices in which a device waits for an acknowledgement of the receipt of data before sending new data.

Hard disk A non-flexible disk used for the magnetic storage of data and programs (see also **Fixed disk**).

Hardware The physical components (e.g. system board, keyboard, etc.) which make up a microcomputer system.

High state The more positive of the two voltage levels used to represent binary logic states. A high state (logic 1) is generally represented by a voltage in the range 2.0–5.0 V.

High memory A legacy term used to describe the first 64 K of extended memory. This area is used by some DOS applications and also by Windows. See **Extended memory**.

IDE IDE (or 'Integrated Drive Electronics') is the forerunner of the EIDE interface used in most modern PC's. See **EIDE**.

Input/output (I/O) Devices and lines used to transfer information to and from external (peripheral) devices.

Integrated circuit An electronic circuit fabricated on a single wafer (chip) and packaged as a single component.

Interface A shared boundary between two or more systems, or between two or more elements within a system. In order to facilitate interconnection of systems, various interface standards are adopted (e.g. RS-232 in the case of asynchronous data communications).

Interleave A system of numbering the sectors on a disk in a non-consecutive fashion in order to optimize data access times.

Interrupt A signal generated by a peripheral device when it wishes to gain the attention of the CPU. The Intel 80×86 family of microprocessors support both software and hardware interrupts. The former provide the means of invoking BIOS and DOS services whilst the latter are generally managed by an interrupt controller chip (e.g. 8259).

ISA ISA (or 'Industry Standard Architecture') is the long-surviving standard for connecting multiple interface adapters to the PC bus. Due to speed limitations, the ISA bus is no longer used for hardware that requires fast data throughput and local bus schemes (such as VL-bus or PCI-bus) are much preferred.

Isochronous data transfer Data transfer (i.e. the movement of digital information from one place to another) is said to be isochronous when the stream of digital information does not require a separate clock or timing signal. In effect, the timing of the isochronous data stream is implied by the rate at which it is delivered. In the Universal Serial Bus, isochronous data transfers provide periodic, continuous communication between a host and a device.

Joystick A device used for positioning a cursor, pointer, or output device using switches or potentiometers which respond to displacement of the stick in the X and Y directions.

Jumper Jumpers, like DIP switches, provide a means of selecting configuration options on adapter cards (see **DIP switch**).

Keyboard buffer A small area in memory which provides temporary storage for keystrokes. See **Buffer**.

Kilobyte (KB) 1024 bytes (note that $2^{10} = 1024$).

Logical device A device which is normally associated with microcomputer I/O, such as the console (which comprises keyboard and display) and printer.

Low state The more negative of the two voltage levels used to represent the binary logic states. A low state (logic 0) is generally represented by a voltage in the range 0–0.8 V.

Megabyte (MB) 1 048 576 bytes (note that $2^{20} = 1\,048\,576$). The basic addressing range of the 8086 processor (which has 20 address bus lines) is 1 MB.

Memory That part of a microcomputer system into which information can be placed and later retrieved. Storage and memory are interchangeable terms. Memory can take various forms including semiconductor (RAM and ROM), magnetic (floppy and hard disks), and optical disks. Note that memory may also be categorized as read-only (in which case data cannot subsequently be written to the memory) or read/write (in which case data can both be read from and written to the memory).

Memory resident program See **TSR**.

Microprocessor A central processing unit fabricated on a single chip.

MIDI The MIDI (or 'Musical Instrument Digital Interface') is the current industry standard for connecting musical instruments to a PC.

Modem A contraction of modulator–demodulator; a communications interface device that enables a serial port to be interfaced to a conventional voice-frequency telephone line.

Modified frequency modulation (MFM) A method of data encoding employed with hard disk storage. This method of data storage is 'self-clocking'.

Motherboard The motherboard (or system board) is the mother printed circuit board which provides the basic functionality of the microcomputer system including processor (or CPU), RAM, and ROM. The system board is fitted with connectors which permit the installation of one, or more, option cards (e.g. graphics adapters, disk controllers, etc.).

Multimedia A combination of various media technologies including sound, video, graphics, and animation.

Multitasking A process in which several programs are running simultaneously.

NAND Inverse of the logical AND function.

Negative acknowledge (NAK) A signal used in serial data communications which indicates that erroneous data has been received.

Network A system which allows two or more computers to be linked via a physical communications medium (e.g. coaxial cable) in order to exchange information and share resources.

Nibble A group of 4 bits which make up one half of a byte. A hexadecimal character can be represented by such a group.

Noise Any unwanted signal component which may appear superimposed on a wanted signal.

NOR Inverse of the logical OR function.

NRZI NRZI (or Non Return to Zero Invert) is a method of encoding serial data in which 1's and 0's are represented by opposite and alternating high- and low- voltage states and where there is no return to the reference (zero) voltage between encoded bits. NRZI is self-clocking and does not require a train of separately transmitted clock pulses.

Operating system A control program which provides a low-level interface with the system hardware. The operating system thus frees the programmer from the need to produce hardware specific I/O routines (e.g. those associated with disk filing). See also **Disk operating system**.

Option card A printed circuit board (adapter card) which complies with the physical and electrical specification for a particular system and which provides the system with additional functionality (e.g. asynchronous communications facilities).

OR Logical function which is asserted (true) when any one or more of its inputs are asserted.

Page A contiguous area of memory of defined size (often 256 bytes but can be larger, see **Expanded memory**).

Paragraph Sixteen consecutive bytes of data. The segment address can be incremented to point to consecutive paragraphs of data.

Parallel interface (parallel port) A communications interface in which data is transferred a byte at a time between a computer and a peripheral device, such as a printer.

PCI The PCI (or 'Peripheral Component Interconnect') standard provides a means of connecting 32- or 64-bit expansion cards to a motherboard. PCI expansion slots are available in most modern PC's.

PCMCIA The PCMCIA (or simply 'PC Card') standard provides a means of connecting a sub-miniature expansion card (such as a memory card or modem) to a laptop or book computer.

Peripheral An external hardware device whose activity is under the control of the microcomputer system.

Port A general term used to describe an interface circuit which facilitates transfer of data to and from external devices (peripherals).

Processor The processor is generally taken to mean the central processing unit (CPU) which is invariably a single very large scale integrated circuit that decodes instructions and controls the other hardware elements of the system (see Appendix D). The CPU comprises a control unit, arithmetic/logic unit and internal storage (see also 'Microprocessor'). Other processors may be dedicated to functions such as graphics and complex I/O functions.

Processor socket The socket (or 'slot') used to mount the processor on the motherboard (see **Appendix F**).

Program A sequence of executable microcomputer instructions which have a defined function. Such instructions are stored in program files having EXE or COM extensions.

Propagation delay The time taken for a signal to travel from one point to another. In the case of logic elements, propagation delay is the time interval between the appearance of a logic state transition at the input of a gate and its subsequent appearance at the output.

Protocol A set of rules and formats necessary for the effective exchange of data between intelligent devices.

Random access An access method in which each word can be retrieved in the same amount of time (i.e. the storage locations can be accessed in any desired order). This method should be compared with sequential access in which access times are dependent upon the position of the data within the memory.

Random access memory (RAM) A term which usually refers to semiconductor read/write memory (in which access time is independent of actual storage address). Note that semiconductor read-only memory (ROM) devices also provide random access.

Read The process of transferring data to a processor from memory or I/O.

Read-only memory (ROM) A memory device which is permanently programmed. Erasable–programmable read only memory (EPROM) devices are popular for storage of programs and data in stand-alone applications, and can be erased under ultraviolet light to permit reprogramming.

Register A storage area within a CPU, controller, or other programmable device, in which data (or addresses) are placed during processing. Registers will commonly hold 8-, 16-, or 32-bit values.

RISC The term RISC refers to a 'Reduced Instruction Set Computer' – a computer based on a processor that accepts only a limited number of basic instructions but which decodes and executes them faster than the alternative technology (CISC). See also **CISC**.

Root directory The principal directory of a disk (either hard or floppy) which is created when the disk is first formatted. The root directory may contain the details of further sub-directories which may themselves contain yet more sub-directories, and so on.

Run length limited (RLL) A method of data encoding employed with hard disk storage. This method is more efficient than conventional MFM encoding.

SCSI The SCSI (or 'Small Computer Systems Interface') provides a means of interfacing up to eight peripheral devices, (such as hard disks, CD-ROM drives and scanners) to a microcomputer system. With its roots in larger minicomputer systems, SCSI tends to be more complex and expensive in comparison with EIDE.

Sector The name given to a section of the circular track placed (during formatting) on a magnetic disk. Tracks are commonly divided into ten sectors (see also **Format**).

Segment 64 KB of contiguous data within memory. The starting address of such a block of memory may be contained within one of the four segment registers (DS, CS, SS, or ES).

Serial interface (serial port) A communications interface in which data is transferred a bit at a time between a computer and a peripheral device, such as a modem. In serial data transfer, a byte of data (i.e. 8 bits) is transmitted by sending a stream of bits, one after another. Furthermore, when such data is transmitted asynchronously (i.e. without a clock), additional bits must be added for synchronization together with further bits for error (parity) checking (if enabled).

Server A computer which provides network accessible services (e.g. hard disk storage, printing, etc.).

Shell The name given to an item of software which provides the principal user interface to a system. The DOS program COMMAND.COM provides a simple DOS shell however later versions of MS-DOS and DR-DOS provide much improved graphical shells (DOSSHELL and VIEWMAX, respectively).

Signal The information conveyed by an electrical quantity.

Signal level The relative magnitude of a signal when considered in relation to an arbitrary reference (usually expressed in volts, V).

SIMM SIMMs (or 'Single In-line Memory Modules') are used to house the DRAM chips used in all modern PCs. The modular packaging and standard pin connections makes memory expansion very straightforward.

Slot A general term used to describe the sockets within a system (such as AGP, PCI, ISA, RAM, etc.) which can be used to expand the system. Each slot is designed to accept a printed circuit card. Note that the term 'slot' is also used to describe the socket used to mount certain types of processor on the motherboard (see Appendix F).

Software A series of computer instructions (i.e. a program).

Sound Card An interface card used to process audio data and provide audio output to external speakers; also typically includes interfaces to a microphone, game controller, and external MIDI devices.

Sub-directory A directory which contains details of a group of files and which is itself contained within another directory (or within the root directory).

Surge suppressor/protector A component or device (placed between the computer and the user's local AC source) that prevents transient voltage surges (*spikes*) from reaching the protected equipment.

Swap file A swap file is a file that resides on a hard disk and is used to provide 'virtual memory'. Swap files may be either 'permanent' or 'temporary' (see also **Virtual memory**).

System board See motherboard.

System file A file that contains information required by DOS. Such a file is not normally shown in a directory listing.

Terminal emulation The ability of a microcomputer to emulate a hardware terminal.

TSR A terminate-and-stay-resident program (i.e. a program which, once loaded, remains resident in memory and which is available for execution from within another application).

UART UART (or 'Universal Asynchronous Transmitter/Receiver') is the name given to the chip that controls the PC's serial interface. Most modern PCs are fitted with 16550 or 16650 UARTs.

Upper memory The 384 K region of memory which extends beyond the 640 K of conventional memory of the original legacy PC specification. This region of memory was unavailable for applications but was reserved for system functions such as the video display memory. Modern PCs are not bound by this restriction (unless booted directly into DOS).

USB USB (or 'Universal Serial Bus') is a medium speed serial interface which provides expansion facilities for modern PCs. The interface is typically used for mice, printers, scanners, and cameras but a wide variety of external hardware for data acquisition and virtual instrumentation is also based on this standard. Note that the number of USB ports provided by a PC can be easily increased with the use of an external hub.

Validation A process in which input data is checked in order to identify incorrect items. Validation can take several forms including range, character, and format checks.

Verification A process in which stored data is checked (by subsequent reading) to see whether it is correct.

Video card (or video graphics accelerator) An interface card with a dedicated video processor and local RAM which is used for processing data for display on a monitor or display screen.

Virtual memory A technique of memory management which uses disk swap files to emulate random-access memory. The extent of RAM can be increased by this technique by an amount which is equivalent to the total size of the swap files on the hard disk.

Visual display unit (VDU) An output device (usually based on a cathode ray tube) on which text and/or graphics can be displayed. A VDU is normally fitted with an integral keyboard in which case it is sometimes referred to as a console.

Volume label A disk name (comprising up to 11 characters). Note that hard disks may be partitioned into several volumes, each associated with its own logical drive specifier (i.e. C:, D:, E:, etc.).

VRAM VRAM (or 'Video Random Access Memory') is a high-speed type of DRAM fitted to a graphics controller card. This type of memory is preferred for the fast throughput of data which is essential when manipulating high-resolution screen images. See also **DRAM**.

Word A data value which comprises a group of 16-bits and which constitutes the fundamental size of data which an 8086 processor can accept and manipulate as a unit.

Write The process of transferring data from a processor (or other bus controlling device) to memory or to an I/O device.

Appendix B SI units

Fundamental units

Quantity	Unit	Symbol
Length	Metre	M
Mass	Kilogram	kg
Time	Second	s
Electric current	Ampere	A
Temperature	Kelvin	K
Energy	Joule	J
Electrical resistance	Ohm	Ω
Luminous intensity	Candela	cd
Amount of substance	Mole	mol

Selected derived units

Quantity	Unit	Symbol
Electric charge	Coulomb	C
Capacitance	Farad	F
Inductance	Henry	H
Frequency	Hertz	Hz
Conductance	Siemen	S
Magnetic flux	Weber	Wb
Magnetic flux density	Tesla	T
Voltage (potential difference)	Volt	V
Area	Square metre	m^2
Volume	Cubic metre	m^3
Volume (fluid capacity)	Litre	l
Force	Pascal	Pa
Velocity	Metre per second	$m\,s^{-1}$ or m/s
Acceleration	Metre per second squared	$m\,s^{-2}$ or m/s^2
Density	Kilogram per metre cubed	$kg\,m^{-3}$ or kg/m^3

Appendix C Multiples and sub-multiples

Factor of 10	Value	Prefix	Symbol
10^{-18}	0.000000000000000001	atto	a
10^{-15}	0.000000000000001	femto	f
10^{-12}	0.000000000001	pico	p
10^{-9}	0.000000001	nano	n
10^{-6}	0.000001	micro	μ
10^{-3}	0.001	milli	m
10^{-2}	0.01	centi	c
10^{-1}	0.1	deci	d
10	10	deca	da
10^{2}	100	hecto	h
10^{3}	1000	kilo	k
10^{6}	1000000	mega	M
10^{9}	1000000000	giga	G
10^{12}	1000000000000	tera	T
10^{15}	1000000000000000	peta	P
10^{18}	1000000000000000000	exa	E
10^{21}	1000000000000000000000	zetta	Z
10^{24}	1000000000000000000000000	yotta	Y

Appendix D Decimal, hexadecimal, binary, and ASCII table

Decimal	Hexadecimal	Binary	ASCII
0	00	00000000	NUL
1	01	00000001	SOH
2	02	00000010	STX
3	03	00000011	ETX
4	04	00000100	EOT
5	05	00000101	ENQ
6	06	00000110	ACK
7	07	00000111	BEL
8	08	00001000	BS
9	09	00001001	HT
10	0A	00001010	LF
11	0B	00001011	VT
12	0C	00001100	FF
13	0D	00001101	CR
14	0E	00001110	SO
15	0F	00001111	SI
16	10	00010000	DLE
17	11	00010001	DC1
18	12	00010010	DC2
19	13	00010011	DC3
20	14	00010100	DC4
21	15	00010101	NAK
22	16	00010110	SYN
23	17	00010111	ETB
24	18	00011000	CAN
25	19	00011001	EM
26	1A	00011010	SUB
27	1B	00011011	ESC
28	1C	00011100	FS
29	1D	00011101	GS
30	1E	00011110	RS

(*continued*)

Decimal	Hexadecimal	Binary	ASCII
31	1F	00011111	US
32	20	00100000	(space)
33	21	00100001	!
34	22	00100010	"
35	23	00100011	#
36	24	00100100	$
37	25	00100101	%
38	26	00100110	&
39	27	00100111	'
40	28	00101000	(
41	29	00101001	)
42	2A	00101010	*
43	2B	00101011	+
44	2C	00101100	,
45	2D	00101101	-
46	2E	00101110	.
47	2F	00101111	/
48	30	00110000	0
49	31	00110001	1
50	32	00110010	2
51	33	00110011	3
52	34	00110100	4
53	35	00110101	5
54	36	00110110	6
55	37	00110111	7
56	38	00111000	8
57	39	00111001	9
58	3A	00111010	:
59	3B	00111011	;
60	3C	00111100	<
61	3D	00111101	=
62	3E	00111110	>
63	3F	00111111	?
64	40	01000000	@
65	41	01000001	A
66	42	01000010	B
67	43	01000011	C
68	44	01000100	D
69	45	01000101	E
70	46	01000110	F
71	47	01000111	G
72	48	01001000	H
73	49	01001001	I
74	4A	01001010	J
75	4B	01001011	K
76	4C	01001100	L
77	4D	01001101	M

(*continued*)

Decimal	Hexadecimal	Binary	ASCII
78	4E	01001110	N
79	4F	01001111	O
80	50	01010000	P
81	51	01010001	Q
82	52	01010010	R
83	53	01010011	S
84	54	01010100	T
85	55	01010101	U
86	56	01010110	V
87	57	01010111	W
88	58	01011000	X
89	59	01011001	Y
90	5A	01011010	Z
91	5B	01011011	[
92	5C	01011100	\
93	5D	01011101	]
94	5E	01011110	^
95	5F	01011111	_
96	60	01100000	‘
97	61	01100001	a
98	62	01100010	b
99	63	01100011	c
100	64	01100100	d
101	65	01100101	e
102	66	01100110	f
103	67	01100111	g
104	68	01101000	h
105	69	01101001	i
106	6A	01101010	j
107	6B	01101011	k
108	6C	01101100	l
109	6D	01101101	m
110	6E	01101110	n
111	6F	01101111	o
112	70	01110000	p
113	71	01110001	q
114	72	01110010	r
115	73	01110011	s
116	74	01110100	t
117	75	01110101	u
118	76	01110110	v
119	77	01110111	w
120	78	01111000	x
121	79	01111001	y
122	7A	01111010	z
123	7B	01111011	{
124	7C	01111100	|

(*continued*)

Decimal	Hexadecimal	Binary	ASCII
125	7D	01111101	}
126	7E	01111110	˜
127	7F	01111111	DEL
128	80	10000000	Ł
129	81	10000001	'
130	82	10000010	,
131	83	10000011	ƒ
132	84	10000100	„
133	85	10000101	…
134	86	10000110	†
135	87	10000111	‡
136	88	10001000	ˆ
137	89	10001001	‰
138	8A	10001010	Š
139	8B	10001011	‹
140	8C	10001100	Œ
141	8D	10001101	Ž
142	8E	10001110	^
143	8F	10001111	–
144	90	10010000	ł
145	91	10010001	‘
146	92	10010010	’
147	93	10010011	“
148	94	10010100	”
149	95	10010101	•
150	96	10010110	–
151	97	10010111	—
152	98	10011000	˜
153	99	10011001	™
154	9A	10011010	š
155	9B	10011011	›
156	9C	10011100	œ
157	9D	10011101	ž
158	9E	10011110	~
159	9F	10011111	Ÿ
160	A0	10100000	(thin space)
161	A1	10100001	¡
162	A2	10100010	¢
163	A3	10100011	£
164	A4	10100100	¤
165	A5	10100101	¥
166	A6	10100110	¦
167	A7	10100111	§
168	A8	10101000	¨
169	A9	10101001	©
170	AA	10101010	ª
171	AB	10101011	«

(*continued*)

Decimal	Hexadecimal	Binary	ASCII
172	AC	10101100	¬
173	AD	10101101	-
174	AE	10101110	®
175	AF	10101111	¯
176	B0	10110000	°
177	B1	10110001	±
178	B2	10110010	^{2}
179	B3	10110011	^{3}
180	B4	10110100	´
181	B5	10110101	μ
182	B6	10110110	¶
183	B7	10110111	·
184	B8	10111000	¸
185	B9	10111001	^{1}
186	BA	10111010	º
187	BB	10111011	»
188	BC	10111100	1/4
189	BD	10111101	1/2
190	BE	10111110	3/4
191	BF	10111111	¿
192	C0	11000000	À
193	C1	11000001	Á
194	C2	11000010	Â
195	C3	11000011	Ã
196	C4	11000100	Ä
197	C5	11000101	Å
198	C6	11000110	Æ
199	C7	11000111	Ç
200	C8	11001000	È
201	C9	11001001	É
202	CA	11001010	Ê
203	CB	11001011	Ë
204	CC	11001100	Ì
205	CD	11001101	Í
206	CE	11001110	Î
207	CF	11001111	Ï
208	D0	11010000	Ð
209	D1	11010001	Ñ
210	D2	11010010	Ò
211	D3	11010011	Ó
212	D4	11010100	Ô
213	D5	11010101	Õ
214	D6	11010110	Ö
215	D7	11010111	×
216	D8	11011000	Ø
217	D9	11011001	Ù

(*continued*)

Decimal	Hexadecimal	Binary	ASCII
218	DA	11011010	Ú
219	DB	11011011	Û
220	DC	11011100	Ü
221	DD	11011101	Ý
222	DE	11011110	Þ
223	DF	11011111	ß
224	E0	11100000	à
225	E1	11100001	á
226	E2	11100010	â
227	E3	11100011	ã
228	E4	11100100	ä
229	E5	11100101	å
230	E6	11100110	æ
231	E7	11100111	ç
232	E8	11101000	è
233	E9	11101001	é
234	EA	11101010	ê
235	EB	11101011	ë
236	EC	11101100	ì
237	ED	11101101	í
238	EE	11101110	î
239	EF	11101111	ï
240	F0	11110000	ð
241	F1	11110001	ñ
242	F2	11110010	ò
243	F3	11110011	ó
244	F4	11110100	ô
245	F5	11110101	õ
246	F6	11110110	ö
247	F7	11110111	÷
248	F8	11111000	ø
249	F9	11111001	ù
250	FA	11111010	ú
251	FB	11111011	û
252	FC	11111100	ü
253	FD	11111101	ý
254	FE	11111110	þ
255	FF	11111111	ÿ

Note: ASCII characters above 127 decimal (FF hex.) are non-standard.

Appendix E Powers of 2

Powers of 2	Number of bytes	Symbol	Name
2^{10}	1 024	KB	kilobyte
2^{20}	1 048 576	MB	Megabyte
2^{30}	1 073 741 824	GB	Gigabyte
2^{40}	1 099 511 627 776	TB	Terabyte
2^{50}	1 125 899 906 843 624	PB	Petabyte
2^{60}	1 152 921 504 607 870 976	EB	Exabyte
2^{70}	1 180 591 620 718 458 879 424	ZB	Zettabyte
2^{80}	1 208 925 819 615 701 892 530 176	YB	Yottabyte

Appendix F Processor sockets

Socket 1	Found on 486 motherboards and supports 486 chips, plus the DX2, DX4 Overdrive.
Socket 2	An upgrade of Socket 1 which has 238 pins and accepts the 486 processor but can also support a Pentium Overdrive.
Socket 3	Similar to Socket 2 but has 237 pins. Operates at 5 V but can be configured for 3.3 V operation.
Socket 4	Supports older Pentium 60–66 and Overdrive processors that operate from a 5 V supply.
Socket 5	Supports Pentium processors from 75 to 133 MHz operating from a 3.3 V supply. Newer chips will not fit because they need an extra pin. Socket 5 has been replaced by Socket 7 although socket converters are available that allow Socket 7 processors to be fitted in Socket 5 motherboards.
Socket 6	A slightly more advanced Socket 3 with 235 pins and 3.3 V operation to suit some 486 chips.
Socket 7	Operating at 2.5 to 3.3 V, Socket 7 is currently the most common motherboard socket still in use. Socket 7 supports Pentium processors from 75 MHz and above, MMX processors, the AMD K5, K6, K6-2, K6-3, 6×86, M2 and M3, and Pentium MMX Overdrives. This socket was the industry standard being suitable for sixth-generation chips by IDT, AMD, and Cyrix. Intel abandoned the socket for its sixth-generation lineup in favour of Slot 1 (see below).
Socket 8	Supports the Pentium Pro. Other modern Pentium Processors do not use Socket 8 but use Slots.
Slot 1	Slot 1 supports the P2, P3, and Celeron processors. A Pentium Pro can be fitted by using a Socket 8 on a daughtercard which is then fitted into the Slot 1.
Slot 2	Slot 2 is a 330 pin version of Slot 1. The Slot 2 design allows the processor to communicate with the level 2 cache at the CPU's full clock speed, in contrast to Slot 1 which communicates at half that speed.
Slot A	Similar to Slot 1, this design suits the AMD Athlon processor. It uses a different bus protocol (EV6) to support a 200 MHz front side bus (FSB).
Socket 370	Socket 370 is a Socket 7 with an extra row of pins on all four sides. Socket 370 supports the Pentium III, Celeron, and Celeron II processors.
Socket 462	Socket 462 is also known as Socket A and is used for AMD's Athlon and Duron processors. It supports the 200 MHz EV6 bus, as well as the new 266 MHz EV6 bus.
Socket 423/478	Socket 423 is the original socket used by Pentium 4 processors. Socket 478 supports the newer 478-pin Pentium 4's.

Appendix G Processor data

Maker	Name	Core	Socket	Process	Voltage (V)	Speed	Effective FSB speed	Level 2 Cache	Internal Bus (bit)	Introduced
Intel	Pentium	P5	Socket 4	0.8 μm	5	60–66 MHz	60–66		64	March 1993
Intel	Pentium	P54C	Socket 5	0.6 μm	3.38–3.52	75–120 MHz	60–66		64	March 1994
Intel	Pentium	P54C	Socket 7	0.35 μm	3.38–3.52	120–200 MHz	60–66		64	March 1995
Cyrix/ IBM	6×86	M1(R)	Socket 7	0.65–0.44 μm	3.3–3.52	90–200 MHz	40–75		64	October 1995
AMD	K5	Model 0-3	Socket 7	0.35 μm	3.52	75–166 MHz	60–66		64	June 1996
Cyrix/ IBM	6×86L	M1L	Socket 7	0.35 μm	2.8	120–200 MHz	50–75		64	January 1997
Intel	Pentium MMX	P55C	Socket 7	0.35 μm	2.8	133–233 MHz	60–66		64	January 1997
AMD	K6	Model 6	Socket 7	0.35 μm	2.9–3.3	166–233 MHz	66		64	April 1997
Cyrix/ IBM	6×86MX/ MII	M2	Socket 7	0.35–0.25 μm	2.9	166–366 MHz	66–83		64	May 1997
AMD	K6	Model 7	Socket 7	0.25 μm	2.2	200–300 MHz	66		64	January 1998
AMD	K6-2 3D	Model 8/[7:0]	Socket 7	0.25 μm	2.2	266–400 MHz	66–100		64	May 1998
AMD	K6-2 3D CXT	Model 8/[F:8]	Socket 7	0.25 μm	2.2–2.4	333–550 MHz	95–100		64	November 1998
AMD	K6-III	Model 9	Socket 7	0.25 μm	2.4	400–450 MHz	100	256 KB 4-way	64	February 1999
AMD	K6-2+		Socket 7	0.18	2	450–550 MHz	100	128 KB	64	April 2000
AMD	K6-III+	Model 13	Socket 7	0.18	2	450–500 MHz	95–100	256 KB 4-way	64	April 2000
Intel	Pentium Pro	P6	Socket 8	0.6–0.35	3.1–3.3	150–200 MHz	60/66	256, 512, 1024 KB	64	November 1995
Intel	Pentium II	Klamath	Slot 1	0.35	2.8	233–300 MHz	66	512 KB	64	May 1997
Intel	Celeron	Covington	Slot 1	0.25	2	266–300 MHz	66		64	April 1998
Intel	Pentium II Xeon	Drake	Slot 2	0.25	2	400–450 MHz	100	512, 1024, 2048 KB	64	June 1998
Intel	Pentium II	Deschutes	Slot 1	0.25	2	333–450 MHz	66–100	512 KB	64	September 1998
Intel	Celeron	Mendocino	Slot 1/ Socket 370	0.25	2	300–533 MHz	66	128 KB	64	August 1998

(*continued*)

Maker	Name	Core	Socket	Process	Voltage (V)	Speed	Effective FSB speed	Level 2 Cache	Internal Bus (bit)	Introduced
Intel	Pentium III	Katmai	Slot 1	0.18–0.25	1.65–2.05	450–600 MHz	100/133	512 KB	64	February 1999
Intel	Pentium III Xeon	Tanner	Slot 2	0.25	2	500–550 MHz	100	512, 1024, 2048 KB	64	March 1999
Intel	Pentium III	Coppermine	Slot 1/ Socket 370	0.18	1.6–1.8	500–1133 MHz	100/133	256 KB	256	October 1999
Intel	Pentium III Xeon	Cascades	Slot 2	0.18	2.8	600–1000 MHz	100/133	256, 1024, 2048 KB	256	October 1999
Intel	Celeron II	Coppermine	Socket 370	0.18	1.5–1.7	533–1100 MHz	66/100	128 KB	256	March 2000
Intel	Pentium III Server	Tualatin	Socket 370	0.13	1.1–1.45	700–1400 MHz	100/133	512 KB	256	June 2001
Intel	Pentium III Desktop	Tualatin	Socket 370	0.13	1.45	1133–1200 MHz	133	256 KB	256	August 2001
Intel	Pentium III Celeron	Tualatin	Socket 370	0.13	1.45	1000–1400 MHz	100	256 KB	256	October 2001
AMD	Athlon	K7	Slot A	0.25	1.6	500–700 MHz	200	512 KB	64	August 1999
AMD	Athlon	K75	Slot A	0.18	1.6–1.8	500 MHz–1 GHz	200	512 KB	64	January 2000
AMD	Duron	Spitfire	Socket A	0.18	1.5–1.6	600–950 MHz	200	64 KB	64	June 2000
AMD	Athlon	Thunderbird	Slot A/ Socket A	0.18	1.75	650 MHz–1.4 GHz	200/266	256 KB	64	June 2000
AMD	Duron	Morgan	Socket A	0.18	1.75	1.0–1.3 GHz	200	64 KB	64	August 2001
AMD	Athlon XP	Palomina	Socket A	0.18	1.75	1.333–1.733 GHz	266	256 KB	64	October 2001
AMD	Athlon XP	Thoroughbred A	Socket A	0.13	1.5–1.65	1.467–1.8 GHz	266	256 KB	64	June 2002
AMD	Athlon XP	Thoroughbred B	Socket A	0.13	1.65	1.8–2.25 GHz	266–322	256 KB	64	August 2002
Intel	Pentium 4	Willamette	Socket 423/ Socket 478	0.18	1.75	1.3–2 GHz	400	256 KB	256	November 2000
Intel	Pentium 4 Celeron	Willamette	Socket 478	0.18	1.75	1.7–1.8 GHz	400	128 KB	256	May 2002
Intel	Pentium 4	Northwood	Socket 478	0.13	1.5	1.6–2.5 GHz	400–533	512 KB	256	January 2002
Intel	Pentium 4	Northwood	Socket 478	0.13	1.5–1.525	2.5–2.8 GHz	400–533	512 KB	256	August 2002
Intel	Pentium 4 Celeron	Northwood	Socket 478	0.13	1.5	2 GHz	400	128 KB	256	October 2002
Intel	Pentium 4	Northwood	Socket 478	0.13	1.5	1.4–2.6 GHz	400	512 KB	256	August 2002
Intel	Pentium 4	Northwood 'A'	Socket 478	0.13	1.5	2.26–3.60 GHz	533	512 KB	256	August 2002
Intel	Pentium 4	Prescott	Socket 478	0.09	1.5	3.60 GHz	664	512 KB	256	Mid-2003

Appendix H Common file extensions

Extension	Type of file
ASC	An ASCII text file
ASM	An assembly language source code file
ASP	An Active Server Page file (see HTM)
BAK	A back-up file (often created automatically by a text editor which renames the source file with this extension and the revised file assumes the original file specification)
BAS	A BASIC program source file
BAT	A batch file which contains a sequence of operating system commands
BIN	A binary file (comprising instructions and data in binary format)
BMP	A bit-mapped picture file
C	A source code file written in the C language
CFG	A configuration file
CLP	A Windows 'clipboard' file
COM	An executable program file in small memory format (i.e. confined to a single 64 KB memory segment)
CPI	A 'code page information' file
CPP	A source code file written in the C++ language
CRD	A Windows 'card index' file
CSS	A Cascading Style Sheet file (used to control the display styles used in HTM and HTML files)
DAT	A data file (usually presented in either binary or ASCII format)
DBG	A DEBUG text file
DLL	A Dynamic Link Library file
DOC	A document file (not normally presented in standard ASCII format)
EXE	An executable program file in large memory format (i.e. not confined to a 64 KB memory model)
GIF	An image file in Graphics Interchange Format
H	A header file containing class names and definitions used in the C language
HEX	A file presented in hexadecimal (an intermediate format sometimes used for object code)
HLP	A help file
HPP	A header file containing class names and definitions used in the C++ language

(*continued*)

Extension	Type of file
HTM	A file in HyperText Markup language format (designed for display using a standard Web browser)
HTML	A file in HyperText Markup Language format (designed for display using a standard Web browser)
INC	An 'include' file used in assembly language programming
INI	An initialization file which may contain a set of inference rules and/or environment variables
JPG	An image saved in JPEG (Joint Photographic Experts Group) format
LIB	A library file (containing multiple object code files)
LST	A listing file (usually showing the assembly code corresponding to each source code instruction together with a complete list of symbols)
MAK	A 'make' file
MAP	A file containing symbol information generated by a compiler and designed for use by an external debugger
MSC	A Microsoft Management Console file
OBJ	An object code file. Object code modules are linked to form executable files
OLD	A back-up file (replaced by a more recent version of the file)
PAS	A source code file written in Pascal
PBC	A PowerBASIC chain file
PBU	A precompiled unit file (PowerBASIC)
PCX	A picture file
PDF	An Adobe Acrobat Document file (requires Adobe Acrobat Reader)
PIF	A Windows 'Program Interchange File'
PRN	A printer file (using dedicated printer codes)
REG	A Windows Registry file
SCR	A DEBUG script file
SYS	A system file
TIF	A tagged image file
TMP	A temporary file
TXT	A text file (usually in ASCII format)
VBP	Visual Basic Project file
WRI	A document file in MS Write format
XLS	A file in MS Excel format
$$$	A temporary file

Appendix I BIOS error codes

IBM BIOS

Indication	Meaning
One short beep	Normal POST – no error
Two short beeps	POST error – see screen for error code
No beeps	Power missing, loose card, or short circuit
Continuous beep	Power missing, loose card, or short circuit
Repeating short beep	Power missing, loose card, or short circuit
One long and one short beep	System board error
One long and two short beeps	Video (mono/CGA display adapter)
One long and three short beeps	Video (EGA display adapter)
Three long beeps	Keyboard error
One beep, blank/incorrect display	Video display circuitry

AMI BIOS

Indication	Meaning
One short beep	DRAM refresh failure
Two short beeps	Parity circuit failure
Three short beeps	Base memory (64 KB) RAM failure
Four short beeps	System timer failure
Five short beeps	CPU failure
Six short beeps	Keyboard controller error
Seven short beeps	Virtual mode exception error
Eight short beeps	Display memory failure
Nine short beeps	ROM BIOS checksum failure
One long and three short beeps	Base/extended memory failure
One long and eight short beeps	Display/retrace test failure

Award BIOS

Indication	Meaning
One short beep	No error during POST
Two short beeps	Any non-fatal error
One long and two short beeps	Video error
One long and three short beeps	Keyboard controller error

Phoenix BIOS

Indication	Meaning
One, one and three beeps	CMOS read/write failure
One, one and four beeps	ROM BIOS checksum failure
One, two and one beep	Programmable interval timer failure
One, two and two beeps	DMA initialization failure
One, two and three beeps	DMA page register read/write failure
One, three and one beep	RAM refresh verification error
One, three and three beeps	First 64 KB RAM chip/data line failure
One, three and four beeps	First 64 KB odd/even logic failure
One, four and one beep	Address line failure first 64 KB RAM
One, four and two beeps	Parity failure first 64 KB RAM
One, four and three beeps	Fail-safe timer feature (EISA only)
One, four and four beeps	Software NMI port failure (EISA only)
Two, one and up to four beeps	First 64 KB RAM chip/data line failure (bits 0 to 3, respectively)
Two, two and up to four beeps	First 64 KB RAM chip/data line failure (bits 4 to 7, respectively)
Two, three and up to four beeps	First 64 KB RAM chip/data line failure (bits 8 to 11, respectively)
Two, four and up to four beeps	First 64 KB RAM chip/data line failure (bits 12 to 15, respectively)
Three, one and one beep	Slave DMA register failure
Three, one and two beeps	Master DMA register failure
Three, one and three beeps	Master interrupt mask register failure
Three, one and four beeps	Slave interrupt register failure
Three, two and four beeps	Keyboard controller test failure
Three, three and four beeps	Screen initialization failure
Three, four and one beep	Screen retrace test failure
Four, two and one beep	Timer tick failure
Four, two and two beeps	Shutdown test failure
Four, two and three beeps	Gate A20 failure
Four, two and four beeps	Unexpected interrupt in protected mode
Four, three and one beep	RAM text address failure
Four, three and three beeps	Interval timer channel 2 failure
Four, three and four beeps	Time of day clock failure
Four, four and three beeps	Maths coprocessor failure

Appendix J Manufacturers, suppliers, and distributors

Expansion systems, embedded controllers, DAQ, and industrial control systems

Alphi Technology Corporation
6202 South Maple Avenue #120
Tempe
AZ 85283
USA
Telephone: 480 838 2428
Fax: 480 838 4477
Web site: www.alphitech.com
E-mail: sales@alphitech.com

Amplicon Liveline Ltd
Hollingdean Road
Centenary Industrial estate
Brighton
East Sussex
BN2 4AW
UK
Telephone: 01273 570220
Fax: 01273 570215
Web site: www.amplicon.co.uk
E-mail: sales@amplicon.co.uk

Arcom Control Systems Ltd
Clifton Road
Cambridge
CB1 7EA
UK
Telephone: 01223 411200
Freephone: 0800 411300
Fax: 01223 410457
Web site: www.arcom.com
E-mail: sales@arcom.com

Biodata Limited
10 Stocks Street
Manchester
M8 8QG
UK
Telephone: 0161 834 6688
Fax: 0161 833 2190
Web site: http://www.microlink.co.uk
E-mail: info@microlink.co.uk

Datel (UK)
Unit 15
Campbell Court Business Park
Campbell Road
Bramley
Tadley
Berkshire
RG26 5EG
UK
Telephone: 01256 880444
Fax: 01256 880706
Web site: www.datel.com
E-mail: datel.ltd@datel.com

Datel Inc
11 Cabot Building
Mansfield MA
02048 1151
USA

Fairchild Semiconductor Ltd
Interface House
Interface Business Park
Wootton Bassett

Swindon
Wiltshire
SN5 8QL
UK
Telephone: 01793 856856
Fax: 01793 856857
Web site: www.fairchildsemi.com
E-mail: sales@fairchildsemi.com

Keithley Instruments
2 Commerce Park
Brunel Road
Theale
Berkshire
RG7 4AB
UK
Telephone: 01189 297500
Fax: 01189 297519
Web site: www.keithley.co.uk
E-mail: info@keithley.co.uk

Keithley Inc
28775 Aurora Road
Cleveland
OH 44139
USA

Measurement Computing Corporation
16 Commerce Boulevard
Middleboro
MA 02346
USA
Telephone: 508 946 5100
Fax: 508 946 9500
Web site: www.measurementcomputing.com
E-mail: info@measurementcomputing.com

Microbus
Treadway Hill
Loudwater
High Wycombe
Buckinghamshire
HP109QL
UK
Telephone: 01628 537333
Fax: 01628 537334
Web site: www.microbus.com
E-mail: sales@microbus.com

National Instruments UK
Measurement House
Newbury Business Park
London Road
Newbury
Berkshire
RG14 2PS
UK
Telephone: 01635 572414
Fax: 01635 523154
Web site: www.digital.ni.com
E-mail: info.uk@ni.com

Semaphore Systems Ltd
Unit 3
Hampstead West
Iverson Road
London
NW6 2HX
UK
Telephone: 020 7625 7744
Fax: 020 7625 7788
Web site: www.semaphore-systems.co.uk
E-mail: sales@semaphore-systems.co.uk

Siemens
Sir William Siemens House
Princess Road
Manchester
M20 2UR
UK
Telephone: 0161 446 6400
Fax: 0161 446 5327
Web site: www.siemens.co.uk

Signametrics Corporation
6073 50th Avenue NE
Seattle
WA 98115
USA
Telephone: 206 524 4074
Fax: 206 525 8578
Web site: www.signametrics.com
E-mail: sales@signametrics.com

Spectrum GmbH
Ahrensfelder Weg 13-17
22927 Grosshandsdorf
Germany

Telephone: 49 4102/6956-0
Fax: 49 4102/6956-66
Web site: www.spectrum-gmbh.com

Strategic Test Corporation
One Broadway
Suite 600
Cambridge
MA 02142
USA
Telephone: 866 898 FAST
Fax: 617 577 1209
Web site: www.strategic-test.com

Talisman Electronics Ltd
2 The Courtyard
Denmark Street
Wokingham
Berkshire
RG40 2AZ
UK
Telephone: 01452 500588
Fax: 01452 513867
Web site: www.talisman-uk.com
E-mail: sales@talisman-uk.com

United Electronic Industries Inc.
611 Neponset Street
Canton
MA 02021
USA
Telephone: 781 821 2890
Fax: 781 821 2891
Web site: www.ueidaq.com

XYCOM Europe Ltd
Pro-face House
8 Orchard Court
Binley Business Park
Coventry
CV3 2TQ
Telephone: 02476 440088
Fax: 02476 440099
Web site: www.xycom.com
E-mail: info@profaceuk.com

XYCOM USA
750 North Maple Road
Saline
MI 48176
USA

Motherboards, memories, processors, drives, and accessories

A and P Computers Ltd
35 Walnut Tree Close Guildford
Surrey
GU1 4UN
UK
Telephone: 01483 841000
Fax: 01483 880011
Web site: www.ap-computers.com
E-mail: sales@ap-computers.com

Evesham Technology
Vale Park
Evesham
Worcestershire
WR11 1TD
UK
Telephone: 08707 1609500
Fax: 01386 769781
Web site: www.evesham.com
E-mail: sales@evesham.com

dabs.com
Direct House
Wingates Industrial Park
Westhoughton
Bolton
UK
Telephone: 0870 4293000
Web site: www.dabs.com.uk

Novatech
Harbour House
Hamilton Road
Cosham
Portsmouth
PO6 4PU
UK
Telephone: 023 923 22522

Scan Computers International Ltd
Unit 27/28
Enterprise Park
Horwich
Bolton
Lancashire
BL6 6PE
UK
Telephone: 01204 474747

Simply.Com
Nimrod House
Enigma Commercial Centre
Malvern
Worcestershire
WR14 1JJ
UK
Telephone: 0870 121 7660 UK
+44 1684 893020 International
Fax: 0870 121 7659 UK
+44 1684 580898 International
Web site: www.simply.com
E-mail: sales@simply.com

Stak Trading Computer Services Ltd
26 Somers Road
Rugby
Warwickshire
CV22 7DH
UK
Telephone: 0870 444 4484
Fax: 0870 444 4485
Web site: www.online.stak.com
E-mail: sales@stak.com

Unimart Computers Ltd.
119 Groveley Road
Sunbury-on-Thames
Middlesex
TW16 7J2
UK
Telephone: 020 8893 2969
Fax: 020 8893 2961
Web site: unimart.co.uk
E-mail: info@unimart.co.uk

Data communication products and accessories

Connexions (UK) PLC
Unit 3
Travellers Close
Welham Lane
Hertfordshire
AL9 7NT
UK
Telephone: 01707 272091
Fax: 01707 269444
Web site: www.cxcxcx.com
E-mail: sales@cxcxcx.com

Quatech
5675 Hudson Industrial Parkway
Hudson
OH 44236 5012
USA
Telephone: 330 655 9000
Fax: 330 655 9010
Web site: www.quatech.com
E-mail: sales@quatech.com

Memory devices

Kingston Technology
Kingston Court
Brooklands Close
Sunbury-on-Thames
Middlesex
TW16 7EP
UK
Telephone: 01932 738888
Fax: 01932 738811
Web site: www.kingston.com

Memory Bank at Powermark
The Powermark Centre
Elstree Road
Elstree
Hertfordshire
WD6 3RP
UK
Telephone: 020 8956 7777
Fax: 020 8956 7878

Electronic components and test equipment

Farnell Electronic Components
Canal Road
Leeds
West Yorkshire
LS12 2TU
UK
Telephone: 0870 1200 200
Fax: 0870 1200 201
Web site: www.farnellinone.co.uk
E-mail: uksales@farnellinone.co.uk

Maplin Electronic Ltd
National Distribution Centre
Valley Road
Wombwell
Barnsley
South Yorkshire
UK
Telephone: 0870 429 6000
Fax: 0870 429 6001
Web site: www.maplin.co.uk
E-mail: sales@maplin.co.uk

RS Components
Birchington Road
Corby
Nothants
NN17 9RS
UK
Telephone: 01536 201201
Fax: 01536 201501
Web site: www.rswww.com
E-mail: sales@rswww.com

Computer supplies

Watford Electronics Computer Supplies
Jessa House
Finway
Luton
LU1 1 TR
UK
Telephone: 0871 666 0200
Fax: 0871 666 5200
Web site: www.savastore.com/watford

Software

Adept Scientific plc
Amor Way
Letchworth
Hertfordshire
SG6 1ZA
UK
Telephone: 01462 480055
Fax: 01462 480213
Web site: www.adeptscience.co.uk
E-mail: info@adeptscience.co.uk

PowerBASIC Inc
1978 Tamiami Trail S #200
Venice
FL 34293
USA
Telephone: 941 408 8700
Fax: 941 408 8820
Web site: www.powerbasic.com
E-mail: sales@powerbasic.com

Virtual instruments

PicoTechnology Ltd
The Mill House
Cambridge Street
St Neots
Cambridgeshire
PE19 1QB
UK
Telephone: 01480 396395
Fax: 01480 396296
Web site: www.picotech.com
E-mail: sales@picotech.com

USB Instruments
EasySync Ltd
373 Scotland Street
Glasgow
G5 8QB
UK
Telephone: 01414 180181
Fax: 01414 180110
Web site: www.usb-instruments.com
E-mail: sales@usb-instruments.com

Appendix K Useful web sites

Company/organization	URL
AMD	http://www.amd.com
Caldera	http://www.caldera.co.uk
Carrera	http://www.carrera.co.uk
Computer Information Centre	http://www.compinfo.co.uk
Dan Technology	http://www.dan.co.uk
Dell	http://www.dell.com/uk
Dr Solomon's Software	http://www.drsolomon.com
Epson	http://www.epson.co.uk
Gateway	http://www.gw2k.co.uk
Geek Hideout	http://www.geekhideout.com
Hewlett Packard	http://www.hp.com
IBM	http://www.ibm.com
Internals	http://www.internals.com
Kingston Technology	http://www.kingston.com
Logix4U	http://www.logix4u.net
Matrox	http://www.matrox.com
Maxtor	http://www.maxtor.com
McAfee	http://www.mcafee.com
Mesh	http://www.meshplc.co.uk
Microsoft	http://www.microsoft.com
NEC	http://www.nec.com
Seagate	http://www.seagate.com
Scientific Software Tools, Inc.	http:// www.sstnet.com
Symantec	http://www.symantec.com
Taxan	http://www.taxan.co.uk
US Robotics	http://www.usr.co.uk
Western Digital	http://www.wdc.com
Zeal SoftStudio	http://www.zealsoft.com
Companion website	http://www.key2control.com (downloadable resources, weblinks, FAQ, source code and other material)

Appendix L Bibliography

Interfacing

James, K. (2000) A practical guide to the programming and interfacing techniques involved in data collection, and the subsequent measurement and control systems. *PC Interfacing and Data Acquisition*, Newnes. ISBN 0 7506 4624 1.

An, P. Provides a practical guide to interfacing with examples of both software and hardware. *PC Interfacing using Centronic, RS-232 and Game Ports*, Newnes, ISBN 0 7506 3637 8.

Hutchings, H., James, M. (2001). An excellent introduction to using C in a wide variety of practical interfacing applications. *Interfacing with C*, 2nd edition, ISBN 0 7506 4831 7.

Smith, G. (1999) Provides a basic introduction to computer interfacing with sample code in C. *Computer Interfacing*, Newnes, ISBN 0 7506 4474 5.

Axelson, J. (2000) An excellent book dedicated to the PC's parallel port. Contains numerous examples using Visual Basic. *Parallel Port Complete*, Lakeview Research, ISBN 0 9650 8191 5.

Electronic circuits

Hickman, I. (1999) Describes and explains a wide variety of analogue electronic circuits. *Analog Circuits Cookbook*, Newnes, 2nd edition, ISBN 0 7506 4234 3.

Tooley, M. (2002) Designed specifically for the newcomer, this book combines comprehensive coverage of the principles of both digital and analogue electronics. *Electronic Circuits: Fundamentals and applications*, Newnes, 2nd edition, ISBN 0 7506 5394 9.

PC hardware

Beales, R.P. (2003) Provides a comprehensive guide to installing and maintaining PC systems. *PC Systems Installation and Maintenance*, Newnes, 2nd edition, ISBN 0 7506 6074 0.

Abbott, D. (2000) A comprehensive reference on the PCI bus (including CompactPCI). *PCI Bus Demystified*, Newnes, ISBN 1 878707 54 X.

Rosch, W. (2003) A comprehensive technical reference on the PC. *The Winn Rosch Hardware Bible*, Sams, 6th edition, ISBN 0789 72859 1.

Programming

Wyatt, A. (1993) A comprehensive introduction to x86 assembly language programming. *Using Assembly Language*, Que, 3rd edition, ISBN 0 8802 2464 9.

Jamsa, K. (2002) Contains a huge selection of code fragments and ideal for those who need commented source code examples. *Jamsa's C/C++ Programmer's Bible*, Jamsa Press, 2nd edition, ISBN 1 884133 25 6.

Hubbard, J. (2000) A student text that provides a rigorous introduction to C++ including a number of more advanced topics. *Programming with C++*, McGraw Hill, 2nd edition, ISBN 0 0713 5346 1.

Perry, G., Hettihewa, S. (1998) Presented as a series of one hour lessons this text aims to get beginners up and running with Visual Basic in the shortest possible time. *Teach Yourself Visual Basic 6*, Sams, ISBN 0 672 31533 5.

Leonik, T. (2000) Shows how a PC with Visual Basic 6 can be used in conjunction with a Programmable Logic Controller (PLC) in a domestic control and monitoring system. *Home Automation Basics*, Prompt, ISBN 0 7906 1214 3.

Appendix M Reference material available from the Web

The following references are available via the World Wide Web:

8086 instruction set
`http://www.ziplib.com/emu8086help/8086_instruction_set.html`

A86/A386 assembler and D86/D386 debugger (Eric Isaacson)
`http://eji.com/a86`

Advanced Configuration and Power Interface Specification, Version 1.0
`http://www.teleport.com/~acpi/`

All BIOS World
`http://www.abios.com`

BIOS
`http://www.biosworld.com`

Motherboards
`http://www.motherboards.org`

'OnNow and ACPI: Introduction and Specifications' and related white papers. Power management specifications for device and bus classes.
`http://www.microsoft.com/hwdev/onnow.htm`

OnNow capabilities and power management
`http://www.microsoft.com/hwdev/pcfuture/onnowwdm.htm`

PC bus standards
`http://www.techfest.com/hardware/bus.htm`

PC standards and data for interface designers
`http://www.interfacebus.com`

PC systems and components
`http://www.pcguide.com`

PC/104 and PCI-104 Specifications
http://www.pc104.org/technology/pc104_tech.html

PCI Local Bus Specification, Revision 2.1 (PCI 2.1)
http://www.pcisig.com

PCI database of devices and systems
http://members.datafast.net.au/dft0802/pcidevs.txt

PCMCIA standards
http://www.pc-card.com

Plug and Play specifications
http://www.microsoft.com/hwdev/specs/

USB Specification, Version 1.0
http://www.usb.org

Windows 2000
http://www.microsoft.com/ntserver/

Windows Hardware Compatibility List (HCL)
http://www.microsoft.com/ntserver/

Index